Approximation Theory IX

Computational Aspects

Innovations in Applied Mathematics

An international series devoted to the latest research in modern areas of mathematics, with significant applications in engineering, medicine, and the sciences.

Series Editor:
Larry L. Schumaker
Stevenson Professor of Mathematics
Vanderbilt University

Previously published titles include

*Mathematical Methods for
Curves and Surfaces* (1995)

*Curves and Surfaces with
Applications in CAGD* (1997)

*Surface Fitting and
Multiresolution Methods* (1997)

*Mathematical Methods for
Curves and Surfaces II* (1998)

*Mathematical Models in
Medical and Health Science* (1998)

Approximation Theory IX

Volume 2. Computational Aspects

EDITED BY

Charles K. Chui
Department of Mathematics
Texas A&M University

Larry L. Schumaker
Department of Mathematics
Vanderbilt University

VANDERBILT UNIVERSITY PRESS
Nashville & London

First Edition 1998
98 99 00 01 02   5 4 3 2 1

CONTENTS

Preface

The *Ninth International Conference on Approximation Theory* was hosted by Vanderbilt University in Nashville, January 3 – 6, 1998. Previous conferences in this series were held in 1973, 1976, 1980, and 1992 in Austin, and 1983, 1986, 1989, and 1995 in College Station. The conference was attended by 190 mathematicians from 21 different countries.

The program included seven invited one-hour survey talks and the one-hour presentation of this year's Popov Prize winner, Arno Kuijlaars. In addition to a large number of contributed talks, there were eight special sessions on interdisciplinary topics of current research interest, arranged by Yang Wang (wavelets), Richard Varga (approximation in the complex plane), Joe Ward (radial basis functions), Boris Shekhtman (abstract approximation), Hrushikesh Mhaskar (neural networks), Günther Nürnberger (splines), Hans Hagen (computer aided geometric design), and Vladymir Temlyakov (nonlinear m-term approximation).

Because of the large number of submitted papers, this proceedings appears in two separate volumes. Volume I covers several areas of classical and modern approximation theory, while Volume II is devoted to recent applied and computational developments.

We are indebted to the National Science Foundation and to Vanderbilt University for their financial support of the conference. We would like to thank the members of the local organizing committee, Tom Hogan, Kirill Kopotun, and Mike Neamtu for their assistance with various aspects of the meeting, and in particular for creating and managing the web site. Thanks are also due to the invited speakers, organizers of the special sessions, presenters, and everyone who attended for making the conference a success. We would like to thank our reviewers who helped select articles for the proceedings, and Robin Campbell for her TEXnical assistance with various papers. Finally, our special thanks to Margaret Chui and Gerda Schumaker who also assisted in the preparation of these volumes.

Charles K. Chui   Sept. 1, 1998
Larry L. Schumaker

CONTRIBUTORS

Numbers in parentheses indicate pages on which authors' contributions begin. Articles marked with * and ** are in Volumes 1 and 2, respectively.

GEORGE A. ANASTASSIOU (*1), *Department of Mathematical Sciences, The University of Memphis, Memphis, TN 38152, USA* [anastasg@hermes.msci.memphis.edu]

ALEXANDER V. ANDRIANOV (*7), *Department of Mathematics, University of South Carolina, Columbia, SC 29208, USA* [andriano@math.sc.edu]

MARTIN BARTELT (*15), *Department of Mathematics, Christopher Newport University, Newport News, VA 23606, USA* [mbartelt@pcs.cnu.edu]

NEERAJ BHARDWAJ (*263), *Department of Mathematics, D.N. (P.G.) College, C.C.S. University, Meerut Uttar Pradesh, India*

KAI BITTNER (**1), *Institute of Biomathematics and Biometry, GSF - National Research Center for Environment and Health, D-85764 Neuherberg, Germany* [bittner@gsf.de]

B. BORDIN (*23), *IMECC-UNICAMP, Caixa Postal 6065, 13081-970, Campinas, SP, Brazil* [bordin@ime.unicamp.br]

PETER BORWEIN (*31,*51), *Department of Mathematics and Statistics, Simon Fraser University, Burnaby, B. C., Canada, V5A 1S6* [pborwein@cecm.sfu.ca]

MICHELE CAMPITI (*59), *Department of Mathematics, Polytechnic of Bari, Via E. Orabona, 4, 70125 Bari, Italy* [campiti@dm.uniba.it]

PETER G. CASAZZA (*67), *Department of Mathematics, University of Missouri, Columbia, MO 65211, USA* [pete@casazza.math.missouri.edu]

JOSÉ E. CASTILHO (**9), *Universidade Federal de Uberlândia, Campus Santa Monica, 38400-902, Uberlândia MG, Brazil* [jecastilho@ufu.br]

BRUCE L. CHALMERS (*75,*83), *Department of Mathematics, University of California, Riverside, CA 92507, USA* [blc@math.ucr.edu]

OLE CHRISTENSEN (*67), *Mathematics Institute, Building 303 Technical University of Denmark, 2800 Lyngby, Denmark* [olechr@mat.dtu.dk]

MAURICE G. COX (*89), *National Physical Laboratory, Teddington, Middlesex, TW11 0LW, UK* [mgc@npl.co.uk]

OLEG DAVYDOV (**17,**25), *Universität Dortmund, Fachbereich Mathematik, Lehrstuhl VIII, 44221 Dortmund, Germany*
[oleg.davydov@math.uni-dortmund.de]

UWE DEPCZYNSKI (*97), *Institut für Angewandte Mathematik und Statistik, Universität Hohenheim, D-70593 Stuttgart, Germany*
[depczyns@uni-hohenheim.de]

FRANK DEUTSCH (*105), *Department of Mathematics, The Pennsylvania State University, University Park, PA 16802, USA*
[deutsch@math.psu.edu]

KAI DIETHELM (*113), *Institut für Mathematik, Universität Hildesheim, Marienburger Platz 22, 31141 Hildesheim, Germany*
[diethelm@informatik.uni-hildesheim.de]

P. DRAGNEV (*119), *Department of Mathematics, Indiana-Purdue University, Fort Wayne, IN 46805, USA* [dragnevp@ipfw.edu]

NIRA DYN (**33), *Tel Aviv University, Sackler Faculty of Exact Sciences, School of Mathematical Sciences, Tel Aviv 69978, Israel*
[niradyn@math.tau.ac.il]

ALAN EGGER (*127), *Department of Mathematics, Idaho State University, Pocatello, Idaho 83209, USA* [eggealan@isu.edu]

STANLEY EINSTEIN-MATTHEWS (*239), *Department of Mathematics, Howard University, Washington, DC 20059, USA*
[smem@scs.howard.edu]

BENEDIKTE ELBEL (**39), *TU Darmstadt, Fachbereich Mathematik, Schlossgartenstrasse 7, D-64289 Darmstadt, Germany*
[elbel@mathematik.tu-darmstadt.de]

KATHLEEN W. FARMER (**47), *Department of Mathematics, Northwestern State University, 3329 University Parkway, Leesville, Louisiana 71446, USA* [kfarmer@cp-tel.net]

GREGORY E. FASSHAUER (**55), *Department of Computer Science and Applied Mathematics, Illinois Institute of Technology, Chicago, IL 60616, USA* [fass@amadeus.csam.iit.edu]

MICHAEL FELTEN (*133), *University of Dortmund, Department of Mathematics, D–44221 Dortmund, Germany*
[felten@zx2.hrz.uni-dortmund.de]

MICHAEL S. FLOATER (**63), *SINTEF Applied Mathematics, Post Box 124, Blindern, 0314 Oslo, Norway*
[Michael.Floater@math.sintef.no]

MICHAEL I. GANZBURG (*141), *Department of Mathematics, Hampton University, Hampton, VA 23668, USA*
[ganzbrgm@fusion.hamptonu.edu]

SÔNIA M. GOMES (**9), *Universidade Estadual de Campinas, IMECC, Caixa Postal 6065, 13081-970 Campinas SP, Brazil* [soniag@ime.unicamp.br]

TIM N. T. GOODMAN (**71), *Department of Mathematics, Dundee University, Dundee, DD1 4HN, Scotland* [tgoodman@mcs.dundee.ac.uk]

CHRISTIAN GOUT (*149), *UPRES A 5033, Laboratoire de Mathematiques Appliquees, Universite de Pau - IPRA, Avenue de l'Universite, 64000 Pau, France* [christian.gout@univ-pau.fr]

HANS HAGEN (**243), *Department of Computer Science, University of Kaiserslautern, Postfach 3049, D-67653 Kaiserslautern, Germany* [hagen@informatik.uni-kl.de]

BIN HAN (**97), *Department of Mathematical Sciences, University of Alberta, Edmonton, Alberta, Canada T6G 2G1* [bhan@math.ualberta.ca]

MATTHEW HE (**105), *Department of Mathematics, Nova Southeastern University, Ft. Lauderdale, FL 33314, USA* [hem@polaris.nova.edu]

TIAN XIAO HE (**113), *Department of Mathematics, Illinois Wesleyan University, Bloomington, IL 61702-2900, USA* [the@sun.iwu.edu]

WENJIE HE (**121), *Department of Mathematics, Texas A&M University, College Station, TX 77843-3368, USA* [wjhe@math.tamu.edu]

THOMAS A. HOGAN (**97), *Department of Mathematics, Vanderbilt University, Nashville, TN 37240, USA* [hogan@math.vanderbilt.edu]

DON HONG (**129), *Department of Mathematics, East Tennessee State University, Johnson City, TN 37614-0663, USA* [hong@etsu.edu]

IVAN V. IVANOV (*157), *Department of Mathematics, University of South Florida, Tampa, FL 33620, USA* [ivanov@chuma.cas.usf.edu]

KURT JETTER (**137), *Institut für Angewandte Mathematik und Statistik, Universität Hohenheim, D–70593 Stuttgart, Germany* [kjetter@uni-hohenheim.de]

HELEN E. JOYCE (*89), *School of Computing and Mathematics, University of Huddersfield, Huddersfield, HD1 3DH, UK* [j.c.mason@hud.ac.uk]

CHANDRA KAMBHAMETTU (**105), *Dept. of Computer and Information Science, University of Delaware, Newark, DE 19716-2712* [chandra@eecis.udel.edu]

J. KOREVAAR (*165), *Department of Mathematics, University of Amsterdam, Plantage Muidergracht 24, 1018 TV Amsterdam, Netherlands* [korevaar@wins.uva.nl]

ANDREI V. KROT (*175), *Department of Mathematics and Mechanics, Belorussian State University, 4 Scorina Ave., Minsk, 220080, Belarus*

ARNO KUIJLAARS (*201), *Department of Mathematics, Katholieke Universiteit Leuven, Celestijnenlaan 200 B, 3001 Leuven, Belgium* [arno@wis.kuleuven.ac.be]

A. K. KUSHPEL (*23,**145), *IMECC-UNICAMP, Caixa Postal 6065, 13081-970, Campinas, SP, Brazil* [ak99@ime.unicamp.br, (ak99@mcs.le.ac.uk)]

MING-JUN LAI (**47,**121,**153), *Department of Mathematics, University of Georgia, Athens, GA 30602, USA* [mjlai@math.uga.edu]

JOSEPH D. LAKEY (**161), *Department of Mathematical Sciences, New Mexico State University, Las Cruces, NM 88003-8001, USA* [jlakey@nmsu.edu]

TERRY L. LAY (*127), *Department of Mathematics, Idaho State University, Pocatello, ID 83209, USA* [layterr@isu.edu]

D. LEGG (*119), *Department of Mathematics, Indiana-Purdue University, Fort Wayne, IN 46805, USA* [legg@ipfw.edu]

J. LEVESLEY (*23,**145), *Department of Mathematics and Computer Science, University of Leicester, Leicester LE1 7RH, UK* [jll@mcs.le.ac.uk]

DANY LEVIATAN (*75,*223), *School of Mathematical Sciences, Sackler Faculty of Exact Sciences, Tel Aviv University, Tel Aviv 69978, Israel* [leviatan@math.tau.ac.il]

JIAN-AO LIAN (**169,**179), *Department of Mathematics, Prairie View A&M University, Prairie View, Texas 77446, USA* [jian-ao_lian@pvamu.edu]

XUE-ZHANG LIANG (**189), *Institute of Mathematics, Jilin University, Changchun, 130023, P. R. China* [xzliang@mail.jlu.edu.cn]

HUAN-WEN LIU (**129), *Department of Mathematics, University of Wollongong, Wollongong, NSW 2522, Australia* [Huanwen_Liu@uow.edu.au]

YONGPING LIU (*231), *Department of Mathematics, Beijing Normal University, Beijing 100875, P. R. China* [ypliu@bnu.edu.cn]

R. A. LORENTZ (**197), *GMD, Schloss Birlinghoven, 53757 St. Augustin, Germany* [lorentz@gmd.de]

CHUN-MEI LU (**189), *Department of Mathematics, University of South Carolina, Columbia, SC 29208, USA* [clu@math.sc.edu]

CLEMENT LUTTERODT (*239), *Department of Mathematics, Howard University, Washington, DC 20059, USA* [clutterodt@fac.howard.edu]

TOM LYCHE (**33), *University of Oslo, Institutt for Informatikk, P.O. Box 1080, Blindern 0316, Oslo* [tom@ifi.uio.no]

W. R. MADYCH (**197), *Department of Mathematics, U-9, University of Connecticut, Storrs, CT 06269, USA* [madych@uconnvm.uconn.edu]

MOHSEN MAESUMI (**205), *Mathematics Department, Lamar University, Beaumont, TX 77710, USA* [maesumi@math.lamar.edu]

MILJENKO MARUŠIĆ (**213), *Dept. of Mathematics, University of Zagreb, Bijenička 30, 10000 Zagreb, Croatia* [miljenko.marusic@math.hr]

JOHN C. MASON (*89), *School of Computing and Mathematics, University of Huddersfield, Huddersfield, HD1 3DH, UK* [j.c.mason@hud.ac.uk]

PETER R. MASSOPUST (**161), *Department of Mathematics, Sam Houston State University, Huntsville, TX 77341, USA* [mth_prm@shsu.edu]

ALLAN W. MCINNES (*247), *Department of Mathematics, University of Canterbury, Christchurch, New Zealand* [A.McInnes@math.canterbury.ac.nz]

G. MIN (*255), *Department of Mathematics and Statistics, Simon Fraser University, Burnaby, B. C., Canada V5A 1S6* [gmin@cecm.sfu.ca]

M. L. MITTAL (*263), *Department of Mathematics, University of Roorkee, Roorkee 247 667, Uttar Pradesh, India*

RAM MOHAPATRA (**129), *Department of Mathematics, University of Central Florida, Orlando, FL 32816, USA* [ramm@pegasus.cc.ucf.edu]

FRANCIS J. NARCOWICH (**221), *Department of Mathematics, Texas A&M University, College Station, TX 77843-3368, USA* [fnarc@math.tamu.edu]

ALEXA NAWOTKI (**243), *Department of Computer Science, University of Kaiserslautern, Postfach 3049, D-67653 Kaiserslautern, Germany* [nawotki@informatik.uni-kl.de]

ERICH NOVAK (**251), *University of Erlangen and Nürnberg, Mathematical Institute, Bismarckstr. 1 1/2, D-91054 Erlangen, Germany* [novak@mi.uni-erlangen.de]

GÜNTHER NÜRNBERGER (**17,**259), *Universität Mannheim, Fakultät für Mathematik und Informatik, 68131 Mannheim, Germany* [nuern@euklid.math.uni-mannheim.de]

TERESA H. O'DONNELL (**267), *ARCON Corporation, 260 Bear Hill Road, Waltham, MA 02154, USA* [terry@arcon.com]

PETER OSWALD (**275), *Bell Laboratories, Lucent Technologies, 600 Mountain Av., Rm. 2C403, Murray Hill, NJ 07974, USA* [poswald@research.bell-labs.com]

K. PAN (*83), *Department of Mathematics, Barry University, Miami Shores, FL 33161, USA* [pan@buvax.barry.edu]

MARIA C. PEREYRA (**161), *Department of Mathematics, University of New Mexico, Albuquerque, NM 87131, USA* [crisp@math.unm.edu]

IGOR E. PRITSKER (*271), *Department of Mathematics, Case Western Reserve University, 10900 Euclid Avenue, Cleveland, OH 44106-7058, USA* [iep@po.cwru.edu]

VASILIY A. PROKHOROV (*175), *Department of Mathematics, University of South Florida, Tampa, Florida 33620, USA* [prokhoro@math.usf.edu]

MICHAEL P. PROPHET (*75), *Department of Mathematics, Murray State University, Murray, KY 42071, USA* [mike@banach.mursuky.edu]

EWALD G. QUAK (**63), *SINTEF Applied Mathematics, Post Box 124, Blindern, 0314 Oslo, Norway* [Ewald.Quak@math.sintef.no]

GEETHA S. RAO (*279), *Ramanujan Institute for Advanced Study in Mathematics, University of Madras, Madras-600 005, India* [geetsr@unimad.ernet.in]

KLAUS RITTER (**251), *Fakultät für Mathematik und Informatik, Universität Passau, 94030 Passau, Germany* [Klaus.Ritter@fmi.uni-passau.de]

GEORGE A. ROBERTS (**179), *Department of Mathematics, Prairie View A&M University, Prairie View, Texas 77446, USA* [groberts@zeno.math.pvamu.edu]

AMOS RON (**283), *Department of Computer Sciences, University of Wisconsin - Madison, 1210 West Dayton, Madison, WI 57311, USA* [amos@cs.wisc.edu]

EDWARD B. SAFF (*175), *Institute for Constructive Mathematics, Department of Mathematics, University of South Florida, Tampa, Florida 33620, USA* [esaff@math.usf.edu]

R. SARAVANAN (*279), *Ramanujan Institute for Advanced Study in Mathematics, University of Madras, Madras-600 005, India* [geetsr@unimad.ernet.in]

THOMAS SAUER (*287), *Mathematical Institute, University Erlangen–Nürnberg, Bismarckstr 1 1/2, D-91054 Erlangen, Germany* [sauer@mi.uni-erlangen.de]

BL. SENDOV (*295), *Bulgarian Academy of Sciences, 1113 Sofia, Bulgaria* [sendov@amigo.acad.bg]

BORIS SHEKHTMAN (*157,*303), *Department of Mathematics, University of South Florida, Tampa, FL 33620, USA* [boris@chuma.cas.usf.edu]

IGOR A. SHEVCHUK (*223), *Institute of Mathematics, NAS of Ukraine, 3, Tereshchenkivska str., Kyiv 252601, Ukraine* [shevchuk@imath.kiev.ua]

M.-R. SKRZIPEK (*309), *FernUniversität-GHS Hagen, Fachbereich Mathematik, Postfach 940, D-58084 Hagen, Germany* [Michael.Skrzipek@FernUni-Hagen.de]

MANFRED SOMMER (**25), *Katholische Universität Eichstätt, Mathematisch-Geographische Fakultät, 85071 Eichstätt, Germany* [manfred.sommer@ku-eichstaett.de]

HUGH SOUTHALL (**267), *Electromagnetics Technology Division, US Air Force Research Laboratory, 31 Grenier Street, Hanscom AFB, MA 01731-3010, USA* [southall@maxwell.rl.plh.af.mil]

FRAUKE SPRENGEL (**319), *CWI, Kruislaan 413, P. O. Box 94079, NL-1090 GB Amsterdam, The Netherlands* [frauke.sprengel@cwi.nl]

GABRIELE STEIDL (**39), *Universität Mannheim, Fakultät für Mathematik und Informatik, A5, D-68131 Mannheim, Germany* [steidl@math.uni-mannheim.de]

ACHIM STEINBAUER (**251), *University of Erlangen and Nürnberg, Mathematical Institute, Bismarckstr. 1 1/2, D-91054 Erlangen, Germany* [steinbau@mi.uni-erlangen.de]

JOACHIM STÖCKLER (*97,**137), *Department of Mathematics and Computer Science, University of Missouri - St. Louis, 8001 Natural Bridge Road, St. Louis, MO 63121-4499* [stockler@uni-hohenheim.dc]

HANS STRAUSS (*317), *Institut für Angewandte Mathematik, Universität Erlangen–Nürnberg, Martensstr. 3, 91058 Erlangen, Germany* [strauss@am.uni-erlangen.de]

GANCHO TACHEV (*325), *Department of Mathematics, University of Architecture, 1 Hristo Smirnenski blvd., 1421 Sofia, Bulgaria*
[gtt_fte@bgace5.uacg.acad.bg]

G. D. TAYLOR (*127), *Department of Mathematics, Colorado State University, Fort Collins, CO 80523, USA*
[taylor@math.colostate.edu]

VLADIMIR N. TEMLYAKOV (*7), *Department of Mathematics, University of South Carolina, Columbia, SC 29208, USA*
[temlyak@math.sc.edu]

JUN TIAN (**327), *Computation Mathematics Laboratory, Rice University, Houston, TX 77005-1892, USA* [juntian@rice.edu]

D. TOWNSEND (*119), *Department of Mathematics, Indiana-Purdue University, Fort Wayne, IN 46805, USA* [townsend@ipfw.edu]

S. A. TOZONI (*23), *IMECC-UNICAMP, Caixa Postal 6065, 13081-970, Campinas, SP, Brazil* [tozoni@ime.unicamp.br]

GENRIKH TS. TUMARKIN (*331), *630 Rossmore Road, Goleta, CA 93117, USA* [tumarkin@gte.net]

SHAYNE WALDRON (*339), *Department of Mathematics, University of Auckland, Private Bag 92019, Auckland, New Zealand*
[waldron@math.auckland.ac.nz]

GUIDO WALZ (**337), *Department of Mathematics and Computer Science, University of Mannheim, D-68131 Mannheim, Germany*
[walz@math.uni-mannheim.de]

JOSEPH D. WARD (**137), *Department of Mathematics, Texas A&M University, College Station, TX 77843, USA*
[jward@math.tamu.edu]

JOE WARREN (**345), *Department of Computer Science, Rice University, P.O. Box 1892, Houston, TX 77005-1892, USA*
[jwarren@rice.edu]

SUMIO WATANABE (*347), *Advanced Information Processing Division, Precision and Intelligence Laboratory, Tokyo Institute of Technology, Tokyo, Japan* [swatanab@pi.titech.ac.jp]

G. ALISTAIR WATSON (**353), *Department of Mathematics, University of Dundee, Dundee DD1 4HN, Scotland*
[gawatson@mcs.dundee.ac.uk]

HENRIK WEIMER (**345), *Department of Computer Science, Rice University, P.O. Box 1892, Houston, TX 77005-1892, USA*
[henrik@rice.edu]

RAYMOND O. WELLS, JR. (**327), *Computation Mathematics Laboratory, Rice University, Houston, TX 77005-1892, USA* [wells@rice.edu]

HOLGER WENDLAND (**361), *Institut für Numerische und Angewandte Mathematik, Universität Göttingen, Lotzestr. 16-18, D-37083 Göttingen, Germany* [wendland@math.uni-goettingen.de]

PAUL WENSTON (**153), *Department of Mathematics, University of Georgia, Athens, GA 30602, USA* [paul@math.uga.edu]

NORMAN WEYRICH (**369), *Synopsys, Inc., Kaiserstr. 100, 52134 Herzogenrath, Germany* [weyrich@synopsys.com]

MARY F. WHEELER (**377), *Texas Institute for Computational and Applied Mathematics, The University of Texas at Austin, Austin, Texas 78712, USA* [mfw@ticam.utexas.edu]

IVAN YOTOV (**377), *Texas Institute for Computational and Applied Mathematics, The University of Texas at Austin, Austin, Texas 78712, USA* [yotov@ticam.utexas.edu]

FRANK ZEILFELDER (**17,**259), *Universität Mannheim, Fakultät für Mathematik und Informatik, 68131 Mannheim, Germany* [zeilfeld@fourier.math.uni-mannheim.de]

PING ZHOU (*51), *Department of Mathematics and Statistics, Simon Fraser University, Burnaby, B. C., Canada, V5A 1S6* [pzhou@cecm.sfu.ca]

On the Reproduction of Linear Functions by Local Trigonometric Bases

Kai Bittner

Abstract. Using an approach of Chui and Shi we construct smooth local trigonometric bases which reproduce linear functions locally by a finite set of basis functions. More precisely, we want to determine window functions w_j such that each linear function can be written locally as a finite linear combination of the functions

$$\psi_{2j}^k(x) = w_{2j}(x) \cos k\pi x,$$
$$\psi_{2j+1}^k(x) = w_{2j+1}(x) \sin k\pi x, \qquad j \in \mathbb{Z}, \quad k = 1, \ldots, N.$$

§1. Introduction

Since Daubechies, Jaffard, and Journé in [6] gave a method to construct an orthonormal basis of

$$L^2 := L^2(\mathbb{R}),$$

consisting of windowed trigonometric functions, local trigonometric bases have been investigated by many authors. In particular, the approach of Coifman and Meyer [5] to consider "two-overlapping" window functions turned out to be useful in many applications. A detailed study of "two-overlapping" orthonormal local trigonometric bases can be found in [1, 2]. To include various desirable features in [7] and [8], orthogonality is replaced by bi-orthogonality. In particular, to improve the approximation properties, bases are considered which reproduce constant functions locally by a finite set of basis functions. A more general approach of Chui and Shi [4] gives necessary conditions for Riesz-stability of "two-overlapping" local trigonometric bases

$$\{\psi_j^k := w_j \, T_j^k \; : \; k \in \mathbb{N}_0, \; j \in \mathbb{Z}\}$$

Approximation Theory IX, Volume 2: Computational Aspects
Charles K. Chui and Larry L. Schumaker (eds.), pp. 1–8.
Copyright ⓒ 1998 by Vanderbilt University Press, Nashville, TN.
ISBN 0-8265-1326-3.

with certain trigonometric functions T_j^k and "two-overlapping" window functions w_j. Here, we call a system of window functions w_j, $j \in \mathbb{Z}$, two-overlapping if $w_j(x)w_k(x) = 0$ for $|j - k| > 1$.

Using the results of Chui and Shi, in [3], smooth window functions w were constructed such that the functions

$$\psi_j^k(x) := w(x - j)\,\sqrt{2}\cos\big((k + \tfrac{1}{2})\pi(x - j)\big), \quad k \in \mathbb{N}_0, \quad j \in \mathbb{Z},$$

form a Riesz basis for L^2 which reproduce each linear function locally by a finite set of basis functions. This improves the approximation properties of the bases. Observe, that these window functions are constructed for trigonometric functions $T_j^k(x) := \cos\big((k + \tfrac{1}{2})\pi(x - j)\big)$. An interesting alternative is given by the functions

$$T_j^k(x) = D_j^k(x) := \begin{cases} \epsilon_k \cos k\pi x, & \text{for even } j, \\ \sqrt{2}\sin(k+1)\pi x, & \text{for odd } j, \end{cases} \tag{1}$$

where $\epsilon_0 = 1$ and $\epsilon_k = \sqrt{2}$ if $k \neq 0$. This system looks more complicated, but the symmetry properties give rise to expect more desirable features (cf. [7]). Therefore, we investigate in this paper how linear functions can be reproduced by local trigonometric functions of the type $\psi_j^k := w_j\,D_j^k$.

§2. Bi-orthogonal Local Trigonometric Bases

In the sequel, we use the approach of Chui and Shi [4] for bi-orthogonal local trigonometric bases in the "two-overlapping" setting. Here, we shortly recall the main results of this approach.

In what follows, we consider functions $\psi_j^k = w_j(x)D_j^k(x)$, $k \in \mathbb{N}_0$, $j \in \mathbb{Z}$, with trigonometric functions D_j^k given by (1). To obtain a two-overlapping setting, we demand that supp $w_j \subset [j - \tfrac{1}{2}, j + \tfrac{3}{2}]$. For the investigation of the basis properties of $\{\psi_j^k\}$, we introduce the matrices

$$M_j(x) := \begin{pmatrix} w_j(x) & w_j(2j - x) \\ -w_{j-1}(x) & w_{j-1}(2j - x) \end{pmatrix}.$$

In many applications one needs a Riesz basis, i.e., we require that there exist positive constants A and B such that each function $f \in L^2$ has the unique representation $f = \sum_{j,k} a_{j,k}\psi_j^k$, where

$$A \sum_{j \in \mathbb{Z}} \sum_{k \in \mathbb{N}_0} |a_{j,k}|^2 \leq \|f\|_{L^2}^2 \leq B \sum_{j \in \mathbb{Z}} \sum_{k \in \mathbb{N}_0} |a_{j,k}|^2.$$

In [4], it is shown that $\{\psi_j^k\}$ is a Riesz basis with Riesz bounds A and B, if and only if

$$A \geq \inf_{j \in \mathbb{Z}} \operatorname*{ess\,inf}_{x \in [j, j+\frac{1}{2}]} \|M_j^{-1}(x)\|_2^{-2} > 0,$$

$$B \leq \sup_{j \in \mathbb{Z}} \operatorname*{ess\,sup}_{x \in [j, j+\frac{1}{2}]} \|M_j(x)\|_2^2 < \infty.$$

where $\|M\|_2 = \sqrt{\rho(M^T M)}$ is the spectral norm of a Matrix M.

If $\{\psi_j^k\}$ is a Riesz basis, then the dual basis is given by $\{\tilde{\psi}_j^k := \tilde{w}_j \, D_j^k\}$ with two overlapping dual windows

$$
\tilde{w}_j(x) := \begin{cases}
\frac{w_{j-1}(2j-x)}{\det M_j(x)}, & \text{if } j - \frac{1}{2} \le x < j + \frac{1}{2}, \\
\frac{w_{j+1}(2(j+1)-x)}{\det M_{j+1}(x)}, & \text{if } j + \frac{1}{2} \le x < j + \frac{3}{2}, \\
0, & \text{otherwise.}
\end{cases}
\tag{2}
$$

Furthermore, for Riesz bases $\{\psi_j^k\}$ it holds that $w_j \in C^m(\mathbb{R})$, $j \in \mathbb{Z}$, if and only if $\tilde{w}_j \in C^m(\mathbb{R})$, $j \in \mathbb{Z}$.

§3. Dual Windows for the Reproduction of Linear Functions

Now, we are looking for a local trigonometric basis which reproduces linear functions locally with a finite number of basis functions. In particular, for the constant function we want to have the representation

$$
1 = \sum_{j \in \mathbb{Z}} \psi_j^0(x), \quad x \in \mathbb{R}.
\tag{3}
$$

Furthermore, for a linear function $p_1(x)$ we want to have

$$
p_1(x) = \sum_{j \in \mathbb{Z}} \sum_{k=0}^{N} a_{j,k} \psi_j^k(x), \quad x \in \mathbb{R}
\tag{4}
$$

with some coefficients $a_{j,k} \in \mathbb{R}$. Since the basis functions ψ_j^k and the dual basis functions $\tilde{\psi}_j^k$ have compact support, one deduces easily that (3) and (4) are equivalent to

$$
\int_{-1/2}^{3/2} \tilde{\psi}_j^k(x+j) \, dx = \delta_{k,0}, \quad j \in \mathbb{Z}
\tag{5}
$$

and

$$
\int_{-1/2}^{3/2} (x - \tfrac{1}{2}) \, \tilde{\psi}_j^k(x+j) \, dx = 0, \quad k > N, \; j \in \mathbb{Z}.
\tag{6}
$$

Therefore, we will investigate the dual window functions in more detail. Since the smoothness of the window functions is an important property of local trigonometric bases, we want to find a smoothest dual window such that (5) and (6) are satisfied for any fixed $N \in \mathbb{N}$.

Lemma 1. *Let $N \in \mathbb{N}$ and $j \in \mathbb{Z}$ be given. There exists one and only one dual window function $\tilde{w}_{N,0} \in C^{N-1}(\mathbb{R})$ with supp $\tilde{w}_{N,0} \subset [-\frac{1}{2}, \frac{3}{2}]$ such that $\tilde{\psi}_j^k(x) := \tilde{w}_{N,0}(x-j)\,\epsilon_k \cos k\pi x$ satisfies (5) and (6) for each $k \in \mathbb{N}_0$. This window is given by*

$$\tilde{w}_{N,0}(x) = \begin{cases} \frac{1}{2} + \dfrac{\frac{1}{2} + \sum_{k=0}^{\lfloor N/2 \rfloor} a_k \cos(2k+1)\pi x}{2x}, & \text{if } x \in [-\frac{1}{2}, \frac{1}{2}]\backslash\{0\}, \\ \frac{1}{2}, & \text{if } x = 0, \\ \tilde{w}_{N,0}(1-x), & \text{if } x \in (\frac{1}{2}, \frac{3}{2}], \\ 0, & \text{otherwise,} \end{cases} \qquad (7)$$

where the coefficients a_k can be computed as the solution of the linear system of equations

$$\sum_{k=0}^{\lfloor N/2 \rfloor} a_k = -\frac{1}{2}, \qquad (8)$$

$$\sum_{k=0}^{\lfloor N/2 \rfloor} (-1)^k (2k+1)^{2\mu+1} a_k = -\frac{\delta_{\mu,0}}{\pi}, \quad \mu = 0, \dots, \lfloor N/2 \rfloor - 1. \qquad (9)$$

Proof: Since $\cos k\pi x = \cos k\pi(-x)$ and $\cos k\pi x = \cos k\pi(2-x)$, we conclude that (5) and (6) are equivalent to

$$\int_0^1 \big(\tilde{w}(-x) + \tilde{w}(x) + \tilde{w}(2-x) \big)\epsilon_k \cos k\pi x \, dx = \delta_{k,0},$$

$$\int_0^1 \big((-x - \tfrac{1}{2})\tilde{w}(-x) + (x - \tfrac{1}{2})\tilde{w}(x) + (\tfrac{3}{2} - x)\tilde{w}(2-x) \big)\epsilon_k \cos k\pi x \, dx = 0,$$

$$k > N.$$

Because $\{\epsilon_k \cos k\pi x : k \in \mathbb{N}_0\}$ is an orthonormal basis of $L^2([0,1])$ these equalities are satisfied if and only if, for $x \in [0, \frac{1}{2}]$,

$$\tilde{w}(x) + \tilde{w}(-x) = 1,$$

$$-\frac{1}{2}\big(\tilde{w}(x) + \tilde{w}(-x) \big) + x\big(\tilde{w}(x) - \tilde{w}(-x) \big) = \sum_{k=0}^{N} u_k \cos k\pi x$$

and, for $x \in [\frac{1}{2}, 1]$,

$$\tilde{w}(x) + \tilde{w}(2-x) = 1,$$

$$\frac{1}{2}\big(\tilde{w}(x) + \tilde{w}(2-x) \big) + (x-1)\big(\tilde{w}(x) - \tilde{w}(2-x) \big) = \sum_{k=0}^{N} u_k \cos k\pi x$$

with some coefficients $u_k \in \mathbb{R}$. From these equalities we obtain

$$\tilde{w}(x) = \begin{cases} \frac{1}{2} + \dfrac{\frac{1}{2} + \sum_{k=0}^{N} u_k \cos k\pi x}{2x}, & \text{if } x \in [-\frac{1}{2}, \frac{1}{2})\backslash\{0\}, \\ \frac{1}{2} + \dfrac{-\frac{1}{2} + \sum_{k=0}^{N} u_k \cos k\pi x}{2(x-1)}, & \text{if } x \in [\frac{1}{2}, \frac{3}{2})\backslash\{1\}. \end{cases}$$

Thus we have $\tilde{w}(x) = \tilde{w}(1-x)$.

For $\tilde{w}$ to be in C^{N-1}, it is necessary that $\tilde{w}$ is bounded. This holds if and only if

$$\sum_{k=0}^{N} u_k = -\tfrac{1}{2} \quad \text{and} \quad \sum_{k=0}^{N} (-1)^k u_k = \tfrac{1}{2}. \tag{10}$$

Then, with $\tilde{w}(0) = \tilde{w}(1) = \tfrac{1}{2}$, the function $\tilde{w}(x)$ is analytic in $(-\tfrac{1}{2}, \tfrac{1}{2})$ and $(\tfrac{1}{2}, \tfrac{3}{2})$. Furthermore, for $\tilde{w} \in C^{N-1}(\mathbb{R})$ it is necessary that

$$\tilde{w}^{(\mu)}(-\tfrac{1}{2} + 0) = 0, \quad \mu = 0, \ldots, N - 1. \tag{11}$$

Obviously, for $x \in [-\tfrac{1}{2}, \tfrac{1}{2})$ we can write $\tilde{w}(x) = \frac{g(x)}{2x}$ with

$$g(x) := x + \frac{1}{2} + \sum_{k=0}^{N} u_k \cos k\pi x.$$

Since

$$\tilde{w}^{(n)}(x) = \sum_{\mu=0}^{n} \binom{n}{\mu} \frac{(-1)^\mu \mu!}{x^{\mu+1}} g^{(n-\mu)}(x), \quad x \in [-\tfrac{1}{2}, \tfrac{1}{2}),$$

we have that (11) is equivalent to

$$g^{(\mu)}(-\tfrac{1}{2}) = 0, \quad \mu = 0, \ldots, N - 1.$$

This, in turn, is equivalent to

$$\sum_{k=0}^{\lfloor (N+1)/2 \rfloor} (-1)^k (2k\pi)^{2\mu} u_{2k} = 0, \quad \mu = 0, \ldots, \lfloor \tfrac{N+1}{2} \rfloor - 1, \tag{12}$$

$$\sum_{k=0}^{\lfloor N/2 \rfloor} (-1)^k ((2k+1)\pi)^{2\mu+1} u_{2k+1} = -\tfrac{\delta_{\mu,0}}{2}, \quad \mu = 0, \ldots, \lfloor \tfrac{N}{2} \rfloor - 1. \tag{13}$$

Now, we rewrite (10) as

$$\sum_{k=0}^{\lfloor (N+1)/2 \rfloor} u_{2k} = 0, \quad \text{and} \quad \sum_{k=0}^{\lfloor N/2 \rfloor} u_{2k+1} = -\tfrac{1}{2}.$$

Thus, the coefficients u_{2k} resp. u_{2k+1} are the solution of the linear systems

$$\boldsymbol{C}\,\boldsymbol{u}^e = \boldsymbol{0} \quad \text{resp.} \quad \boldsymbol{D}\,\boldsymbol{u}^o = \boldsymbol{r},$$

where $\quad \boldsymbol{u}^e := (u_{2k}), \quad \boldsymbol{u}^o := (u_{2k+1}), \quad \boldsymbol{r} := (-\tfrac{1}{\pi}\delta_{\mu,0} - \tfrac{1}{2}\delta_{\mu,\lfloor N/2 \rfloor}),$
$\boldsymbol{C} := (c_{k\mu})_{k,\mu=0}^{\lfloor (N+1)/2 \rfloor}$ and $\boldsymbol{D} := (d_{k\mu})_{k,\mu=0}^{\lfloor N/2 \rfloor}$ with

$$c_{k\mu} = \begin{cases} 1, & \text{if } \mu = \lfloor (N+1)/2 \rfloor, \\ (-1)^k (2k)^\mu, & \text{otherwise}, \end{cases}$$

$$d_{k\mu} = \begin{cases} 1, & \text{if } \mu = \lfloor N/2 \rfloor, \\ (-1)^k (2k+1)^{2\mu+1}, & \text{otherwise}. \end{cases}$$

Analogous to [3, Proof of Theorem 7] one can show $\det C \neq 0$ and $\det D \neq 0$. Hence, the coefficients u_k are uniquely determined with $u_{2k} = 0$ and $u_{2k+1} = a_k$. We have still to show that the window is in fact contained in $C^{N-1}(\mathbb{R})$. Indeed, one shows by easy computations that, for $u_{2k} = 0$, (11) implies

$$\tilde{w}^{(\mu)}(\tfrac{1}{2} - 0) = \tilde{w}^{(\mu)}(\tfrac{1}{2} + 0) \quad \text{and} \quad \tilde{w}^{(\mu)}(\tfrac{3}{2} - 0) = 0, \quad \mu = 0, \dots, N - 1.$$

This completes the proof of the lemma. $\square$

Lemma 2. *Let $N \in \mathbb{N}$ and $j \in \mathbb{Z}$ be given. There exists one and only one dual window function $\tilde{w}_{N,1} \in C^N(\mathbb{R})$ with supp $\tilde{w}_{N,1} \subset [-\tfrac{1}{2}, \tfrac{3}{2}]$ such that $\tilde{\psi}_j^k(x) := \tilde{w}_{N,1}(x - j) \sin(k + 1)\pi x$ satisfies (5) and (6) for each $k \in \mathbb{N}_0$. This window is given by*

$$\tilde{w}_{N,1}(x) = \begin{cases} \dfrac{\sin \pi x}{\sqrt{2}} + \dfrac{\frac{\sin \pi x}{\sqrt{2}} + \sum_{k=1}^{\lfloor (N+1)/2 \rfloor} b_k \sin 2k\pi x}{2x}, & \text{if } x \in [-\tfrac{1}{2}, \tfrac{1}{2}] \backslash \{0\}, \\[2ex] \dfrac{\pi}{2\sqrt{2}} + \pi \sum_{k=1}^{\lfloor (N+1)/2 \rfloor} k\, b_k, & \text{if } x \in \{0\}, \\[2ex] \tilde{w}_{N,1}(1 - x) & \text{if } x \in (\tfrac{1}{2}, \tfrac{3}{2}], \\[1ex] 0, & \text{otherwise}, \end{cases} \tag{14}$$

where the coefficients b_k are the solution of the linear system of equations

$$\sum_{k=1}^{\lfloor (N+1)/2 \rfloor} (-1)^k (2k)^{2\mu+1} b_k = \sqrt{2}\,\frac{2\mu + 1}{\pi}, \quad \mu = 0, \dots, \lfloor (N+1)/2 \rfloor - 1. \tag{15}$$

Proof: Analogous to the proof of Lemma 1 we show that (5) and (6) hold if and only if

$$\tilde{w}(x) = \begin{cases} \dfrac{\sin \pi x}{\sqrt{2}} + \dfrac{\frac{\sin \pi x}{\sqrt{2}} + \sum_{k=1}^{N+1} u_k \sin k\pi x}{2x}, & \text{if } x \in [-\tfrac{1}{2}, \tfrac{1}{2}) \backslash \{0\}, \\[2ex] \dfrac{\sin \pi x}{\sqrt{2}} + \dfrac{-\frac{\sin \pi x}{\sqrt{2}} + \sum_{k=1}^{N+1} u_k \sin k\pi x}{2(x-1)}, & \text{if } x \in [\tfrac{1}{2}, \tfrac{3}{2}) \backslash \{1\}. \end{cases}$$

Obviously, with

$$\tilde{w}(0) = \tilde{w}(1) = \frac{\pi}{2\sqrt{2}} + \pi \sum_{k=1}^{N+1} \frac{k}{2} u_k,$$

$\tilde{w}$ is analytic in $(-\tfrac{1}{2}, \tfrac{1}{2})$ and $(\tfrac{1}{2}, \tfrac{3}{2})$. One easily checks that $\tilde{w}(x) = \tilde{w}(1 - x)$. By analogous reflections as in the proof of Lemma 1, we obtain that

$$\tilde{w}^{(\mu)}(-\tfrac{1}{2} + 0) = 0, \quad \mu = 0, \dots, N - 1,$$

if and only if

$$\sum_{k=0}^{\lfloor N/2 \rfloor} (-1)^k (2k + 1)^{2\mu} u_{2k+1} = 0, \quad \mu = 0, \dots, \lfloor \tfrac{N}{2} \rfloor,$$

$$\sum_{k=1}^{\lfloor (N+1)/2 \rfloor} (-1)^k (2k)^{2\mu+1} u_{2k} = \sqrt{2}\,\frac{2\mu + 1}{\pi}, \quad \mu = 0, \dots, \lfloor \tfrac{N-1}{2} \rfloor.$$

Finally, one shows again that this system of equations has a unique solution with $u_{2k+1} = 0$ and $u_{2k} = b_k$ and that for this solution the window $\tilde{w}$, given by (14), is contained in $C^N(\mathbb{R})$. $\square$

§4. Smooth Local Trigonometric Bases

From Lemmas 1 and 2 we see that $\tilde{w}_{2N,0} = \tilde{w}_{2N+1,0} \in C^{2N}(\mathbb{R})$ and $\tilde{w}_{2N,1} = \tilde{w}_{2N-1,1} \in C^{2N}(\mathbb{R})$. In general, one wants all basis functions to have the same smoothness. Therefore, for each $N \in \mathbb{N}$ we choose dual basis functions $\tilde{\psi}_j^k \in C^{2N}(\mathbb{R})$ in the following way

$$\tilde{\psi}_j^k(x) := \begin{cases} \tilde{w}_{2N,0}(x-j)\,\epsilon_k \cos k\pi x & \text{for even } j, \\ \tilde{w}_{2N,1}(x-j)\,\sqrt{2}\sin k\pi x & \text{for odd } j. \end{cases}$$

Applying (2), we obtain windows

$$w_{2N,0}(x) := \begin{cases} \frac{\tilde{w}_{2N,1}(x)}{\tilde{w}_{2N,0}(x)\tilde{w}_{2N,1}(x)+\tilde{w}_{2N,0}(-x)\tilde{w}_{2N,1}(-x)}, & \text{if } -\frac{1}{2} \leq x < \frac{1}{2}, \\ \frac{\tilde{w}_{2N,1}(x)}{\tilde{w}_{2N,0}(x)\tilde{w}_{2N,1}(x)+\tilde{w}_{2N,0}(2-x)\tilde{w}_{2N,1}(2-x)}, & \text{if } \frac{1}{2} \leq x < \frac{3}{2}, \qquad (16) \\ 0, & \text{otherwise}, \end{cases}$$

and

$$w_{2N,1}(x) := \begin{cases} \frac{\tilde{w}_{2N,0}(x)}{\tilde{w}_{2N,0}(x)\tilde{w}_{2N,1}(x)+\tilde{w}_{2N,0}(-x)\tilde{w}_{2N,1}(-x)}, & \text{if } -\frac{1}{2} \leq x < \frac{1}{2}, \\ \frac{\tilde{w}_{2N,0}(x)}{\tilde{w}_{2N,0}(x)\tilde{w}_{2N,1}(x)+\tilde{w}_{2N,0}(2-x)\tilde{w}_{2N,1}(2-x)}, & \text{if } \frac{1}{2} \leq x < \frac{3}{2}, \qquad (17) \\ 0, & \text{otherwise}. \end{cases}$$

If $\{\tilde{\psi}_j^k\}$ is a Riesz basis with Riesz bounds A and B we know that the functions

$$\psi_j^k(x) := \begin{cases} w_{2N,0}(x-j)\,\epsilon_k \cos k\pi x & \text{for even } j, \\ w_{2N,1}(x-j)\,\sqrt{2}\sin k\pi x & \text{for odd } j, \end{cases}$$

form also a Riesz basis with Riesz bounds B^{-1} and A^{-1}. From Lemmas 1 and 2 we obtain further that $\psi_j^k \in C^{2N}(\mathbb{R})$ and

$$c\,x + d = \sum_{j \in \mathbb{Z}} \left(\left(c\left(2j + \tfrac{1}{2}\right) + d \right)\psi_{2j}^0 + \sum_{k=0}^{N} \frac{c}{\sqrt{2}}\,a_k\,\psi_{2j}^{2k+1} \right.$$

$$\left. - \left(c\left(2j + \tfrac{3}{2}\right) + d \right)\psi_{2j+1}^0 + \sum_{k=1}^{N} \frac{c}{\sqrt{2}}\,b_k\,\psi_{2j+1}^{2k-1} \right).$$

Now, we have for each $N \in \mathbb{N}_0$ a local trigonometric basis $\{\psi_j^k\} \subset C^{2N}(\mathbb{R})$ such that each linear function can be reproduced by using $N+2$ (resp. $N+1$) basis functions for each even (resp. odd) time index j. This is an improvement of the result in [3], where basis functions have been constructed which are contained in $C^{2N+1}(\mathbb{R})$ and reproduce linear functions with $2N + 3$ basis functions for each time index $j \in \mathbb{Z}$.

In Table 1, we give finally the coefficients a_k and b_k as well as the Riesz bounds for some small, even N.

N	$a_k, \quad k=0,\ldots,\lfloor\frac{N}{2}\rfloor$	$b_k, k=1,\ldots,\lfloor\frac{N+1}{2}\rfloor$	A	B
0	$-\frac{1}{2}$		$\frac{1}{2}$	$2.5088\ldots$
2	$\frac{-2-3\pi}{8\pi}, \frac{2-\pi}{8\pi}$	$\frac{-1}{\sqrt{2}\pi}$	$\frac{(\pi-2)^2}{4}$	2
4	$\frac{-19-15\pi}{48\pi}, \frac{31-15\pi}{96\pi}, \frac{7-3\pi}{96\pi}$	$\frac{-13}{12\sqrt{2}\pi}, \frac{-1}{24\sqrt{2}\pi}$	$\frac{(3\pi-7)^2}{36}$	2
6	$\frac{-757-420\pi}{1536\pi}, \frac{2629-1260\pi}{7680\pi}, \frac{1007-420\pi}{7680\pi}, \frac{149-60\pi}{7680\pi}$	$\frac{-425}{384\sqrt{2}\pi}, \frac{-29}{480\sqrt{2}\pi}, \frac{-3}{640\sqrt{2}\pi}$	$\frac{(60\pi-149)^2}{14400}$	2
8	$\frac{-5767-2520\pi}{10240\pi}, \frac{2629-1260\pi}{7680\pi}, \frac{307-126\pi}{1792\pi}, \frac{6383-2520\pi}{143360\pi}, \frac{2161-840\pi}{430080\pi}$	$\frac{-1715}{1536\sqrt{2}\pi}, \frac{-539}{7680\sqrt{2}\pi}, \frac{-159}{17920\sqrt{2}\pi}, \frac{-5}{7168\sqrt{2}\pi}$	$\frac{(840\pi-2161)^2}{2822400}$	2

Table 1. Coefficients a_k and b_k and corresponding Riesz bounds for small N.

References

1. Auscher, P., Remarks on the local Fourier bases, in *Wavelets: Mathematics and Applications*, J. Benedetto and M. Frazier (eds.), CRC Press, Boca Raton, FL, 1993, pp. 203–218.

2. Auscher, P., G. Weiss, and M. V. Wickerhauser, Local sine and cosine bases of Coifman and Meyer and the construction of smooth wavelets, in *Wavelets - A Tutorial in Theory and Applications*, C. K. Chui (ed.), Academic Press, Boston, 1992, pp. 237–256.

3. Bittner, K., Error estimates and polynomial reproduction for biorthogonal local trigonometric bases, Appl. Comput. Harmonic Anal., to appear.

4. Chui, C. K. and X. Shi, Characterization and construction of biorthogonal cosine wavelets, J. Fourier Anal. Appl. **32** (1997), 559–575.

5. Coifman, R. R. and Y. Meyer, Remarques sur l'analyse de Fourier á fenêtre, C. R. Acad. Sci. Paris **312** (1991), 259–261.

6. Daubechies, I., S. Jaffard, and J. L. Journé, A simple Wilson orthonormal basis with exponential decay, SIAM J. Math. Anal. **22** (1991), 554–572.

7. Jawerth, B. and W. Sweldens, Biorthogonal smooth local trigonometric bases, J. Fourier Anal. Appl. **2** (1995), 109–133.

8. Matviyenko, G., Optimized local trigonometric bases, Appl. Comput. Harmonic Anal. **3** (1996), 301–323.

Kai Bittner
Institute of Biomathematics and Biometry
GSF - National Research Center for Environment and Health
D-85764 Neuherberg, Germany
bittner@gsf.de

Discretization of Nonlinear Terms Using Biorthogonal Wavelets: Hybrid Formulations

José E. Castilho and Sônia M. Gomes

Abstract. In applications of wavelets to PDEs, the discretization of non-linear terms is usually done in physical space, with the help of functionals defined in terms of point values. These functionals may be related to interpolation or quasi-interpolation. We adopt here an abstract approach to formulate several strategies in a unified framework. We analyze the truncation error for the discretization of $L(u, v) = uv_x$. In the context of biorthogonal splines, we give a precise description of the interaction of different Fourier modes. We also briefly discuss an efficient implementation of these schemes in a multilevel setting.

§1. Introduction

The term pseudo-spectral method appears in numerical analysis of nonlinear evolution equations, and it is associated with hybrid strategies usually used for updating the approximate solution at each time step: the linear part is calculated in the Fourier space, and the nonlinear terms are evaluated in the physical domain. In wavelet analysis, similar pseudo-wavelet methods have been adopted [2, 3, 8, 10]. In these formulations, the evaluation of the nonlinear terms in physical space uses functionals defined in terms of point values. These functionals are associated with interpolation or quasi-interpolation operators onto the multiresolution spaces.

Our purpose in this paper is to give a common formulation for these methods in order to have a unified framework for the analysis of the truncation error. This kind of abstract formulation uses the concepts of discretization and reconstruction operators, and was introduced in [5] in the context of piecewise linear finite elements. We were also inspired by the methodology suggested by A. Harten [9], which also uses these concepts in the definition of a general approach for the construction of multiresolution analyses of data.

Precisely, for differential operators $L(u, v)$, we shall consider a general form of discretization $L^j(u^j, v^j)$ given by

$$L^j(u^j, v^j) = D^j_c \left[L(\mathcal{R}^j u^j, \mathcal{R}^j v^j) \right].$$

Approximation Theory IX, Volume 2: Computational Aspects   9
Charles K. Chui and Larry L. Schumaker (eds.), pp. 9–16.
Copyright ⓒ 1998 by Vanderbilt University Press, Nashville, TN.
ISBN 0-8265-1326-3.

Let us describe the main ingredients in this formulation. The functions u and v, which are in the domain of L, are supposed to be in a functional space V. For the present case, where $L(u, v) = uv_x$, V can be taken as $H^1(\mathbf{R})$. u^j and v^j are vectors containing discrete values and belong to discrete vector spaces V^j. $\mathcal{D}^j$ is a discretization operator that maps V onto V^j, $\mathcal{D}^j : V \to V^j$, and gives the discrete values $u^j = \mathcal{D}^j u$ and $v^j = \mathcal{D}^j v$. Conversely, there is also a reconstruction operator: $\mathcal{R}^j : V^j \to V$ associating discrete vectors in V^j with functions in V. In Petrov-Galerkin formulations the discretization and reconstruction operators are defined in terms of test and trial functions, e.g.,

$$(\mathcal{D}^j u)_s = 2^j \int u(x)\phi^*_{j,s}(x)dx, \quad (\mathcal{R}^j u^j)(x) = \sum_s u^j_s \phi_{j,s}(x).$$

In this paper we have $\phi^*_{j,s}(x) = \phi^*(2^j x - s)$ and $\phi_{j,s}(x) = \phi(2^j x - s)$. To complete the description of our formulation it remains to specify $\mathcal{D}^j_c$, which is also a discretization operator. Having $\mathcal{D}^j_c$ equal to $\mathcal{D}^j$ would mean a pure Petrov-Galerkin method, which usually does not give an efficient strategy for the evaluation of nonlinear terms. Instead, hybrid formulations use two different discretizations $\mathcal{D}^j \neq \mathcal{D}^j_c$. While $\mathcal{D}^j$ is used in the discretization of u and v, $\mathcal{D}^j_c$ is used to perform the nonlinear step. The whole idea consists in having efficient discretizations $\mathcal{D}^j_c$ defined in terms of node values. By efficiency we mean that only a limited number of point evaluations are required for the computation of each functional $(\mathcal{D}^j_c u)_s$. Furthermore, we also expect to get the best possible order of accuracy for the approximation scheme $\mathcal{R}^j \mathcal{D}^j_c$.

In this paper we are concerned with algorithms designed in the framework of biorthogonal spline multiresolution analyses. However, we should mention that they can be extended straightforward to more general settings.

§2. The Biorthogonal Framework

We say that the approximation scheme $\{\mathcal{D}^j, \mathcal{R}^j\}$ is conservative if $\mathcal{R}^j$ is a right-inverse of $\mathcal{D}^j$, i.e., $\mathcal{D}^j(\mathcal{R}^j u^j) = u^j$. In other words, $\mathcal{R}^j \mathcal{D}^j$ is a projection operator. In the present case, conservation means biorthogonality, i.e., $\int \phi^*(x-k)\phi(x-l)dx = \delta_{k,l}$. We have particular interest in approximation schemes designed in terms of the biorthogonal spline families [4], where $\phi^*(x) = \varphi_{N^*}(x)$ are the B-splines of order N^*, and $\phi(x) = \varphi_{N,N^*}(x)$ are their dual functions. The numbers N and N^* are positive integers with same parity, i.e., $N + N^* = M$ is always an even integer. In the limit case $N^* = 0$, $\phi^*(x) = \delta(x)$ is the Dirac distribution, and $\phi = \varphi_{0,M}(x)$ are the Delauriers and Dubuc interpolatory scaling functions. These functions satisfy scale relations

$$\phi(x) = 2\sum_{k \in \mathbb{Z}} h(k)\phi(2x - k), \quad \phi^*(x) = 2\sum_{k \in \mathbb{Z}} h^*(k)\phi^*(2x - k), \quad (1)$$

which can be expressed as $\hat{\phi}(\xi) = H(\xi/2)\hat{\phi}(\xi/2)$ and $\hat{\phi}^*(\xi) = H^*(\xi/2)\hat{\phi}^*(\xi/2)$, where $\hat{\phi}$ stands for the Fourier transform of ϕ, and $H(\xi) = \sum_{k \in \mathbb{Z}} h(k)e^{-ik\xi}$, and $H^*(\xi) = \sum_{k \in \mathbb{Z}} h^*(k)e^{-ik\xi}$, are the filters associated with ϕ and ϕ^*.

It is well known that the best order of accuracy in approximations from shift invariant spaces is characterized by the Strang-Fix condition. We recall that ϕ satisfies the Strang-Fix condition of order p if $\hat\phi(0) \neq 0$ and $\hat\phi(\xi)$ has zeros of order $p + 1$ at all points $\xi = 2\pi k$, $0 \neq k \in \mathbb{Z}$. In this case, all the polynomials up to degree p can be locally reproduced by linear combinations of the trial functions $\phi_{j,s}(x)$, given an accuracy of order $p + 1$. For the biorthogonal splines, the order of the Strang-Fix condition is $N - 1$ for ϕ, and $N^* - 1$ for ϕ^*, corresponding to the order of the zeros of $H(\xi)$ and $H^*(\xi)$ at $\xi = \pi$. This means that for $\phi_{N^*,N}$ the best order of accuracy is N, which is also realized by the biorthogonal projection $\mathcal{R}^j \mathcal{D}^j$, as well as by all alternative approximations $\mathcal{R}^j \mathcal{D}_c^j$ presented in this paper [7, 11].

§3. Hybrid Discretizations

For the three hybrid formulations described in this section we shall assume that the basic approximation scheme $\{\mathcal{D}^j, \mathcal{R}^j\}$ comes from a biorthogonal spline framework. The hybrid formulations will be distinguished by different discretizations $\mathcal{D}_c^j$. All of them will be defined in terms of discrete convolutions of point values with some specific filter coefficients, which are chosen in order to fit some prescribed constraints.

3.1 Interpolation

For some $0 \leq \alpha < 1$, consider the discretization

$$(\mathcal{D}_c^j f)_k = \sum_n \gamma(n) f((k - n + \alpha)2^{-j}). \tag{2}$$

The requirement is that $(\mathcal{D}_c^j \phi_{j,k})_l = \delta_{k,l}$, which means that the operator $I^j(x; f) = \sum_k (\mathcal{D}_c^j f)_k \phi_{j,k}(x)$ interpolates f at the nodes $\mu = (n + \alpha)2^{-j}$. This can be realized provided that $\tilde\phi_\alpha(\xi) = \sum_k \phi(k + \alpha)e^{-ik\xi} \neq 0$. In this case, the filter coefficients $\gamma(k)$ are determined from the relation

$$\tilde\gamma(\xi)\tilde\phi_\alpha(\xi) = 1, \tag{3}$$

where $\tilde\gamma(\xi) = \sum_k \gamma(k)e^{-ik\xi}$. Numerical experiments suggest that the interpolatory condition (3) is satisfied by $\phi = \phi_{N,N^*}$ with $\alpha = 0$ for even N, and with $\alpha = 1/2$ for odd N [1]. But, except for the case $N^* = 0$, where $\phi(x)$ is itself an interpolatory function, this interpolating constraint can only be achieved with infinitely many coefficients $\gamma(k) \neq 0$. Therefore, the implementation of D_c^j in physical space requires truncated filter coefficients [8].

3.2 Quasi-Interpolation Scheme

Here the discretization has the same form as in (2), but the constraint is weaker. We only require $I^j(x; f)$ to be a quasi-interpolation in the sense of exact reproduction of polynomials up to degree $N - 1$: $I^j(f; x) = f(x)$ for $f(x) = x^m$, $m = 0, 1, \ldots, N - 1$. Then a moment relation is necessary:

$$\tilde\gamma(\xi)\tilde\phi_\alpha(\xi) = 1 + \mathcal{O}(\xi^N). \tag{4}$$

Table 1. Quasi Interpolation – $\gamma(k) \neq 0$, $k \geq 0$

(N^*, N)	$k = 0$	1	2
$(1,3)$	$\frac{11}{12}$	$\frac{1}{24}$	
$(1,5)$	$\frac{863}{960}$	$\frac{77}{1440}$	$-\frac{17}{5760}$
$(2,4)$	$\frac{5}{6}$	$\frac{1}{12}$	

Note that for scaling functions having zero moments $\int x^m \phi(x)dx = 0, m = 1, \ldots, N-1$, this relation is satisfied with $\alpha = 0$ and $\tilde{\gamma}(\xi) \equiv 1$. In such cases, the discretization (2) reduces to point values. See [7, 10] for applications. For biorthogonal splines, finitely many nonzero and symmetric coefficients $\gamma(k)$ can be chosen in order to satisfy (4). See some examples in Table 1. Again, we use $\alpha = 0$ if N is even, and $\alpha = 1/2$ if N is odd.

3.3 Discrete Projection

In contrast to the biorthogonal projection, and to the interpolation operator, quasi-interpolation schemes are not conservative in general. Discrete projection is a concept introduced in [11] with the idea of having a conservative quasi-interpolation scheme $\{\mathcal{D}_c^j, \mathcal{R}^j\}$, where the discretization is performed with finitely many nonzero coefficients. Oversampling is then required:

$$(\mathcal{D}_c^j f)_k = \sum_n \gamma(n) f((2k - n)2^{-j-1}).$$

Conservation occurs if $(\mathcal{D}_c^j \phi_{j,l})_k = \delta_{l,k}$. This can be expressed as

$$\tilde{\gamma}_{even}(\xi)\tilde{\phi}_0(\xi) + \tilde{\gamma}_{odd}(\xi)\tilde{\phi}_{1/2}(\xi) = 1, \tag{5}$$

where $\tilde{\gamma}_{even}(\xi) = \sum_k \gamma(2k)e^{-ik\xi}$ and $\tilde{\gamma}_{odd}(\xi) = \sum_k \gamma(2k+1)e^{-ik\xi}$ are the filters coming from the split of $\gamma(k)$ into two parts of even and odd index coefficients. Bezout's Theorem [6] gives a condition that implies the solvability of (5) with finitely many nonzero coefficients. Namely, the symbols $\tilde{\phi}_0(\xi)$ and $\tilde{\phi}_{1/2}(\xi)$ should not have common zeros. This property is known to be valid for the B-splines. According to our numerical experiments, this property also results to be true for several $\phi_{N^*,N}$ listed in the literature. In Table 2 are the coefficients $\gamma(k)$ for some tested cases.

§4. Truncation Error

Our purpose is to analyze the accuracy of the discretization L^j by means of the truncation error $\mathcal{D}^j L(u, v) - L^j(u^j, v^j)$. As used in previous works [5, 7, 10], Fourier analysis is the main tool. In this sense, an important first

Table 2. Discrete Projection: the coefficients $\gamma(k) \neq 0$

(N^*, N)	$k = 0$	1	2	3	4	5	6	7
$(1,3)$	$\frac{7}{9}$	$\frac{55}{576}$	$\frac{1}{72}$	$\frac{1}{576}$				
$(1,5)$	$\frac{1151}{1624}$	$\frac{107}{792}$	$\frac{142}{12067}$	$-\frac{19}{25832}$	$-\frac{27}{63649}$	$-\frac{4}{54441}$	$-\frac{7}{934226}$	$\frac{18}{107334566}$
$(2,4)$	$\frac{535}{1346}$	$\frac{2605}{10013}$	$\frac{231}{4475}$	$-\frac{94}{16799}$	$-\frac{33}{7450}$	$-\frac{9}{17374}$		

step is the understanding of the interactions of Fourier modes. Therefore, we shall focus our analysis on functions $u(x) = e^{-i\eta x}$ and $v(x) = e^{-i\zeta x}$. In all the schemes considered above, the truncation error has the general form $i\zeta e^{i(w+z)}TE(w,z)$. For interpolation and quasi-interpolation

$$|TE| = \left| \hat{\phi}^*(w+z) - \frac{i}{z}\hat{\phi}^*(w)\overline{\hat{\phi}_\alpha(w)} \; \hat{\phi}^*(z) \; \overline{\tilde{\beta}_\alpha(z)} \; \overline{\tilde{\gamma}(w+z)} \right|,$$

where $w = \eta 2^{-j}, z = \zeta 2^{-j}$, $\tilde{\beta}_\alpha(z) = \sum_k \phi'(k+\alpha)e^{-ikz}$, and $\tilde{\gamma}(\xi)$ is the filter satisfying (3) for interpolation, and (4) for quasi-interpolation. On the other hand, for the discrete projection we obtain

$$|TE| = \left| \hat{\phi}^*(w+z) - \frac{i}{z}\hat{\phi}^*(w)\hat{\phi}^*(z)\tilde{\Gamma}(w,z) \right|,$$

where $\tilde{\Gamma}(w,z) = \overline{\tilde{\gamma}_{even}(w+z)}\hat{\phi}_0(w)\tilde{\beta}_0(z) + \overline{\tilde{\gamma}_{odd}(w+z)}\hat{\phi}_{1/2}(w)\tilde{\beta}_{1/2}(z)$.

Theorem 1. *The truncation error for the three formulations satisfies*
(a) *For even N, $|TE| = \sum_{j=0}^{N} \mathcal{O}(w^j z^{N-j})$.*
(b) *For odd N, $|TE| = \mathcal{O}(z)^{N-1} + \sum_{j=0}^{N+1} \mathcal{O}(w^j z^{N+1-j})$.*

The main ingredients for the proof are the Strang Fix conditions for ϕ and ϕ^*, and the biorthogonal relation. The improvement in the order of accuracy, characteristic of splines of even degree, is a consequence of symmetry arguments combined with the fact that only first derivative is involved. In Table 3 we report some numerical experiments that confirm the estimates of Theorem 1, and give the asymptotic constants for the families $(N^*, N) = (1,5), (2,4)$. We used the software Mathematica and Matlab. At this point, some comparison remarks are in order. In favor to quasi-interpolation are the few nonzero coefficients required: $\gamma(k) = 0$ for $|k| > (N-1)/2$ for odd N, and $|k| > (N-2)/2$ for even N. In addition to oversampling, discrete projections have $\gamma(k) = 0$ for $|k| > 2N - 3$. However, the leading constants in quasi-interpolation are bigger than those in discrete projections. Quasi-interpolation has another drawback: it is not conservative. As described in the next section, this property is important in a multi-level setting.

Table 3. Leading terms in the truncation error

scheme	$(1,5)$	$(2,4)$
I	$\frac{23}{180}z^4$	$\frac{691}{3150}z^4 + \frac{184}{315}z^3w + \frac{92}{105}z^2w^2 + \frac{184}{315}zw^3$
QI	$\frac{23}{180}z^4$	$\frac{83}{1200}z^4 - \frac{1}{60}z^3w - \frac{1}{40}z^2w^2 - \frac{1}{60}zw^3 - \frac{757}{5040}w^4$
DP	$-\frac{46}{315}z^4$	$\frac{83}{7996}z^4 + \frac{123}{4450}z^3w + \frac{100}{2629}z^2w^2 + \frac{123}{4450}zw^3$

§5. Multiscale Framework

The scaling relations (1) imply that biorthogonal splines have a multiresolution structure. This means that the spaces $\mathcal{V}^j$, which are spanned by the basis $\phi_{j,k}$, form a ladder of spaces $\mathcal{V}^{j-1} \subset \mathcal{V}^j$. Furthermore, the difference of information between two consecutive approximations, $\mathcal{R}^j u^j - \mathcal{R}^{j-1}u^{j-1}$ is given by a projection operator $\mathcal{Q}^{j-1}d^{j-1}$. Now the discretization $d^{j-1} = \mathcal{G}^{j-1}u$ is obtained by testing u with $\psi^*_{j-1,k}(x)$, and the reconstruction $\mathcal{Q}^{j-1}$ uses an expansion in terms of trial functions $\psi_{j-1,k}(x)$, where $\psi^*(x)$ and $\psi(x)$ are the associated biorthogonal wavelets. Therefore, multilevel representations $\mathcal{R}^j u^j = \mathcal{R}^J u^J + \sum_{l=J}^{j-1} \mathcal{Q}^l d^l$ hold. The one-to-one relation between the one-level discretization u^j and the multi-level discretization $u^j_{ML} := \{u^J, d^J, \ldots d^{j-1}\}$ is known as wavelet transform, and it corresponds to the change of basis $\{\phi_{j,k}\} \leftrightarrow \{\phi_{J,k}\} \cup \{\psi_{J,k}\} \cup \ldots \{\psi_{j-1,k}\}$.

In wavelet based methods for PDE's, the idea is to use discretization and reconstruction operators in the multilevel form. However, in the presence of nonlinearities, hybrid formulations are also required. Therefore, our attention is again directed to the search of efficient alternative dicretizations $\mathcal{G}^l_c$, also defined in terms of a finite number of point evaluations. Noting that $d^l_k = (\mathcal{G}^l u)_k = \sum_m g^*_{m-2k} u^{l+1}_m$, with $g^*_k = (-1)^k h_{1-k}$, we may follow the same pattern and define $(\mathcal{G}^l_c u)_k = \sum_m g^*_{m-2k}(\mathcal{D}^{l+1}_c u)_m$. This may be a good option for high levels of discretization. However, at coarse levels the results are very poor. Another alternative was suggested in [8] for interpolation, and explored in [11] for discrete projection. The main idea is to use at each level the information already computed at higher levels. Precisely, a modified multiresolution discretization $\hat{u}^j_{ML}$ is defined by the application of the following algorithm

- For $l = j - 1 : (-1) : J$
$$\hat{d}^l \leftarrow \mathcal{G}^l_c(u)$$
$$u(x) \leftarrow u(x) - \sum_k \hat{d}^l_k \psi_{l,k}$$
- $\hat{u}^J_k \leftarrow \mathcal{D}^J_c(u)$

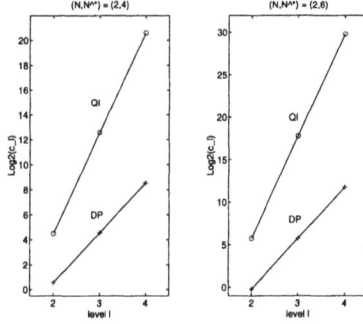

The conservation property plays an important role in the design of this algorithm. Since the quasi interpolation is just approximately conservative, we do not expect a performance as good as in the discrete projection. Applying the algorithm to $u(x) = e^{-i\eta x}$ gives $\hat{d}_k^l = e^{-i2^{j-l}kz}S_{j-l}(z)$, $z = 2^{-j}\eta$, where the symbols $S_m(z)$ may be calculated iteratively. For quasi-interpolation, $S_1 = \tilde{\gamma}(z)\tilde{g}^*(z)$, and for $m > 1$ the next formula holds

$$S_m(z) = S_1(2^{m-1}z)\left\{1 - \left[\sum_{\lambda=1}^{m-1} S_\lambda(z)\overline{\tilde{\psi}(2^\lambda z)}\right]\right\}.$$

In the case of discrete projection, $S_1(z) = \tilde{g}^*(z)\overline{\tilde{\gamma}(z/2)}$, $S_2(z) = S_1(2z) - S_1(z)T(2z)$, with $T(z) = \tilde{g}^*(z)\left[\overline{\tilde{\gamma}_{even}(z)\tilde{\psi}(z)} + e^{iz}\overline{\tilde{\gamma}_{odd}(z)\tilde{\psi}_{1/2}(z)}\right]$, and

$$S_m(z) = S_1(2^{m-1}z)\left\{1 - \left[\sum_{\lambda=1}^{m-2} S_\lambda(z)\overline{\tilde{\psi}(2^\lambda z)}\right]\right\} - S_{l-1}(z)T(2^{l-1}z), \quad m > 2.$$

Comparing with respect to the exact wavelet coefficients $d_k^l = e^{-i2^l kz}\hat{\psi}^*(2^l z)$, the error results to be $e_k^l = |\hat{d}_k^{j-l} - d_k^{j-l}| \sim c(l)z^{2N}$, for the tested cases $(2,4)$ and $(2,6)$. We found $c_{DP}(l) = 2^{Nl}$ for the discrete projection, and $c_{QI}(l) = 2^{2Nl}$. See the figure above.

The use of wavelet based methods for PDEs seems to be adequate when the target solution has isolated point singularities allowing sparse wavelet representation. As a final remark, we point out another important feature of the above algorithm. It has the capability to be easily adapted to compute a lacunary multiresolution representation, with computation effort proportional to the number of presciled wavelet coefficients [8, 11]. Given the set Λ_l of wavelet coefficients indices, there are sample points X^l required to implement $(\mathcal{G}_c^l f)_k$, $k \in \Lambda^l$. Since these functionals use finitely many filters coefficients, the size $|X^l|$ is of the same order as $|\Lambda^l|$. Therefore, the total number of operations is $O(\sum |\Lambda^l|)$.

Acknowledgments. The first author was partially supported by PICD-CAPES. This paper was completed while the second author was visiting the Institute for Advanced Study, Princeton. She is grateful for the hospitality received during her stay there. Her research was also partially supported by FAPESP (Grant 97/2248-5) and CNPq (Grant 302714/88-0).

References

1. Castilho, J. E., M. O. Domingues, and S. M. Gomes, Estimation of truncation error in discretization of nonlinear differential operators using biorthogonal wavelets, 1997, preprint.

2. Charton, P. and V. Perrier, A pseudo-wavelet scheme for the two dimensional Navier-Stokes equation, Mat. Aplic. Comp. **15** (2), 1996.

3. Cohen, A., Wavelets in numerical analysis, in *Handbook of Numerical Analysis*, Ph. Ciarlet and J. L. Lions (eds.), to appear.

4. Cohen, A., I. Daubechies, and J. C. Feauveau, Biorthogonal bases of compactly supported wavelets, Comm. Pure and Appl. Math. **45** (1992), 485–560.

5. Cullen, M. J. P. and K. W. Morton, Analysis of evolutionary error in finite element and other methods, J. Comput. Phys. **34** (1980), 245–267.

6. Daubechies I., *Ten Lectures on Wavelets*, CBMS-NSF **61**, SIAM, Philadelphia, 1992.

7. Gomes, S., Convergence estimate for the wavelet-Galerkin method: superconvergence at the node points, Advances in Comp. Math. **4** (1995), 261–282.

8. Fröhlich, J. and K. Schneider, An adaptive wavelet-vaguelette algorithm for the solution of PDEs, J. Comput. Phys. **130**(2) (1997), 174–190.

9. Harten, A., Multiresolution representation of data: A general framework, SIAM J. Numer. Anal. **33** (1996), 1205–1256.

10. Liandrat, J. and Ph. Tchamitchian, On the fast approximation of some nonlinear operators in non regular wavelet bases, Advances in Comp. Math. **8** (1998), 179–192.

11. Ware, A., Discrete projections onto wavelet subspaces, R NA-97/04, Durham University, UK, preprint.

J. E. Castilho
Universidade Federal de Uberlândia
38400-902, Uberlândia MG, Brasil
jecastilho@ufu.br

S. M. Gomes
Universidade Estadual de Campinas - IMECC
Caixa Postal 6065
13081-970 Campinas SP, Brasil
soniag@ime.unicamp.br

Interpolation by Cubic Splines
on Triangulations

Oleg Davydov, Günther Nürnberger, and Frank Zeilfelder

Abstract. We describe an algorithm for constructing point sets which admit unique Lagrange and Hermite interpolation from the space $S_3^1(\Delta)$ of C^1 splines of degree 3 defined on a general class of triangulations Δ. The triangulations Δ consist of nested polygons whose vertices are connected by line segments. In particular, we have to determine the dimension of $S_3^1(\Delta)$ which is not known for arbitrary triangulations Δ. Numerical examples are given.

§1. Introduction

In the literature, point sets which admit unique Lagrange and Hermite interpolation from spaces $S_q^r(\Delta)$ of splines of degree q and smoothness r were constructed for crosscut partitions Δ, in particular for Δ^1- and Δ^2-partitions. Results on the approximation order of these interpolation methods were also proved. (Because of space limitations, we refer to the references of our paper [5] in this volume.) Hermite interpolation schemes for $S_q^1(\Delta), q \geq 5$, where Δ is an arbitrary triangulation, were given in [1, 3].

An inductive method for constructing Lagrange and Hermite interpolation points for $S_q^1(\Delta), q \geq 5$, where Δ is an arbitrary triangulation, was developed in [2]. Here, in each step, one vertex is added to the subtriangulation considered before. For $q = 4$, this method works under certain assumptions on Δ.

The most complex case is $q = 3$, since even the dimension of $S_3^1(\Delta)$ is not known for arbitrary triangulations Δ. In this paper, we develop Lagrange and Hermite interpolation methods for $S_3^1(\Delta)$. The triangulations Δ consist of nested polygons whose vertices are connected in a natural way. The interpolation points are constructed inductively by passing through the vertices of the nested polygons, where in contrast to [2], the choice of these vertices is unique.

Approximation Theory IX, Volume 2: Computational Aspects
Charles K. Chui and Larry L. Schumaker (eds.), pp. 17–24.
Copyright © 1998 by Vanderbilt University Press, Nashville, TN.
ISBN 0-8265-1326-3.

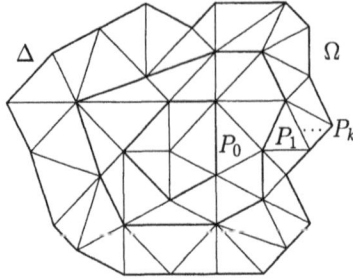

Fig. 1. Triangulation Δ (nested polygons).

§2. Main Results

Let Δ be a regular triangulation of a simply connected polygonal domain Ω in $\mathbb{R}^2$. Given an integer $q \geq 2$, we denote by $S_q^1(\Delta) = \{s \in C^1(\Omega) : s|_T \in \Pi_q \text{ for all } T \in \Delta\}$ the space of bivariate splines of degree q and smoothness one (with respect to Δ). Here $\Pi_q = \text{span}\{x^\alpha y^\beta : \alpha, \beta \geq 0, \ \alpha + \beta \leq q\}$ denotes the space of bivariate polynomials of total degree q. We investigate the following problem. Construct sets $\{z_1, \ldots, z_d\}$ in Ω, where $d = \dim S_q^1(\Delta)$, such that for each function $f \in C(T)$, a unique spline $s \in S_q^1(\Delta)$ exists which satisfies the Lagrange interpolation conditions $s(z_\nu) = f(z_\nu), \nu = 1, \ldots, d$. If we consider not only function values of f but also partial derivatives, then we speak of Hermite interpolation conditions.

The Class of Triangulations. We consider the following general type of triangulations Δ. The vertices of Δ are the vertices of closed simple polygons $P_0, P_1, \ldots, P_k$ which are nested and one vertex inside P_0. This means that $\Omega_{\mu-1} \subset \Omega_\mu$, where Ω_μ is the closed (not necessarily convex) polyhedron with boundary $P_\mu, \mu = 0, \ldots, k$, and Δ is a triangulation of $\Omega := \Omega_k$ (see Figure 1). To be more precise, we note that the vertices of P_μ are connected by line segments with the vertices of $P_{\mu+1}, \mu = 0, \ldots, k - 1$. On the other hand, for each closed simple polygon P_μ, there is no additional line segment connecting two vertices of $P_\mu, \mu = 0, \ldots, k$. In order to construct interpolation points for $S_3^1(\Delta)$, we assume that the triangulation Δ has the following properties:

(T1) Each vertex of P_μ is connected with at least two vertices of $P_{\mu+1}, \mu = 0, \ldots, k - 1$.

(T2) There exist vertices w_μ of $P_\mu, \mu = 0, \ldots, k$, such that w_μ and $w_{\mu+1}$ are connected, and each vertex w_μ is connected with at least three vertices of $P_{\mu+1}, \mu = 0, \ldots, k - 1$.

Remark 1. (*i*) Since the polygons P_μ grow with increasing index μ, it is natural to assume that the number of vertices of $P_{\mu+1}$ is greater than the number of vertices of $P_\mu, \mu = 0, \ldots, k - 1$. Then it is natural to connect the vertices of the polygons in such a way that the properties (T1) and (T2) are satisfied. (*ii*) Moreover, the properties (T1) and (T2) of Δ remain valid if Δ is deformed, i.e., the location of the vertices of Δ are changed but the

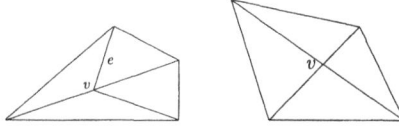

Fig. 2. Degenerate edge, respectively singular vertex.

connection of the vertices remain unchanged. (In other words, the graphs of the triangulation Δ and the deformed triangulation are the same.)

Decomposition of the Domain. In order to construct interpolation points, we decompose the domain Ω into finitely many sets $V_0 \subset V_1 \subset \cdots \subset V_m = \Omega$, where each set V_i is the union of closed triangles of Δ, $i = 0, \ldots, m$. Let V_0 be an arbitrary closed triangle of Δ in Ω_0. We define the sets $V_1 \subset \cdots \subset V_m$ by induction according to the following rule: If V_{i-1} is defined, then we choose a vertex v_i of Δ with the following property: Let $T_{i,1}, \ldots, T_{i,n_i} (n_i \geq 1)$ be all the triangles of Δ with vertex v_i having a common edge with V_{i-1}. (Since Δ satisfies property $(T1)$, we have $n_i \leq 2$.) We set $V_i = V_{i-1} \cup \overline{T}_{i,1} \cup \ldots \cup \overline{T}_{i,n_i}$. (Note that we choose the vertex v_i in such a way that at least one such triangle exists.)

The vertices $v_i, i = 1, \ldots, m$, are chosen as follows. After choosing V_0 to be an arbitrary closed triangle of Δ in Ω_0, we pass through the vertices of P_0 in clockwise order by applying the above rule. (It is clear that the choice of these vertices is unique.) Now, we assume that we have passed through the vertices of $P_{\mu-1}$. Then w.r.t. clockwise order, we choose the first vertex of P_μ greater than w_μ which is connected with at least two vertices of $P_{\mu-1}$. Then we pass through the vertices of P_μ in clockwise order until w_μ^- and pass through the vertices of P_μ in counterclockwise order until w_μ^+ by applying the above rule. (Here w_μ^+ denotes the vertex next to w_μ in clockwise order and w_μ^- denotes the vertex next to w_μ in counterclockwise order.) Finally, we choose the vertex w_μ. (It is clear that the choice of the vertices is unique.) In this way, we obtain the sets $V_0 \subset V_1 \subset \ldots \subset V_m = \Omega$.

Construction of Interpolation Sets. The choice of interpolation points depends on the following properties of the triangulation Δ.

Definition 2. *(i) An interior edge e with vertex v of the triangulation Δ is called degenerate at v if the edges with vertex v adjacent to e lie on a line. (ii) An interior vertex v of Δ is called singular if v is a vertex of exactly four edges and these edges lie on two lines. (iii) An interior vertex v of Δ on the boundary of a given subtriangulation Δ' of Δ is called semi-singular of type 1 w.r.t. Δ' if exactly one edge with endpoint v is not contained in Δ' and this edge is degenerate at v. (iv) An interior vertex v of Δ on the boundary of a given subtriangulation Δ' of Δ is called semi-singular of type 2 w.r.t. Δ' if exactly two edges with endpoint v are not contained in Δ' and these edges are degenerate at v. (v) A vertex v of Δ is called semi-singular w. r. t. Δ' if v satisfies (iii) or (iv).*

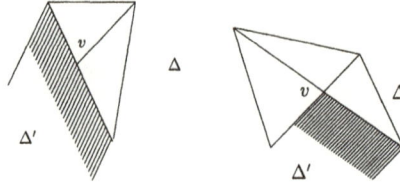

Fig. 3. Semi-singular vertex.

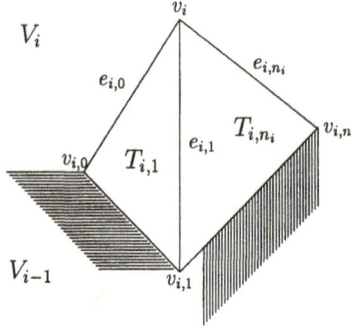

Fig. 4. The set $V_i \setminus V_{i-1}$.

Now, we construct interpolation sets for $S_3^1(\Delta)$ inductively as follows. First, we choose interpolation points on V_0 and then on $V_i \setminus V_{i-1}, i = 1, \ldots, m$. In the first step, we choose 10 different points (respectively 10 Hermite interpolation conditions) on V_0 which admit unique Lagrange interpolation by the space Π_3. (For example, we may choose four parallel line segments l_ν in V_0 and ν different points on each $l_\nu, \nu = 1, 2, 3, 4$.)

Now, we assume that we have already chosen interpolation points on V_{i-1}. Then we choose interpolation points on $V_i \setminus V_{i-1}$ as follows. By the above decomposition of Ω, $V_i \setminus V_{i-1}$ is the union of consecutive triangles $T_{i,1}, \ldots, T_{i,n_i}$ with vertex v_i having common edges with V_{i-1}. We denote the consecutive endpoints of these edges by $v_{i,0}, v_{i,1}, \ldots, v_{i,n_i}$. Moreover, the edges $[v_{i,j}, v_i]$ are denoted by $e_{i,j}, j = 0, \ldots, n_i$ (see Figure 4).

The choice of interpolation points on $V_i \setminus V_{i-1}$ depends on the following properties of the subtriangulation $\Delta_i = \{T \in \Delta : T \subset V_i\}$ at the vertices $v_{i,0}, \ldots, v_{i,n_i}$: (i) $e_{i,j}$ is non-degenerate at $v_{i,j}$. (ii) $e_{i,j}$ is non-degenerate at $v_{i,j}$ and in addition, $v_{i,j}$ is semi-singular w.r.t. Δ_i.

For $j \in \{0, n_i\}$, we set $c_{i,j} = 1$ if (ii) holds; and $c_{i,j} = 0$ otherwise. For $0 < j < n_i$, we set $c_{i,j} = 1$ if (i) holds; and $c_{i,j} = 0$ otherwise. Moreover, we set $c_i = \sum_{j=0}^{n_i} c_{i,j}$ and note that $0 \le c_i \le 3$. For Lagrange interpolation, we choose the following points on $V_i \setminus V_{i-1}$: If $c_i = 3$, then no point is chosen. If $c_i = 2$, then we choose v_i. If $c_i = 1$, then we choose v_i and one additional point on some edge $e_{i,j}$ with $c_{i,j} = 0$. If $c_i = 0$, then we choose v_i and two additional points on two different edges. For Hermite interpolation, we require the following interpolation conditions for $s \in S_3^1(\Delta)$ at the vertex v_i: If $c_i = 3$, then no interpolation condition is required at v_i. If $c_i = 2$, then we require $s(v_i) = f(v_i)$. If $c_i = 1$, then we require $s(v_i) = f(v_i)$ and $\frac{\partial s}{\partial e_{i,j}}(v_i) = \frac{\partial f}{\partial e_{i,j}}(v_i)$, where $e_{i,j}$ is some edge with $c_{i,j} = 0$. If $c_i = 0$, then we require $s(v_i) = f(v_i)$, $\frac{\partial s}{\partial x}(v_i) = \frac{\partial f}{\partial x}(v_i)$ and $\frac{\partial s}{\partial y}(v_i) = \frac{\partial f}{\partial y}(v_i)$. (For simplicity, we have denoted the derivative in direction of a unit vector parallel to the edge $e_{i,j}$ by $\frac{\partial}{\partial e_{i,j}}$.) By the above construction, we obtain a set of points for Lagrange interpolation (respectively a set of Hermite interpolation conditions).

Theorem 3. *If the triangulation Δ satisfies the properties (T1) and (T2), then there exists a unique spline in $S_3^1(\Delta)$ which satisfies the above Lagrange (respectively Hermite) interpolation conditions. In particular, the total number of interpolation conditions is equal to the dimension of $S_3^1(\Delta)$.*

Proof: Let $s \in S_3^1(\Delta)$ be a given spline which satisfies the homogenous Lagrange (respectively Hermite) interpolation conditions. We will show that $s = 0$ on Ω and that the total number of interpolation conditions on Ω is equal to the dimension of $S_3^1(\Delta)$.

First, we show by induction that $s = 0$ on $V_i, i = 0, \ldots, m$. It is clear that the interpolation conditions on V_0 imply $s = 0$ on V_0. Now, we assume that $s = 0$ on V_{i-1} for some $i \in \{1, \ldots, m\}$ and consider V_i. Set $\tilde{e}_{i,j} = [v_{i,j-1}, v_{i,j}]$ and $p_{i,j} = s|_{T_{i,j}} \in \Pi_3, j = 1, \ldots, n_i$. For simplicity, we omit the index i of $v_i, v_{i,j}, e_{i,j}, \tilde{e}_{i,j}, T_{i,j}, p_{i,j}$ and n_i. Since $s \in C^1(\Omega)$, it follows from the induction hypothesis (i.e., $s = 0$ on V_{i-1}) that for all $j \in \{1, \ldots, n\}$,

$$\frac{\partial^{\alpha+\beta} p_j}{\partial^\alpha \tilde{e}_j \partial^\beta e_{j-1}} = 0 \quad \text{and} \quad \frac{\partial^{\alpha+\beta} p_j}{\partial^\alpha \tilde{e}_j \partial^\beta e_j} = 0 \quad \text{on } \tilde{e}_j \tag{1}$$

for all $\alpha \geq 0, \beta = 0, 1$ and $\alpha + \beta \leq 3$. We consider the following cases.

Case 1. $c_i = 0$. For Lagrange interpolation, we may assume that the three interpolation points are chosen on the edges e_0 and e_1. Since p_1 is zero at these points and p_1 satisfies the zero properties (1), it follows that $p_1 = 0$. The same arguments hold for Hermite interpolation. Since $s = 0$ on T_1 and $s \in C^1(\Omega)$, we obtain $p_2(v) = 0$, $\frac{\partial p_2}{\partial \tilde{e}_2}(v) = 0$, and $\frac{\partial p_2}{\partial e_2}(v) = 0$. This together with (1) for p_2 implies $p_2 = 0$, and therefore $s = 0$ on V_i.

Case 2. $c_i = 1$. First, we consider the case $c_{i,1} = 1$, where $n = 2$. In this case e_1 is non-degenerate at v_1. Hence, we have $e_1 = \gamma_1 \tilde{e}_1 + \gamma_2 \tilde{e}_2$, where $\gamma_1, \gamma_2 \in \mathbb{R} \setminus \{0\}$. Thus, $\frac{\partial^2 p_1}{\partial^2 e_1}(v_1) = \gamma_1 \frac{\partial^2 p_1}{\partial \tilde{e}_1 \partial e_1}(v_1) + \gamma_2 \frac{\partial^2 p_1}{\partial \tilde{e}_2 \partial e_1}(v_1)$. It follows from (1) that $\frac{\partial^2 p_1}{\partial^2 e_1}(v_1) = 0$. Since v is an interpolation point, we obtain that $p_1 = 0$ on e_1. Analogously as in Case 1, we conclude $s = 0$ on V_i.

Now, we consider $c_{i,0} = 1$, i.e., v_0 is semi-singular of type 2 w.r.t. Δ_i and e_0 is non-degenerate at v_0. Let e be the edge outside of Δ_i attached to v_0, which is not lying on the same line as e_0. Denote by $\tilde{T}_j, j = 1, \ldots, 3$, the triangles with vertex v_0 outside of Δ_i in counterclockwise order and set

$\tilde{p}_j = s|_{\tilde{T}_j} \in \Pi_3, j = 1, \ldots, 3$. Since $s = 0$ on V_{i-1}, we have

$$\frac{\partial^2 p_1}{\partial e \partial e_0}(v_0) = \frac{\partial^2 \tilde{p}_1}{\partial e \partial e_0}(v_0) = -\frac{\partial^2 \tilde{p}_2}{\partial e \partial(-e_0)}(v_0) = \frac{\partial^2 \tilde{p}_3}{\partial(-e)\partial(-e_0)}(v_0) = 0 \ .$$

Since e_0 is non-degenerate at v_0, analogously as above, it follows from (1) that $\frac{\partial^2 p_1}{\partial^2 e_0}(v_0) = 0$. As in the above case, we conclude from the interpolation conditions that $s = 0$ on V_i.

Finally, we consider $c_{i,n} = 1$, i.e., v_n is semi-singular w.r.t. Δ_i and e_n is non-degenerate at v_n. We may assume that v_n is semi-singular of type 1 w.r.t. Δ_i, since the remaining case can be treated analogous to the above case $c_{i,0} = 1$. Let e be the edge outside of Δ_i attached to v_n, denote by $\tilde{T}_j, j = 1, 2$, the triangles with vertex v_n outside of Δ_i in clockwise order and set $\tilde{p}_j = s|_{\tilde{T}_j} \in \Pi_3, j = 1, 2$. Since $s = 0$ on V_{i-1}, we have

$$\frac{\partial^2 p_n}{\partial e \partial e_n}(v_n) = \frac{\partial^2 \tilde{p}_1}{\partial e \partial e_n}(v_n) = -\frac{\partial^2 \tilde{p}_2}{\partial e \partial(-e_n)}(v_n) = 0 \ .$$

Since e_n is non-degenerate at v_0, analogously as above, it follows from (1) that $\frac{\partial^2 p_n}{\partial^2 e_n}(v_n) = 0$. As in the above case, we conclude from the interpolation conditions that $s = 0$ on V_i.

Case 3. $c_i = 2$. Here, we have three cases which can be treated by analogous arguments as in Case 2.

Case 4. $c_i = 3$. By analogous arguments as in Case 2, we obtain $\frac{\partial^2 p_1}{\partial^2 e_0}(v_0) = 0$, $\frac{\partial^2 p_1}{\partial^2 e_1}(v_1) = 0$ and $\frac{\partial^2 p_2}{\partial^2 e_2}(v_2) = 0$. It is well known (cf. [4], p. 124) that each univariate polynomial p of degree 3 on an interval $[a, b]$ satisfies $6p(a) + 2(b - a)p'(a) = 6p(b) - 4(b - a)p'(b) + (b - a)^2 p''(b)$. It follows that

$$3p_1(v) + \alpha_j \frac{\partial p_1}{\partial e_j}(v) = 0, \qquad j = 0, \ldots, 2, \tag{2}$$

where α_j is the length of $e_j, j = 0, \ldots, 2$. If e_0 and e_2 lie on a line, then it is easy to see that these equations imply

$$p_1(v) = \frac{\partial p_1}{\partial e_0}(v) = \frac{\partial p_1}{\partial e_1}(v) = \frac{\partial p_1}{\partial e_2}(v) = 0. \tag{3}$$

If e_0 and e_2 do not lie on a line, then we have $\sin(\theta_1 + \theta_2)e_1 = \sin(\theta_2)e_0 + \sin(\theta_1)e_2$, where $\theta_j \in (0, \pi)$ is the angle in $T_j, j = 1, 2$ at v. Thus,

$$\sin(\theta_1 + \theta_2)\frac{\partial p_1}{\partial e_1}(v) = \sin(\theta_2)\frac{\partial p_1}{\partial e_0}(v) + \sin(\theta_1)\frac{\partial p_1}{\partial e_2}(v) \ .$$

This and (2) lead to a homogenous linear system with corresponding determinant $3(-\alpha_1\alpha_2 \sin(\theta_2) - \alpha_0\alpha_1 \sin(\theta_1) + \alpha_0\alpha_2 \sin(\theta_1 + \theta_2))$. It is obvious that for $\theta_1 + \theta_2 > \pi$, this determinant is nonzero. Moreover, this also holds for

$\theta_1 + \theta_2 < \pi$, since the area of the triangle with vertices v, v_0, and v_2 is different from the sum of the areas of the triangles T_1 and T_2. It follows that (3) holds in all cases. Now, analogous to Case 1, we obtain $s = 0$ on V_i. This shows $s = 0$ on V_i.

Finally, we show that the total number $M(\Delta)$ of interpolation conditions for $S_3^1(\Delta)$ is equal to the dimension of $S_3^1(\Delta)$. It follows from the above proof that $\dim S_3^1(\Delta) \le M(\Delta)$. On the other hand, it was shown in [6] that $L(\Delta) = 3V_B(\Delta) + 2V_I(\Delta) + \sigma(\Delta) + 1 \le \dim S_3^1(\Delta)$, where $V_B(\Delta)$ (respectively $V_I(\Delta)$) is the number of boundary (respectively interior) vertices of Δ and $\sigma(\Delta)$ is the number of singular vertices of Δ. Thus, it remains to show $L(\Delta) = M(\Delta)$. We prove by induction that $L(\Delta_i) = M(\Delta_i), i = 0, \ldots, m$. Since we have chosen $\dim \Pi_3 = 10$ interpolation conditions on Δ_0, $L(\Delta_0) = M(\Delta_0)$. We assume $L(\Delta_{i-1}) = M(\Delta_{i-1})$ and show $L(\Delta_i) = M(\Delta_i)$. If $n_i = 1$, then $V_B(\Delta_i) = V_B(\Delta_{i-1}) + 1, V_I(\Delta_i) = V_I(\Delta_{i-1})$. Thus, $L(\Delta_i) = L(\Delta_{i-1}) + 3$. By the choice of interpolation conditions, we have $M(\Delta_i) = M(\Delta_{i-1}) + 3$. Thus, $L(\Delta_i) = M(\Delta_i)$. If $n_i = 2$, then $V_B(\Delta_i) = V_B(\Delta_{i-1}), V_I(\Delta_i) = V_I(\Delta_{i-1}) + 1$. Thus, $L(\Delta_i) = L(\Delta_{i-1}) + 2$. If $c_{i,1} = 1$, then by the choice of interpolation conditions, $M(\Delta_i) = M(\Delta_{i-1}) + 2$. Thus, $L(\Delta_i) = M(\Delta_i)$. Now, we consider $c_{i,1} = 0$. We assume that $v_{i,1}$ is not singular. (Note that by property (T1) of Δ, the vertex inside P_0 is the only vertex which may be singular.) We claim that there exist a unique integer $i_0 \le i - 1$ and $j \in \{0, \ldots, n_{i_0}\}$ such that $v_{i,1} = v_{i_0}$ is semi-singular w.r.t Δ_{i_0} and $c_{i_0,j} = 1$. First, we note that it follows from properties (T1) and (T2) of Δ that $v_{i_1} := v_{i,1}$ is not semi-singular w.r.t. Δ_{i_1}. We consider two cases.

Case 1. Suppose the vertices $v_{i,1}$ and v_{i-2} are connected by a line segment e.

If the edges $e_{i,1}$ and e do not lie on a line, then $i_0 = i - 1$ and $c_{i_0,1} = 1$. (In this case, $v_{i,1}$ is semi-singular of type 1 w.r.t Δ_{i-1}.) If the edges $e_{i,1}$ and e do lie on a line, then $i_0 = i - 2$ and, since $v_{i,1}$ is not singular, $c_{i_0,n_{i_0}} = 1$. (In this case $v_{i,1}$ is semi-singular of type 2 w.r.t Δ_{i-2}.) Moreover, $v_{i,1}$ is not semi-singular w.r.t Δ_{i_2}, where $v_{i_2} := v_{i,2}$ since at least three edges of Δ_i outside of Δ_{i_2} are attached to $v_{i,1}$.

Case 2. Suppose the vertices $v_{i,1}$ and v_{i-2} are not connected by a line segment.

Let e be the edge which connects $v_{i,1}$ with the vertex on its polygon in counter-clockwise order. If the edges $e_{i,1}$ and e do not lie on a line, then we also have $i_0 = i - 1$ and $c_{i_0,1} = 1$. Moreover, since $\tilde{e}_1$ is non-degenerate at $v_{i,1}$, it follows that $v_{i,1}$ is not semi-singular w.r.t Δ_{i_2}, where $v_{i_2} := v_{i,2}$. If the edges $e_{i,1}$ and e do lie on a line, then $v_{i,1}$ is semi-singular of type 2 w.r.t Δ_{i_2}, where $v_{i_2} := v_{i,2}$, and $c_{i_2,0} = 1$. Moreover, in this case, $v_{i,1}$ is semi-singular of type 1 w.r.t Δ_{i-1}, but $c_{i-1,1} = 0$. This shows that if $c_{i,1} = 0$ (and $v_{i,1}$ not singular), then there exists a unique integer $i_0 \le i - 1$ such that $c_{i_0,j} = 1$. By the choice of interpolation conditions, it follows that $M(\Delta_i) = (M(\Delta_{i-1}) - 1) + 3 = M(\Delta_{i-1}) + 2$. Thus, $L(\Delta_i) = M(\Delta_i)$. This proves Theorem 3. $\square$

If a triangulation consists of subrectangles by adding one diagonal (of the same direction), then we speak of a Δ^1-partition. The next result on Δ^1-partitions which are deformed (see Remark 1) follows from Theorem 3.

Corollary 4. *Let Δ be a deformed Δ^1-partition. Then there exists a unique spline in $S_3^1(\Delta)$ which satisfies the Lagrange (respectively Hermite) interpolation conditions obtained by the above method.*

Numerical examples. We interpolate Franke's test function (see [5] in this volume) by splines in $S_3^1(\Delta)$, where Ω is somewhat larger than $[0, 1] \times [0, 1]$ and Δ is a uniform triangulation of Ω consisting of nested polygons. By using 3747 (resp. 14403) interpolation points, we obtain an error of $1.61 * 10^{-4}$ (resp. $2.03 * 10^{-5}$) in the uniform norm. (In the case of non-uniform Δ, our method may be modified. If $V_i \setminus V_{i-1}$ is a convex (respectively non-convex) quadrangle with one diagonal, then the second diagonal is added (respectively one triangle of the two is subdivided into three subtriangles). In this case, the interpolation points are obtained easily by combining the methods in this paper and in [5].) The interpolating splines are computed by passing through the triangles and by solving several small systems instead of one large system.

Finally, we note that our basic principle of passing through the vertices of the nested polygons of Δ can also be applied to the space $S_q^1(\Delta), q \geq 4$, in combination with the algorithm for constructing interpolation points in [2]. Then, in contrast to [2], the choice of the vertices is unique.

Acknowledgments. Oleg Davydov was supported in part by a research fellowship from the Alexander von Humboldt Foundation.

References

1. Davydov, O. V., Locally linearly independent basis for C^1 bivariate splines of degree $q \geq 5$, in *Mathematical Methods for Curves and Surfaces II*, M. Daehlen, T. Lyche, and L. L. Schumaker (eds.), Vanderbilt University Press, 1998, pp. 1–7.

2. Davydov, O. V. and G. Nürnberger, in preparation.

3. Morgan, J. and R. Scott, A nodal basis for C^1 piecewise polynomials of degree $n \geq 5$, Math. Comp. **29** (1975), 736–740.

4. Nürnberger, G., *Approximation by Spline Functions*, SpringerVerlag, Berlin, Heidelberg, New York, 1989.

5. Nürnberger, G. and F. Zeilfelder, Spline interpolation on convex quadranqulations, in *Approximation Theory IX, Volume 2: Computational Aspects*, C. K. Chui and L. L. Schumaker (eds.), Vanderbilt University Press, Nashville, 1998, 259–266.

6. Schumaker, L. L., Bounds on the dimension of spaces of multivariate piecewise polynomials, Rocky Mountain J. Math. **14** (1984), 251–264.

Interpolation and Almost Interpolation
by Weak Chebyshev Spaces

Oleg Davydov and Manfred Sommer

Abstract. Some new results on univariate interpolation by weak Chebyshev spaces, using conditions of Schoenberg-Whitney type and the concept of almost interpolation sets, are given.

§1. Introduction

Let U denote a finite-dimensional subspace of real-valued functions defined on some set K. We are interested in describing those configurations $T = \{t_1, \ldots, t_s\} \subset K$, $s \leq n = \dim U$, such that

$$\dim U_{|T} = s.$$

T is called an interpolation set (I-set) w.r.t. U. If $s = n$, then it is clearly equivalent to the condition that for any given data $\{y_1, \ldots, y_n\}$ there exists a unique $u \in U$ such that

$$u(t_i) = y_i, \qquad i = 1, \ldots, n.$$

It is well known that in the case of univariate polynomial spline spaces all interpolation sets can be characterized by the Schoenberg-Whitney condition (see, e.g., [3, 4]).

A new approach to multivariate interpolation has been found by Sommer and Strauss using the concept of almost interpolation. A set $T = \{t_1, \ldots, t_s\} \subset K$, $s \leq n$, is called an almost interpolation set (AI-set) w.r.t. U if for any system of neighborhoods B_i of t_i, $i = 1, \ldots, s$, there exist points $t_i' \in B_i$ such that $T' = \{t_1', \ldots, t_s'\}$ is an I-set w.r.t. U. They have shown that for a wide class of generalized spline spaces defined on polyhedral partitions AI-sets can be characterized by conditions of Schoenberg-Whitney type (for detail see [3]).

Approximation Theory IX, Volume 2: Computational Aspects   25
Charles K. Chui and Larry L. Schumaker (eds.), pp. 25–32.
Copyright © 1998 by Vanderbilt University Press, Nashville, TN.
ISBN 0-8265-1326-3.

Davydov [1] has considered AI-sets in the case of an arbitrary topological space K. Using the concept of local dimension (Definition 1) he has shown that under some minor additional hypotheses on K, every subspace U has a spline-like structure and every AI-set w.r.t. U can be characterized by a Schoenberg-Whitney type condition.

In this paper we apply the concept of almost interpolation and local dimension to the case when K is a real subset and U denotes a weak Chebyshev space of dimension n; i.e., every $u \in U$ has at most $n - 1$ sign changes.

§2. Characterizations of Interpolation Sets in the Weak Chebyshev Case

In the sequel we shall suppose that $K \subset \mathbb{R}$ and shall use the notations I-set and AI-set w.r.t. a space U, respectively, as defined above. We denote by $F(K)$ the linear space of all real-valued functions defined on K, and by $C(K)$, its subspace consisting of all continuous functions. Moreover, we denote the number of elements of a finite set M by card M.

We set, for a subspace U of $F(K)$,

$$Z(U) := \{t \in K : u(t) = 0 \text{ for all } u \in U\}.$$

For a space U and a subset M of K, we set

$$U(M) := \{u \in U : u = 0 \text{ on } M\}.$$

Moreover, we define for any subset $\{t_1, \ldots, t_s\} \subset K$ such that $t_1 < \ldots < t_s$,

$$t_{s+1} := t_1, [t_i, t_{i+1}] := \{t \in K : t_i \le t \le t_{i+1}\}, \ i = 1, \ldots, s - 1,$$
$$[t_s, t_1] := \{t \in K : t \ge t_s \ \text{ or } \ t \le t_1\}.$$

The following characterization of interpolation sets has been obtained in [2].

Theorem 1. *Let U be an n-dimensional weak Chebyshev subspace of $F(K)$. Moreover, suppose that $T = \{t_1, \ldots, t_n\} \subset K \setminus Z(U)$ such that $t_1 < \ldots < t_n$. The following conditions are equivalent:*

1) *T is an I-set w.r.t. U;*
2) *For all $P \subset \{1, \ldots, n\}$,*

$$\text{card}\,(T \cap \bigcup_{i \in P} [t_i, t_{i+1}]) \le \dim U_{|\bigcup_{i \in P} [t_i, t_{i+1}]}.$$

The assumption that U is a weak Chebyshev space cannot be removed.

Example 1. Let $K = [0, 3] \subset \mathbb{R}$ and assume that $U = \text{span}\,\{u_1, u_2\}$, where $u_1 = 1$ on K and

$$u_2(t) = \begin{cases} 1 - t, & \text{if } 0 \le t \le 1 \\ 0, & \text{if } 1 < t < 2 \\ t - 2, & \text{if } 2 \le t \le 3. \end{cases}$$

Set $\tilde{u} = 1/2u_1 - u_2$. Then it is obvious that $\tilde{u}$ has two sign changes at $t_1 = 1/2$ and $t_2 = 5/2$, respectively. This implies that U fails to be a weak Chebyshev space and, in particular, that T fails to be an I-set w.r.t. U, where $T = \{t_1, t_2\}$. On the other hand, T satisfies condition 2 in Theorem 1.

We also need a spline-like behaviour of each finite-dimensional subspace U of $F(K)$ as shown in [1].

Definition 1. *Let t be any element of K. The* local dimension *of U on $\{t\}$ is defined by*

$$\phi(t) := \text{l-dim}_t U := \inf\{\dim U_{|B} : t \in B, \ B \text{ open}\}.$$

It is easily seen that ϕ is an upper semicontinuous function on K. Moreover, some additional conditions hold.

Theorem 2. [3, p. 53] *Denote by $G_U \subset K$ the set of all points of continuity of local dimension. Then*

1) *G_U is an open and everywhere dense subset of K;*

2) *$G_U = \bigcup_{i \in I} K_i$, where K_i are disjoint connected components of G_U and I is a countable set;*

3) *If $Z(U) = \emptyset$ and U is a weak Chebyshev space, then $U_{|\text{int } K_i}$ is a Haar space for each $i \in I$ such that int $K_i \neq \emptyset$.*

Hence we consider the set $\{K_i\}_{i \in I}$ as a *partition* of K with *cells* K_i, and U as a "piecewise Haar" space.

Moreover, we need a "local" property of the elements of K. We say that a point $t \in K$ has V-property if either $(t - \varepsilon, t + \varepsilon) \cap K = \{t\}$ for some $\varepsilon > 0$ or

$$t = \sup\{x \in K : x < t\} = \inf\{x \in K : x > t\}.$$

We are now ready to state our first main result, which characterizes interpolation sets in terms of their locations with respect to the cells K_i.

Theorem 3. *Suppose that U denotes an n-dimensional weak Chebyshev subspace of $C(K)$. Let $K = \overline{\bigcup_{i \in I} K_i}$ be as above. Moreover, suppose that $T = \{t_1, \ldots, t_n\} \subset K \setminus Z(U)$ such that $t_1 < \cdots < t_n$, and every t_i has V-property. Then the following conditions are equivalent:*

1) *T is an I-set w.r.t. U;*

2) *For all $P \subset I$ we have*

$$\text{card}\,(T \cap M_P) \leq \dim U_{|M_P},$$

where $M_P := \overline{\bigcup_{i \in P} K_i}$.

Example 1 also shows that the assumption that U is a weak Chebyshev space cannot be removed from the statement of Theorem 3.

By definition it is obvious that every interpolation set represents an almost interpolation set. We are, therefore, interested in the converse; i.e., under what conditions an AI-set is already an I-set, and we obtain the following result for the case of weak Chebyshev spaces.

Theorem 4. *Let U denote an n-dimensional weak Chebyshev subspace of $F(K)$. Suppose that $T = \{t_1, \ldots, t_s\} \subset G_U \setminus Z(U)$ such that $s \leq n$, and every t_i has V-property. The following conditions are equivalent:*

1) *T is an AI-set w.r.t. U;*

2) *T is an I-set w.r.t. U.*

This statement is no longer true if one omits the assumption that U is a weak Chebyshev space.

Example 2. Let us consider the subspace $U = \text{span} \{u_1, u_2\}$ of $C[0, 3]$ as has been defined in Example 1. Recall that U fails to be a weak Chebyshev space. Moreover, it is easily seen that the local dimension ϕ of U on $\{t\}$ is given by

$$\phi(t) = \begin{cases} 2 & \text{if } 0 \leq t \leq 1 \text{ and } 2 \leq t \leq 3 \\ 1 & \text{if } 1 < t < 2. \end{cases}$$

Set $T = \{t_1, t_2\}$ where $t_1 = 1/2$ and $t_2 = 5/2$. Then $T \subset G_U \setminus Z(U)$, and it is obvious that t_i has V-property for $i = 1, 2$. By the arguments in Example 1, we know that T fails to be an I-set w.r.t. U. On the other hand, T is an AI-set w.r.t. U, because it is easily verified that for each sufficiently small $\varepsilon > 0$ the set $T_\varepsilon = \{\tilde{t}_1, \tilde{t}_2\}$ is an I-set w.r.t. U, where $\tilde{t}_1 = t_1$ and $\tilde{t}_2 = t_2 + \varepsilon$, respectively.

Remark. Theorems 3 and 4 together extend a statement on interpolation by generalized splines [5, Theorem 4.6]. In fact, it is shown that every weak Chebyshev space satisfies a "weak SSW-property" in terminology of [5].

§3. Proofs

In this section we shall give the proofs of the Theorems 3 and 4. To prove the first one, we need the following lemma.

Lemma 1. *Let U denote an n-dimensional subspace of $C(K)$. Moreover, suppose that $T = \{t_1, \ldots, t_s\} \subset K$, where $s \leq n$. Let $K = \bigcup_{i \in I} K_i$ be decomposed as in Section 2. If for all $P \subset I$,*

$$\text{card} \, (T \cap M_P) \leq \dim U_{|M_P}, \tag{1}$$

where $M_P = \overline{\bigcup_{i \in P} K_i}$, and every point t_j has V-property, then for all $Q \subset \{1, \ldots, s\}$,

$$\text{card} \, (T \cap R_Q) \leq \dim U_{|R_Q}, \tag{2}$$

where $R_Q = \bigcup_{j \in Q} [t_j, t_{j+1}]$.

Proof: Suppose that (2) fails. Then, for some $\tilde{Q} \subset \{1, \ldots, s\}$, it follows that $\dim U_{|R_{\tilde{Q}}} < \text{card} \, (T \cap R_{\tilde{Q}})$. Let us consider the set

$$\tilde{P} = \{i \in I : K_i \cap R_{\tilde{Q}}^o \neq \emptyset\} \cup \{i \in I : K_i = \{t_j\} \text{ or } K_i = \{t_{j+1}\} \text{ for some } j \in \tilde{Q}\},$$

where

$$R_{\tilde{Q}}^o := \bigcup_{j \in \tilde{Q}} (t_j, t_{j+1}).$$

Then, because $\bigcup_{i \in I} K_i$ is everywhere dense in K, and every t_j has V-property, we have

$$R_{\tilde{Q}} \subset M_{\tilde{P}}. \tag{3}$$

In order to reach a contradiction to (1), it is now enough to show that

$$\dim U_{|M_{\tilde{P}}} = \dim U_{|R_{\tilde{Q}}}. \tag{4}$$

On the contrary, assume that (4) is not true. Then there exists $u \in U$ such that

$$u_{|R_{\tilde{Q}}} = 0 \quad \text{and} \quad u_{|M_{\tilde{P}}} \neq 0. \tag{5}$$

Since $u \in C(K)$, it follows from (5) that $u_{|\bigcup_{i \in \tilde{P}} K_i} \neq 0$ which implies that $u_{|K_{i_0}} \neq 0$ for some $i_0 \in \tilde{P}$. Let $x \in K_{i_0}$ be such that $u(x) \neq 0$. In view of (5), we have $x \notin R_{\tilde{Q}}$. Consequently, by the definition of $\tilde{P}$, we see that $K_{i_0} \cap R_{\tilde{Q}}^o \neq \emptyset$. Therefore, K_{i_0} also contains a point y such that $y \in R_{\tilde{Q}}^o$. Assume, without loss of generality, that $x < y$. Since K_{i_0} is connected, we have $(x, y) \subset K_{i_0}$. Moreover, there exists j_0 such that $x < t_{j_0} < y$ and $(t_{j_0}, y) \subset R_{\tilde{Q}}^o$. Then $u_{|(t_{j_0}, y)} = 0$. However, by [3, Theorem 4.7], $U_{|(x,y)}$ is an almost Haar space. Hence, $u_{|(x,y)} = 0$, and, by continuity, $u(x) = 0$, a contradiction. $\square$

Proof of Theorem 3: Suppose first that T is an I-set w.r.t. U and

$$c := \operatorname{card}(T \cap M_{\tilde{P}}) > \dim U_{|M_{\tilde{P}}} =: \tilde{c}$$

for some $\tilde{P} \subset I$. Thus we could interpolate arbitrary data $\{y_1, \ldots, y_c\}$ by $U_{|M_{\tilde{P}}}$ which contradicts $c > \tilde{c}$.

Suppose now that for all $P \subset I$, we have

$$\operatorname{card}(T \cap M_P) \leq \dim U_{|M_P}.$$

Then, in view of Lemma 1, it is obvious that for all $Q \subset \{1, \ldots, n\}$,

$$\operatorname{card}(T \cap R_Q) \leq \dim U_{|R_Q}.$$

Hence it follows from Theorem 1 that T is an I-set w.r.t. U. $\square$

To prove Theorem 4, we first introduce some notations. Suppose that U denotes a subspace of $F(A)$. A finite set $T = \{t_1, \ldots, t_s\}$ is said to be an NI-set w.r.t. U, if T fails to be an I-set w.r.t. U. Every minimal NI-set is called a C-set w.r.t. U.

It is easily verified that for every C-set $T = \{t_1, \ldots, t_s\}$ there exists a (up to a factor ± 1) unique signature $\varepsilon^T = \{\varepsilon_1^T, \ldots, \varepsilon_s^T\}$, $|\varepsilon_i^T| = 1$ for all i such that no function $u \in U$ satisfies $\varepsilon_i^T u(t_i) > 0$, $i = 1, \ldots, s$. By definition, we have $\dim U_{|T} = s - 1$.

The proof of Theorem 4 is based on the following three lemmas.

Lemma 2. *Let* U *denote an* n-*dimensional weak Chebyshev subspace of* $F(K)$. *Suppose that* $T = \{t_1, \ldots, t_s\} \subset K$ *is an NI-set w.r.t.* U, *where* $t_1 < \ldots < t_s$, $s \leq n$, *and* $X = \{x_1, \ldots, x_p\} \subset K$ *is an I-set w.r.t.* $U(T)$, *where* $x_1 < \ldots < x_p$ *and* $p = n - s + 1$. *Then there does not exist any function* $u \in U$ *such that*

$$\varepsilon_i(X)u(t_i) > 0, \quad i = 1, \ldots, s, \tag{6}$$

where $\varepsilon(X) = \{\varepsilon_1(X), \ldots, \varepsilon_s(X)\}$ *is defined by*

$$\varepsilon_1(X) := 1, \quad \varepsilon_{i+1}(X) := (-1)^{\gamma_i + 1}\varepsilon_i(X), \quad i = 1, \ldots, s - 1 \tag{7}$$

with $\gamma_i := \text{card } (X \cap [t_i, t_{i+1}])$, $i = 1, \ldots, s - 1$.

Proof: On the contrary, assume that there exists $u \in U$ such that (6) holds. Since X is an I-set w.r.t. $U(T)$, it is clear that $X \cap T = \emptyset$. Hence, for each $j \in \{1, \ldots, p\}$ there exists $i \in \{0, \ldots, s\}$ such that $x_j \in (t_i, t_{i+1})$. (We set $t_0 := -\infty$, $t_{s+1} := \infty$.) Let

$$\delta_j := \begin{cases} (-1)^{\nu_j}, & \nu_j = \text{card } (X \cap [x_j, t_1]) & \text{if } i = 0 \\ (-1)^{\nu_j}\varepsilon_i(X), & \nu_j = \text{card } (X \cap [t_i, x_j]) & \text{if } i \geq 1. \end{cases}$$

Because X is an I-set w.r.t. $U(T)$, there exists a function $v \in U(T)$ such that $v(x_j) = \delta_j$, $j = 1, \ldots, p$. Then for a sufficiently small $\alpha > 0$, the function $\tilde{u} := \alpha u + v$ has n sign changes on $T \cup X \subset K$ which contradicts the hypothesis on U to be a weak Chebyshev space. $\square$

Lemma 3. *Let* U *denote an* n-*dimensional weak Chebyshev subspace of* $F(K)$. *Suppose that* $T = \{t_1, \ldots, t_s\} \subset K$ *is a C-set w.r.t.* U, *where* $t_1 < \cdots < t_s$ *with* $s \leq n$. *Then*

$$U(T) = U(K \cap [t_1, t_s]) \oplus \bigoplus_{i=1}^{s-1} U(K \setminus (t_i, t_{i+1})). \tag{8}$$

Moreover, the signature ε^T *is determined by the following equations*

$$\varepsilon_i^T \varepsilon_{i+1}^T = (-1)^{\mu_i + 1}, \quad i = 1, \ldots, s - 1, \tag{9}$$

where $\mu_i := \dim U(K \setminus (t_i, t_{i+1}))$.

Proof: Because T is a C-set, we have $\dim U(T) = n - \dim U_{|T} = n - s + 1$. Hence there exists an I-set $X = \{x_1, \ldots, x_p\}$ w.r.t. $U(T)$ such that $p = n - s + 1$. In view of Lemma 2, the sign vector $\varepsilon(X) = \{\varepsilon_1(X), \ldots, \varepsilon_s(X)\}$ defined by (7) is a signature of T.

Let the functions $u_1, \ldots, u_p \in U(T)$ be given by the conditions $u_j(x_i) = \delta_{ij}$, $i, j = 1, \ldots, p$. If (8) is not true, then there exists $j_0 \in \{1, \ldots, p\}$ such that

$$u_{j_0} \notin U(K \cap [t_1, t_s]) \cup \bigcup_{i=1}^{s-1} U(K \setminus (t_i, t_{i+1})). \tag{10}$$

Let $i_0 \in \{0, \ldots, s\}$ such that $x_{j_0} \in (t_{i_0}, t_{i_0+1})$. By (10) we can find x'_{j_0} such that $u_{j_0}(x'_{j_0}) \neq 0$ and $x'_{j_0} \in (t_i, t_{i+1})$ with $i \neq i_0$ (and $i \notin \{0, s\}$ if $i_0 \in \{0, s\}$). Consider $X' := (X \setminus \{x_{j_0}\}) \cup \{x'_{j_0}\}$ which is evidently an I-set w.r.t. $U(T)$. Let $\varepsilon(X') = \{\varepsilon_1(X'), \ldots, \varepsilon_s(X')\}$ be defined by (7), replacing X by X'. It follows from Lemma 2 that $\varepsilon(X')$ is also a signature of T. However, it is clear that $\varepsilon(X') \notin \{\varepsilon(X), -\varepsilon(X)\}$. This contradicts the uniqueness of the signature up to a factor ± 1. Thus (8) holds.

In order to check (9), it is sufficient to notice that $\varepsilon^T = \pm \varepsilon(X)$ and, in view of (8), $\mu_i = \text{card}\,(X \cap [t_i, t_{i+1}])$, $i = 1, \ldots, s-1$. $\quad\square$

Lemma 4. *Let K be a topological space, and let U denote a locally finite-dimensional subspace of $F(K)$ (i.e., for every $t \in K$ there exists a neighborhood $B(t)$ such that $U_{|B(t)}$ is finite-dimensional). Suppose that V is any linear subspace of U. Then*

$$G_V \supset G_U,$$

where G_U and G_V denote the sets of points of continuity of local dimension of U and V, respectively.

Proof: Define $\phi(t) := \text{l-dim}_t\,U$ and assume that $\bar{t} \in G_U$. Since ϕ takes only integer values, there exists an open set B, with $\bar{t} \in B \subset K$, such that $\phi(t) = c := \dim U_{|B}$ for all $t \in B$. Let $U_{|B} = V_{|B} \oplus \tilde{V}$. We set $\bar{\phi}(t) := \text{l-dim}_t\,V$ and $\tilde{\phi}(t) := \text{l-dim}_t\,\tilde{V}$, $t \in B$. Then $c = \dim V_{|B} + \dim \tilde{V}$, which implies that

$$\phi(t) \geq \bar{\phi}(t) + \tilde{\phi}(t), \quad t \in B.$$

We want to show the equality. Assume that $\phi(\tilde{t}) > \bar{\phi}(\tilde{t}) + \tilde{\phi}(\tilde{t})$ for some $\tilde{t} \in B$. Let $\hat{B} \subset B$, $\hat{B}$ open, $\tilde{t} \in \hat{B}$, such that $\bar{\phi}(\tilde{t}) = \dim V_{|\hat{B}}$ and $\tilde{\phi}(\tilde{t}) = \dim \tilde{V}_{|\hat{B}}$. Then $c = \dim U_{|\hat{B}} > \dim V_{|\hat{B}} + \dim \tilde{V}_{|\hat{B}}$, a contradiction.

Since both $\bar{\phi}$ and $\tilde{\phi}$ are upper semicontinuous, we have for some $\tilde{B} \subset B$, with $\bar{t} \in \tilde{B}$,

$$\bar{\phi}(t) \leq \bar{\phi}(\bar{t}), \quad \tilde{\phi}(t) \leq \tilde{\phi}(\bar{t}), \qquad t \in \tilde{B},$$

which, together with $\bar{\phi}(t) + \tilde{\phi}(t) = c$, $t \in B$, implies that each of $\bar{\phi}$ and $\tilde{\phi}$ is constant on $\tilde{B}$. In particular, $\bar{\phi}$ is continuous at $\bar{t}$, and hence $\bar{t} \in G_V$. $\quad\square$

Proof of Theorem 4: Assume that T is an AI-set but fails to be an I-set w.r.t. U. Without loss of generality, let T be a C-set with $s \geq 2$, since $T \cap Z(U) = \emptyset$. Thus $\dim U_{|T} = s - 1$. It is easy to check that $T \subset Z(U(T))$ and $\dim U_{|T} = \dim U_{|Z(U(T))}$.

If $T \subset$ int $Z(U(T))$, then considering the local dimension of U on T, we obtain

$$\text{l-dim}_T\, U := \inf\{\dim U_{|B} : B \supset T,\ B\ \text{open}\} \leq \dim U_{|\text{int}\, Z(U(T))} = \dim U_{|T}$$

which implies that $\text{l-dim}_T\, U \leq s - 1 <$ card T contradicting the hypothesis on T to be an AI-set. (We have used here a characterization of AI-sets given in [1, Theorem 3.3].)

Otherwise, there exists $i_0 \in \{1, \ldots, s\}$ such that $t_{i_0} \in$ bd $Z(U(T))$. This implies that t_{i_0} fails to be an isolated point of K, because otherwise $\{t_{i_0}\}$ would be open in K. Hence by the V-property, $t_{i_0} = \sup\{x \in K : x < t_{i_0}\} = \inf\{x \in K : x > t_{i_0}\}$. Then it follows from Lemma 3 that t_{i_0} is a point of discontinuity of the local dimension of $U(T)$. By Lemma 4, t_{i_0} is also a point of discontinuity of local dimension of U contradicting the hypothesis that $T \subset G_U$. $\square$

Acknowledgments. O. Davydov was supported in part by a Research Fellowship from the Alexander von Humboldt Foundation

References

1. Davydov, O., On almost interpolation, J. Approx. Theory **91** (1997), 398–418.
2. Davydov, O. and M. Sommer, Interpolation by weak Chebyshev spaces, preprint.
3. Davydov, O., M. Sommer, and H. Strauss, On almost interpolation by multivariate splines, in *Multivariate Approximation and Splines*, G. Nürnberger, J. W. Schmidt, and G. Walz (eds.), Birkhäuser, Basel, 1997, pp. 45–58.
4. Schumaker, L. L., *Spline Functions: Basic Theory*, Wiley, New York, 1981.
5. Sommer, M. and H. Strauss, Interpolation by uni- and multivariate generalized splines, J. Approx. Theory **83** (1995), 423–447.

Oleg Davydov
Universität Dortmund
Fachbereich Mathematik
Lehrstuhl VIII
44221 Dortmund
Germany
oleg.davydov@math.uni-dortmund.de

Manfred Sommer
Katholische Universität Eichstätt
Mathematisch-Geographische Fakultät
85071 Eichstätt
Germany
manfred.sommer@ku-eichstaett.de

A Hermite Subdivision Scheme
for the Evaluation of the
Powell-Sabin 12-Split Element

Nira Dyn and Tom Lyche

Abstract. It is observed that the Powell-Sabin 12-split triangle is refinable since the same split of the 4 similar subtriangles of a triangle contains the lines of split of the original triangle. This property of the split is the key to the existence of a subdivision scheme for the evaluation of the C^1 quadratic spline on the split which interpolates function and gradient values at the 3 vertices of the triangle, and normal derivatives at the midpoints of the edges. Explicit formulae for the Hermite subdivision step are given. For rendering the interpolant it is suggested to use the triangulation and the function values at the vertices obtained after a small number of subdivision iterations, and to use the known values of the gradient at the vertices to obtain the normals to the surface at the vertices of the triangulation. The shading of the 3D triangulation can then be done by Gouraud shading. It is further suggested to perturb the C^1-Hermite subdivision scheme which evaluates the above interpolant on the Powell-Sabin 12-split triangle to obtain other C^1 schemes with a shape parameter.

§1. Introduction

For bivariate smooth spline spaces of low degree on triangulations it is difficult to construct basis functions with local support. One approach which leads to good results is the splitting of each triangle into subtriangles according to the same rule of splitting. Among the known splits is the Powell-Sabin 12-split (PS-12 split). Here each triangle is divided into 12 subtriangles by connecting each vertex of the triangle to the midpoint of the opposite edge and connecting the midpoints, see Figure 1. On this split there is a unique quadratic C^1-spline interpolant to function values and gradients at the vertices and cross derivatives at the midpoints of the three edges ([9]). This interpolant is called the PS-12 split element.

Due to the large number of subtriangles in this split, it is hard to compute these elements (see [1]). Yet this split has the following advantage ([8, 7]):

Approximation Theory IX, Volume 2: Computational Aspects
Charles K. Chui and Larry L. Schumaker (eds.), pp. 33–38.
Copyright © 1998 by Vanderbilt University Press, Nashville, TN.
ISBN 0-8265-1326-3.

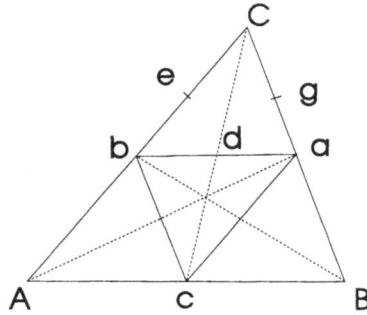

Fig. 1. A triangle divided into 12 subtriangles by the PS-12 split.

Suppose we subdivide the original triangle into 4 similar triangles (called refined triangles) by connecting midpoints of edges. Then using the PS-12 split on each of the refined triangles we obtain a spline space of C^1 piecewise quadratic polynomials which contains the spline space of C^1 piecewise quadratic polynomials defined on the PS-12 split of the original triangle as a subspace. This property is called *refinability* and is important for multiresolution analysis. Other well-known splits do not have this property. This refinability property allows us to compute the PS-12 split element in a simple way which ignores the split and generates function values and gradients on a dense set of points. By repeatedly subdividing the refined triangles in this way, we obtain the values and gradients of the PS-12 split element on the vertices of the refined triangulation. We can stop this process after a few iterations of the subdivision (4-6 iterations) and display the piecewise linear surface on the current refined triangulation, using the function values. To improve the rendering of the surface, one can use the gradients for evaluating the normals at the vertices of the current triangulation, and perform Gouraud shading of the piecewise linear surface. (Our figures of the subdivision surfaces were obtained without the Gouraud shading, which generates the same quality of rendered surfaces after 2-3 iterations).

In this paper we show how to compute the PS-12 split element by a Hermite subdivision scheme on triangles. A different C^1-Hermite subdivision scheme was designed in [3], with a limit surface which is not a spline. Univariate Hermite subdivision schemes were studied in [2, 5, 6]. Special cases of these schemes lead to univariate splines.

By perturbing the formula we obtain a one-parameter family of new C^1-Hermite subdivision schemes with the perturbation parameter acting as a *tension parameter*. The resulting surfaces are no longer spline surfaces.

§2. Computing the PS-12 Split Element by Subdivision

Given function values and gradients at the vertices A, B, C of a triangle T, and cross derivatives at midpoints a, b, c of edges opposite A, B, C (see Figure 1), a unique PS-12 split element which interpolates this data is determined ([9]).

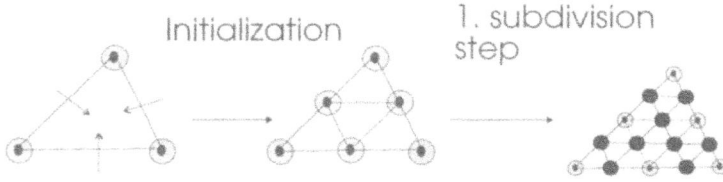

Fig. 2. Subdividing the PS-12 split element. A circle around a vertex means that both the function value and the gradient are known at that vertex.

2.1. Initialization

The first step in the computation of such an element involves the computation of its value and gradient at the midpoints a, b, c of the triangle T (see Figure 1). Here we use the formula

$$f_b = (f_A + f_C)/2 - (\nabla f_A - \nabla f_C) \cdot (A - C)/8$$

for the function value at the midpoint b of AC. For the gradient we first compute the directional derivative in the direction AC at b

$$(A - C) \cdot \nabla f_b = 2(f_A - f_C) - (\nabla f_A + \nabla f_C) \cdot (A - C)/2.$$

Combining this value with the given value of the cross-derivative at b, we can calculate ∇f_b. For the other midpoints we use similar formulae.

These formulae are obtained from the observation that along each side of T the PS-12 split element is a piecewise quadratic C^1-spline with a knot at the midpoint.

2.2. The General Subdivision Step

For the first subdivision step (see Figure 1) we use the following formulae:

$$
\begin{aligned}
f_e &= (f_b + f_C)/2 - (\nabla f_b - \nabla f_C) \cdot (b - C)/8 \\
f_g &= (f_a + f_C)/2 - (\nabla f_a - \nabla f_C) \cdot (a - C)/8 \\
f_d &= (f_b + f_a)/2 - (\nabla f_b - \nabla f_a) \cdot (b - a)/8 \\
\nabla f_e &= (\nabla f_b + \nabla f_C)/2 \\
\nabla f_g &= (\nabla f_a + \nabla f_C)/2 \\
(a - b) \cdot \nabla f_d &= 2(f_a - f_b) - (\nabla f_a + \nabla f_b) \cdot (a - b)/2 \\
(C - d) \cdot \nabla f_d &= 2(f_C - f_d) - \nabla f_C \cdot (C - d).
\end{aligned}
\tag{1}
$$

From the last two values we can solve for ∇f_d. Similar formulae are used for the two other corner triangles Abc and Bca, and we obtain the values and gradients at locations shown to the right in Figure 2. This process can now be continued for as many levels of refinement as desired.

Figure 3 displays a PS-12 split element obtained from random initial data. The implementation was done using Mathematica.

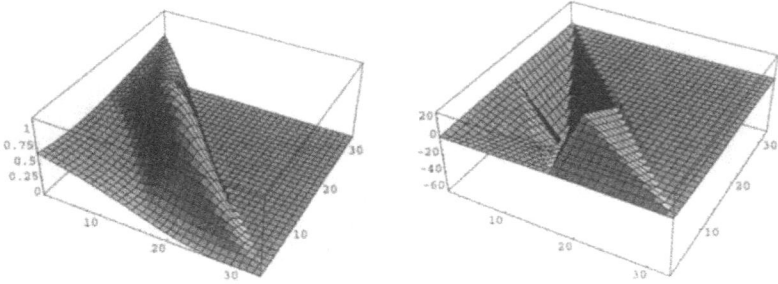

Fig. 3. A PS-12 split element (left) constructed from random data using 5 levels of subdivision. The partial derivative in the x-direction (right).

2.3. Computation of Interpolants on a Given Triangulation

The above procedure can be extended to any initial triangulation Δ with data consisting of function values and gradient values at the vertices, and cross derivatives at midpoints of edges. The data determines a unique interpolant f in the space $S_2^1(\Delta_{PS})$, where Δ_{PS} is the triangulation obtained from Δ by splitting each triangle of Δ into 12 subtriangles according to the PS-12 split. Here $S_d^r(\Delta)$ denotes the spline space of C^r-piecewise polynomials of degree at most d on the triangulation Δ.

With the subdivision procedure, one can get values and gradients of the interpolant f at the vertices of a refined triangulation of the initial one, for any level of refinement.

§3. Tension Parameters

Since the formulae (1) generate a C^1-surface by subdivision, small changes in parameters will also generate a C^1-surface. This follows from the observation that a subdivision scheme which depends on a parameter and generates a C^1-surface will have this property in a neighborhood of that parameter value (see [4]).

Suppose in (1) we keep the formulae for the gradient values, but change the formulae for function values to

$$f_e = (f_b + f_C)/2 - (\nabla f_b - \nabla f_C) \cdot (b - C)\alpha$$
$$f_g = (f_a + f_C)/2 - (\nabla f_a - \nabla f_C) \cdot (a - C)\alpha \qquad (2)$$
$$f_d = (f_b + f_a)/2 - (\nabla f_b - \nabla f_a) \cdot (b - a)\alpha.$$

Then for $\alpha = 1/8$ we get a C^1-surface. Therefore, for α in a neighborhood of $1/8$ we still get a C^1-surface.

We observe that for $\alpha = 0$ we obtain a piecewise linear interpolant to the data of function values at the vertices, see Figure 4. Therefore, we expect to get a "tighter" surface in the left neighborhood of $1/8$ and a "looser" surface in the right neighborhood of $1/8$. Figures 5 and 6 are in agreement with this expectation. This can be seen by comparing the range of function values of the partial derivative in the x-direction.

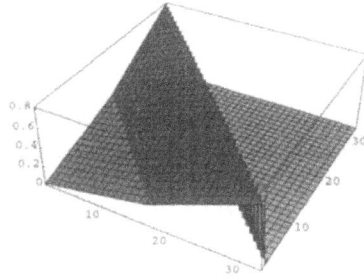

Fig. 4. Subdivision surface with tension parameter $\alpha = 0$ using the same initial data as in Figure 3.

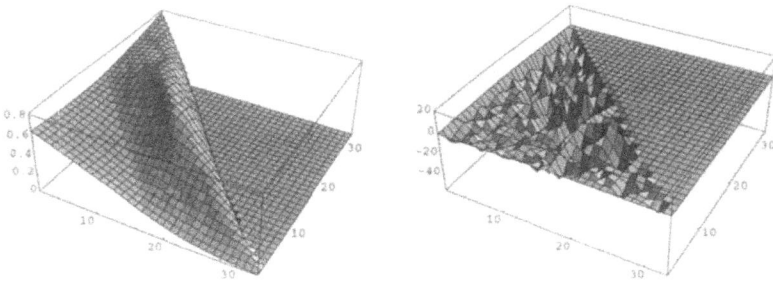

Fig. 5. Subdivision surface (left) with tension parameter $\alpha = 1/16$ using the same initial data as in Figure 3. The partial derivative in the x-direction (right).

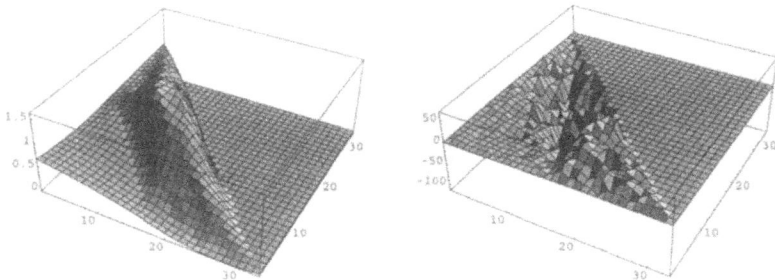

Fig. 6. Subdivision surface (right) with tension parameter $\alpha = 3/16$ using the same initial data as in Figure 3. The partial derivative in the x-direction (right).

Acknowledgments. Part of the work of the second author was carried out during a stay at "Institut National des Sciences Appliquées" and "Laboratoire Approximation et Optimisation" of "Université Paul Sabatier," Toulouse, France.

References

1. Chui, C. K. and T.-X. He, Bivariate C^1 quadratic finite elements and vertex splines, Math. Comp. **54** (1990), 169–187.

2. Merrien, J.-L., A family of Hermite interpolants by bisection algorithms, Numer. Algorithms **2** (1992), 187–200.

3. Merrien, J.-L., Dyadic Hermite interpolants on a triangulation, Numer. Algorithms **7** (1994), 391–410.

4. Dyn, N., Subdivision schemes in computer-aided geometric design, in *Advances in Numerical Analysis - Volume II, Wavelets, Subdivision Algorithms and Radial Basis Functions*, W. Light (ed.), Clarendon Press, Oxford, 1992, pp. 36–104.

5. Dyn, N. and D. Levin, Analysis of Hermite-type subdivision schemes, in *Approximation Theory VIII, Vol. 2: Wavelets*, Charles K. Chui and Larry L. Schumaker (eds.), World Scientific Publishing Co., Inc., Singapore, 1995, pp. 117–124.

6. Dyn, N. and D. Levin, Analysis of Hermite-interpolatory subdivision schemes, in *Spline Functions and the Theory of Wavelets*, S. Dubuc (ed.), AMS series CRM Proceedings and Lecture Notes, 1998, to appear.

7. Dæhlen, M., T. Lyche, K. Mørken, R. Schneider, and H-P. Seidel, Multiresolution analysis based on quadratic Hermite interpolation, Part 2: Piecewise polynomial surfaces over triangles, preprint.

8. Oswald, P., Hierarchical conforming finite element methods for the biharmonic equation, SIAM J. Numer. Anal. **29** (1992), 1610–1625.

9. Powell, M. J. D. and M. A. Sabin, Piecewise quadratic approximation on triangles, ACM Trans. Math. Software **3** (1977), 316–325.

Nira Dyn
Tel Aviv University
Sackler Faculty of Exact Sciences
School of Mathematical Sciences
Tel Aviv, 69978
ISRAEL
niradyn@math.tau.ac.il

Tom Lyche
University of Oslo
Institutt for informatikk
P.O. Box 1080, Blindern, 0316 Oslo
NORWAY
tom@ifi.uio.no

Fast Fourier Transforms for Nonequispaced Data

Benedikte Elbel and Gabriele Steidl

Abstract. In this paper, we propose a general efficient method for the fast approximative computation of discrete Fourier transforms on nonequispaced grids in the time domain and the frequency domain. We derive estimates for the approximation error which depend on the arithmetic complexity of the algorithm.

§1. Introduction

Let $\Pi^d := [-\frac{1}{2}, \frac{1}{2}]^d$, $I_N := \{k = (k_j)_{j=1}^d \in \mathbb{Z}^d : -\frac{N}{2} \le k_j < \frac{N}{2}\}$. For $x_k \in \Pi^d$ and $v_j \in N\Pi^d$, we are interested in the fast and stable computation of the M^d sums

$$f(v_j) = \sum_{k \in I_N} f_k \, e^{-2\pi i x_k v_j}, \qquad j \in I_M. \tag{1}$$

For equispaced nodes $x_k := \frac{k}{N}$, $k \in I_N$ and $v_j := j$, $j \in I_N$, the values $f(v_j)$ can be computed by the well-known fast Fourier transform (FFT) with only $\mathcal{O}(N^d \log N)$ arithmetical operations. However, the FFT requires sampling on an equally spaced grid which represents a significant limitation for many applications. Unfortunately, for arbitrary nodes, the direct evaluation of (1) takes $\mathcal{O}(N^d M^d)$ arithmetical operations, too much for practical purposes. Therefore, based on local series expansions of the complex exponentials (Taylor series [1], series of prolate spheroidal functions [8]), algorithms for the fast approximate computation of (1) were developed which are significantly better than the direct evaluation. Further progress was achieved by algorithms based on local Lagrange interpolation [3, 10]. Finally, these algorithms were outperformed by algorithms recently introduced by Dutt and Rokhlin [5] and Beylkin [2] for the computation of the discrete Fourier transform on nonequispaced nodes either in the time domain or in the frequency domain, i.e., for the computation of

$$f(v_j) = \sum_{k \in I_N} f_k \, e^{-2\pi i k v_j / N}, \qquad j \in I_M \tag{2}$$

Approximation Theory IX, Volume 2: Computational Aspects   39
Charles K. Chui and Larry L. Schumaker (eds.), pp. 39–46.
Copyright © 1998 by Vanderbilt University Press, Nashville, TN.
ISBN 0-8265-1326-3.

or

$$f(k) = \sum_{j \in I_N} f_j \, e^{-2\pi i k v_j / N}, \qquad k \in I_M. \tag{3}$$

Note that an algorithm for the fast computation of (2) implies an algorithm for the fast computation of the "transposed" problem (3) in a straightforward way by reversing the order of the computation steps. Similar algorithms were always known in image processing contexts as so-called "gridding" [9, 11]. However, only the papers [2] and [5] investigate the dependence of speed and accuracy of the algorithms on the choice of the approximation scheme and the oversampling factor. While Dutt and Rokhlin approximate the complex exponentials by suitable series of Gaussian bells, Beylkin uses interpolation by translates of B-splines and special approximations involving Lemarié-Battle scaling functions.

In this paper, we propose a new unified approach to the fast computation of (1), which incorporates the algorithms in [2] and [5] for the special setting (2). Our point of view allows simple error estimates. In particular, we improve the estimates given in [5].

Finally, let us remark that in the *univariate case*, there exists a quite different type of algorithms which works also for the inverse transform. These algorithms are based on a suitable factorization of the transform matrix followed by multipole techniques [4, 6].

§2. FFT for Nonequispaced Grids in Time or Frequency Domain

Let us begin with the computation of (2). For notational reasons, we restrict our attention to the case $M = N$. We have to evaluate the 1-periodic trigonometric polynomial

$$f(v) := \sum_{k \in I_N} f_k \, e^{-2\pi i k v} \tag{4}$$

at the nodes $w_j := v_j / N \in \Pi^d$ ($j \in I_N$). We introduce the oversampling factor $\alpha > 1$ and set $n := \alpha N$. Let φ be a continuous 1-periodic function of bounded variation. We approximate f by

$$s_1(v) := \sum_{l \in I_n} g_l \, \varphi\left(v - \frac{l}{n}\right). \tag{5}$$

Switching to the frequency domain, we obtain

$$s_1(v) = \sum_{k \in \mathbb{Z}^d} \hat{g}_k \, c_k(\varphi) \, e^{-2\pi i k v}$$

$$= \sum_{k \in I_n} \hat{g}_k \, c_k(\varphi) \, e^{-2\pi i k v} + \sum_{r \in \mathbb{Z}^d \setminus \{0\}} \sum_{k \in I_n} \hat{g}_k \, c_{k+nr}(\varphi) \, e^{-2\pi i (k+nr) v} \tag{6}$$

with

$$\hat{g}_k := \sum_{l \in I_n} g_l \, e^{2\pi i k l/n}, \tag{7}$$

$$c_k(\varphi) := \int_{\Pi^d} \varphi(v) e^{2\pi i k v} \, dv, \quad k \in \mathbb{Z}^d.$$

If the Fourier coefficients c_k become sufficiently small for $k \in \mathbb{Z}^d \backslash I_n$ and if $c_k(\varphi) \neq 0$, $k \in I_N$, then we suggest by comparing (4) with (6) to set

$$\hat{g}_k := \begin{cases} f_k/c_k(\varphi) & k \in I_N, \\ 0 & k \in I_n \backslash I_N. \end{cases} \tag{8}$$

Now the values g_l can be obtained from (7) by the reduced inverse d-variate FFT of size αN. If φ is also well-localized in time domain such that it can be approximated by a 1-periodic function ψ with $\operatorname{supp} \psi \subseteq \frac{2m}{n} \Pi^d$, $m \ll N$, then

$$f(w_j) \approx s_1(w_j) \approx s(w_j) = \sum_{l \in I_{n,m}(w_j)} g_l \, \psi\left(w_j - \frac{l}{n}\right) \tag{9}$$

with $I_{n,m}(w) := \{l \in I_N : \lceil n w_s \rceil - m \leq l_s \leq \lfloor n w_s \rfloor + m\}$. For fixed w_j, the above sum contains at most $(2m+1)^d$ nonzero summands.

In summary, we obtain the following algorithm for the fast computation of (2) with $\mathcal{O}((\alpha N)^d \log(\alpha N))$ arithmetical operations:

Algorithm 1. (Computation of (2))
    Precomputation: $n := \alpha N$,
    $c_k(\varphi), \quad k \in I_N$,
    $\psi(w_j - \frac{l}{n}), \quad j \in I_N, l \in I_{n,m}(w_j)$.

    1) $\hat{g}_k := f_k/c_k(\varphi), \quad k \in I_N$.

    2) $g_l := n^{-d} \sum_{k \in I_N} \hat{g}_k \, e^{-2\pi i k l/n}, \quad l \in I_n$.

    3) $s(w_j) := \sum_{l \in I_{n,m}(w_j)} g_l \, \psi(w_j - \frac{l}{n}), \quad j \in I_N$.

If only few nodes v_j differ from the equispaced nodes j, then the approximating function s_1 of f can be alternatively constructed by requiring exact interpolation at the nodes j/n, i.e., $f(j/n) = s_1(j/n)$, $j \in I_n$. Then we have only to replace $c_k(\varphi)$ in step 1 of the algorithm by the discrete Fourier sum of $(\varphi(l/n))_{l \in I_n}$. (See also [2].) Moreover, there is a relationship of the first part of Algorithm 1 (steps 1 and 2) to the gridded interpolation algorithms introduced in [7].

The approximation error introduced by Algorithm 1 splits as follows:

$$\begin{aligned} E(w_j) &:= |f(w_j) - s(w_j)| \\ &\leq |f(w_j) - s_1(w_j)| + |s_1(w_j) - s(w_j)| = E_1(w_j) + E_2(w_j). \end{aligned}$$

Let

$$E_\infty := \max_{j \in I_N} E(w_j)$$

and

$$\|f\|_1 := \sum_{k \in I_N} |f_k|.$$

Then we obtain immediately by (4), (6), and (8) that

$$E_1(w_j) \le \|f\|_1 \max_{k \in I_N} \sum_{r \in \mathbb{Z}^d \setminus \{0\}} \left| \frac{c_{k+nr}(\varphi)}{c_k(\varphi)} \right|$$

and by (5), (7), and (9) that

$$E_2(w_j) \le \|f\|_1 n^{-d} \left(\max_{k \in I_N} |c_k(\varphi)|^{-1} \right) \sum_{l \in I_n} \left| \varphi\left(w_j - \frac{l}{n}\right) - \psi\left(w_j - \frac{l}{n}\right) \right|.$$

To fill these error estimates with life, we consider special functions φ. Using the integral estimates in [12], we can prove

Theorem 1. *Let* (2) *be computed by Algorithm 1 with the tensor product* φ *of the dilated, periodized Gaussian bell* $(\pi b)^{-1/2} \sum_{r \in \mathbb{Z}} e^{-(n(v+r))^2/b}$ *and with the tensor product* ψ *of* $(\pi b)^{-1/2} \sum_{r \in \mathbb{Z}} e^{-(n(v+r))^2/b} \chi_{(-m,m)}(n(v+r))$.

Let $\frac{3}{2} \le \alpha$ *and* $\frac{3}{2} \le b \le \frac{2\alpha m}{\pi(2\alpha-1)\sqrt{1+(d-1)/(2\alpha-1)^2}}$. *Then*

$$E_\infty \le \|f\|_1 \, d \, 2^{d+1} \, e^{-b\pi^2(1-\frac{1}{\alpha})}.$$

Using a different approach (estimates of the complex exponentials e^{ixv}), Dutt and Rokhlin [5] have estimated the error introduced by Algorithm 1 in the univariate case by

$$E_\infty \le (b+9) \|f\|_1 \, e^{-b\pi^2(1-\frac{1}{\alpha^2})/4}$$

under the assumption that $2 < b \le \frac{2(m-1)}{\pi}$. Our estimate is sharper. Moreover, in contrast to $b = \frac{2(m-1)}{\pi}$, our suggestion $b = \frac{2\alpha m}{(2\alpha-1)\pi}$ provides a good choice for b as function of α and m. It follows from our estimates that the optimal parameter b lies slightly above $b = \frac{2\alpha m}{(2\alpha-1)\pi}$. Note that up to now suitable parameters b were determined by trial computations. In contrast to [5], our approach emphasizes the periodic point of view which will also play a significant role in the solution of the more general problem (1).

Theorem 2. *Let* (2) *be computed by Algorithm 1 with the tensor product* φ *of dilated, periodized, centered cardinal B-splines* $\sum_{r\in\mathbb{Z}} M_{2m}(n(v+r))$ *of order $2m$ ($m \geq 1$) and let $\psi = \varphi$. Then, for $\alpha \geq \frac{3}{2}$,*

$$E_\infty \leq \|f\|_1 \, d \, 2^{d+1} \, (2\alpha - 1)^{-2m}.$$

As in Theorem 1, the restriction $\alpha \geq \frac{3}{2}$ instead of $\alpha > 1$ is only necessary to obtain the constant $d\,2^{d+1}$. From our point of view, Beylkin [2] suggested an interpolatory Algorithm 1 with B-splines (without error estimates) and a "transposed" Algorithm 1 with B-splines, which contains a modified first step based on properties of Lemarié-Battle scaling functions. The following table contains the relative approximation error

$$\max_{j\in I_N} |f(w_j) - s(w_j)| / \max_{j\in I_N} |f(w_j)| \tag{10}$$

of Algorithm 1 with tensor products of Gaussian bells and cardinal B-splines as φ, respectively. The nodes w_j are randomly distributed in Π^2, $N = 256$, $\alpha = 2$, and $b = \frac{2\alpha m}{(2\alpha-1)\pi}$. The algorithm was implemented in C and tested on HP9000 (710). While the exact computation of (2) takes approximately 10000 seconds of CPU-time, our algorithm requires less than 6 seconds to achieve single precision ($m = 5$) and about 20 seconds to achieve double precision ($m = 13$).

m	Gaussian	B-Splines
5	2.93838e-06	2.33224e-06
7	5.65913e-08	2.15342e-08
9	6.70520e-10	2.08683e-10
11	8.32639e-12	2.12740e-12
13	9.74287e-14	2.19395e-14
15	2.69682e-14	1.78535e-14

Tab. 1. Approximation error of Algorithm 1 for $N = 256$ and $\alpha = 2$.

§3. FFT for Nonequispaced Grids in Time and Frequency Domain

Let us turn to the fast computation of (1). By $\varphi_1 \in L_2(\mathbb{R}^d)$ we denote a sufficiently smooth function with Fourier transform $\hat{\varphi}_1(v) := \int_{\mathbb{R}^d} \varphi_1(x)e^{-2\pi i v x}dx \neq 0$ for all $v \in N\Pi^d$. Then we obtain for

$$G(x) := \sum_{k\in I_N} f_k\,\varphi_1(x - x_k)$$

that

$$\hat{G}(v) = \sum_{k\in I_N} f_k\,e^{-2\pi i x_k v}\,\hat{\varphi}_1(v),$$

and, consequently,

$$f(v_j) = \frac{\hat{G}(v_j)}{\hat{\varphi}_1(v_j)} \qquad (j \in I_N).$$

Thus, given $\hat{\varphi}_1(v_j)$, it remains to compute $\hat{G}(v_j)$.

Let $n_1 := \alpha_1 N$ $(\alpha_1 > 1)$, $m_1 \in \mathbb{N}$ $(m_1 \ll N)$. We introduce the parameter $a := 1 + \frac{2m_1}{n_1}$ and rewrite $\hat{G}(v)$ as

$$\hat{G}(v) = \sum_{k \in I_N} f_k \int_{\mathbb{R}^d} \varphi_1(x - x_k) e^{-2\pi i x v} \, dx$$

$$= \sum_{k \in I_N} f_k \int_{a\Pi^d} \sum_{r \in \mathbb{Z}^d} \varphi_1(x + ra - x_k) e^{-2\pi i(x+ra)v} \, dx. \qquad (11)$$

Discretization of the integral by the rectangular rule leads to

$$\hat{G}(v) \approx S_1(v) = \sum_{k \in I_N} f_k \, n_1^{-d} \sum_{j \in I_{an_1}} \sum_{r \in \mathbb{Z}^d} \varphi_1\left(\frac{j}{n_1} + ra - x_k\right) e^{-2\pi i(\frac{j}{n_1}+ra)v}.$$

$$(12)$$

Now we replace the function φ_1 by a function ψ_1 with supp $\psi_1 \subseteq \frac{2m_1}{n_1}\Pi^d$. Then the third sum in (12) contains only one nonzero summand for $r = 0$. Changing the order of the summation, we obtain

$$S_1(v) \approx S_2(v) = n_1^{-d} \sum_{t \in I_{an_1}} \left(\sum_{k \in I_N} f_k \psi_1\left(\frac{t}{n_1} - x_k\right)\right) e^{-2\pi i t v/n_1}.$$

After the computation of the inner sum for all $t \in I_{an_1}$, we arrive at computation problem (2), which can be solved in a fast way by Algorithm 1. We summarize:

Algorithm 2. (Computation of (1))

Precomputation: $n_1 := \alpha_1 N$, $a := 1 + \frac{2m_1}{n_1}$, $n_2 := \alpha_2 \, a \, n_1$,
$\hat{\varphi}_1(v_j)$, $j \in I_N$,
$\psi_1(\frac{t}{n_1} - x_k)$, $k \in I_N$; $t \in I_{an_1, m_1}(x_k)$,
$\hat{\varphi}_2(t) = c_t(\varphi_2)$, $t \in I_{an_1}$,
$\psi_2(\frac{v_j}{n_1} - \frac{l}{n_2})$, $j \in I_N$, $l \in I_{n_2, m_2}(\frac{v_j}{n_1})$.

1) $F(t) := \sum\limits_{k \in I_N} f_k \, \psi_1(\frac{t}{n_1} - x_k)$, $t \in I_{an_1}$.

2) $\hat{g}_t := F(t)/c_t(\varphi_2)$, $t \in I_{an_1}$.

3) $g_l := n_2^{-d} \sum\limits_{t \in I_{an_1}} \hat{g}_t \, e^{-2\pi i t l/n_2}$, $l \in I_{n_2}$.

4) $s(v_j) := n_1^{-d} \sum\limits_{l \in I_{n_2, m_2}(v_j/n_1)} g_l \, \psi_2(\frac{v_j}{n_1} - \frac{l}{n_2})$, $j \in I_N$.

5) $S(v_j) = s(v_j)/\hat{\varphi}_1(v_j)$, $j \in I_N$.

The algorithm requires $\mathcal{O}((\alpha_1\alpha_2 aN)^d \log(\alpha_1\alpha_2 aN))$ arithmetical operations. The approximation error is given by

$$
\begin{aligned}
E(v_j) &:= |f(v_j) - S(v_j)| \\
&= \left| f(v_j) - \frac{S_1(v_j)}{\hat{\varphi}_1(v_j)} \right| + \left| \frac{S_1(v_j) - S_2(v_j)}{\hat{\varphi}_1(v_j)} \right| + \left| \frac{S_2(v_j) - s(v_j)}{\hat{\varphi}_1(v_j)} \right| \\
&= E_1(v_j) + E_2(v_j) + E_3(v_j).
\end{aligned}
$$

The error E_3 is the product of the error of Algorithm 1 and $|\hat{\varphi}_1(v_j)|^{-1}$. The cut-off error E_2 behaves like the error E_2 in Algorithm 1. The error E_1 arising from the discretization of the integral (11) can be estimated by the following aliasing argument: Assuming that the a-periodic function

$$
\varphi_{k,v}(x) := \sum_{r\in\mathbb{Z}^d} \varphi_1(x + ra - x_k)e^{-2\pi i(x+ra)v}
$$

has a uniformly convergent Fourier series

$$
\varphi_{k,v}(x) := \sum_{l\in\mathbb{Z}^d} c_l(\varphi_{k,v})e^{2\pi i l x/a}, \quad c_l(\varphi_{k,v}) = a^{-d}e^{-2\pi i x_k(v+l/a)}\hat{\varphi}_1\left(v + \frac{l}{a}\right),
$$

we obtain

$$
n_1^{-d} \sum_{t\in I_{an}} \varphi_{k,v}\left(\frac{t}{n_1}\right) = a^d c_0(\varphi_{k,v}) + \sum_{r\in\mathbb{Z}^d\setminus\{0\}} e^{-2\pi i x_k(v+rn)}\hat{\varphi}_1(v + rn).
$$

The left-hand side coincides with the discretized integral (12). Since on the other hand, the integral in (11) is equal to $a^d c_0(\varphi_{k,v})$, the error E_1 can be estimated by

$$
E_1(v_j) \le \|f\|_1 \sum_{r\in\mathbb{Z}^d\setminus\{0\}} \left| \frac{\hat{\varphi}_1(v_j + rn)}{\hat{\varphi}_1(v_j)} \right|.
$$

Therefore, it can be handled as the error E_1 of Algorithm 1.

The following table contains the relative approximation error of type (10) introduced by Algorithm 2 for tensor products of Gaussian bells φ_1 and φ_2, for $N = 128$, $m = m_1 = m_2$, $a = \frac{N}{N-m}$, $\alpha_1 = \frac{2}{a}$, $\alpha_2 = 2$, and for randomly distributed nodes x_j and v_k/N in Π^2. By the choice of α_1, α_2, and a, the main effort of the algorithm consists in the bivariate FFT of size $4N = 1024$. The third column of Table 2 contains the error of Algorithm 2 if we simply set $a := 1$ and $\alpha_1 = \alpha_2 = 2$. This change of parameters influences only the first step of the algorithm. (A similar error occurs if we consider ψ_1 as 1-periodic function.) The table demonstrates the significance of the parameter a.

m	$a = \frac{N}{N-m}$	$a = 1$
5	5.96608e-06	0.0180850
7	5.44728e-08	0.0318376
9	1.07677e-09	0.0541445
11	3.31061e-11	0.0906439
13	1.26030e-12	0.1507300
15	2.16694e-13	0.2500920

Tab. 2. Approximation error of Algorithm 2 for $N = 128$, $a = \frac{N}{N-m}$, $\alpha_1 = \frac{2}{a}$, $\alpha_2 = 2$, and for $a := 1$, $\alpha_1 = \alpha_2 = 2$, respectively.

References

1. Anderson, C. and M. D. Dahleh, Rapid computation of the discrete Fourier transform, SIAM J. Sci. Comput. **17** (1996), 913–919.

2. Beylkin, G., On the fast Fourier transform of functions with singularities, Appl. Comput. Harmon. Anal. **2** (1995), 363–381.

3. Brandt, A., Multilevel computations of integral transforms and particle interactions with oscillatory kernels, Comput. Phys. Comm. **65** (1991), 24–38.

4. Droese, J. O., Verfahren zur schnellen Fourier-Transformation mit nicht-äquidistanten Knoten, Diplomarbeit, TH Darmstadt 1996.

5. Dutt, A. and V. Rokhlin, Fast Fourier transforms for nonequispaced data, SIAM J. Sci. Statist. Comput. **14** (1993), 1368–1393.

6. Dutt, A. and V. Rokhlin, Fast Fourier transforms for nonequispaced data II, Appl. Comput. Harmon. Anal. **2** (1995), 85–100.

7. Jetter, K. and J. Stöckler, Algorithms for cardinal interpolation using box splines and radial basis functions, Numer. Math. **60** (1991), 97–114.

8. Newsam, G. private communication.

9. Pelt, J. Fast computation of trigonometric sums with application to frequency analysis of astronomical data, 1997, preprint.

10. Press, W. H. and G. B. Rybicki, Fast algorithms for spectral analysis of unevenly sampled data, Astrophys. J. **338** (1989), 227–280.

11. Sramek, R. A. and F. R. Schwab, Imaging, in *Astronomical Society of the Pacific Conference* Vol. 6, R. Perley, F. R. Schwab, and A. Bridle (eds.), 1988, pp. 117–138.

12. Steidl, G., A note on fast Fourier transforms for nonequispaced grids, Adv. Comp. Math., in print.

Scattered Data Interpolation by C^2 Quintic Splines Using Energy Minimization

Kathleen W. Farmer and Ming-Jun Lai

Abstract. We use a C^2 quintic spline space over a refinement of Powell-Sabin's type to find scattered data interpolants via energy minimization. We implemented the spline space and the scattered data interpolation via energy minimization in MATLAB. We present numerical evidence to verify that if the scattered data increase uniformly and the data are obtained from a C^2 function f, then the C^2 spline interpolants converge to f.

§1. Introduction

A problem encountered in many numerical applications is that of finding a smooth surface that interpolates a set of scattered data points in $\mathbb{R}^3$. That is, given a set of scattered data $V = \{(x_i, y_i), i = 1, \ldots, n)\} \in \mathbb{R}^2$ over a polygonal domain Ω and given a corresponding set of real numbers $\{z_i\}_{i=1}^n$, we wish to find a surface s that has continuous derivatives everywhere (typically $s \in C^1$ or C^2) and such that

$$s(x_i, y_i) = z_i, \qquad i = 1, \ldots, n. \tag{1}$$

One approach to the solution of this problem is to construct the bivariate function s as a smooth piecewise polynomial surface over Ω. This approach developed naturally from the theory of univariate splines that were studied extensively in the 1960s and 70s. The study of bivariate splines began in the early 70s, but since they are a lot more complex than univariate splines there are still many unexplored areas (cf. [13]). Although there are many methods such as *tensor-product splines*, radial basis functions, etc. available in the literature and in practice, bivariate splines defined on triangulated domains have a greater flexibility (see e.g., [1, 2, 15, 4]).

Approximation Theory IX, Volume 2: Computational Aspects
Charles K. Chui and Larry L. Schumaker (eds.), pp. 47–54.
Copyright © 1998 by Vanderbilt University Press, Nashville, TN.
ISBN 0-8265-1326-3.

So far most of the spline surfaces used for scattered data interpolation in practice have been C^1 spline surfaces (cf. [8] and [10]). Although many C^2 spline schemes have been investigated, very few of them have been implemented (cf. [7] and [11]). The purpose of this paper is to report both our implementation of a C^2 spline scheme that interpolates a set of scattered data and minimizes the energy functional, and also our experiments for the approximation and convergence properties of the scheme.

Specifically, the polynomial spline surfaces being considered herein all belong to the space of C^2 quintic spline functions over triangulated domains which was introduced in [7]. The construction of a locally supported basis $\{\phi_i\}_{i=1}^m$ where m is the dimension of the space was also given in [7]. Using the locally supported basis functions, any spline function s in the space can be represented by

$$s = \sum_{i=1}^m C_i \phi_i, \tag{2}$$

for some set of coefficients $\{C_i\}_{i=1}^m$. Thus any spline surface s is uniquely defined by its particular set of coefficients $\{C_i\}_{i=1}^m$. Note that the basis functions ϕ_i may be so arranged that the coefficients $\{C_i\}_{i=1}^n$ are fixed by the given function values $\{z_i\}_{i=1}^n$.

The method utilized in finding the remaining coefficients $\{C_i\}_{i=n+1}^m$ is a process that minimizes the energy functional $E(s)$, which is defined as

$$E(s) = \int \int_\Omega [s_{xx}^2 + 2s_{xy}^2 + s_{yy}^2] dx dy. \tag{3}$$

It represents the amount of potential energy in a thin elastic plate that passes through the data points (cf. e.g., [4]). Then $E(s)$ can be expressed as a function of the $m - n$ coefficients $\{C_i\}_{i=n+1}^m$. The particular values for the coefficients $\{C_i\}_{i=n+1}^m$ may be found so that the energy functional is minimized. This involves solving a system of $m - n$ equations. As will be shown in Section 2, the system always has a solution since the matrix A associated with this system is a symmetric, positive definite matrix. Our next aim is to investigate if the interpolating spline surface constructed by the energy minimization resembles the unknown target function, and how well the spline surface approximates the unknown function.

Computer programs have been written in MATLAB to implement the C^2 quintic spline space and to construct the interpolating spline of energy minimization (cf. [3]). Many numerical experiments have been performed. Several functions were tested with six data sets, with each data set being created from the previous one by increasing its size in a uniform manner. Table 3.1 shows that the interpolating spline surfaces created by the energy minimization are converging to the known functions as the number of data points is increased.

§2. Main Results

Suppose we are given a set of points $P = \{(x_i, y_i, z_i), i = 1, \ldots, n\} \in I\!R^3$, where the points $V = \{\mathbf{v}_i = (x_i, y_i)\}_{i=1}^n$ are over a polygonal domain Ω. Let $\triangle$ be a regular triangulation of Ω with the points V as the vertices for the triangulation. Although a number of algorithms exist for triangulations, the best triangulations are those that avoid long thin triangles (cf. [5] and [15]). It is well understood that this is necessary for both aesthetic reasons and approximation power since longer and thinner triangles yield a greater approximation error (cf. [9] and [15]). After an initial triangulation $\triangle$ of scattered data points $\{(x_i, y_i), i = 1, \cdots, n\}$ is found, a refinement Δ of $\triangle$ is found by subdividing each triangle $t \in \triangle$ into six triangles by using a Powell-Sabin type refinement scheme (cf. [7]). The scheme utilizes the center $\mathbf{u}_t$ of the inscribed circle of each triangle $t \in \triangle$ as follows: $\mathbf{u}_t$ is connected to each of the three vertices of t and is also connected to the midpoint of any boundary edge of t (if such exists) and to the center $\mathbf{u}_{t'}$ of any triangle $t' \in \triangle$ which shares a common edge with t.

Given any triangulation $\triangle$ of Ω, let

$$S_5^2(\triangle) = \{s \in C^2(\Omega) : s \mid_t \in I\!P_5, \forall t \in \triangle\},$$

be the C^2 quintic spline space, where $I\!P_5 = \{p(x, y) = \sum_{0 \leq i+j \leq 5} c_{ij} x^i y^j\}$.

The quintic spline space that is utilized in this paper is the super spline space $SS_5^2(\Delta)$ which was investigated in [7], where Δ is the Powell-Sabin type refinement of $\triangle$ as above. The super spline space $SS_5^2(\Delta)$ is defined as follows:

$$SS_5^2(\Delta) = \{s \in S_5^2(\Delta) : s \in C^3 \; \forall \; \mathbf{v} \in V \; \text{ and } \; s \in C^3 \text{ at } \mathbf{u}_t, \; \forall \; t \in \triangle\},$$

where $\mathbf{u}_t$ is the center of the inscribed circle of the triangle.

It has been shown in [7] that

$$\dim(SS_5^2(\Delta)) = 10(V_I + V_B) + N_T + E + V_B,$$

where V_I and V_B denote the number of interior and boundary vertices of $\triangle$, respectively, N_T the number of triangles in $\triangle$, and E the number of edges in $\triangle$. We let $m = 10(V_I + V_B) + N_T + E + V_B$. Clearly, $m \gg n$, where n is equal to the number of scattered data points V. Note that we can so arrange the locally supported basis functions ϕ_i constructed in [7] that

$$\phi_j(x_i, y_i) = \begin{cases} 1, & \text{if } i = j, \\ 0, & \text{otherwise}, \end{cases}$$

for $i, j = 1, \cdots, n$ and $\phi_j(x_i, y_i) = 0$ for $i = 1, \cdots, n$ and $j = n+1, \cdots, m$. Thus, for a given set of scattered data, many interpolating spline surfaces s can be found by letting

$$s = \sum_{i=1}^n z_i \phi_i + \sum_{i=n+1}^m C_i \phi_i$$

for any choice of $\{C_i : i = n+1, \cdots, m\}$. We wish to find the specific coefficients $\{C_i : i = n+1, \cdots, m\}$ that are the result of the energy minimization method to be described below.

Recall that the energy functional $E(s)$ is an expression for the amount of potential energy in a thin elastic plate that passes through the data points P. It is generally accepted that a surface is smooth if the potential energy of the thin plate is minimal (see e.g., [4]). The potential energy of the thin plate is given by

$$E_p = \int \int_S [a(\kappa_1^2 + \kappa_2^2) + 2(1-b)\kappa_1\kappa_2]dS, \tag{4}$$

where κ_1 and κ_2 are the principle curvatures of the plate (and will thus depend on the second derivatives of the surface), and a and b are constants which depend on the material. Note that minimizing the potential energy amounts to minimizing the curvatures or second derivatives. In fact, "kinking" occurs when the second derivative becomes infinite (cf. [16]).

An approximation for the thin plate energy is obtained from (4) by letting $a = 1$ and $b = 0$ and making a commonly accepted assumption that the plate has only small deflections. This leads to the energy functional (3).

In the space $SS_5^2(\Delta)$, (3) can be represented as

$$E(s) = \int_\Omega \left[\left(\frac{\partial^2}{\partial x^2} s \right)^2 + \left(\frac{\partial^2}{\partial y^2} s \right)^2 + 2 \left(\frac{\partial^2}{\partial x \partial y} s \right)^2 \right] dx dy$$

$$= \int_\Omega \left[\left(\sum_{i=1}^m C_i \frac{\partial^2}{\partial x^2} \phi_i \right)^2 + \left(\sum_{i=1}^m C_i \frac{\partial^2}{\partial y^2} \phi_i \right)^2 + 2 \left(\sum_{i=1}^m C_i \frac{\partial^2}{\partial x \partial y} \phi_i \right)^2 \right] dx dy.$$

Since $C_i = z_i$ for all $i \in \{1, \ldots, n\}$, the set of n coefficients $\{C_i\}_{i=1}^n$ is fixed. Thus $E(s)$ is a function of the $m - n$ coefficients $\{C_j\}_{j=n+1}^m$. That is,

$$E(s) = E(C_{n+1}, \ldots, C_m).$$

We now find $C_{n+1}, \ldots, C_m$ by minimizing the energy functional. That is, we need to find a spline $s_* \in SS_5^2(\Delta)$ such that

$$E(s_*) = \min\{E(s) : s(x_i, y_i) = z_i, \ i = 1, \ldots, n; \ s \in SS_5^2(\Delta)\}.$$

Since $E(s)$ is a function of $\{C_{n+1}, \ldots, C_m\}$, in order to minimize $E(s)$ we let $\frac{\partial}{\partial C_j} E(s) = 0$ for each $C_j \in \{C_{n+1}, \ldots, C_m\}$. Thus taking the partial of $E(s)$ with respect to each of the $m - n$ unknown coefficients, we have

$$\frac{\partial}{\partial C_j} E(s) = 2 \sum_{t \in \Delta} \int_t \left[\left(\sum_{i=1}^m C_i \frac{\partial^2}{\partial x^2} \phi_i \right) \frac{\partial^2}{\partial x^2} \phi_j + \left(\sum_{i=1}^m C_i \frac{\partial^2}{\partial y^2} \phi_i \right) \frac{\partial^2}{\partial y^2} \phi_j + \right.$$

$$2 \left(\sum_{i=1}^m C_i \frac{\partial^2}{\partial x \partial y} \phi_i \right) \frac{\partial^2}{\partial x \partial y} \phi_j \right] dx dy$$

$$= 2 \sum_{i=1}^{m} C_i \sum_{t \in \Delta} \int_t \left(\frac{\partial^2}{\partial x^2} \phi_i \frac{\partial^2}{\partial x^2} \phi_j + \frac{\partial^2}{\partial y^2} \phi_i \frac{\partial^2}{\partial y^2} \phi_j \right.$$

$$\left. + 2 \frac{\partial^2}{\partial x \partial y} \phi_i \frac{\partial^2}{\partial x \partial y} \phi_j \right) dx dy$$

$$= 2 \sum_{i=n+1}^{m} C_i \langle \oplus \phi_i, \oplus \phi_j \rangle + 2 \sum_{i=1}^{n} C_i \langle \oplus \phi_i, \oplus \phi_j \rangle,$$

where

$$\oplus \phi_i = \left(\frac{\partial^2}{\partial x^2} \phi_i, \sqrt{2} \frac{\partial^2}{\partial x \partial y} \phi_i, \frac{\partial^2}{\partial y^2} \phi_i \right)^T$$

is a vector of three components and

$$\langle \oplus \phi_i, \oplus \phi_j \rangle = \int_\Omega \left(\frac{\partial^2}{\partial x^2} \phi_i \frac{\partial^2}{\partial x^2} \phi_j + \frac{\partial^2}{\partial y^2} \phi_i \frac{\partial^2}{\partial y^2} \phi_j + 2 \frac{\partial^2}{\partial x \partial y} \phi_i \frac{\partial^2}{\partial x \partial y} \phi_j \right) dx dy.$$

To minimize $E(s)$, we want $\frac{\partial}{\partial C_j} E(s) = 0$ for all $j \in \{n+1, \ldots, m\}$. From the calculation of $\frac{\partial}{\partial C_j} E(s)$ above, it follows that

$$\sum_{i=n+1}^{m} C_i \langle \oplus \phi_i, \oplus \phi_j \rangle = - \sum_{i=1}^{n} C_i \langle \oplus \phi_i, \oplus \phi_j \rangle$$

for each $j \in \{n+1, \ldots, m\}$. Thus we have a system of $m - n$ equations in $m - n$ unknown coefficients $\{C_{n+1}, \ldots, C_m\}$. The coefficient matrix A associated with the linear system is an $m - n$ square matrix composed of the inner-products $\langle \oplus \phi_i, \oplus \phi_j \rangle$.

If $\mathbf{c}$ contains the unknown coefficients, the system $A\mathbf{c} = \mathbf{b}$ becomes:

$$\begin{bmatrix} \langle \oplus \phi_{n+1}, \oplus \phi_{n+1} \rangle & \langle \oplus \phi_{n+2}, \oplus \phi_{n+1} \rangle & \cdots & \langle \oplus \phi_m, \oplus \phi_{n+1} \rangle \\ \vdots & & & \vdots \\ \langle \oplus \phi_{n+1}, \oplus \phi_m \rangle & \langle \oplus \phi_{n+2}, \oplus \phi_m \rangle & \cdots & \langle \oplus \phi_m, \oplus \phi_m \rangle \end{bmatrix} \begin{bmatrix} C_{n+1} \\ \vdots \\ C_m \end{bmatrix}$$

$$= - \begin{bmatrix} \sum_{i=1}^{n} C_i \langle \oplus \phi_i, \oplus \phi_{n+1} \rangle \\ \vdots \\ \sum_{i=1}^{n} C_i \langle \oplus \phi_i, \oplus \phi_m \rangle \end{bmatrix}.$$

Clearly, A is symmetric. We can further show A is positive definite as follows.

Theorem 1. *Let V be a set of n vertices over a polygonal domain Ω, $\triangle$ a triangulation of V, and $\triangle$ the Powell-Sabin's refined triangulation of $\triangle$. Let $\dim SS_5^2(\triangle) = m$. Suppose $\{\phi_i\}_{i=n+1}^m$ are all locally supported basis functions of $SS_5^2(\triangle)$ such that $\phi_i = 0$ at all vertices of V for all $i \in \{n+1,\ldots,m\}$. Then the matrix $A = [\langle \oplus\phi_i, \oplus\phi_j \rangle]_{n+1 \le i,j \le m}$ is positive definite.*

Proof: Suppose $\mathbf{c}$ is an $(m-n) \times 1$ non-zero vector. In order to prove that A is positive definite, we must show that $\mathbf{c}^T A\mathbf{c} > 0$. Now

$$\mathbf{c}^T A\mathbf{c} = \sum_{j=n+1}^m \sum_{i=n+1}^m c_i c_j \langle \oplus\phi_i, \oplus\phi_j \rangle$$

$$= \sum_{t \in \triangle} \int_t \left[\sum_{i=n+1}^m \sum_{j=n+1}^m c_i c_j \frac{\partial^2}{\partial x^2}\phi_i \frac{\partial^2}{\partial x^2}\phi_j + \sum_{i=n+1}^m \sum_{j=n+1}^m c_i c_j \frac{\partial^2}{\partial y^2}\phi_i \frac{\partial^2}{\partial y^2}\phi_j \right.$$

$$\left. + \sum_{i=n+1}^m \sum_{j=n+1}^m c_i c_j \frac{\partial^2}{\partial x \partial y}\phi_i \frac{\partial^2}{\partial x \partial y}\phi_j \right] dxdy$$

$$= \sum_{t \in \triangle} \int_t \left[\left(\sum_{i=n+1}^m c_i \frac{\partial^2}{\partial x^2}\phi_i \right)^2 + \left(\sum_{i=n+1}^m c_i \frac{\partial^2}{\partial y^2}\phi_i \right)^2 + 2\left(\sum_{i=n+1}^m c_i \frac{\partial^2}{\partial x \partial y}\phi_i \right)^2 \right] dxdy.$$

The equation above implies that $\mathbf{c}^T A\mathbf{c} \ge 0$. Suppose $\mathbf{c}^T A\mathbf{c} = 0$. Then each of the three terms in the equation above must be 0 over each triangle $t \in \triangle$. This implies

$$\sum_{i=n+1}^m c_i \frac{\partial^2}{\partial x^2}\phi_i = 0, \qquad \sum_{i=n+1}^m c_i \frac{\partial^2}{\partial y^2}\phi_i = 0, \qquad \sum_{i=n+1}^m c_i \frac{\partial^2}{\partial x \partial y}\phi_i = 0.$$

Letting $f(x,y) = \sum_{i=n+1}^m c_i \phi_i(x,y)$, then the above three equations become

$$\frac{\partial^2}{\partial x^2} f(x,y) = 0, \qquad \frac{\partial^2}{\partial y^2} f(x,y) = 0, \qquad \frac{\partial^2}{\partial x \partial y} f(x,y) = 0,$$

which implies $f(x,y)$ is a linear polynomial on each triangle $t \in \triangle$. Fix $\tilde{t} \in \triangle$, the initial triangulation. Since $f(x,y) \in SS_5^2(\triangle)$, then $f(x,y) \in C^2$ and hence $f(x,y)$ is a linear polynomial on $\tilde{t}$. Since for any vertex $\mathbf{v}$, $\phi_i(\mathbf{v}) = 0$ for all $i \in \{n+1,\ldots,m\}$, then $f(x,y) = 0$ at the three vertices of $\tilde{t}$. So $f(x,y) = 0$ for all $(x,y) \in \tilde{t}$ and hence $f(x,y) = 0$ for all $(x,y) \in \Omega$. But since $\{\phi_i\}_{i=n+1}^m$ are basis functions and therefore are linearly independent, this implies $c_i = 0$ for all $i \in \{n+1,\ldots,m\}$. Thus $\mathbf{c} = 0$, a contradiction. Therefore $\mathbf{c}^T A\mathbf{c} > 0$ for all non-zero vectors $\mathbf{c}$, which implies A is positive definite. $\square$

Thus since A is a symmetric, positive definite matrix, A is non-singular and the system $A\mathbf{c} = \mathbf{b}$ has a unique solution. So there exists a unique solution S_* of the energy minimization satisfying the interpolation conditions. By the uniqueness of the solution, we also have the following:

Theorem 2. *Suppose that f is a linear polynomial. Let $S_f \in SS_5^2(\Delta)$ be the solution of the energy minimization satisfying the interpolation condition:*

$$S_f(x_i, y_i) = f(x_i, y_i), \qquad i = 1, \cdots, n.$$

Then $S_f(x, y) \equiv f(x, y)$.

§3. Numerical Experiments

The purpose of this section is to demonstrate how well the interpolating spline functions created by the energy minimization approximate known functions as the number of data points for a given application is increased in a uniform manner. By a uniform manner, we mean that each data set is refined uniformly. Then resulting triangulations will be in a class of almost uniform triangulations.

The set of three points $A1 = \{(0,0), (2,0), (1,2)\}$ was used as the initial set of points. Hence the initial triangulation consisted of one triangle T. The data set A1 was refined by augmenting it with the mid-points of the sides of T. Thus the set A2 consisted of A1 and the midpoints of the three sides of T for a total of 6 vertices. Six refinements, A1–A6, were created in this manner.

Each of the data sets described above was tested with four known functions. One of them is the well-known Franke function (cf. [6]). All the calculations were done on a Pentium PC and on a Sun Ultra-II workstation. Table 3.1 shows the maximum error for each of the four functions implemented with data sets A1-A6. It is evident that in general the error decreases with each refinement of the data.

Tab. 3.1. Convergence over Data Sets A1-A6.

Number of Points	$z = e^{x+y}$ Errors	$z = x^6 + y^6$ Errors	Franke(x,y) Errors	$\sin(x^2 + y^3)$ Errors
3	6.4093	54.085	0.5734	1.4848
6	1.7713	16.017	0.7185	1.6491
15	0.5553	5.2891	0.6232	1.2812
45	0.1508	1.4000	0.2105	0.6723
153	0.0376	0.3459	0.0466	0.0600
561	0.0084	0.0827	0.0062	0.0130

References

1. Chui, C. K., *Multivariate Splines*, SIAM Publications, Philadelphia, 1988.
2. Dierckx, P., *Curve and Surface Fitting with Splines*, Oxford Science Publications, Oxford, 1995.

3. Farmer, K., *Scattered Data Interpolation by C^2 Quintic Splines Using Energy Minimization*, Master Thesis, Univ. of Georgia, 1997.

4. Fasshauer, G. and L. L. Schumaker, Minimal energy surfaces using parametric splines, Computer Aided Geometric Design **13** (1996), 45–79.

5. Foley, T. A., Scattered data interpolation and approximation with error bounds, Computer Aided Geometric Design **3** (1986), 163–177.

6. Franke, R., Scattered data interpolation: tests of some methods, Math. Comp. **38** (1982), 181–200.

7. Lai, M. J., On C^2 quintic spline functions over triangulations of Powell-Sabin's type, J. Comput. Appl. Math. **73** (1996), 135–155.

8. Lai, M. J., Scattered data interpolation and approximation using bivariate C^1 piecewise cubic polynomial, Computer Aided Geometric Design **13** (1996), 81–88.

9. Lai, M. J. and L. L. Schumaker, On the approximation power of bivariate splines, submitted.

10. Peters, J., C^1-surface splines, SIAM J. Numer. Anal. **32** (1995), 645–666.

11. Peters, J., Curvature continuous spline surfaces over irregular meshes, Computer Aided Geometric Design **13** (1996), 101–131.

12. Powell, M. J. D. and M. A. Sabin, Piecewise quadratic approximations on triangles, ACM Transactions on Mathematical Software **3** (1977), 316–325.

13. Schumaker, L. L., Applications of multivariate splines, Proc. Symposia in Applied Mathematics **48** (1994), 177–203.

14. Schumaker, L. L., Fitting surfaces to scattered data, in *Approximation Theory II*, G. G. Lorentz, C. K. Chui, and L. L. Schumaker (eds.), Academic Press, 1976, pp. 203–268.

15. Schumaker, L. L. Triangulation methods, in *Topics in Multivariate Approximation*, C. K. Chui, L. L. Schumaker, and F. I. Uteras (eds.), Academic Press, New York, 1987, pp. 219–232.

16. Zienkiewicz, O. C., *The Finite Element Method in Engineering Science*, McGraw-Hill, London, 1971.

Kathleen Farmer
Department of Mathematics
Northwestern State University
3329 University Parkway
Leesville, Louisiana 71446
kfarmer@cp-tel.net

Ming-Jun Lai
Department of Mathematics
University of Georgia
Athens, GA 30602
mjlai@math.uga.edu

On Smoothing for Multilevel Approximation with Radial Basis Functions

Gregory E. Fasshauer

Abstract. In a recent paper with Jerome we have suggested the use of a smoothing operation at each step of the basic multilevel approximation algorithm to improve the convergence rate of the algorithm. In our original paper the smoothing was defined via convolution, and its actual implementation was done via numerical quadrature. In this paper we suggest a different approach to smoothing, namely the use of a precomputed hierarchy of smooth functions. This essentially reduces the cost of the smoothing to zero.

§1. Background and Motivation

Multilevel approximation with radial basis functions (RBFs) was first suggested in [4], and since then also investigated theoretically in [2, 6]. The basic idea is to work with locally supported basis functions at different levels of resolution. This ensures stability and accuracy, something which is difficult to achieve with globally supported RBFs (cf. the well-known *trade-off principle* [7]). One starts with a coarse-level approximation, and then approximates residuals at increasingly finer levels. These residuals are subsequently added to the coarse-level fit resulting in an improved approximation.

In this paper we will illustrate our ideas with an application of the multilevel approach to the solution of a linear differential equation of the form $Lu = f$ with homogeneous boundary conditions. We point out, however, that the same principle can also be applied in the general framework used in [2]. If we use RBFs as approximating functions and attempt to solve the differential equation via collocation, then the approximant $\tilde{u}$ of u should be constructed via the following *Ansatz*

$$\tilde{u}(x) = \sum_{j=1}^{n_B} c_j \Phi_j(x) + \sum_{j=1}^{n_I} d_j L^{(2)} \Phi_j(x). \tag{1}$$

Approximation Theory IX, Volume 2: Computational Aspects
Charles K. Chui and Larry L. Schumaker (eds.), pp. 55–62.
Copyright ℗ 1998 by Vanderbilt University Press, Nashville, TN.
ISBN 0-8265-1326-3.

Here $\Phi_j(x) = \varphi(\|x - x_j\|)$ is a typical basis function, and the superscript on $L^{(2)}$ indicates that the differential operator L treats the function Φ as a function of the second variable, *i.e.*, we are differentiating with respect to the knot locations. The symbols n_B and n_I denote the number of boundary and interior collocation points, respectively. For a more detailed discussion of collocation with RBFs and its connection to Hermite interpolation (in the framework of globally supported RBFs on one single computational grid) we refer the reader to [1]. The main reason for the use of expansion (1) is that the resulting collocation matrix will be symmetric and positive definite, and thus invertible. Error bounds for (single-grid) collocation with globally and locally supported RBFs can be found in [5].

The basic multilevel algorithm for the solution of the problem $Lu = f$ on $\Omega \subset \mathbb{R}^d$ by collocation can be described as follows:

1) Generate a nested sequence of computational grids $\mathcal{X}_0 \subset \mathcal{X}_1 \subset \cdots \subset \mathcal{X}_K \subset \Omega$.

2) Compute a coarse-level approximation of the form

$$\tilde{u}_0(x) = \sum_{j=1}^{n_B^0} c_j^0 \Phi_j^0(x) + \sum_{j=1}^{n_I^0} d_j^0 L^{(2)} \Phi_j^0(x),$$

where $n_B^0 + n_I^0$ is the number of collocation points in $\mathcal{X}_0$, and the coefficients c_j^0 and d_j^0 are determined by solving $L\tilde{u}_0 = f$ on the grid $\mathcal{X}_0$ via collocation.

3) For i from 1 to K do
   a) Compute the residual $r_i = f - L\tilde{u}_{i-1}$ on the grid X_i.
   b) Approximate the residual r_i with a function of the form

$$g_i(x) = \sum_{j=1}^{n_B^i} c_j^i \Phi_j^i(x) + \sum_{j=1}^{n_I^i} d_j^i L^{(2)} \Phi_j^i(x),$$

   where the coefficients are found by solving $Lg_i = r_i$ on the grid X_i (which consists of $n_B^i + n_I^i$ collocation points).
   c) Compute the level-i approximation as

$$\tilde{u}_i(x) = \tilde{u}_0(x) + \sum_{k=1}^{i} g_k(x).$$

In [2] the multilevel approximation algorithm was interpreted as a *Newton method*, and it was shown that a smoothing of the residual at each level can improve the rate of convergence. Specifically, this means that step 3c above has to be modified as follows:

c'. Compute a smoothing of the approximation to the residual at level i as

$$s_i(x) = S_i g_i(x),$$

and then compute the level-i approximation as

$$\tilde{u}_i(x) = \tilde{u}_0(x) + \sum_{k=1}^{i} s_k(x).$$

In [2] the smoothing was defined as a convolution with a suitable kernel, *i.e.*, $S_i s = \phi_i * s$. The actual implementation of the smoothing used for the numerical examples in [2] was done via a quadrature approximation of the convolution integral. Besides the amount of work involved in computing this integral, this approach has another considerable drawback: after the quadrature the smoothed residual s_i is known only on a discrete set of points (the grid X_i), and therefore one needs to compute a representation of s_i which can be differentiated. This is necessary since the computation of the residual for the following iteration requires an application of L (see 3a). This procedure is inefficient and introduces additional errors.

We are therefore motivated to search for a more efficient implementation of the smoothing. The linearity of L along with the fact that $L^{(2)}$ acts on the second variable, the linearity of convolution, and the expansion of the approximation to the residual, s_i, in the form (1) imply

$$S_i s_i(x) = (\phi_i * s_i)(x)$$

$$= \int s_i(y)\phi_i(x-y)dy$$

$$= \int \left[\sum_{j=1}^{n_B^i} c_j^i \Phi_j^i(y) + \sum_{j=1}^{n_I^i} d_j^i L^{(2)} \Phi_j^i(y) \right] \phi_i(x-y)dy$$

$$= \sum_{j=1}^{n_B^i} c_j^i \int \Phi_j^i(y)\phi_i(x-y)dy + \sum_{j=1}^{n_I^i} d_j^i L^{(2)} \int \Phi_j^i(y)\phi_i(x-y)dy$$

$$= \sum_{j=1}^{n_B^i} c_j^i (\Phi_j^i * \phi_i)(x) + \sum_{j=1}^{n_I^i} d_j^i L^{(2)} (\Phi_j^i * \phi_i)(x)$$

$$= \sum_{j=1}^{n_B^i} c_j^i \tilde{\Phi}_j^i(x) + \sum_{j=1}^{n_I^i} d_j^i L^{(2)} \tilde{\Phi}_j^i(x).$$

This means that instead of actually performing the convolution to obtain the smoothed residual s_i at every level of the iteration, we can a priori determine a family of smoother basis functions $\{\tilde{\Phi}_j^i\}$ and then simply evaluate the expansion of g_i in step 3.b. by replacing the basis functions Φ_j^i with the smoother ones $\tilde{\Phi}_j^i$. The coefficients *remain the same*, and therefore – except for the one-time investment in the construction of the suitable set of smoother basis functions $\{\widetilde{\Phi}^i_j\}$ – the smoothing can be implemented at virtually no additional cost.

§2. Supporting Numerical Evidence

A family $\{\psi_{\ell,k}\}$ of compactly supported RBFs was introduced in [8]. The functions are strictly positive definite in $\mathbb{R}^d$ for all d less than or equal to some fixed value d_0, and can be constructed to have any desired amount of smoothness $2k$. For example, for $d_0 = 3$ we can compute

$$
\begin{aligned}
\psi_{4,2}(r) &\doteq (1-r)_+^6 (35r^2 + 18r + 3), \\
\psi_{5,3}(r) &\doteq (1-r)_+^8 (32r^3 + 25r^2 + 8r + 1), \\
\psi_{6,4}(r) &\doteq (1-r)_+^{10} (429r^4 + 450r^3 + 210r^2 + 50r + 5), \\
\psi_{7,5}(r) &\doteq (1-r)_+^{12} (2048r^r + 2697r^4 + 1644r^3 + 566r^2 + 108r + 9), \\
\psi_{8,6}(r) &\doteq (1-r)_+^{14} (46189r^6 + 73206r^5 + 54915r^4 + 24500r^3 + 6755r^2 \\
&\quad + 1078r + 77).
\end{aligned}
\tag{2}
$$

Here $r = \|x\|$, and the symbol $\doteq$ is used to express equality up to a positive multiplicative constant. The functions $\psi_{\ell,k}$ are polynomials in r on $[0,1]$ and are in $C^4(\mathbb{R})$, $C^6(\mathbb{R})$, $C^8(\mathbb{R})$, $C^{10}(\mathbb{R})$, and $C^{12}(\mathbb{R})$, respectively. In view of our observations at the end of the previous section, we are motivated to investigate their use as the smoother basis functions $\widetilde{\Phi}^i_j$ used to evaluate the residuals s_i in the multilevel algorithm.

For our numerical illustrations we choose the following very simple one-dimensional boundary value problem:

$$
\begin{aligned}
-u''(x) + \pi^2 u(x) &= 2\pi^2 \sin \pi x, \qquad x \in (0,1), \\
u(0) &= u(1) = 0,
\end{aligned}
\tag{3}
$$

which has solution $u(x) = \sin \pi x$.

We now compare collocation based on the C^4-function $\psi_{4,2}$. In the left half of Table 1 we do not apply the smoothing, in the right half we do, *i.e.*, we use the smoother functions listed above to evaluate the residuals. The nested sequence of computational grids is generated by taking $\mathcal{X}_i$ to consist of $2^{i+2} + 1$ equally spaced points in $[0,1]$.

	no smoothing		smoothing	
# $\mathcal{X}_i$	ℓ_2 error	rate	ℓ_2 error	rate
5	.1373313245		.1373313245	
9	.04127396164	1.734356908	.03828218880	1.842915509
17	.007501933148	2.459897609	.004390590470	3.124186439
33	.0009036081816	3.053493167	.0003109226600	3.819787305
65	.00008189092187	3.463921899	.00002873418154	3.435715902

Tab. 1. ℓ_2 errors for the approximate solution of (3) using hierarchy (2).

Using a different family of compactly supported functions (to be introduced in the next section) the results for the smoothing are quite remarkable:

# X_i	no smoothing		smoothing	
	ℓ_2 error	rate	ℓ_2 error	rate
5	.3652893821		.3652893820	
9	.2981395993	.2930518106	.2846817499	.3596898097
17	.1851993745	.6869087828	.1351373266	1.074923789
33	.07734116173	1.259770886	.02942014441	2.199549987
65	.01956947148	1.982631642	.001985448971	3.889267140

Tab. 2. ℓ_2 errors for the approximate solution of (3) using hierarchy (6).

We point out that we have made no attempt to optimize our method. In fact, due to a very poor initial fit the errors in our examples are all quite large. The purpose of these examples is only to illustrate the *effect* of the smoothing, and thus give some numerical justification to the alternative approach to smoothing described in Section 1.

§3. A Second Look at Wendland's Functions $\psi_{\ell,k}$

The basic definition of the functions $\psi_{\ell,k}$ in [8] is

$$\psi_{\ell,k}(r) = I^k \psi_{\ell,0}(r), \tag{4}$$

where

$$\psi_{\ell,0}(r) = (1-r)_+^\ell, \quad \text{and} \quad If(r) = \int_r^\infty sf(s)ds. \tag{5}$$

We now prove

Theorem 1. *The functions $\psi_{\ell,k}$, $k \geq 1$, can be computed recursively as*

$$\psi_{\ell,k}(r) = \int_r^1 s\psi_{\ell,k-1}(s)ds.$$

Proof: The proof is by induction on k. The induction step is easily obtained using (4), (5), and the fact that all of the functions $\psi_{\ell,k}$ are supported on $[0,1]$:

$$\psi_{\ell,k}(r) = I^k \psi_{\ell,0}(r) = II^{k-1}\psi_{\ell,0}(r) = I\psi_{\ell,k-1}(r)$$

$$= \int_r^\infty s\psi_{\ell,k-1}(s)ds = \int_r^1 s\psi_{\ell,k-1}(s)ds. \quad \square$$

Theorem 1 has a different flavor than the recursion formulae given in [8, 9]. It is particularly helpful for our purposes since it sheds some light on the smoothness relations of the functions $\psi_{\ell,k}$. On the one hand we see that the functions in the family (2), which we used for our first set of numerical experiments, are *not* smoother versions of one another. Rather they are all a smoother version (in the sense of Theorem 1) of a function from a different family. For example, hierarchy (2) suggests that the C^6-function $\psi_{5,3}$ is a smoothed version of the C^4-function $\psi_{4,2}$. Theorem 1, however, implies that $\psi_{5,3}$ is a smoother version of $\psi_{5,2}$ (which is C^4, but strictly positive definite in $\mathbb{R}^5$ and not $\mathbb{R}^3$, as the other functions in (2)).

On the other hand Theorem 1 tells us how to select a family of functions which are obtained by repeated smoothing of one basic function. In particular, for the example in Table 2 we used the functions

$$
\begin{aligned}
\psi_{7,2}(r) &\doteq (1-r)_+^9 (80r^2 + 27r + 3),\\
\psi_{7,3}(r) &\doteq (1-r)_+^{10}(320r^3 + 197r^2 + 50r + 5),\\
\psi_{7,4}(r) &\doteq (1-r)_+^{11}(128r^4 + 121r^3 + 51r^2 + 11r + 1),\\
\psi_{7,5}(r) &\doteq (1-r)_+^{12}(2048r^5 + 2697r^4 + 1644r^3 + 566r^2 + 108r + 9),\\
\psi_{7,6}(r) &\doteq (1-r)_+^{13}(4096r^6 + 7059r^5 + 5751r^4 + 2782r^3 + 830r^2\\
&\quad + 143r + 11).
\end{aligned}
\tag{6}
$$

As a by-product we mention that the recursive formula of Theorem 1 can be used to obtain *explicit* formulae for the functions $\psi_{\ell,k}$ for fixed values of k.

Corollary 2. *The functions $\psi_{\ell,k}$, $k = 1,2,3$, have the representation*

$$
\begin{aligned}
\psi_{\ell,1}(r) &\doteq (1-r)_+^{\ell+1}\left[(\ell+1)r + 1\right],\\
\psi_{\ell,2}(r) &\doteq (1-r)_+^{\ell+2}\left[(\ell^2 + 4\ell + 3)r^2 + (3\ell+6)r + 3\right],\\
\psi_{\ell,3}(r) &\doteq (1-r)_+^{\ell+3}\left[(\ell^3 + 9\ell^2 + 23\ell + 15)r^3\right.\\
&\quad \left.+(6\ell^2 + 36\ell + 45)r^2 + (15\ell+45)r + 15\right].
\end{aligned}
$$

Corollary 2 gives explicit formulae for Wendland's C^2, C^4, and C^6 functions which can be made strictly positive definite in any $\mathbb{R}^d$ by choosing $\ell = \lfloor \frac{d}{2} \rfloor + k + 1$ (see [8]). Formulae for higher values of k (and thus higher smoothness) can also be computed with the help of symbolic manipulation software.

§4. The Functions $\psi_{\ell,k}$ and Convolution

In order to establish a connection with the general smoothing theory given in [2] we investigate the relation of the functions $\psi_{\ell,k}$ to convolution.

Theorem 3. *The functions $\psi_{\ell,k}$, $k \geq 1$, have the representation*

$$
\psi_{\ell,k}(r) = r\psi_{\ell,k-1}(r) * \chi_{[r,\infty)}.
$$

Proof: We show that the right-hand side is equal to $\psi_{\ell,k}(r)$. Due to the definition of convolution, the support of the characteristic function, and the compact support of $\psi_{\ell,k-1}$ we get

$$
\begin{aligned}
r\psi_{\ell,k-1}(r) * \chi_{[r,\infty)} &= \int_{-\infty}^{\infty} s\psi_{\ell,k-1}(s)\chi_{[r-s,\infty)}ds \\
&= \int_{r}^{\infty} s\psi_{\ell,k-1}(s)ds \\
&= \int_{r}^{1} s\psi_{\ell,k-1}(s)ds \\
&= \psi_{\ell,k}(r). \quad \square
\end{aligned}
$$

We now see that the smoothing S is defined via

$$
Su(r) = ru(r) * \chi_{[r,\infty)} = \int_{r}^{\infty} su(s)ds,
$$

and the functions in family (6) are given by

$$
\psi_{7,k+2} = S^{k}\psi_{7,2}, \qquad k = 0, \dots, 4.
$$

According to the general theory in [2], the smoothing is required to be an approximate identity and to satisfy a Bernstein and Jackson type inequality. The verification of these properties is beyond the scope of this paper. It will be the subject of a future paper [3].

§5. Concluding Remarks

We have suggested two hierarchies of functions for the use in an efficient implementation of smoothing within the basic multilevel approximation algorithm. Simple numerical examples indicate the potential of both families. However, a rigorous proof of the properties of the smoothing operators, including a detailed analysis of the parameters involved, is deferred to [3].

For practical purposes a hierarchy as in (2) has the advantage that it is "open ended." By this we mean that one can start with some basic function ψ_{ℓ_0,k_0}, say, and then use $\{\psi_{\ell_0+i,k_0+i}\}$ as the smoother functions for the evaluation of the residuals in the multilevel algorithm as illustrated in the introductory section. The only real cost for implementing this scheme is the effort involved in precomputing the functions with a symbolic manipulation program such as MAPLE, and hardcoding them into the software.

On the other hand, it seems that a hierarchy such as (6) is better suited to the theoretical analysis required. Also, according to our simple numerical experiments, family (6) seems to be more effective than family (2). The drawback of (6) in practice is that one has to have an idea of how many iterations will be required, say K. Then the family of functions to be used is $\{\psi_{K+1,k}\}$. Again, all functions can be determined a priori symbolically, and then be coded into the software.

In any case, the construction of a suitable set of "smoother" basis functions seems to be preferential to the direct use of convolution for the smoothing at every level of the iteration (realized via fast Fourier transform, numerical quadrature, or similar means).

References

1. Fasshauer, G. E., Solving partial differential equations by collocation with radial basis functions, in *Surface Fitting and Multiresolution Methods*, A. Le Méhauté, C. Rabut, and L. L. Schumaker (eds.), Vanderbilt Univ. Press, Nashville, 1997, pp. 131–138.

2. Fasshauer, G. E. and J. W. Jerome, Multistep approximation algorithms: Improved convergence rates through postconditioning with smoothing kernels, Advances in Comp. Math., to appear.

3. Fasshauer, G. E. and J. W. Jerome; A family of compactly supported radial basis functions for postconditioning of multilevel approximation algorithms, in preparation.

4. Floater, M. S. and A. Iske, Multistep scattered data interpolation using compactly supported radial basis functions, J. Comput. Applied Math. **73** (1996), 65–78.

5. Franke, C. and R. Schaback, Convergence orders of meshless collocation methods using radial basis functions Advances in Comp. Math., to appear.

6. Narcowich, F. J., R. Schaback, and J. D. Ward, Multilevel interpolation and approximation, 1997, preprint.

7. Schaback, R., On the efficiency of interpolation by radial basis functions, in *Surface Fitting and Multiresolution Methods*, A. Le Méhauté, C. Rabut, and L. L. Schumaker (eds.), Vanderbilt Univ. Press, Nashville, 1997, pp. 309–318.

8. Wendland, H., Piecewise polynomial, positive definite and compactly supported radial functions of minimal degree, Advances in Comp. Math.4 (1995), 389–396.

9. Wendland, H., Error estimates for interpolation by compactly supported radial basis functions of minimal degree, J. Approx. Theory, to appear.

Gregory E. Fasshauer
Department of Computer Science and Applied Mathematics
Illinois Institute of Technology
Chicago, IL 60616
fass@amadeus.csam.iit.edu

A Semi-Prewavelet Approach to Piecewise Linear Prewavelets on Triangulations

Michael S. Floater and Ewald G. Quak

Abstract. It is shown how bases of piecewise linear prewavelets can be constructed on arbitrary triangulations by summing pairs of 'semi-prewavelets.' As an example we derive basis elements for the wavelet space on type-1 triangulations.

§1. Introduction

In a recent paper [1], piecewise linear prewavelets of small support were found on arbitrary triangulations. The aim of this paper is to introduce a simpler approach to constructing prewavelets which generates the most symmetric of those found in [1]. These include a set Ψ of prewavelets which were shown in [1] to form a basis for the wavelet space when all vertex degrees are at most 21. As an example, we state the elements of Ψ on a bounded type-1 triangulation, thus generalizing results from [2]. Since the maximum vertex degree in a type-1 triangulation is six, these elements form a basis.

§2. The Wavelet Space

Let $[X]$ denote the convex hull of a subset X of $\mathbb{R}^2$. We will refer to the convex hull of three non-collinear points in $\mathbb{R}^2$ as a *triangle*. Let $\mathcal{T} = \{T_1, \ldots, T_M\}$ be a set of triangles and let $\Omega = \bigcup_{i=1}^{M} T_i$ be their union. Then $\mathcal{T}$ is a triangulation if

(i) $T_i \cap T_j$ is either empty or a common vertex or a common edge, $i \neq j$;
(ii) the number of boundary edges incident on a boundary vertex is two;
(iii) Ω is simply connected.

We denote by V the set of all vertices $v \in \mathbb{R}^2$ of triangles in $\mathcal{T}$ and by E the set of all edges $e = [v, w]$ of triangles in $\mathcal{T}$. By a boundary vertex or boundary edge we mean a vertex or edge contained in the boundary of Ω. All other vertices and edges will be called interior vertices and interior edges.

Approximation Theory IX, Volume 2: Computational Aspects
Charles K. Chui and Larry L. Schumaker (eds.), pp. 63–70.
Copyright © 1998 by Vanderbilt University Press, Nashville, TN.
ISBN 0-8265-1326-3.

For a vertex $v \in V$, the *set of neighbors of v in V* is

$$V_v = \{w \in V : [v, w] \in E\}.$$

Given data values $f_v \in \mathbb{R}$ for $v \in V$, there is a unique function $f : \Omega \to \mathbb{R}$ which is linear on each triangle in $\mathcal{T}$ and interpolates the data: $f(v) = f_v$, $v \in V$. The function f is piecewise linear and the set of all such f constitute a linear space S with dimension $|V|$.

For each $v \in V$, let $\phi_v : \Omega \to \mathbb{R}$ be the unique ('hat') function in S such that for all $w \in V$,

$$\phi_v(w) = \begin{cases} 1, & w = v; \\ 0, & \text{otherwise.} \end{cases}$$

The set of functions $\Phi = \{\phi_v\}_{v \in V}$ is a basis for the space S and for any function $f \in S$,

$$f(x) = \sum_{v \in V} f(v)\phi_v(x), \qquad x \in \Omega. \tag{2.1}$$

The support of ϕ_v is the union of all triangles which contain v, called the molecule of v, which we denote by $M_v := \bigcup_{v \in T \in \mathcal{T}} T$.

Given a triangulation $\mathcal{T}^0$ we next wish to consider a uniform refinement $\mathcal{T}^1$. For a given triangle $[x_1, x_2, x_3]$ in $\mathcal{T}^0$ let y_1, y_2, y_3 denote the midpoints of the edges $[x_2, x_3]$, $[x_3, x_1]$, $[x_1, x_2]$, respectively. Then we set

$$\mathcal{T}^1 = \bigcup_{[x_1,x_2,x_3] \in \mathcal{T}^0} \{[x_1, y_2, y_3], [y_1, x_2, y_3], [y_1, y_2, x_3], [y_1, y_2, y_3]\},$$

which is a triangulation. A refined type-1 triangulation is shown in Figure 4.

Let V^j be the set of vertices in $\mathcal{T}^j$, $j = 0, 1$, and define E^j, V_v^j, S^j, ϕ_v^j, and M_v^j, accordingly. Then S^0 is a subspace of S^1, as

$$\phi_v^0 = \phi_v^1 + \frac{1}{2} \sum_{w \in V_v^1} \phi_w^1, \qquad v \in V^0.$$

Let $\langle \cdot, \cdot \rangle$ be the following inner product, defined for continuous functions on Ω,

$$\langle f, g \rangle = \sum_{T \in \mathcal{T}^0} \frac{1}{\mathrm{a}(T)} \int_T f(x)g(x)\,dx, \qquad f, g \in C(\Omega), \tag{2.2}$$

where $\mathrm{a}(T)$ is the area of triangle T. We observe that when the area of every triangle in $\mathcal{T}^0$ is some constant a, as is the case for the type-1 triangulations in Section 4, the inner product reduces to the scaled L_2 inner product $\langle f, g \rangle = \frac{1}{a} \int_\Omega f(x)g(x)\,dx$.

With respect to the inner product (2.2), the spaces S^0 and S^1 become Hilbert spaces. Let W^0 denote the relative orthogonal complement of the coarse space S^0 in the fine space S^1, so that

$$S^1 = S^0 \oplus W^0.$$

The space W^0 is called the wavelet space, and its elements are called prewavelets. The dimension of W^0 is $|V^1| - |V^0| = |E^0|$. Our goal is to describe a simple construction of a *basis* Ψ for W^0 consisting of prewavelets with small support.

§3. From Semi-Prewavelets to Prewavelets

Let v be a vertex in V^0 and w a vertex in V^1. Then the inner product $\langle \phi_v^0, \phi_w^1 \rangle$ can only be non-zero if v and w belong to a common triangle T in T^0. In this case four different values of the inner product are possible. Let $t(v)$ denote the number of triangles containing the vertex $v \in V^0$, and $t(e)$ the number of triangles (one or two) in T^0 containing the edge $e \in E^0$. Then (see [1])

$$96\langle \phi_v^0, \phi_w^1 \rangle = \begin{cases} 6t(v), & \text{if } w = v, \\ 10t(e), & \text{if } w \in V_v^1, \ w \text{ is midpoint of } e \in E^0, \\ t(e), & \text{if } w \in V_v^0, \ e = [v,w] \in E^0, \\ 4, & \text{if } w \notin V^0 \text{ and } w \notin V_v^1. \end{cases} \tag{3.1}$$

Let $u \in V^1 \setminus V^0$ be a 'new' vertex, namely the midpoint of an edge $e = [a_1, a_2]$ in E^0 with endpoints a_1 and a_2 in V^0. In order to lie in W^0, a function $\psi_u \in S^1$ must satisfy the orthogonality conditions

$$\langle \phi_v^0, \psi_u \rangle = 0, \quad v \in V^0. \tag{3.2}$$

Our approach to finding such an element will be to take ψ_u to be the sum of two functions in S^1 which are 'almost' orthogonal to S^0 and whose supports are just the molecules in T^0 of the edge endpoints a_1 and a_2, respectively. Specifically we will search for an element of W^0 of the form

$$\psi_u(x) = \sigma_{a_1,u}(x) + \sigma_{a_2,u}(x), \tag{3.3}$$

where $\sigma_{a_1,u}$ and $\sigma_{a_2,u}$ are semi-prewavelets.

Definition 3.1. *For any $a_1 \in V^0$ and $u \in V_{a_1}^1$, we call the function*

$$\sigma_{a_1,u}(x) = s_{a_1}\phi_{a_1}^1(x) + \sum_{w \in V_{a_1}^1} s_w \phi_w^1(x), \qquad s_{a_1}, s_w \in \mathbb{R}, \tag{3.4}$$

a semi-prewavelet if it satisfies the orthogonality conditions

$$\langle \phi_v^0, \sigma_{a_1,u} \rangle = 0, \quad v \in V_{a_1}^0 \setminus \{a_2\}, \tag{3.5}$$

where $a_2 \in V_{a_1}^0$ is the neighbor of a_1 such that u is the midpoint of the edge $[a_1, a_2]$ in E^0.

We think of a semi-prewavelet $\sigma_{a_1,u}$ as 'centred' at a_1, though it also depends on u. Due to Equation (3.4), the support of $\sigma_{a_1,u}$ is included in the molecule $M_{a_1}^0$, and so due to condition (3.5), $\sigma_{a_1,u}$ is clearly orthogonal to all basis functions in S^0, except $\phi_{a_1}^0$ and $\phi_{a_2}^0$.

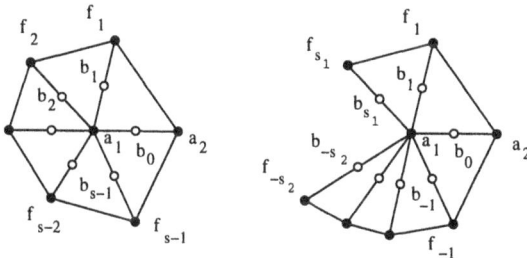

Fig. 1. Two kinds of semi-prewavelets.

There are two kinds of semi-prewavelets $\sigma_{a_1,u}$, depending on whether a_1 is an interior or a boundary vertex.

We first suppose that a_1 is an interior vertex and that its degree is $s := |V_{a_1}^0| = |V_{a_1}^1|$. Following the notation of Figure 1, with $A_1 = s_{a_1}$ and $B_i = s_{b_i}$, and $b_0 = u$, we have

$$\sigma_{a_1,u}(x) = A_1\phi_{a_1}^1(x) + \sum_{i=0}^{s-1} B_i\phi_{b_i}^1(x).$$

Lemma 3.1. *If a_1 is an interior vertex, then any semi-prewavelet $\sigma_{a_1,u}$ satisfies the equations*

$$B_i = -\frac{1}{14}A_1 + L(\lambda^i + \lambda^{s-i}), \qquad i = 0, \ldots s - 1,$$

for some arbitrary constant L, and $\lambda = -5/2 + \sqrt{21}/2 = -0.208712\ldots$

Proof: We get one orthogonality condition (3.5) for each vertex $f_1, \ldots, f_{s-1}$. Using the inner product formulae (3.1), we find that these conditions are

$$2A_1 + 4B_{i-1} + 20B_i + 4B_{i+1} = 0, \qquad i = 1, \ldots, s - 2, \qquad (3.6)$$

and

$$2A_1 + 4B_{s-2} + 20B_{s-1} + 4B_0 = 0. \qquad (3.7)$$

The general solution of the three-term difference equation (3.6) is given by $B_i = -\frac{1}{14}A_1 + L\lambda^i + M\lambda^{-i}$, but Equation (3.7) implies that $M = \lambda^s L$. □

Except for the trivial solution, $\sigma_{a_1,u}$ cannot itself belong to W^0 because a further calculation shows that

$$96\langle\phi_{a_1}^0, \sigma_{a_1,u}\rangle = \frac{32}{7}sA_1 + \frac{20}{7}\sqrt{21}\,(1 - \lambda^s)L, \qquad (3.8)$$

$$96\langle\phi_{a_2}^0, \sigma_{a_1,u}\rangle = 4\sqrt{21}\,(1 - \lambda^s)L. \qquad (3.9)$$

In the second case, a_1 is a boundary vertex and let us suppose that there are s_1 neighbors to the left of the directed edge $[a_1, a_2]$ and s_2 to the right. Let us assume first that $s_1 \geq 1$ and $s_2 \geq 1$, so that $[a_1, a_2]$ is an interior edge. Then, following again the notation of Figure 1, with $A_1 = s_{a_1}$ and $B_i = s_{b_i}$, and $b_0 = u$,

$$\sigma_{a_1,u}(x) = A_1\phi_{a_1}^1(x) + \sum_{i=-s_2}^{s_1} B_i\phi_{b_i}^1(x).$$

Lemma 3.2. *If a_1 is a boundary vertex, then any semi-prewavelet $\sigma_{a_1,u}$ satisfies the equations*

$$
\begin{aligned}
B_i &= -\frac{1}{14}A_1 + K(\lambda^{s_1} + \lambda^{-s_1})(\lambda^{s_2+i} + \lambda^{-s_2-i}), \qquad i = -s_2, \ldots, 0, \\
B_i &= -\frac{1}{14}A_1 + K(\lambda^{s_2} + \lambda^{-s_2})(\lambda^{s_1-i} + \lambda^{i-s_1}), \qquad i = 1, \ldots, s_1,
\end{aligned}
$$
$$(3.10)$$

for some arbitrary constant K, with λ as in Lemma 3.1.

Proof: We get one orthogonality condition for each vertex $f_{-s_2}, \ldots, f_{-1}$, $f_1, \ldots, f_{s_1}$. Using the inner product formulae (3.1), we obtain in this situation

$$2A_1 + 4B_{i-1} + 20B_i + 4B_{i+1} = 0, \qquad i = -s_2+1, \ldots, -1, 1, \ldots, s_1-1, \quad (3.11)$$

$$A_1 + 4B_{-s_2+1} + 10B_{-s_2} = 0, \qquad (3.12)$$

and

$$A_1 + 4B_{s_1-1} + 10B_{s_1} = 0. \qquad (3.13)$$

Equations (3.11) for $i = -s_2 + 1, \ldots, -1$ and Equation (3.12) yield $B_i = -\frac{1}{14}A_1 + L(\lambda^{s_2+i} + \lambda^{-s_2-i})$, and the solution to Equations (3.11) for $i = 1, \ldots, s_1 - 1$ and (3.13) is $B_i = -\frac{1}{14}A_1 + M(\lambda^{s_1-i} + \lambda^{i-s_1})$. In order that the solutions B_0 agree, L and M must satisfy $L(\lambda^{s_2} + \lambda^{-s_2}) = M(\lambda^{s_1} + \lambda^{-s_1})$. $\square$

As in the previous case, $\sigma_{a_1,u}$ will not in general belong to W^0, for

$$96\langle \phi_{a_1}^0, \sigma_{a_1,u} \rangle = \frac{32}{7}(s_1 + s_2)A_1 + \frac{20}{7}\sqrt{21}\left(\lambda^{-s_1-s_2} - \lambda^{s_1+s_2}\right)K, \quad (3.14)$$

$$96\langle \phi_{a_2}^0, \sigma_{a_1,u} \rangle = 4\sqrt{21}\left(\lambda^{-s_1-s_2} - \lambda^{s_1+s_2}\right)K. \qquad (3.15)$$

We remark that the solutions B_i and the inner products above remain valid when $e = [a_1, a_2]$ is a boundary edge in which case either $s_1 = 0$ or $s_2 = 0$ (but not both).

Now we turn our attention to the construction of a prewavelet ψ_u. If we take as our candidate $\psi_u = \sigma_{a_1,u} + \sigma_{a_2,u}$ as in (3.3), ψ_u will be orthogonal to all ϕ_v^0 except for $\phi_{a_1}^0$ and $\phi_{a_2}^0$. So we need to choose specific semi-prewavelets $\sigma_{a_1,u}$ and $\sigma_{a_2,u}$ such that

$$\langle \phi_{a_1}^0, \sigma_{a_1,u} \rangle + \langle \phi_{a_1}^0, \sigma_{a_2,u} \rangle = 0, \qquad (3.16)$$

$$\langle \phi_{a_2}^0, \sigma_{a_1,u} \rangle + \langle \phi_{a_2}^0, \sigma_{a_2,u} \rangle = 0. \qquad (3.17)$$

We achieve this by choosing the two variables determining $\sigma_{a_1,u}$, and the two determining $\sigma_{a_2,u}$, appropriately, using (3.8) and (3.9) for interior vertices and (3.14) and (3.15) for boundary vertices. Let us only consider one of the three possible cases, interior/interior, boundary/interior, and boundary/boundary in detail. Specifically let a_1 be a boundary vertex and a_2 an interior one. Then following the notation of Figure 2,

$$\sigma_{a_1,u}(x) = A_1 \phi_{a_1}^1(x) + \sum_{i=-s_2}^{-1} B_i \phi_{b_i}^1(x) + B_0 \phi_u^1(x) + \sum_{i=1}^{s_1} B_i \phi_{b_i}^1(x)$$

and

$$\sigma_{a_2,u}(x) = A_2 \phi_{a_2}^1(x) + C_0 \phi_u^1(x) + \sum_{i=1}^{t-1} C_i \phi_{c_i}^1(x).$$

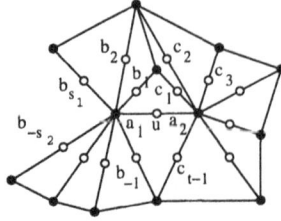

Fig. 2. An $((s_1, s_2), t)$ edge connecting a boundary and an interior vertex.

Applying Lemma 3.2 to $\sigma_{a_1,u}$ implies that the coefficients B_i, $i = -s_2, \ldots, s_1$, satisfy Equation (3.10). Meanwhile, applying Lemma 3.1 to $\sigma_{a_2,u}$ implies that $C_i = -A_2/14 + L(\lambda^i + \lambda^{t-i})$, $i = 0, \ldots, t-1$. Then after some algebra, the two equations, (3.16) and (3.17), in the four unknowns, A_1, A_2, K, and L yield the solutions

$$A_1 = -\frac{\sqrt{21}}{8(s_1 + s_2)}\left(5(\lambda^{-s_1-s_2} - \lambda^{s_1+s_2})K + 7(1 - \lambda^t)L\right),$$

$$A_2 = -\frac{\sqrt{21}}{8t}\left(7(\lambda^{-s_1-s_2} - \lambda^{s_1+s_2})K + 5(1 - \lambda^t)L\right).$$

Let us choose a specific prewavelet $\psi_u^1 = \sigma_{a_1,u}^b + \sigma_{a_2,u}^i$ by choosing the values $K = K_1 := 1/(\lambda^{-s_1-s_2} - \lambda^{s_1+s_2})$ and $L = L_1 := 1/(1 - \lambda^t)$. The "boundary" semi-prewavelet $\sigma_{a_1,u}^b$ then has coefficients $A_1 = -3\sqrt{21}/(2(s_1 + s_2))$, $B_i = 3\sqrt{21}/(28(s_1 + s_2)) + (\lambda^{s_1} + \lambda^{-s_1})(\lambda^{s_2+i} + \lambda^{-s_2-i})/(\lambda^{-s_1-s_2} - \lambda^{s_1+s_2})$, for $i = -s_2, \ldots, 0$, and $B_i = 3\sqrt{21}/(28(s_1 + s_2)) + (\lambda^{s_2} + \lambda^{-s_2})(\lambda^{s_1-i} + \lambda^{-s_1+i})/(\lambda^{-s_1-s_2} - \lambda^{s_1+s_2})$ for $i = 1, \ldots, s_1$. The "interior" semi-prewavelet $\sigma_{a_2,u}^i$ has the coefficients $A_2 = -3\sqrt{21}/(2t)$ and $C_i = 3\sqrt{21}/(28t) + (\lambda^i + \lambda^{t-i})/(1 - \lambda^t)$ for $i = 0, \ldots, t-1$.

Interestingly, it can further be shown that due to these choices of K and L, $\psi_u^1 = \sigma_{a_1,u}^i + \sigma_{a_2,u}^i$ is a prewavelet when a_1 and a_2 are both interior vertices, and $\psi_u^1 = \sigma_{a_1,u}^b + \sigma_{a_2,u}^b$ is a prewavelet when a_1 and a_2 are both boundary vertices.

The prewavelets ψ_u^1 were also found in [1] and according to Theorem 8.3 of [1] the whole set $\Psi^1 = \{\psi_u^1\}_{u \in V^1 \setminus V^0}$ is a basis of W^0 if the degrees of all vertices in T^0 are at most 21.

A second set of prewavelets, $\Psi^2 = \{\psi_u^2\}_{u \in V^1 \setminus V^0}$ can be generated in the same way by setting instead $K = K_2 := \pm K_1$ and $L = L_2 := \mp L_1$, and it was shown in [1] that the two prewavelets ψ_u^1 and ψ_u^2 are linearly independent for each vertex $u \in V^1 \setminus V^0$.

It turns out that in general Ψ^2 is not a basis of W^0. Consider the simplest situation where T^0 contains only one triangle. Then $\dim S^0 = \dim W^0 = 3$, and Figure 3 (with $a = -5\sqrt{21}/24$ and $b = \sqrt{21}/4$) shows the function values on the vertex set V^1 of the three prewavelets ψ_u^2. These prewavelets are clearly linearly dependent, as their sum is identically zero.

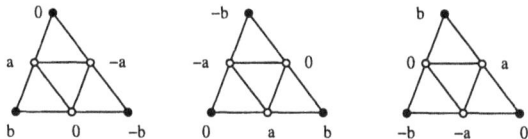

Fig. 3. Three linearly dependent prewavelets.

It was further established in [1] that the space of all prewavelets with supports contained in the union of the molecules in T^0 of a_1 and a_2 has in all cases dimension $k + 1$, where k is the number of neighbours in T^0 common to both a_1 and a_2. When the edge $[a_1, a_2]$ is an interior one, $k \geq 2$ and so the dimension is at least three. Therefore the semi-prewavelet approach (3.3) does not generate all possible prewavelets of this support, even though it does generate the important one ψ_u^1.

§4. Type-1 Triangulations

We illustrate our findings by treating the important case when T^0 is a rectangular triangulation of type I. Let $[a_1, b_1]$ and $[a_2, b_2]$ be two real intervals and for given integers $m, n \geq 1$, let $x_i = (1 - i/m)a_1 + (i/m)b_1$, and $y_j = (1 - j/n)a_2 + (j/n)b_2$, for $i = 0, 1, \ldots, m$ and $j = 0, 1, \ldots, n$. Then defining the points $v_{ij} = (x_i, y_j)$, the set of triangles

$$T^0 = \{[v_{ij}, v_{i+1,j}, v_{i+1,j+1}], [v_{ij}, v_{i,j+1}, v_{i+1,j+1}] : 0 \leq i \leq m-1, 0 \leq j \leq n-1\}$$

is a type-1 triangulation. Up to symmetry, there are only seven different prewavelets ψ_u^1, labeled 1 to 7 in Figure 4.

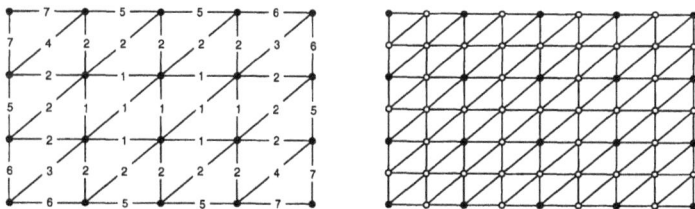

Fig. 4. A type-1 triangulation and its refinement.

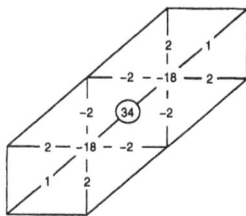

Fig. 5. Case 1, (6, 6).

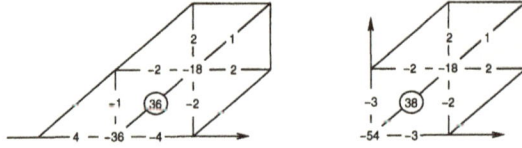

Fig. 6. Cases 2 and 3, $((2,1),6)$ and $((1,1),6)$.

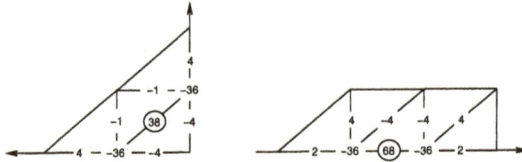

Fig. 7. Cases 4 and 5, $((2,1),(2,1))$ and $((3,0),(3,0))$.

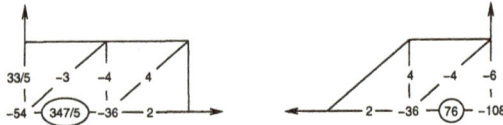

Fig. 8. Cases 6 and 7, $((2,0),(3,0))$ and $((3,0),(1,0))$.

Figures 5–8 show the respective supports and nonzero function values (after multiplying by $72/\sqrt{21}$) at the vertices of T^1 of the seven types of prewavelet ψ_u^1, computed from our formulae, using the notation introduced in [1]. The vertex u itself is circled in each case. The coefficients in Case 1 (in Figure 5) were found by Kotyczka and Oswald [2]. Due to Theorem 8.3 in [1], the set $\Psi^1 = \{\psi_u^1\}_{u \in V^1 \backslash V^0}$ is a basis of the wavelet space W^0.

References

1. Floater, M. S. and E. G. Quak, Piecewise linear prewavelets on arbitrary triangulations, Numer. Math., to appear.

2. Kotyczka, U. and P. Oswald, Piecewise linear prewavelets of small support, in *Approximation Theory VIII*, C. K. Chui and L. L. Schumaker (eds.), World Scientific, Singapore, vol. **2**, 1995, pp. 235–242.

Michael S. Floater and Ewald G. Quak
SINTEF Applied Mathematics
Post Box 124, Blindern
0314 Oslo, NORWAY
Michael.Floater@math.sintef.no
Ewald.Quak@math.sintef.no

Refinable Spline Functions

Tim N. T. Goodman

Abstract. We study piecewise polynomials which are finite linear combinations of dilates of integer shifts of themselves.

§1. Introduction

This paper attempts to give a survey of refinable spline functions. It is intended to be accessible to those without any background in this area and so we include details, examples and simple proofs. However, neither space nor the author's abilities allow this basic approach to cover all the work in this area, and so it is necessary to deal with certain topics in much greater detail than others. The selection is not intended to reflect the importance of the topics but merely the author's interests and limitations.

We shall say a function $\phi : \mathbb{R} \to \mathbb{R}$ is refinable if it satisfies an equation of the form

$$\phi(x) = \sum_{j=-\infty}^{\infty} a_j \phi(2x - j), \quad x \in \mathbb{R}, \tag{1.1}$$

for real numbers (a_j). In this paper we shall be mostly interested in the case when both ϕ and the sequence (a_j) have bounded support. We refer to an equation of the form (1.1) as a refinement equation. Sometimes such equations are called two-scale equations and refinable functions are called scaling functions. To give a simple example, consider the "hat function,"

$$\phi(x) = \begin{cases} x, & 0 \le x \le 1, \\ 2 - x, & 1 < x \le 2, \\ 0, & \text{otherwise.} \end{cases} \tag{1.2}$$

Then it is easily seen that ϕ satisfies the refinement equation

$$\phi(x) = \frac{1}{2}\phi(2x) + \phi(2x - 1) + \frac{1}{2}\phi(2x - 2), \quad x \in \mathbb{R}. \tag{1.3}$$

Approximation Theory IX, Volume 2: Computational Aspects
Charles K. Chui and Larry L. Schumaker (eds.), pp. 71–96.
Copyright © 1998 by Vanderbilt University Press, Nashville, TN.
ISBN 0-8265-1326-3.

In this example the refinable function is defined explicitly as a spline function, i.e. a piecewise polynomial. Thus, properties such as smoothness and approximation order can be derived directly from the definition of the function. It is this approach, which we call the explicit approach, which is considered in this paper. An alternative approach, which we term the implicit approach, is to start with the refinement equation (1.1) and consider properties of the coefficients (a_j) which will ensure that it has a (unique) solution ϕ of a certain type and then study construction and properties of ϕ in terms of the coefficients. In this case the refinable function ϕ will not in general have an explicit representation. This approach was used, for example, by Daubechies in the construction of her refinable functions with orthonormal integer shifts which she used to construct her famous wavelets. This approach will not be considered in this paper, except to mention that certain results hold for refinable functions in general, whether defined explicitly or implicitly.

The simplest refinable spline functions are the uniform B-splines, of which the "hat function" (1.2) is a special case, and these are considered in Section 2. In Section 3 we study the connection between B-splines and spline wavelets, whether non-orthogonal (prewavelets), orthogonal or biorthogonal. Section 4 characterises all spline functions satisfying a refinement equation of form (1.1) and shows, in particular, that B-splines are the only such functions to have stable integer shifts.

This last result means that our above definition of a refinable spline function is very restrictive. So in Section 5 we consider a refinable vector of spline functions, where in the refinement equation (1.1), ϕ denotes a column vector of functions and each coefficient (a_j) is a matrix. As an example, we study B-splines with multiple knots at the integers. However, such B-splines are by no means the only refinable vectors of spline functions and such functions are studied in greater generality in Section 6, while Section 7 gives examples whose integer shifts are orthogonal.

We then turn attention to higher dimension. A generalisation of uniform B-splines is box splines, which are considered in Section 8, and their corresponding wavelets in Section 9. Some related refinable pairs of bivariate splines are studied in Section 10. Finally, in Section 11, we mention briefly work on spline functions in multiresolution methods not covered in the previous sections.

§2. Uniform B-Splines

In [65], Schoenberg defined B-splines with simple knots at the integers $\mathbb{Z}$. Our method of introducing these B-splines will be different from the usual ones as it will be convenient for providing a unified approach to much of this paper. In Section 5 we shall extend our method to define basic splines with multiple knots at the integers and these spline functions will then be used in Section 6 to define very general refinable splines. The same approach will be used in Section 9 to introduce box splines and in Section 11 for "multi-box-splines."

Take an integer $n \geq 0$ and let $\mathcal{S}_n$ denote the space of compactly supported C^{n-1} functions f on $\mathbb{R}$ such that for all integers j, f coincides on $[j, j+1)$

with a polynomial of degree n, i.e. f is a spline function of degree n with simple knots at $\mathbb{Z}$. It is easy to see by integration by parts that the Fourier transform of f in $\mathcal{S}_n$ is of the form

$$\hat{f}(u) = \frac{P(z)}{(iu)^{n+1}}, \quad z = e^{-iu}, \quad u \in \mathbb{R}, \tag{2.1}$$

where P is a Laurent polynomial $P(z) = \sum_{j=r}^{s} a_j z^j$, for some constants (a_j) and integers $r \le s$. Since f has compact support, $\hat{f}$ must be continuous. So P must have a zero of multiplicity $n + 1$ when $u = 0$, i.e.

$$P(z) = Q(z)(1 - z)^{n+1}, \tag{2.2}$$

where Q is also a Laurent polynomial. From (2.1) and (2.2) we may write

$$\hat{f}(u) = Q(z)\hat{N}_n(z)(u), \quad u \in \mathbb{R}, \tag{2.3}$$

where

$$\hat{N}_n(u) := \left(\frac{1-z}{iu}\right)^{n+1}, \quad u \in \mathbb{R}. \tag{2.4}$$

By integration by parts again, we can see that N_n is in $\mathcal{S}_n$ and has support $[0, n+1]$. It is a B-spline of degree n. Writing $Q(z) = \sum_{j=r}^{s} b_j z^j$, we see from (2.3) that $f(x) = \sum_{j=r}^{s} b_j N_n(x - j)$, $x \in \mathbb{R}$. Thus, every function f in $\mathcal{S}_n$ is a finite linear combination of integer shifts of N_n. We find it convenient to describe this property by saying that N_n is a basic function for $\mathcal{S}_n$. We note that this property alone shows that N_n is refinable. The function $N_n(./2)$ has simple knots at $2\mathbb{Z}$ and so lies in $\mathcal{S}_n$. Since N_n is a basic function for $\mathcal{S}_n$,

$$N_n\left(\frac{x}{2}\right) = \sum_{j=-\infty}^{\infty} c_j N_n(x - j), \quad x \in \mathbb{R}, \tag{2.5}$$

for some sequence (c_j) with bounded support. Equation (2.5) is of form (1.1) (replacing x by $2x$).

The coefficients (c_j) in (2.5) can be found explicitly on noting from (2.4) that

$$\hat{N}_n(2u) = \left(\frac{1-z^2}{2iu}\right)^{n+1} = \left(\frac{1+z}{2}\right)^{n+1} \hat{N}_n(u). \tag{2.6}$$

Taking inverse Fourier transforms then gives (2.5), where $\sum_{j=-\infty}^{\infty} c_j z^j = 2\left(\frac{1+z}{2}\right)^{n+1}$, and so for $j \in \mathbb{Z}$, $c_j = \frac{1}{2^n}\binom{n+1}{j}$. The function N_1 equals ϕ in (1.2) and the refinement equation (1.3) is (2.5), (2.6) with $n = 1$.

§3. Wavelets

It is not the main purpose of this paper to study wavelets, but it would be perverse to ignore what is probably the main application of refinable functions. Suppose that $\phi \in L^2(\mathbb{R})$ satisfies the refinement equation (1.1). For $k \in \mathbb{Z}$ we define V_k to be the closed linear subspace of $L^2(\mathbb{R})$ spanned by $\phi(2^k \cdot - j)$, $j \in \mathbb{Z}$. What the refinement equation (1.1) tells is that ϕ is in V_1 and hence $V_0 \subset V_1$. Replacing ϕ by $\phi(2^k \cdot)$ shows that $V_k \subset V_{k+1}$, $k \in \mathbb{Z}$. Thus, we have a nested sequence of spaces (V_k). If in addition we have

$$\cap_{k=-\infty}^{\infty} V_k = \{0\}, \qquad \text{closure } \cup_{k=-\infty}^{\infty} V_k = L^2(\mathbb{R}), \qquad (3.1)$$

then the sequence (V_k) is called a multiresolution analysis. This concept, introduced by Mallat [54], has become basic to the theory of wavelets. Now let W_0 denote the orthogonal complement of V_0 in V_1. More generally, defining $W_k = \{f(2^k \cdot) : f \in W_0\}$, we see that W_k is the orthogonal complement of V_k in V_{k+1}. Then for integers $m < n$,

$$V_n = W_{n-1} \oplus V_{n-1} = \ldots = W_{n-1} \oplus W_{n-2} \ldots \oplus W_m \oplus V_m.$$

Letting $m \to -\infty$, $n \to \infty$, we see that if (V_k) forms a multiresolution analysis, then $L^2(\mathbb{R})$ is the orthogonal sum of W_k, $k \in \mathbb{Z}$.

We now see how to generate the spaces (W_k). We shall assume that ϕ is in $L^2(\mathbb{R})$ and has compact support, and that the sequence (a_j) in (1.1) has bounded support. For $j \in \mathbb{Z}$, let

$$b_j = \int_{-\infty}^{\infty} \phi(x)\phi(2x - j)\,dx \qquad (3.2)$$

and define ψ in V_1 by

$$\psi(x) = \sum_{k=-\infty}^{\infty} (-1)^k b_{1-k}\phi(2x - k), \qquad x \in \mathbb{R}. \qquad (3.3)$$

Then for $j \in \mathbb{Z}$,

$$\int_{-\infty}^{\infty} \phi(x - j)\psi(x)\,dx = \sum_{k=-\infty}^{\infty} (-1)^k b_{1-k} \int_{-\infty}^{\infty} \phi(x - j)\phi(2x - k)\,dx$$

$$= \sum_{k=-\infty}^{\infty} (-1)^k b_{1-k} b_{k-2j} = \sum_{l=-\infty}^{\infty} b_{1-2l} b_{2l-2j} - \sum_{m=-\infty}^{\infty} b_{-2m} b_{2m+1-2j} = 0.$$

Thus, ψ is orthogonal to V_0, i.e. $\psi \in W_0$. Clearly this implies $\psi(\cdot - j) \in W_0$ for all integers j. Indeed we have the following result.

Theorem 3.1. *Let ϕ be a refinable function, as above, and suppose that $\int_{-\infty}^{\infty} \phi(x)\,dx \neq 0$. If ψ is defined by (3.2), (3.3), then W_0 is the closed linear subspace of $L^2(\mathbb{R})$ generated by $\psi(. - j)$, $j \in \mathbb{Z}$, (and hence for $k \in \mathbb{Z}$, W_k is the closed linear span of $\psi(2^k. - j)$, $j \in \mathbb{Z}$). Moreover, $L^2(\mathbb{R})$ is the orthogonal sum of W_k, $k \in \mathbb{Z}$, and if $\{\phi(. - j) : j \in \mathbb{Z}\}$ forms a Riesz basis, then $\{2^{k/2}\psi(2^k. - j) : j, k \in \mathbb{Z}\}$ forms a Riesz basis for $L^2(\mathbb{R})$.*

The function ψ in Theorem 3.1 is called a **wavelet**, though sometimes it is called a **prewavelet** and the term wavelet is reserved for the orthogonal case considered shortly. Theorem 3.1 was proved in [2, 11, 12, 55], see also the survey paper [43].

As a special case we can take ϕ to be the B-spline N_n of the previous section. In this case the wavelet ψ was studied in a series of papers [10–13], see also [5]. Since N_n has support on $[0, n+1]$, the sequence (b_j) in (3.2) has support on $\{-n, \ldots, 2n+1\}$ and so, from (3.3), the wavelet ψ has support $[-n, n+1]$. Since N_n is symmetric about $\frac{1}{2}(n+1)$, we have $b_j = b_{n+1-j}$, $j = -n, \ldots, 2n+1$, and it follows that $\psi(1-x) = (-1)^{n+1}\psi(x)$, $x \in \mathbb{R}$.

We now suppose that ϕ is a refinable function in $L^2(\mathbb{R})$ with orthonormal integer shifts. In this case, combining (3.2) with (1.1) gives, for $j \in \mathbb{Z}$,

$$b_j = \sum_{k=-\infty}^{\infty} a_k \int_{-\infty}^{\infty} \phi(2x-k)\phi(2x-j)\,dx = \frac{1}{2}a_j.$$

Thus, instead of (3.3) we take

$$\psi(x) = \sum_{k=-\infty}^{\infty} (-1)^k a_{1-k}\phi(2x-k). \tag{3.4}$$

Then for $j \in \mathbb{Z}$,

$$\int_{-\infty}^{\infty} \psi\psi(. - j) = \sum_{k,l=-\infty}^{\infty} (-1)^{k+l} a_{1-k}a_{1-l} \int_{-\infty}^{\infty} \phi(2. - k)\phi(2. - l - 2j)$$

$$= \frac{1}{2} \sum_{k=-\infty}^{\infty} a_{1-k}a_{1-k+2j} = \sum_{k,l=-\infty}^{\infty} a_k a_l \int_{-\infty}^{\infty} \phi(2. - k)\phi(2. - l + 2j)$$

$$= \int_{-\infty}^{\infty} \phi\phi(. + j) = \delta_{j,0},$$

on applying (1.1). Thus, ψ has orthonormal integer shifts and from Theorem 3.1, $\{2^{k/2}\psi(2^k. - j) : j, k \in \mathbb{Z}\}$ forms an orthonormal basis for $L^2(\mathbb{R})$.

In fact this orthonormal case predates Theorem 3.1, and is due to Mallat [54]. The B-spline N_n does not have orthogonal integer shifts except for $n = 0$. However, we may construct spline functions with orthonormal integer shifts as follows. From (2.4) it can be seen that the 2π-periodic function

$$f_n(u) := \sum_{j=-\infty}^{\infty} |\hat{N}_n(u + 2j\pi)|^2 \tag{3.5}$$

is bounded away from zero and so we can define a function ϕ_n in $L^2(\mathbb{R})$ by

$$\hat{\phi}_n := \frac{\hat{N}_n}{f_n}. \tag{3.6}$$

Then from (3.5) and (3.6), $\sum_{j=-\infty}^{\infty} |\hat{\phi}_n(u + 2j\pi)|^2 = 1$, $u \in \mathbb{R}$, and it follows that ϕ_n has orthonormal integer shifts. It can be seen from Poisson's summation formula, and the fact that N_n has compact support, that $f_n(u)$ is a polynomial in $z = e^{-iu}$ and so we can write

$$\frac{1}{f_n(u)} = \sum_{j=-\infty}^{\infty} c_j z^j, \quad u \in \mathbb{R}, \tag{3.7}$$

where the sequence (c_j) of coefficients (depending on n) decays exponentially. Then from (3.4), $\phi_n = \sum_{-\infty}^{\infty} c_j N_n(. - j)$ and so ϕ_n is a spline function with infinite support but exponential decay. In a similar manner to (3.7), we can deduce

$$\frac{\hat{\phi}_n(2u)}{\hat{\phi}_n(u)} = \left(\frac{1+z}{2}\right)^{n+1} \frac{f_n(u)}{f_n(2u)} = \sum_{j=-\infty}^{\infty} a_j z^j, \quad z = e^{-iu}, \quad u \in \mathbb{R},$$

where (a_j) decays exponentially. Taking inverse Fourier transforms then gives

$$\phi_n(\frac{x}{2}) = 2 \sum_{j=-\infty}^{\infty} a_j \phi_n(x - j), \quad x \in \mathbb{R}.$$

Thus, ϕ_n is refinable. Although neither the function ϕ_n nor the sequence (a_j) have bounded support, the above procedure of Mallat gives a corresponding wavelet ψ_n with orthonormal integer shifts. These wavelets were introduced in [1, 50] before the prewavelets with compact support discussed above.

To finish this section we mention briefly the biothogonal spline wavelets of [14]. Take $n \geq 0$ and define the space S_n as before. Then it is shown in [14] that there is a function ψ in S_n and a function $\tilde{\psi}$ in $L^2(\mathbb{R})$ so that $\{2^{k/2}\psi(2^k. - j) : j, k \in \mathbb{Z}\}$ and $\{2^{k/2}\tilde{\psi}(2^k. - j) : j, k \in \mathbb{Z}\}$ are both Riesz bases for $L^2(\mathbb{R})$, and

$$\int_{-\infty}^{\infty} 2^{k/2}\psi(2^k x - j)2^{m/2}\tilde{\psi}(2^m x - l) = \delta_{k,m}\delta_{j,l}.$$

Such bases are said to be biorthogonal. For $n \geq 1$, the function $\tilde{\psi}$ is not a spline function, indeed it is defined implicitly, and so we do not study these further.

§4. Characterising Refinable Splines

We saw in the previous section that for $n \geq 1$ the only refinable spline functions with orthogonal integer shifts have both infinite support and an infinite refinement equation, i.e. with a sequence of coefficients of infinite support. Other refinable splines with infinite support or with infinite refinement equations are considered in [74]. However, our interest in this paper is where both function and equation have bounded support. Examples of such splines which do not equal the B-splines N_n are considered, together with corresponding wavelets, in [39], but the choice of such functions is very limited, as we see in the following result.

Theorem 4.1. [48] *For any piecewise polynomial ϕ with compact support, the following are equivalent.*

(a) *The function ϕ satisfies a refinement equation of form (1.1), where (a_j) has bounded support.*

(b) *For some $n \geq 0$,*

$$\phi = \sum_{j=-\infty}^{\infty} c_j N_n(. - j), \qquad (4.1)$$

where the mapping $z \to z^2$ sends the (non-zero) roots of the Laurent polynomial $(z-1)^{n+1} \sum_{j=-\infty}^{\infty} c_j z^j$ into themselves.

Moreover, if (a) holds and the integer shifts of ϕ form a Riesz basis, then $\phi = c N_n(. - j)$, for some $n \geq 0$, constant c and integer j.

Now assume that (a) holds. We shall not prove here that ϕ must be in S_n, some n, but we shall show that if ϕ is in S_n, then (b) follows. Since N_n is basic for S_n, (4.1) must hold for some sequence (c_j) with bounded support. Taking Fourier transforms of (4.1) gives

$$\hat{\phi}(u) = P(z)\hat{N}_n(u) = P(z) \left(\frac{1-z}{iu} \right)^{n+1}, \quad z = e^{-iu}, \quad u \in \mathbb{R}, \qquad (4.2)$$

where $P(z) = \sum_{j=-\infty}^{\infty} c_j z^j$. Taking Fourier transforms of (1.1) gives

$$\hat{\phi}(2u) = A(z)\hat{\phi}(u), \quad u \in \mathbb{R}, \qquad (4.3)$$

where $A(z) = \frac{1}{2} \sum_{j=-\infty}^{\infty} a_j z^j$. Combining (4.2) and (4.3) gives

$$P(z^2)(z^2 - 1)^{n+1} = 2^{n+1} A(z) P(z)(z-1)^{n+1}, \qquad (4.4)$$

which holds for $|z| = 1$ and hence for all z in $\mathbb{C}$. Thus, the polynomial $P(z)(z-1)^{n+1}$ has its roots mapped into themselves (with multiplicity) by the mapping $z \to z^2$, and so (b) holds.

In addition to (a) we now assume that the integer shifts of ϕ form a Riesz basis. It follows that $\sum_{k=-\infty}^{\infty} |\hat{\phi}(u + 2\pi k)|^2 > 0$, $u \in \mathbb{R}$. Then from (4.2), $P(z)$ has no zeros on $|z| = 1$. But we have already seen that the roots of

$P(z)(z-1)^{n+1}$ map into themselves by the mapping $z \to z^2$ and hence lie on $|z| = 1$. So $P(z)$ has no roots, i.e. $P(z) = cz^j$ for some constant c and integer j. Then taking Fourier transforms of (4.2) gives $\phi(x) = c_n N_n(x - j)$, $x \in \mathbb{R}$.

Finally we assume that (b) holds and deduce (a). As above (4.1) gives (4.2). Since the roots of $(z-1)^{n+1}P(z)$ are mapped into themselves by the $z \to z^2$, $(z+1)^{n+1}P(z^2)/P(z)$ is a Laurent polynomial. But by (4.2),

$$\hat{\phi}(2u) = \frac{(z+1)^{n+1}}{2^{n+1}} \frac{P(z^2)}{P(z)} \hat{\phi}(u), \quad z = e^{-iu}, \quad u \in \mathbb{R},$$

and taking inverse Fourier transforms shows that ϕ satisfies a refinement equation of form (1.1), where (a_j) has bounded support.

In most applications it is useful that a refinable function has integer shifts which form a Riesz basis. Theorem 4.1 shows that this is only possible for a spline ϕ when ϕ is a constant multiple of an integer shift of N_n. In order to have other refinable spline functions we must generalise the refinement equation (1.1). One possibility might appear to be to replace the factor 2 in (1.1) by an integer $m \geq 3$. However, it is shown in [48] that Theorem 4.1 still holds, with $N_n(x - j)$ in (4.1) replaced by $N_n(x - \frac{k}{m-1} - j)$ for some fixed integer k, and with the mapping $z \to z^2$ replaced by $z \to z^m$. In particular, if (a) holds and the integer shifts of ϕ form a Riesz basis, then $\phi = cN_n(\cdot - \frac{j}{m-1})$ for some constant c and integer j.

Thus, to have refinable spline functions which are not essentially B-splines N_n, we must employ a more far-reaching generalisation of the refinement equation (1.1). This will be studied in the next three sections.

§5. Splines with Uniform Multiple Knots

For $r \geq 1$, let $\phi = (\phi_1, \ldots, \phi_r)^T$, for real-valued functions $\phi_1, \ldots, \phi_r$ on $\mathbb{R}$. We consider again the equation

$$\phi(x) = \sum_{j=-\infty}^{\infty} a_j \phi(2x - j), \quad x \in \mathbb{R}, \tag{5.1}$$

but this time the coefficients a_j are $r \times r$ matrices of real numbers. We shall call ϕ a refinable vector of functions and the equation (5.1) a vector refinement equation. As a simple example, consider the piecewise linear functions with support on $[0, 1]$, given by

$$\begin{aligned} \phi_1(x) &= 1 - x, \quad 0 \leq x \leq 1, \\ \phi_2(x) &= x, \quad 0 \leq x \leq 1. \end{aligned} \tag{5.2}$$

It is easily seen that $\phi = (\phi_1, \phi_2)^T$ satisfies the vector refinement equation

$$\phi(x) = \begin{bmatrix} 1 & \frac{1}{2} \\ 0 & \frac{1}{2} \end{bmatrix} \phi(2x) + \begin{bmatrix} \frac{1}{2} & 0 \\ \frac{1}{2} & 1 \end{bmatrix} \phi(2x - 1), \quad x \in \mathbb{R}. \tag{5.3}$$

As a class of examples of refinable vectors of spline functions we shall now define splines with multiple knots at the integers in an analogous manner to the uniform B-splines in Section 2. Take integers $n \geq 0$, $1 \leq r \leq n + 1$, and let $\mathcal{S}_{n,r}$ denote the space of compactly supported C^{n-r} functions f on $\mathbb{R}$ such that for all integers j, f coincides with a polynomial of degree n on $[j, j + 1)$, i.e. f is a spline function of degree n with knots of multiplicity r at $\mathbb{Z}$.

By integration by parts, any function f in $\mathcal{S}_{n,r}$ satisfies

$$\hat{f}(u) = \sum_{j=0}^{r-1} \frac{P_j(z)}{(iu)^{n+1-j}}, \quad z = e^{-iu}, \quad u \in \mathbb{R}, \tag{5.4}$$

for Laurent polynomials $P_0, \ldots, P_{r-1}$. Because $\hat{f}$ is continuous at $u = 0$, $\sum_{j=0}^{r-1} (iu)^j P_j(z)$ must have a zero of order $n + 1$ at $u = 0$.

Now for $1 \leq j \leq n$, let $P_{n,j}(z)$ denote the Taylor polynomial of degree n at $z = 1$ for $(\log z)^j$. Then $(-iu)^j - P_{n,j}(z)$ has a zero of order $n + 1$ at $u = 0$. Thus, $P_0(z) + \sum_{j=1}^{r-1} (-1)^j P_{n,j}(z) P_j(z)$ has a zero of multiplicity $n + 1$ when $u = 0$, i.e.

$$P_0(z) + \sum_{j=1}^{r-1} (-1)^j P_{n,j}(z) P_j(z) = Q(z)(1 - z)^{n+1}, \tag{5.5}$$

for a Laurent polynomial Q. Then substituting for P_0 from (5.5) into (5.4) gives, after some simplification, for $u \in \mathbb{R}$,

$$\hat{f}(u) = Q(z)\hat{N}_n(u) + \sum_{j=1}^{r-1} P_j(z)\hat{B}_{n,j}(u), \tag{5.6}$$

$$\hat{B}_{n,j} := \frac{(iu)^j - (-1)^j P_{n,j}(z)}{(iu)^{n+1}}. \tag{5.7}$$

By integration by parts we can see that $B_{n,j}$ is uniquely defined by the following properties.

(a) The function $B_{n,j}$ is a piecewise polynomial of degree n with support on $[0, n]$.

(b) The only discontinuities of derivatives of $B_{n,j}$ are in $B_{n,j}^{(n)}$ at $0, \ldots, n$ and in $B_{n,j}^{(n-j)}$ at 0, where $B_{n,j}^{(n-j)}(0^+) = 1$.

From (5.4) we see that every function f in $\mathcal{S}_{n,r}$ is a finite linear combination of integer shifts of $N_n, B_{n,1}, \ldots, B_{n,r-1}$. We describe this property by saying that $\{N_n, B_{n,1}, \ldots, B_{n,r-1}\}$ is a basic set for $\mathcal{S}_{n,r}$. As for the case $r = 1$ in Section 2, this property shows that the vector $\phi = (N_n, B_{n,1}, \ldots, B_{n,r-1})^T$ satisfies a vector refinement equation. From (2.4), (5.7), we have

$$\hat{\phi}(u) = \frac{1}{(iu)^{n+1}} R(z)[1, iu, \ldots, (iu)^{r-1}]^T, \tag{5.8}$$

where R is an $r \times r$ lower triangular matrix of Laurent polynomials with $\det R(z) = (1 - z)^{n+1}$. Thus, the Fourier transform of the vector refinement equation satisfied by ϕ is $\hat{\phi}(2u) = A(z)\hat{\phi}(u)$, where $A(z)$ is an $r \times r$ lower triangular matrix of Laurent polynomials given by

$$A(z) = \frac{1}{2^{n+1}} R(z^2) \operatorname{diag}[1, 2, \ldots, 2^{r-1}] R^{-1}(z),$$

and so $\det A(z) = 2^{\frac{1}{2}r(r-1)-r(n+1)}(1 + z)^{n+1}$.

Theorem 5.1. [33] Let $\phi = (N_n, B_{n,1}, \ldots, B_{n,r-1})^T$ for $n \geq 0$, $1 \leq r \leq n+1$. Then for $s \leq r$, $\eta = (\eta_1, \ldots, \eta_s)^T$ is a basic set for $\mathcal{S}_{n,r}$ if and only if $s = r$ and

$$\hat{\eta}(u) = N(z)\hat{\phi}(u), \quad u \in \mathbb{R}, \quad z = e^{-iu}, \qquad (5.9)$$

for some $r \times r$ matrix of Laurent polynomials with $\det N(z) = cz^j$ for $c \neq 0$ and $j \in \mathbb{Z}$.

Proof: Suppose that η satisfies (5.9). Take any f in $\mathcal{S}_{n,r}$. Since ϕ is basic for $\mathcal{S}_{n,r}$, there is a matrix M of Laurent polynomials with

$$\hat{f}(u) = M(z)\hat{\phi}(u) = M(z)N^{-1}(z)\hat{\eta}(u).$$

Since N^{-1} is a matrix of Laurent polynomials, it follows that η is a basic set for $\mathcal{S}_{n,r}$.

Conversely, suppose that η is a basic set for $\mathcal{S}_{n,r}$. Then for matrices M, N of Laurent polynomials,

$$\hat{\eta}(u) = N(z)\hat{\phi}(u), \quad \hat{\phi}(u) = M(z)\hat{\eta}(u) = M(z)N(z)\hat{\phi}(u).$$

From (5.8) we see that for $u \neq 0$, the matrix $[\hat{\phi}(u)\hat{\phi}(u + 2\pi) \ldots \hat{\phi}(u + 2\pi(r - 1))]$ is non-singular and thus $M(z)N(z)$ is the identity. So $s = r$ and $\det N(z) = cz^j$ for some $c \neq 0$ and $j \in \mathbb{Z}$. $\square$

We remark that in [33] it was not made clear that the above result holds only for $s \leq r$. Clearly we can obtain basic sets with more than r elements by simply adding any elements of $\mathcal{S}_{n,r}$ to ϕ.

We note that any basic function for $\mathcal{S}_n = \mathcal{S}_{n,1}$ is of the form $cN_n(. - j)$ for $c \neq 0$, $j \in \mathbb{Z}$. An example of a basic set for $\mathcal{S}_{1,2}$ is (ϕ_1, ϕ_2) given by (5.2). It is easily seen that $\phi_1 = B_{1,1}$, $\phi_2 = N_1 - B_{1,1}(. - 1)$. In fact the functions (5.2) are examples of B-splines, which for general knots were introduced in [16]. (Although published in 1966, this paper was written in 1946, see [15].) For $n \geq 0$ and $t_0 \leq t_1 \leq \ldots \leq t_{n+1}$, $t_0 < t_{n+1}$, there is a B-spline N of degree n, unique up to a constant multiple, with support on $[t_0, t_{n+1}]$, with knots $t_1, \ldots, t_{n+1}$. This means that for any $t_i < t_{i+1}$, N coincides with a polynomial of degree n on $[t_i, t_{i+1}]$, while if t_j has multiplicity m in $(t_0, \ldots, t_{n+1})$, then N is C^{n-m} at t_j. We are concerned here with the case of knots at $\mathbb{Z}$ with multiplicity r, $1 \leq r \leq n+1$. For $i \in \mathbb{Z}$, we take $t_i = [\frac{i}{r}]$ (the integer part of $\frac{i}{r}$), and define $N_i^{n,r}$ to be the B-spline with knots $t_i, \ldots, t_{i+n+1}$, normalised so that $\sum_{i=-\infty}^{\infty} N_i^{n,r} = 1$. Then $(N_0^{n,r}, \ldots, N_{r-1}^{n,r})$ is a basic set for $\mathcal{S}_{n,r}$ and hence satisfies a vector refinement equation. The functions in (5.2) are $\phi_1 = N_0^{1,2}$, $\phi_2 = N_1^{1,2}$.

We now turn to a consideration of wavelets, corresponding to those considered in Section 3. Take $n \geq 0$, $1 \leq r \leq n+1$, and let V_0 denote the closure of $\mathcal{S}_{n,r}$ in $L^2(\mathbb{R})$. More generally for $k \in \mathbb{Z}$, let $V_k = \{f(2^k.) : f \in V_0\}$. Since $V_0 \subset V_1$, we may define W_0 to be the orthogonal complement of V_0 in V_1. For $k \in \mathbb{Z}$ we let $W_k = \{f(2^k.) : f \in W_0\}$, so that W_k is the orthogonal complement of V_k in V_{k+1}. As in Section 3, we can show that $L^2(\mathbb{R})$ is the orthogonal sum of W_k, $k \in \mathbb{Z}$.

We call $\psi = (\psi_1, \ldots, \psi_r)^T$ a multi-wavelet for $\mathcal{S}_{n,r}$ if W_0 is the closed linear span of $\{\psi_i(. - j) : i = 1, \ldots, r, j \in \mathbb{Z}\}$. It is shown in [34], following work in [37], that there is a multiwavelet ψ for $\mathcal{S}_{n,r}$ with compact support satisfying the following properties. For $i = 1, \ldots, r$, ψ_i must lie in $W_0 \subset V_1$ and so is a spline function of degree n with knots at multiplicity r at $\frac{1}{2}\mathbb{Z}$. With $t_i = [\frac{i}{r}]$, as before, ψ_i has support on $[t_{i-1}, \ldots, t_{i+2n+1-r}]$ and on no smaller interval. Its knots (with multiplicity) which are integers lie in $(t_{i-1}, \ldots, t_{i+2n+1-r})$ and lie in neither $(t_i, \ldots, t_{i+2n+1-r})$ nor in $(t_{i-1}, \ldots, t_{i+2n-r})$. Moreover, these properties determine ψ_i uniquely up to a constant multiple.

There are different choices of multi-wavelets for $\mathcal{S}_{n,r}$ with compact support. For example in [30], multiwavelets are given for $\mathcal{S}_{2r-1,r}$, which correspond, in a sense, to the basic set for $\mathcal{S}_{2r-1,r}$ introduced in [66] and further studied in [49]. In [56], multi-wavelets for $\mathcal{S}_{n,r}$ are constructed with an explicit representation, while in [58], all multi-wavelets for $\mathcal{S}_{n,r}$ are characterised. This characterisation is derived from earlier work in [57], where a recursive scheme is given for determining the coefficients in the vector refinement equation for $(N_0^{n,r}, \ldots, N_{r-1}^{n,r})$.

§6. General Refinable Splines

So far we have studied only refinable splines with knots at the integers. We now extend this work to consider univariate refinable splines with very general knots, i.e. discontinuities in derivatives. As with $\mathcal{S}_{n,r}$ we shall consider a space $\mathcal{S}$ of spline functions of degree n with compact support which is shift invariant, i.e. $f \in \mathcal{S}$ implies $f(. - 1) \in \mathcal{S}$, and refinable, i.e. $f \in \mathcal{S}$ implies $f(\frac{.}{2}) \in \mathcal{S}$. Then the knots are determined by the knots in $[0, 1)$, which we assume to be finite in number. The only other assumption is that if we allow a discontinuity in $f^{(j)}(\alpha)$ for some j, $0 \leq j < n$, then we also allow a discontinuity in $f^{(n)}(\alpha)$.

Now we take $n \geq 0$ and define our space $\mathcal{S}$. Let S_0 be a finite subset of $[0, 1)$ and for $k = 1, \ldots, n$, let S_k be a subset of S_0. Then $\mathcal{S}$ comprises all piecewise polynomials of degree n with compact support whose only discontinuities in derivatives are in $f^{(n-k)}$ at $S_k + \mathbb{Z}$, $k = 0, \ldots, n$. We let r denote the total number of knots in $[0, 1)$, i.e. $r = |S_0| + \ldots + |S_n|$. Clearly $\mathcal{S}$ is shift invariant. We also assume that $\mathcal{S}$ is refinable. Since $f^{(k)}$ has a discontinuity at α if and only if $f(\frac{.}{2})^{(k)}$ has a discontinuity at 2α, $\mathcal{S}$ is refinable if and only if for $k = 0, \ldots, n$, $\alpha \in S_k$ implies $2\alpha (\text{mod} \mathbb{Z}) \in S_k$. In [35] are characterised the choices of S_k which satisfy this condition but this is not discussed here.

The space $\mathcal{S}_{n,r}$ of the last section is the special case $S_k = \{0\}$, $k = 0, \ldots, r-1$, (where we mention only non-empty sets S_k). For $r = 2$, the only spaces $\mathcal{S}$ are the following.

1. $S_0 = S_j = \{0\}$ for some j, $1 \leq j \leq n$. 2. $S_0 = \{0, \frac{1}{2}\}$.   3. $S_0 = \{\frac{1}{3}, \frac{2}{3}\}$.

Theorem 6.1. [35] *Let* $S_0 = \{t_1, \ldots, t_l\}$, $0 \leq t_1 < \ldots < t_l < 1$ *and extend to a sequence* (t_j) *so that* $t_{j+l} = t_j + 1$, $j \in \mathbb{Z}$. *For* $j = 1, \ldots, l$, *let* N_j *denote a B-spline of degree* n *with support* $[t_j, t_{j+n+1}]$ *and simple knots* $t_j, \ldots, t_{j+n+1}$. *Then* $\{N_1, \ldots, N_l\} \cup \{B_{n,k}(. - \alpha) : \alpha \in S_k, k = 1, \ldots, n\}$ *is a basic set for* $\mathcal{S}$.

Proof: Take f in $\mathcal{S}$. Define

$$g = \sum_{k=1}^{n} \sum_{\alpha \in S_k} \sum_{j=-\infty}^{\infty} (f^{(n-k)}(\alpha + j^+) - f^{(n-k)}(\alpha + j^-))B_{n,k}(. - \alpha - j).$$

Then for $1 \leq k \leq n$, $\alpha \in S_k$, $j \in \mathbb{Z}$,

$$g^{(n-k)}(\alpha + j^+) - g^{(n-k)}(\alpha + j^-) = f^{(n-k)}(\alpha + j^+) - f^{(n-k)}(\alpha + j^-),$$

and so $f - g$ has discontinuities only in $(f-g)^{(n)}$ at $S_0 + \mathbb{Z}$, i.e. in (t_j). Thus, $f - g$ is a finite linear combination of integer translates of $N_1, \ldots, N_l$. $\square$

We note that the basic set in Theorem 6.1 comprises r functions. For the case $S_k = \{0\}$, $k = 0, \ldots, r-1$, the basic set is the set $\{N_n, B_{n,1}, \ldots, B_{n,r-1}\}$ considered in the previous section.

As always, a basic set for $\mathcal{S}$ is vector refinable. We shall characterise all refinable vectors of r functions in $\mathcal{S}$ which have linearly independent integer shifts, in a similar manner to Theorem 5.1. By linear independence we mean that if an arbitrary linear combination (possibly infinite) of the functions equals the zero function, then all coefficients are zero.

Theorem 6.2. [35] *The functions* $\{N_1, \ldots, N_l\} \cup \{B_{n,k}(. - \alpha) : \alpha \in S_k, k = 1, \ldots, n\}$ *have linearly independent integer shifts.*

Proof: Suppose that

$$f := \sum_{i=1}^{l} \sum_{j=-\infty}^{\infty} a_{i,j} N_i(. - j) + \sum_{k=1}^{n} \sum_{j=-\infty}^{\infty} \sum_{\alpha \in S_k} b_{k,j,\alpha} B_{n,k}(. - \alpha - j) = 0.$$

Then for $1 \leq k \leq n$, $j \in \mathbb{Z}$, $\alpha \in S_k$, $0 = f^{(n-k)}(\alpha + j^+) - f^{(n-k)}(\alpha + j^-) = b_{k,j,\alpha}$. The integer shifts of $N_1, \ldots, N_l$ are the B-splines for the knot sequence (t_j) of Theorem 6.1 and hence linearly independent. Thus, $a_{i,j} = 0$, $i = 1, \ldots, l$, $j \in \mathbb{Z}$. $\square$

Theorem 6.3. [35] *Let* $\phi = (\phi_1, \ldots, \phi_r)^T$ *denote the basic set of Theorem 6.1. Then* $\eta = (\eta_1, \ldots, \eta_r)^T$ *is a refinable vector of functions in* $\mathcal{S}$, *whose integer shifts are linearly independent, if and only if* $\hat{\eta}(u) = N(z)\hat{\phi}(u)$, $u \in \mathbb{R}$, $z = e^{-iu}$, *for some* $r \times r$ *matrix* N *of Laurent polynomials with* $\det N(z) = cz^j$ *for* $c \neq 0$ *and* j *in* $\mathbb{Z}$.

We shall not prove this here. It follows from a characterisation in terms of the Fourier transform in [45] for the integer shifts of a vector of functions to be linearly independent.

The result is not true, even for $r = 2$, if the condition of linear independence is relaxed to the requirement that the integer shifts form a Riesz basis. In [36] a characterisation is given of all refinable vectors of functions in $\mathcal{S}$ with $r = 2$ whose integer shifts form a Riesz basis.

Note that Theorem 6.3 considers only refinable vectors of r functions. It is shown in [35] that if $\eta = (\eta_1, \ldots, \eta_s)^T$ is a refinable vector of functions in $\mathcal{S}$ whose integer translates form a Riesz basis, then $s \leq r$. In [36], results are given concerning possible values of s and concerning linear independence of integer shifts, which include the following.

Theorem 6.4. [36] *Suppose that* $\eta = (\eta_1, \ldots, \eta_s)^T$ *is a refinable vector of functions in* $\mathcal{S}$.

(a) *If all elements of* S_0 *have odd denominators and the integer shifts of* η *form a Riesz basis, then the integer shifts of* η *are linearly independent and* $s = r$.

(b) *If* $s \leq 2$, *then* $r \leq s$.

For the case $s = 1$, part (b) tells us that $r = 1$ and so any refinable spline function with compact support must lie in $\mathcal{S}_n$ for some $n \geq 0$, as is stated in Theorem 4.1. For $s = 2$, (b) shows that $r = 1$ or 2 and so any refinable pair of spline functions with compact support must lie in $\mathcal{S}_n$ or in $\mathcal{S}$ for one of the three cases given before Theorem 6.1, while if the integer shifts of η form a Riesz basis, then $r = 2$ and so the case $\mathcal{S}_n$ is not allowed. If the situation of (a) holds, or the situation of (b) holds and the integer shifts of η are linearly independent, then we have the characterisation of Theorem 6.3. However, if $s \geq 3$ and some knots have even denominators, we may have $s < r$ and so Theorem 6.3 does not apply. Some interesting cases of this situation are considered in the next section.

§7. Orthogonal Refinable Splines

It is interesting and potentially useful for applications to construct refinable spline functions of compact support whose integer shifts are orthonormal. This is done in [26, 27] by using a general procedure for generating refinable functions with orthonormal shifts.

Suppose that $\phi = (\phi_1, \ldots, \phi_r)^T$ is a refinable vector of functions with compact support. We define a multiresolution analysis as for the case $r = 1$ at the beginning of Section 3. For $k \in \mathbb{Z}$, let V_k denote the closed linear span of $\{\phi_i(2^k. - j) : j \in \mathbb{Z}, i = 1, \ldots, r\}$. Then the refinement equation shows that $V_k \subset V_{k+1}$, $k \in \mathbb{Z}$. We assume that (3.1) holds and in addition that $\{\phi_i(. - j) : j \in \mathbb{Z}, i = 1, \ldots, r\}$ forms a Riesz basis for V_0.

Theorem 7.1. [26] *Under the above conditions, there are integers q, m and a refinable vector of functions $\tilde{\phi} = (\tilde{\phi}_1, \ldots, \tilde{\phi}_s)$ with compact support such that $\{\tilde{\phi}_i(. - j) : j \in \mathbb{Z}, i = 1, \ldots, s\}$ is orthonormal and its closed linear span $\tilde{V}_0$ satisfies*

$$V_q \subset \tilde{V}_0 \subset V_{q+m}. \tag{7.1}$$

Since $\tilde{\phi}$ is vector refinable, we have $\tilde{V}_k \subset \tilde{V}_{k+1}$, $k \in \mathbb{Z}$. Since $\tilde{V}_0 \subset V_{q+m}$, then $\cap_{k=-\infty}^{\infty} \tilde{V}_k \subset \cap_{k=-\infty}^{\infty} V_k = \{0\}$, and since $\tilde{V}_0 \supset V_q$, $\cup_{k=-\infty}^{\infty} \tilde{V}_k \supset \cup_{k=-\infty}^{\infty} V_k$, which is dense in $L^2(\mathbb{R})$. Thus, $(\tilde{V}_k)$ also forms a multiresolution analysis. In [26], (V_k) and $(\tilde{V}_k)$ are called intertwining multiresolution analyses.

We shall not prove Theorem 7.1 here but we mention briefly a useful technique which is essential to the proof. Take $k \in \mathbb{Z}$ and let $\psi_l = \phi(2^k. - l)$, $l = 0, \ldots, 2^k - 1$. Now V_k is the closed linear span of $\{\phi_i(2^k. - j) : i = 1, \ldots, r, j \in \mathbb{Z}\}$ and hence of $\{(\psi_l)_i(. - j) : i = 1, \ldots, r, l = 0, \ldots, 2^k - 1, j \in \mathbb{Z}\}$. Thus, the multiresolution generated by ϕ equals that generated by the refinable vector of $2^k r$ functions $(\psi_l)_i$, $l = 0, \ldots, 2^k - 1$, $i = 1, \ldots, r$. By taking a suitable integer shift, we may assume that ϕ has support in $[0, N]$ for some $N > 0$ and so ψ_l has support in $[2^{-k}l, 2^{-k}(l+N)] \subset [0, 1 + 2^{-k}(N-1)]$ for $l = 0, \ldots, 2^k - 1$. Thus by taking k large enough we may assume, without loss of generality, that our refinable vector of spline functions ϕ has support in $[0, 2]$.

It is further shown in [26] that, by taking suitable linear combinations, we may assume that $\phi = (\phi_1, \ldots, \phi_s, \phi_{s+1}, \ldots, \phi_t)^T$, where $\phi_{s+1}, \ldots, \phi_t$ have support in $[0, 1]$, and on $[0, 1]$ the functions $\phi_1, \ldots, \phi_s, \phi_1(. + 1), \ldots, \phi_s(. + 1), \phi_{s+1}, \ldots, \phi_t$ are linearly independent. As this technique is used in proving Theorem 7.1, the resulting functions $\tilde{\phi}$, with orthonormal integer shifts, have support in $[0, 2]$.

As a special case of Theorem 7.1 we can take the multiresolution (V_k) to be that generated by $S_{n,r}$ for some $n \geq 0$, $1 \leq r \leq n + 1$. It is shown in [26] that in this case we can take $m = 1$ in (7.1). The following examples are calculated explicitly in [26].

Case 1. $S_{1,1}$.

Here $V_1 \subset \tilde{V}_0 \subset V_2$, $\tilde{\phi} = (\tilde{\phi}_1, \tilde{\phi}_2, \tilde{\phi}_3)$, where $\tilde{\phi}_1$ has support $[0, 2]$ and $\tilde{\phi}_2$, $\tilde{\phi}_3$ have support $[0, 1]$. For the space $\tilde{V}_0$, in the notation of Section 6 we have $n = 1$, $S_0 = \{0, \frac{1}{4}, \frac{1}{2}, \frac{3}{4}\}$, $r = 4$, $s = 3$.

Case 2. $S_{1,1}$.

The functions $\tilde{\phi}_1$, $\tilde{\phi}_2$ in Case 1 have no symmetry properties. To gain symmetry we take $V_2 \subset \tilde{V}_0 \subset V_3$, $\tilde{\phi} = (\tilde{\phi}_1, \ldots, \tilde{\phi}_4)$, where $\tilde{\phi}_1$ has support $[0, 2]$, $\tilde{\phi}_2, \tilde{\phi}_3, \tilde{\phi}_4$ have support $[0, 1]$, $\tilde{\phi}_1, \tilde{\phi}_2, \tilde{\phi}_3$ are symmetric, and $\tilde{\phi}_4$ is antisymmetric. For the space $\tilde{V}_0$, $n = 1$, $S_0 = \{\frac{i}{8} : 1 = 0, \ldots, 7\}$, $r = 8$, $s = 4$.

Case 3. $S_{3,2}$.

Here $V_1 \subset \tilde{V}_0 \subset V_2$, $\tilde{\phi} = (\tilde{\phi}_1, \ldots, \tilde{\phi}_6)$, where $\tilde{\phi}_1, \tilde{\phi}_2$ have support $[0,2]$ and $\tilde{\phi}_3, \ldots, \tilde{\phi}_6$ have support $[0,1]$. For the space $\tilde{V}_0$, $n = 3$, $S_0 = S_1 = \{0, \frac{1}{4}, \frac{1}{2}, \frac{3}{4}\}$, $r = 8$, $s = 6$.

For the above refinable vectors of spline functions, corresponding wavelets are calculated in [26] by using the technique of [29]. Further examples of orthonormal refinable vectors of spline functions are given in [27], including a single example of C^2 spline functions, comprising a vector of 10 splines of degree 11, 7 of which have support $[0,1]$

§8. Box Splines

We now extend the definition in the introduction to say that for $d \geq 1$, a function $\phi : \mathbb{R}^d \to \mathbb{R}$ is refinable if it satisfies a refinement equation

$$\phi(x) = \sum_{j \in \mathbb{Z}^d} a_j \phi(2x - j), \quad x \in \mathbb{R}^d, \tag{8.1}$$

for real numbers (a_j). Again we are interested in the case that both ϕ and (a_j) have bounded support.

Now suppose that for $i = 1, \ldots, d$, the function $\phi_i : \mathbb{R} \to \mathbb{R}$ satisfies the refinement equation

$$\phi_i(x) = \sum_{j=-\infty}^{\infty} a_{i,j} \phi_i(2x - j), \quad x \in \mathbb{R}, \tag{8.2}$$

and define their tensor-product

$$\phi(x) = \phi_1(x_1) \ldots \phi_d(x_d), \quad x \in \mathbb{R}^d. \tag{8.3}$$

Then for $x \in \mathbb{R}^d$,

$$\phi(x) = \prod_{i=1}^{d} \sum_{j=-\infty}^{\infty} a_{i,j} \phi_i(2x_i - j) = \sum_{j \in \mathbb{Z}^d} b_j \phi(2x - j), \tag{8.4}$$

where $b_j = \prod_{i=1}^{d} a_{i,j_i}$, $j \in \mathbb{Z}^d$.

Thus, the function ϕ in (8.3) satisfies the refinement equation (8.4). Although this generates a large class of refinable multivariate functions, and in some respects these functions may have essentially different behaviour form the univariate case, for our purposes the study of tensor-product refinable functions reduces essentially to the univariate case already considered. Therefore we do not discuss the tensor-product case (also called the separable case) any further.

We now generalise the procedure of Section 2 for defining uniform B-splines. Take $n \geq 0$ and vectors $v_1, \ldots, v_{n+d}$ in $\mathbb{Z}^d$ which span $\mathbb{R}^d$. By analogy with the characterisation (2.1) of the space $\mathcal{S}_n$, we shall define the space $\mathcal{S} = \mathcal{S}(v_1, \ldots, v_{n+d})$ to comprise all integrable functions $f : \mathbb{R}^d \to \mathbb{R}$ with compact support for which

$$\hat{f}(u) = \frac{P(z)}{(iv_1 u) \ldots (iv_{n+d} u)}, \quad z = e^{-iu}, \quad u \in \mathbb{R}^d, \tag{8.5}$$

where P is a Laurent plynomial in $z = (z_1, \ldots, z_d) = (e^{-iu_1}, \ldots, e^{-iu_d})$. Here for $w, u \in \mathbb{R}^d$, wu denotes their scalar product $w_1 u_1 + \ldots + w_d u_d$. We note that the definition of $\mathcal{S}$ is independent of any scaling of $v_1, \ldots, v_{n+d}$ and so we may assume that for $j = 1, \ldots, n + d$, the components of v_j are coprime. (For $d = 1$, we assume $v_1 = \ldots v_{n+1} = 1$.)

Now take $z \in \mathbb{C}^d$ with $z^{v_1} = 1$. Since the components of v_1 are coprime we can write $z = e^{-iu}$, where $v_1 u = 0$. Since the denominator of (8.5) vanishes, we must have $P(z) = 0$. Thus, the Laurent polynomial $P(z)$ must be divisible by $1 - z^{v_1}$. Factoring out $\frac{1 - z^{v_1}}{iv_1 u}$ and repeating this procedure, we see that $P(z)$ must be divisible by $(1 - z^{v_1}) \ldots (1 - z^{v_{n+d}})$. So for some Laurent polynomial Q, $\hat{f}(u) = Q(z)\hat{M}_n(u)$, $u \in \mathbb{R}^d$, where

$$\hat{M}_n(u) = \frac{1 - z^{v_1}}{iv_1 u} \cdots \frac{1 - z^{v_{n+d}}}{iv_{n+d} u}, \quad u \in \mathbb{R}^d. \tag{8.6}$$

Thus, every function in $\mathcal{S}$ is a finite linear combination of integer shifts of M_n. In our previous terminology, M_n is a basic function for $\mathcal{S}$.

To see the nature of the function M_n we first consider the case $n = 0$. We denote by V the parallelopiped $\{t_1 v_1 + \ldots t_d v_d : 0 \le t_i \le 1\}$. Since $v_1, \ldots, v_d$ span $\mathbb{R}^d$, the d-dimensional measure $m_d(V)$ of V is non-zero. Then it can be seen from (8.6) that

$$m_0 = \frac{1}{m_d(V)} \chi_V, \tag{8.7}$$

where χ_V denotes the characteristic function of V. For $n \ge 1$, we can also see from (8.6) that

$$M_n(x) = \int_0^1 M_{n-1}(x - tv_{n+d})\, dt, \quad x \in \mathbb{R}^d. \tag{8.8}$$

It follows by induction that M_n is a piecewise polynomial of degree n with support $\{t_1 v_1 + \ldots + t_{n+d} v_{n+d} : 0 \le t_i \le 1\}$. The function M_n is called a box spline. Box splines were introduced in [3]. Since M_n is a basic function for $\mathcal{S}$, every function in $\mathcal{S}$ is a piecewise polynomial of degree n. In general the space $\mathcal{S}$, restricted to a region on which its elements are polynomials, does not coincide with all piecewise polynomials of degree n, see [3, 22].

Since M_n is basic for $\mathcal{S}$, it follows that M_n is refinable. To find the refinement equation we note that for $u \in \mathbb{R}^d$,

$$\hat{M}_n(2u) = \frac{1 - z^{2v_1}}{2iv_1 u} \cdots \frac{1 - z^{2v_{n+d}}}{2iv_{n+d} u} = \frac{1 + z^{v_1}}{2} \cdots \frac{1 + z^{v_{n+d}}}{2} \hat{M}_n(u).$$

Taking inverse Fourier transforms gives (8.1) for $\phi = M_n$, where

$$\sum_{j \in \mathbb{Z}^d} a_j z^j = 2^d \frac{1 + z^{v_1}}{2} \cdots \frac{1 + z^{v_{n+d}}}{2}. \tag{8.9}$$

We can see from (8.6) that M_n is symmetric about $c = \frac{1}{2} \sum_{j=1}^{n+d} v_j$.

The integer shifts of M_n are linearly independent (in fact locally linearly independent) if and only if for any elements $w_1, \ldots, w_d$ in $\{v_1, \ldots, v_{n+d}\}$ which are linearly independent, the $d \times d$ matrix with rows $w_1, \ldots, w_d$ has determinant 1 or -1. This result is shown in [22, 42].

For $d = 1$, we have $v_1 = \ldots = v_{n+1} = 1$, and then (8.6) and (2.4) show that $M_n = N_n$, the uniform B-spline. Now take $d = 2$. If $n = 0$, $v_1 = (1, 0)$, $v_2 = (0, 1)$, then from (8.2) M_0 is the characteristic function of the unit square $[0, 1]^2$, which satisfies the refinement equation

$$M_0(x) = M_0(2x) + M_0(2x - (1, 0)) + M_0(2x - (0, 1)) + M_0(2x - (1, 1)), \quad x \in \mathbb{R}^2.$$

Now add the vector $v_3 = (1, 1)$. Then from (8.8), $M_1(x) = m_1(\{t : x - (t, t) \in [0, 1]^2\})$, $x \in \mathbb{R}^2$, i.e. M_1 is the continuous piecewise linear function on the triangulation formed by the lines through $\mathbb{Z}^2$ parallel to $(1, 0), (0, 1), (1, 1)$, satisfying $M_1(j) = \delta_{j,(1,1)}$, $j \in \mathbb{Z}^2$. In this case S comprises all continuous piecewise linear functions with compact support on the above triangulation. From (8.1), (8.9), (or direct inspection), M_1 satisfies the refinement equation

$$M_1 = M_1(2. - (1, 1)) +$$
$$\frac{1}{2} \sum \{M(2. - j) : j = (0, 0), (1, 0), (0, 1), (2, 1), (1, 2), (2, 2)\}.$$

It is easily seen that M_1 is symmetric about $(1, 1)$ and the integer shifts of M_1 are linearly independent. However, if we add the further vector $(1, -1)$, the integer shifts of M_2 are not linearly independent, since the matrix with rows $(1, 1)$ and $(1, -1)$ has determinant -2.

§9. Multivariate Wavelets

We now consider the extension of Section 3 to the multivariate case. Suppose that $\phi \in L^2(\mathbb{R}^d)$ satisfies the refinement equation (8.1). For $k \in \mathbb{Z}$ we define V_k to be the closed linear subspace of $L^2(\mathbb{R}^d)$ spanned by $\phi(2^k . - j)$, $j \in \mathbb{R}^d$. As for the univariate case, the nested sequence (V_k) is called a multiresolution analysis if (3.1) holds. In this case $L^2(\mathbb{R}^d)$ is the orthogonal sum of W_k, $k \in \mathbb{Z}$, where W_k denotes the orthogonal complement of V_k in V_{k+1}.

Now let E denote the set of vertices of $[0, 1]^d$. While V_0 is spanned by the integer shifts of ϕ, V_1 is spanned by the integer shifts of the 2^d functions $\{\phi(2. - j) : j \in E\}$. This suggests that W_0 should be spanned by the integer shifts of $2^d - 1$ functions. Such a result was shown in [61], as we now describe.

We shall need to assume that ϕ is symmetric about a point $c \in \mathbb{R}^d$, i.e.

$$\phi(c + x) = \phi(c - x), \quad x \in \mathbb{R}^d. \tag{9.1}$$

88   *T. N. T. Goodman*

It follows from (9.1) that c is in $\frac{1}{2}\mathbb{Z}^d$ and

$$a_{2c-j} = a_j, \quad j \in \mathbb{Z}^d. \tag{9.2}$$

We shall also need a function $\eta : E \to E$ satisfying

$$\eta(0) = 0, \quad (\eta(\mu) + \eta(\nu))(\mu + \nu) \text{ is odd for } \mu \neq \nu. \tag{9.3}$$

For $d = 1$, such a map is given by $\eta(0) = 0$, $\eta(1) = 1$, while for $d = 2$ it can be given by $\eta(0) = 0$, $\eta(0,1) = (0,1)$, $\eta(1,0) = (1,1)$, $\eta(1,1) = (1,0)$. Such a mapping for $d = 3$ is given in [61] but, as remarked there, no such maps exist for $d > 3$.

Theorem 9.1. [61] *Suppose that $d \leq 3$ and ϕ in $L^2(\mathbb{R}^d)$ satisfies the refinement equation (8.1) for $(a_j) \in l^2(\mathbb{Z}^d)$, is symmetric about c in $\frac{1}{2}\mathbb{Z}^d$, and has orthonormal integer shifts. For $\mu \in E$, let*

$$\psi_\mu(x) = \sum_{j \in \mathbb{Z}^d} (-1)^{j\mu} a_{(-1)^{2c\mu}(j+\eta(\mu))} \phi(2x - j), \quad x \in \mathbb{R}^d. \tag{9.4}$$

Then W_0 is the closed linear subspace of $L^2(\mathbb{R}^d)$ generated by $\{\psi_\mu(. - j) : \mu \in E \setminus \{0\}, j \in \mathbb{Z}^d\}$ and $\{2^{\frac{1}{2}kd}\psi_\mu(2^k. - j) : \mu \in E \setminus \{0\}, k \in \mathbb{Z}, j \in \mathbb{Z}^d\}$ forms an orthonormal basis for $L^2(\mathbb{R}^d)$.

If $d = 1$, then $\mu = \eta(\mu) = 1$. If $c \in \mathbb{Z}^d$, then $(-1)^{2c\mu}(j + \eta(\mu)) = j + 1$ and, by (9.2), $a_{j+1} = a_{2c-j-1}$. If $c \notin \mathbb{Z}^d$, then $(-1)^{2c\mu}(j + \eta(\mu)) = -j - 1$. So in either case, ψ_μ in (9.4) is an integer shift of ψ in (3.4).

The box splines M_n do not have orthogonal integer shifts for $n \geq 1$, but by a similar trick to (3.6), corresponding refinable splines with orthonormal integer shifts are constructed in [61]. As for the univariate case these spline functions do not have compact support. To gain wavelets from refinable functions without orthogonal integer shifts, we have the following partial generalisation of Theorem 3.1.

Theorem 9.2. [62] *Suppose that $d \leq 3$ and ϕ in $L^2(\mathbb{R}^d)$ has compact support, satisfies the refinement equation (8.1) for (a_j) with bounded support, and is symmetric about c in $\frac{1}{2}\mathbb{Z}^d$. Suppose further that the integer shifts of ϕ form a Riesz basis. For $\mu \in E$, let*

$$\psi_\mu(x) = \sum_{j \in \mathbb{Z}^d} (-1)^{j\mu} b_{(-1)^{2c\mu}(j+\eta(\mu))} \phi(2x - j), \quad x \in \mathbb{R}^d,$$

where $b_j = \int_{\mathbb{R}^d} \phi(x)\phi(2x - j)\,dx$. Then W_0 is the closed linear subspace of $L^2(\mathbb{R}^d)$ generated by $\{\psi_\mu(. - j) : \mu \in E \setminus \{0\}, j \in \mathbb{Z}^d\}$. Moreover, $L^2(\mathbb{R}^d)$ is the orthogonal sum of W_k, $k \in \mathbb{Z}$, and $\{2^{\frac{1}{2}kd}\psi_\mu(2^k. - j) : \mu \in E \setminus \{0\}, k \in \mathbb{Z}, j \in \mathbb{Z}^d\}$ forms a Riesz basis for $L^2(\mathbb{R}^d)$.

Under the conditions for linear independence at the end of Section 8, the box spline M_n satisfies the conditions of Theorem 9.2 and so we can define corresponding wavelets. This was also done in [7]. For $d \geq 4$, this construction does not work. A result on the existence of wavelets in arbitrary dimensions is given in [44]. There the proof is not constructive. Explicit constructions for wavelets from box splines in general dimensions are given in [69]. Further results on box spline wavelets appear in [51, 52, 76].

§10. Vector Refinable Bivariate Splines

The characterisation of univariate refinable splines in Section 4 has been partially extended in [72] to show that a multivariate refinable spline must in general be a linear combination of integer shifts of box splines. In this section we shall give an analogue of the vector refinable splines in Section 5 for $r = 2$ in 2 dimensions. However, the situation is richer in two dimensions than in one, and an indication of the difference is that these bivariate splines of degree n have maximal continuity C^{n-1}.

Take $n \geq 0$ and pairwise linearly independent vectors $v_1, \ldots, v_{n+3}$ in $\mathbb{Z}^2$, where the components of each v_j are coprime. We define the space $S = S_2(v_1, \ldots, v_{n+3})$ to comprise all integrable functions $f : \mathbb{R}^2 \to \mathbb{R}$ with compact support for which

$$\hat{f}(u) = \frac{P(z)u}{(v_1 u) \cdots (v_{n+3} u)}, \quad z = e^{-iu}, \quad u \in \mathbb{R}^2, \tag{10.1}$$

where P is a 1×2 matrix of Laurent polynomials.

Theorem 10.1. [32] *A function f lies in S if and only if it is a C^{n-1} piecewise polynomial of degree n with compact support on the triangulation formed by lines through $\mathbb{Z}^2$ parallel to $v_1, \ldots, v_{n+3}$, and such that the jumps in the nth derivatives of f across mesh lines remain constant except across points in $\mathbb{Z}^2$.*

By following a similar procedure to that for the construction of box splines, it is shown in [32] that one can construct, recursively on n, a pair of functions $\phi = (\phi_1, \phi_2)^T$ which are basic for S and hence satisfy a vector refinement equation. As in Theorem 5.1, a pair $\eta = (\eta_1, \eta_2)^T$ is basic for S if and only if $\hat{\eta}(u) = N(z)\hat{\phi}(u)$, for a 2×2 matrix of Laurent polynomials with $\det N(z) = cz^j$ for $c \neq 0$ and $j \in \mathbb{Z}^2$.

As a first example we take $n = 0$, $v_1 = (1, 0)$, $v_2 = (0, 1)$, $v_3 = (1, 1)$. Then S comprises all piecewise constant functions with compact support on the triangulation formed by the lines through $\mathbb{Z}^2$ parallel to v_1, v_2, v_3. A basic pair for S is (ϕ_1, ϕ_2), where ϕ_1 is the characteristic function of the triangle with vertices $(0, 0)$, $(1, 0)$, $(1, 1)$, while ϕ_2 is the characteristic function of the triangle with vertices $(0, 0)$, $(0, 1)$, $(1, 1)$.

Next we add the vector $v_4 = (1, -1)$. Then S comprises all continuous, piecewise linear functions with compact support on the triangulation formed by the lines through $\mathbb{Z}^2$ parallel to $v_1, \ldots, v_4$. The functions in S are determined uniquely by their values on the mesh points $\mathbb{Z}^2 \cup (\mathbb{Z}^2 + (\frac{1}{2}, \frac{1}{2}))$. A basic pair for $\mathcal{S}$ is $\phi = (\phi_1, \phi_2)^T$, where $\phi_1(0) = 1$, $\phi_2(\frac{1}{2}, \frac{1}{2}) = 1$, while ϕ_1 and ϕ_2 vanish at all other mesh points.

Now return to the general case and suppose that M is a non-singular 2×2 integer matrix such that for $j = 1, \ldots, n+3$, Mv_j is a scalar multiple of a vector in $\{v_1, \ldots, v_{n+3}\}$. Then from (10.1) or from Theorem 10.1, it is easily seen that $f \in \mathcal{S}$ implies $f(M^{-1}.) \in \mathcal{S}$. If ϕ is a basic pair of functions for $\mathcal{S}$, then $\phi(M^{-1}.)$ is in $\mathcal{S}$ and so ϕ satisfies an equation of form

$$\phi(x) = \sum_{j \in \mathbb{Z}^2} a_j \phi(Mx - j), \quad x \in \mathbb{R}^2, \tag{10.2}$$

for a sequence (a_j) of constant 2×2 matrices with bounded support. If $M = 2I$, then (10.2) reduces to the usual vector refinement equation. If $\lim_{n \to \infty} M^{-n} = 0$, then we still refer to (10.2) as a vector refinement equation.

As an example, consider again the 4-direction mesh $v_1 = (1, 0)$, $v_2 = (0, 1)$, $v_3 = (1, 1)$, $v_4 = (1, -1)$ and the basic pair ϕ above. Let $M = \begin{bmatrix} 1 & 1 \\ -1 & 1 \end{bmatrix}$. Then $\lim_{n \to \infty} M^{-n} = 0$ and $Mv_1 = v_4$, $Mv_2 = v_3$, $Mv_3 = 2v_1$, $Mv_4 = -2v_2$. It is easily checked that ϕ satisfies, for $x \in \mathbb{R}^2$,

$$\phi_1(x) = \phi_1(Mx) + \phi_2(Mx + (0, 1)) + \phi_2(Mx + (0, 1)) + \phi_2(Mx + (1, 1)),$$
$$\phi_2(x) = \phi_1(Mx - (1, 0)).$$

Wavelets corresponding to this example are constructed in [31, 38].

Now recall the condition in Section 8 for the integer shifts of a box spline M_n to be linearly independent. If $d = 2$, then the condition is equivalent to requiring that any two distinct lines through $\mathbb{Z}^2$ parallel to vectors in $\{v_1, \ldots, v_{n+2}\}$ can intersect only in $\mathbb{Z}^2$. In [32], a corresponding condition is derived for any basic pair ϕ for $\mathcal{S}$: the integer shifts of ϕ form a Riesz basis if and only if at any point not in $\mathbb{Z}^2$, no more than two lines through $\mathbb{Z}^2$ parallel to vectors in $\{v_1, \ldots, v_{n+3}\}$ can intersect. For the 4-direction mesh considered above, the only intersections not in $\mathbb{Z}^2$ are at points in $\mathbb{Z}^2 + (\frac{1}{2}, \frac{1}{2})$, at each of which exactly two such lines intersect. Thus, in this case the integer translates of ϕ form a Riesz basis.

It is shown in [32] that the condition for a Riesz basis can be satisfied only for $n \leq 3$. For $n = 3$, there is an infinite choice of vectors $\{v_1, \ldots, v_6\}$ for which it is satisfied, which include the choices $(1, 0)$, $(0, 1)$, $(1, 1)$, $(1, -1)$, $(2j, 1)$, $(-1, 2j)$, for any non-zero integer j.

The following results are shown in [33]. If the condition for a Riesz basis is satisfied and $n \geq 1$, then the condition on the jumps on the nth derivatives in Theorem 10.1 is automatically satisfied, i.e. $\mathcal{S}$ comprises all C^{n-1} piecewise polynomials of degree n with compact support on the triangulation formed by the lines through $\mathbb{Z}^2$ parallel to $v_1, \ldots, v_{n+3}$. If, in addition, $\{v_1, \ldots, v_{n+3}\}$ contains $(1, 0)$, $(0, 1)$ and $(1, 1)$, then any C^{n-1} piecewise polynomial of degree n on the above triangulation (with arbitrary support) can be written as a linear combination (possibly infinite) of the integer shifts of any basic pair ϕ for $\mathcal{S}$.

§11. Final Remarks

Biorthogonal spline wavelets for the univariate vector case are considered in [20, 21, 70, 71], while biorthogonal wavelets corresponding to box splines are considered in [18, 40, 41, 63]. Box spline wavelet packets and frames are studied in [64, 67]. Related to the orthogonal refinable splines of Section 7, some bivariate examples are given in [28].

There is an elegant theory for periodic refinable splines [46, 59, 73], which can also be considered as splines on a finite interval with periodic boundary conditions. For other "refinable" splines on a finite interval, see [6, 9, 20, 21, 60]. Corresponding work for splines on $\mathbb{R}$ with non-periodic knots appears in [4, 8, 19, 23, 53]. The bivariate case for general triangulations is considered in [17, 24, 47, 68] and applications of such work to computer aided geometric design are reviewed in [25, 75].

Even if we restrict our attention to spline functions satisfying a vector refinement equation of form (8.1), there is still much work to be done, especially for $d \geq 2$. It is hoped that this paper will give a stimulus for further work.

References

1. Battle, G., A block spin construction of ondelettes, Part 1: Lemarie functions, Comm. Math. Phys. **110** (1987), 601–615.

2. de Boor, C., R. DeVore, and A. Ron, On the construction of multivariate (pre)wavelets, Constr. Approx. **9** (1993), 123–166.

3. de Boor, C. and K. Höllig, B-splines from parallelopipeds, J. Analyse Math. **42** (1982), 99–115.

4. Buhmann, M. D. and C. A. Micchelli, Spline prewavelets for non-uniform knots, Numer. Math. **61** (1992), 455–474.

5. Chui, C. K., *An Introduction to Wavelets*, Academic Press, Boston, 1992.

6. Chui, C. K. and E. Quak, Wavelets on a bounded interval, in *Numerical Methods of Approximation Theory, Vol. 9*, D. Braess and L. L. Schumaker (eds.), Birkhäuser, Basel, 1992, pp. 53–75.

7. Chui, C. K., J. Stöckler, and J. Ward, Compactly supported box spline wavelets, Approx. Theory Appl. **8** (1992), 77–100.

8. Chui, C. K. and J. de Villiers, Applications of optimally local interpolation to constructing interpolatory approximants and compactly supported wavelets, Math. Comp. **65**(213) (1996), 99–114.

9. Chui, C. K. and J. de Villiers, Spline-wavelets with arbitrary knots on a bounded interval: orthogonal decomposition and computational algorithms, Comm. Appl. Math. **2** (1998), to appear.

10. Chui, C. K. and J. Z. Wang, A cardinal spline approach to wavelets, Proc. Amer. Math. Soc. **113** (1991), 785–793.

11. Chui, C. K. and J. Z. Wang, On compactly supported spline wavelets and a duality principle, Trans. Amer. Math. Soc. **330** (1992), 77–100.

12. Chui, C. K. and J. Z. Wang, A general framework of compactly supported splines and wavelets, J. Approx. Theory **71** (1992), 263–304.

13. Chui, C. K. and J. Z. Wang, An analysis of cardinal spline wavelets, J. Approx. Theory **72** (1993), 54–68.

14. Cohen, A., I. Daubechies, and J.-C. Feauveau, Biorthogonal bases of compactly supported wavelets, Comm. Pure Appl. Math. **45** (1992), 485–560.

15. Curry, H. B. and I. J. Schoenberg, On spline distributions and their limits: the Pólya distribution functions, Abstract 380t, Bull. Amer. Math. Soc. **53** (1947), 1114.

16. Curry, H. B. and I. J. Schoenberg, On Pólya frequency functions IV. The fundamental spline functions and their limits, J. Analyse Math. **17** (1966), 71–107.

17. Dæhlen, M., T. Lyche, K. Mørken, R. Schneider, and H.-P. Seidel, Multiresolution analysis based on quadratic Hermite interpolation on triangles, preprint.

18. Dahlke, S., K. Gröchenig, and V. Latour, Biorthogonal box spline wavelet bases, in *Surface Fitting and Multiresolution Methods*, A. Le Méhauté, C. Rabut, and L. L. Schumaker (eds.), Vanderbilt Univ. Press, Nashville, 1997, pp. 83–92.

19. Dahmen, W., J. M. Carnicer, and J. M. Peña, Local decomposition of refinable spaces and wavelets, Appl. Comput. Harmonic Anal. **3** (1996), 127–153.

20. Dahmen, W., B. Han, R. Q. Jia, and A. Kunoth, Biorthogonal multiwavelets on the interval: Cubic Hermite splines, preprint.

21. Dahmen, W., A. Kunoth, and K. Urban, Biorthogonal spline wavelets on the interval - Stability and moment conditions, preprint.

22. Dahmen, W. and C. A. Micchelli, Translates of multivariate splines, Linear Algebra Appl. **52** (1983), 217–234.

23. Dahmen, W. and C. A. Micchelli, Banded matrices with banded inverses II: Locally finite decomposition of spline spaces, Constr. Approx. **9** (1993), 263–281.

24. Dahmen, W. and R. P. Stevenson, Element-by-element construction of wavelets satisfying stability and moment conditions, preprint.

25. DeRose, T. D., Applications of multiresolution surfaces, in *The Mathematics of Surfaces VII*, T. Goodman and R. Martin (eds.), Information Geometers, Winchester, 1997, pp. 1–15.

26. Donovan, G. C., J. S. Geronimo, and D. P. Hardin, Intertwining multiresolution analyses and the construction of piecewise polynomial wavelets, SIAM J. Math. Anal. **27** (1996), 1791–1815.

27. Donovan, G. C., J. S. Geronimo, and D. P. Hardin, Orthogonal polynomials and the construction of piecewise smooth wavelets, preprint.

28. Donovan, G. C., J. S. Geronimo, and D. P. Hardin, Non-separable scaling functions in two dimensions, preprint.

29. Donovan, G. C., J. S. Geronimo, D. P. Hardin, and P. R. Massopust, Construction of orthogonal wavelets using fractal interpolation functions, SIAM J. Math. Anal. **27** (1996), 1158–1192.

30. Goodman, T. N. T., Interpolatory Hermite spline wavelets, J. Approx. Theory **78** (1994), 174–189.

31. Goodman, T. N. T., Construction of wavelets with multiplicity, Rendiconti di Mathematica (Serie VIII) **14** (1994), 665–691.

32. Goodman, T. N. T., Constructing pairs of refinable bivariate spline functions, in *Surface Fitting and Multiresolution Methods*, A. Le Méhauté, C. Rabut, and L. L. Schumaker (eds.), Vanderbilt Univ. Press, Nashville, 1997, pp. 145–162.

33. Goodman, T. N. T., Properties of bivariate refinable spline pairs, in *Multivariate Approximation, Recent Trends and Results*, W. Haussmann, K. Jetter, and M. Reimer (eds.), Akademie Verlag, Berlin, 1997, pp. 63–82.

34. Goodman, T. N. T. and S. L. Lee, Wavelets of multiplicity r, Trans. Amer. Math. Soc. **342** (1994), 307–324.

35. Goodman, T. N. T. and S. L. Lee, Refinable vectors of spline functions, in *Mathematical Methods for Curves and Surfaces II*, M. Dæhlen, T. Lyche, and L. L. Schumaker (eds.), Vanderbilt Univ. Press, Nashville & London, 1998, pp. 213–220.

36. Goodman, T. N. T. and S. L. Lee, Linear independence and stability of refinable vectors of spline functions, preprint.

37. Goodman, T. N. T., S. L. Lee, and W. S. Tang, Wavelets in wandering subspaces, Trans. Amer. Math. Soc. **338** (1993), 639–654.

38. Goodman, T. N. T., S. L. Lee, and W. S. Tang, Wavelet bases for a set of commuting unitary operators, Advances in Comp. Math. **1** (1993), 109–125.

39. He, T. X., Spline interpolation and its wavelet analysis, in *Approximation Theory VIII, Vol. 2: Wavelets*, Charles K. Chui and Larry L. Schumaker (eds.), World Scientific Publishing Co., Inc., Singapore, 1995, pp. 143–150.

40. He, W. J. and M. J. Lai, Construction of trivariate compactly supported biorthogonal box spline wavelets, preprint.

41. Ji, H., S. D. Riemenschneider, and Z. W. Shen, Multivariate compactly supported fundamental refinable functions, duals and biorthogonal wavelets, preprint.

42. Jia, R. Q., Local linear independence of the translates of a box spline, Constr. Approx. **1** (1985), 175–182.

43. Jia, R. Q., Refinable shift-invariant spaces: from splines to wavelets, in *Approximation Theory VIII, Vol. 2: Wavelets*, Charles K. Chui and Larry

L. Schumaker (eds.), World Scientific Publishing Co., Inc., Singapore, 1995, pp. 179–208.

44. Jia, R. Q. and C. A. Micchelli, Using the refinement equation for the construction of pre-wavelets II: Powers of two, in *Curves and Surfaces*, P.-J. Laurent, A. Le Méhauté, and L. L. Schumaker (eds.), Academic Press, New York, 1991, pp. 209–246.

45. Jia, R. Q. and C. A. Micchelli, On linear independence of integer translates of a finite number of functions, Proc. Edinburgh Math. Soc. **36** (1992), 69–85.

46. Koh, Y. W., S. L. Lee, and H. H. Tan, Periodic orthogonal splines and wavelets, Appl. Comput. Harmonic Anal. **2** (1995), 201–212.

47. Kotycka, U. and P. Oswald, Piecewise linear prewavelets of small support, in *Approximation Theory VIII, Vol. 2: Wavelets*, Charles K. Chui and Larry L. Schumaker (eds.), World Scientific Publishing Co., Inc., Singapore, 1995, pp. 235–242.

48. Lawton, W., S. L. Lee, and Z. W. Shen, Characterization of compactly supported refinable splines, Adv. Comp. Math. **3** (1995), 137–145.

49. Lee, S. L., *B*-splines for cardinal Hermite interpolation, Linear Algebra Appl. **12** (1975), 269–280.

50. Lemarie, Ondelettes a localisation exponentielles, J. de Math. Pures et Appl. **67** (1988), 227–236.

51. Lorentz, R. A. and W. R. Madych, Wavelets and generalised box splines, Appl. Anal. **44** (1992), 51–76.

52. Lorentz, R. A. and P. Oswald, Nonexistence of compactly supported box spline prewavelets in Sobolev spaces, *Surface Fitting and Multiresolution Methods*, A. Le Méhauté, C. Rabut, and L. L. Schumaker (eds.), Vanderbilt Univ. Press, Nashville, 1997, pp. 235–244.

53. Lyche, T. and K. Mørken, Spline wavelets of minimal support, in *Numerical Methods in Approximation Theory, Vol. 9*, D. Braess and L. L. Schumaker (eds.), Birkhauser Verlag, Basel, 1992, pp. 177–194.

54. Mallat, S. G., Multiresolution approximations and wavelet orthonormal basis of $L^2(\mathbb{R})$, Trans. Amer. Math. Soc. **315** (1989), 69–87.

55. Micchelli, C. A., Using the refinement equation for the construction of pre-wavelets, Numer. Algorithms **1** (1991), 75–116.

56. Plonka, G., Spline wavelets with higher defect, in *Wavelets, Images, and Surface Fitting*, P.-J. Laurent, A. Le Méhauté, and L. L. Schumaker (eds.), A. K. Peters, Wellesley MA, 1994, pp. 387–398.

57. Plonka, G., Two-scale symbol and autocorrelation symbol for B-splines with multiple knots, Advances in Comp. Math. **3** (1995), 1–22.

58. Plonka, G., Generalized spline wavelets, Constr. Approx. **12** (1996), 127–155.

59. Plonka, G. and M. Tasche, On the computation of periodic spline wavelets, Appl. Comput. Harmonic Anal. **2** (1995), 1–14.

60. Quak, E., On a spline multiresolution analysis with homogeneous boundary conditions, *Surface Fitting and Multiresolution Methods*, A. Le Méhauté, C. Rabut, and L. L. Schumaker (eds.), Vanderbilt Univ. Press, Nashville, 1997, pp. 301–308.

61. Riemenschneider, S. D. and Z. W. Shen, Box splines, cardinal series and wavelets, in *Approximation Theory and Functional Analysis*, C. K. Chui (ed.), Academic Press, New York, 1991, pp. 133–149.

62. Riemenschneider, S. D. and Z. W. Shen, Wavelets and pre-wavelets in low dimensions, J. Approx. Theory **71** (1992), 18–38.

63. Riemenschneider, S. D. and Z. W. Shen, Construction of compactly supported biorthogonal wavelets in $L_2(\mathbb{R}^d)$, preprint.

64. Ron, A. and Z. W. Shen, Compactly supported tight affine spline frames in $L_2(\mathbb{R}^d)$, preprint.

65. Schoenberg, I. J., Contributions to the problem of approximation of equidistant data by analytic functions, Quart. Appl. Math. **4** (1946), 45–99, 112–141.

66. Schoenberg, I. J. and A. Sharma, Cardinal interpolation and spline functions. V. The B-splines for cardinal Hermite interpolation, Linear Algebra Appl. **7** (1973), 1–42.

67. Shen, Z. W., Non-tensor product wavelet packets in $L_2(\mathbb{R}^s)$, SIAM J. Math. Anal. **26** (1995), 1061–1074.

68. Stevenson, R. P., Piecewise linear (pre)-wavelets on non-uniform meshes, preprint.

69. Stöckler, J., Multivariate wavelets, in *Wavelets: a tutorial in theory and applications*, C. K. Chui (ed.), Academic Press, San Diego, 1992, pp. 325–355.

70. Strela, V., A note on construction of biorthogonal multi-scaling functions, preprint.

71. Strela, V. and G. Strang, Pseudo-biorthogonal multiwavelets and finite elements, preprint.

72. Sun, Q., Refinable functions with compact support, J. Approx. Theory **86** (1996), 240–252.

73. Tasche, M., Orthogonal periodic spline wavelets, in *Wavelets, Images and Surface Fitting*, P.-J. Laurent, A. LeMéhauté, and L. L. Schumaker (eds.), A. K. Peters, Wellesley MA, 1994, pp. 37 50.

74. de Villiers, J., On the construction of a multiresolution based on spline interpolation, in *Approximation Theory IX*, Charles K. Chui and Larry L. Schumaker (eds.), Vanderbilt University Press, Nashville, TN, 1998.

75. Warren, J., Sparse filter banks for binary subdivision schemes, in *The Mathematics of Surfaces VII*, T. Goodman and R. Martin (eds.), Information Geometers, Winchester, 1997, pp. 427–438.

76. Zhou, D. X., Construction of real-valued wavelets by symmetry, J. Approx. Theory **81** (1995), 323–331.

Tim N. T. Goodman
Department of Mathematics
Dundee University
Dundee, DD1 4HN, Scotland
`tgoodman@mcs.dundee.ac.uk`

How is a Vector Pyramid Scheme Affected by Perturbation in the Mask?

Bin Han and Thomas A. Hogan

Abstract. When implementing a pyramid scheme on a computer, it is not possible to use the exact values of the terms of the refinement mask if some are irrational. The irrational terms must be perturbed. When this is the case, it is important to know that the pyramid scheme associated with the perturbed mask converges and that the limit function is close to the solution of the original refinement equation. In this paper, we provide sharp L_p $(1 \leq p \leq \infty)$ error estimates for univariate vector pyramid schemes when the terms of a finitely supported mask are perturbed slightly.

§1. Introduction

A refinable function vector ϕ is an $r \times 1$ vector of functions satisfying the following vector refinement equation:

$$\phi = \sum_{k \in \mathbb{Z}} a(k)\phi(2 \cdot - k), \tag{1}$$

where $a : \mathbb{Z} \to \mathbb{R}^{r \times r}$ is a finitely supported sequence of $r \times r$ matrices called the (matrix refinement) mask. Vector refinement equations and refinable function vectors have been well-studied recently (see e.g., [1, 2, 3, 4, 5, 8, 9, 11]).

We investigate how perturbation of a finitely supported mask affects the corresponding refinable function vector and associated pyramid scheme. As in [11], we assume throughout that the mask a satisfies the following condition:

$$\sum_{k \in \mathbb{Z}} a(k) = \begin{pmatrix} 2 & 0 \\ 0 & \Lambda \end{pmatrix} \quad \text{with} \quad \lim_{n \to \infty} (\Lambda/2)^n = 0. \tag{2}$$

Under this condition, Heil and Colella proved in [8] that there exists a unique vector ϕ of compactly supported distributions such that ϕ satisfies the vector

Approximation Theory IX, Volume 2: Computational Aspects
Charles K. Chui and Larry L. Schumaker (eds.), pp. 97–104.
Copyright © 1998 by Vanderbilt University Press, Nashville, TN.
ISBN 0-8265-1326-3.

refinement equation (1) and $\widehat{\phi}(0) = (1, 0, \ldots, 0)^T$. We call this solution the normalized solution of (1) associated with the mask a and denote it by ϕ_a.

The notation $\|\cdot\|_p$ will denote norms in several different spaces. First, it denotes the standard p-norm on $\mathbb{R}^m$ for any positive integer m. In addition,

$$\|A\|_p := \max_{\|v\|_p = 1} \|Av\|_p \text{ for } A \text{ in } \mathbb{R}^{m \times m},$$

$$\|a\|_p^p := \left(\sum_{j \in \mathbb{Z}} \|a(j)\|_p^p \right)^{1/p} \quad \text{for } a \text{ in } (\ell_0(\mathbb{Z}))^{m \times m} \text{ or } (\ell_0(\mathbb{Z}))^m,$$

$$\text{and } \|f\|_p^p := \int_{\mathbb{R}} \|f(x)\|_p^p \, dx \text{ for } f \text{ in } (L_p(\mathbb{R}))^m.$$

The particular interpretation should always be clear from context.

There is generally no explicit expression for the normalized solution ϕ_a of Equation (1). Instead, the solution is approximated by iterating the operator

$$Q_a : (L_p(\mathbb{R}))^r \to (L_p(\mathbb{R}))^r : f \mapsto Q_a f := \sum_{k \in \mathbb{Z}} a(k) f(2 \cdot - k). \qquad (3)$$

This process is called the vector pyramid scheme associated with the matrix mask a (see [1]). Note that the fixed points of Q_a are the solutions to Equation (1). It was proved in [11] that if f is a compactly supported function vector and the sequence $(Q_a^n f, n \in \mathbb{N})$ converges to ϕ_a in the L_p norm for some $1 \leq p \leq \infty$, then f must satisfy the following moment conditions of order 1:

$$e_1^T \widehat{f}(0) = 1 \quad \text{and} \quad e_1^T \widehat{f}(2k\pi) = 0 \qquad \forall k \in \mathbb{Z} \backslash \{0\}, \qquad (4)$$

where, for $i = 1, \ldots, r$, e_i denotes the ith standard unit vector in $\mathbb{R}^r$.

Now, let h be the hat function

$$h(x) := \max\{0, 1 - |1 - x|\}, \qquad x \in \mathbb{R}$$

and define the function vector ϕ_0 by

$$\phi_0 := \left(h, h(2r \cdot - 2), h(2r \cdot - 4), \cdots, h(2r \cdot - 2(r-1)) \right)^T. \qquad (5)$$

That is, the first component of ϕ_0 is h, and the jth component of ϕ_0 is $h(2r \cdot - 2(j-1))$ for $j = 2, \ldots, r$. Then ϕ_0 satisfies the moment conditions in (4). If there exists $\phi \in (L_p(\mathbb{R}))^r$ such that $\lim_{n \to \infty} \|Q_a^n \phi_0 - \phi\|_p = 0$, we say that the pyramid scheme associated with a converges to ϕ in the L_p norm.

It is well known (see [4, 8, 11]) that if a satisfies condition (2) and the pyramid scheme associated with a converges in L_p norm for some $1 \leq p \leq \infty$, then

$$e_1^T \sum_{k \in \mathbb{Z}} a(2k) = e_1^T \sum_{k \in \mathbb{Z}} a(1 + 2k) = e_1^T. \qquad (6)$$

Also, the limit function vector ϕ is the normalized solution of Equation (1).

For any positive integers m and n, let $(\ell_0(\mathbb{Z}))^{m \times n}$ denote the linear space of all sequences of $m \times n$ matrices on $\mathbb{Z}$ with finite support. We say that the shifts of an $r \times 1$ function vector $f \in (L_p(\mathbb{R}))^r$ $(1 \leq p \leq \infty)$ are stable if there exist two positive constants C_1 and C_2 such that

$$C_1\|\lambda\|_p \leq \left\|\sum_{k \in \mathbb{Z}} \lambda(k)f(\cdot - k)\right\|_p \leq C_2\|\lambda\|_p \qquad \forall \lambda \in (\ell_0(\mathbb{Z}))^{1 \times r}. \qquad (7)$$

Suppose f is a compactly supported function vector with stable shifts that satisfies the moment conditions of order 1. If the sequence $(Q_a^n f, n \in \mathbb{N})$ converges to ϕ_a in the L_p norm, then for any compactly supported function vector g satisfying the moment conditions of order 1, the sequence $(Q_a^n g, n \in \mathbb{N})$ also converges to ϕ_a in the L_p norm (see [11]). In this paper, we only use the initial function ϕ_0 given in (5). However, our results and proofs are valid for general initial function vectors that satisfy (4).

It is natural to raise the following question: how are vector pyramid schemes and refinable function vectors affected by a perturbation of the mask? For example, if some of the entries of the mask coefficient matrices are irrational then, in most applications, it is necessary to perturb those entries. Daubechies and Huang [6] first investigated what effect this might have on univariate scalar $(r = 1)$ refinable functions for the case $p = \infty$. More recently, Han [7] provided a sharp error estimate for multivariate scalar functions in any L_p norm, $1 \leq p \leq \infty$. In this paper, we provide results along the lines of [7] for univariate refinable function vectors.

Of course this is only a small aspect of the effect of round-off error on a pyramid scheme. Another important question to be answered regards possible accumulation of errors in any single iteration of the scheme which are inevitable when using floating point arithmetic. These considerations are very interesting and should be addressed in the future.

For any positive integer N, denote by $(\ell([-N, N]))^{r \times r}$ the set of all sequences in $(\ell_0(\mathbb{Z}))^{r \times r}$ which are supported in $[-N, N] \cap \mathbb{Z}$.

Main Theorem. *Let a be an element of $(\ell([-N, N]))^{r \times r}$ for some positive integer N. Suppose that a satisfies condition (2) and the pyramid scheme associated with a converges in the L_p norm $(1 \leq p \leq \infty)$. Then there exist positive constants η and C such that, for any b in $(\ell([-N, N]))^{r \times r}$ satisfying*

$$e_1^T \sum_{k \in \mathbb{Z}} b(2k) = e_1^T \sum_{k \in \mathbb{Z}} b(1 + 2k) = e_1^T, \qquad (8)$$

and $\|a - b\|_1 < \eta$, the pyramid scheme associated with b converges in the L_p norm and

$$\|Q_a^n \phi_0 - Q_b^n \phi_0\|_p \leq C\|a - b\|_1 \qquad \forall n \in \mathbb{N}. \qquad (9)$$

Consequently, $\|\phi_a - \phi_b\|_p \leq C\|a - b\|_1$ where ϕ_a and ϕ_b are the normalized solutions of the refinement equation with mask a and b, respectively.

Remark. Suitable constants η and C could be obtained constructively by looking into the details of the proofs.

§2. Auxiliary Results

Our results rely on the characterization provided in [11] of the convergence of a vector pyramid scheme in terms of the ℓ_p-norm joint spectral radius of two matrices associated with the mask. The uniform joint spectral radius was introduced by Rota and Strang in [12], and the mean spectral radius was studied by Wang in [13]. Their definitions were extended to the ℓ_p-norm joint spectral radius for $1 \leq p \leq \infty$ by Jia in [10]. We state that and other relevant definitions now. Throughout, a is assumed to be an element of $(\ell_0(\mathbb{Z}))^{r \times r}$.

Let $\mathcal{A}$ be a finite collection of square matrices and $\|\cdot\|$ be a matrix norm. For a positive integer n, define $\mathcal{A}^n := \{(A_1, \cdots, A_n) : A_1, \cdots, A_n \in \mathcal{A}\}$ and

$$\|\mathcal{A}^n\|_p := \begin{cases} \left(\sum_{(A_1,\cdots,A_n)\in\mathcal{A}^n} \|A_1 \cdots A_n\|^p\right)^{1/p}, & 1 \leq p < \infty, \\ \max\{\|A_1 \cdots A_n\| : (A_1, \cdots, A_n) \in \mathcal{A}^n\}, & p = \infty. \end{cases}$$

Then the ℓ_p-norm joint spectral radius of $\mathcal{A}$ is

$$\rho_p(\mathcal{A}) := \lim_{n\to\infty} \|\mathcal{A}^n\|_p^{1/n} = \inf_{n\geq 1} \|\mathcal{A}^n\|_p^{1/n}. \tag{10}$$

For $\varepsilon \in \{0,1\}$, let A_ε denote the following linear operator on $(\ell_0(\mathbb{Z}))^r$:

$$A_\varepsilon \lambda(i) := \sum_{k\in\mathbb{Z}} a(\varepsilon + 2i - k)\lambda(k) \qquad i \in \mathbb{Z}, \lambda \in (\ell_0(\mathbb{Z}))^r;$$

and for any positive integer N, let

$$U_N := \{\lambda \in (\ell([-N,N]))^r : e_1^T \sum_{k\in\mathbb{Z}} \lambda(k) = 0\}.$$

As proved in [11], for $a \in (\ell([-N,N]))^{r\times r}$ satisfying (2), the pyramid scheme associated with a converges in L_p norm ($1 \leq p \leq \infty$) if and only if

1) a satisfies Equations (6) (hence U_N is invariant under A_0 and A_1); and

2) $\rho_p(\mathcal{A}) < 2^{1/p}$, where $\mathcal{A}$ contains matrices representing $A_0|_{U_N}$ and $A_1|_{U_N}$.

We assume that the perturbed mask b satisfies condition 1). Then the idea of the proof of our Main Theorem is to deal with condition 2) by showing that the ℓ_p-norm joint spectral radius is continuous under perturbations of the mask.

Define the symbol of a to be

$$\widetilde{a}(z) := \sum_{k\in\mathbb{Z}} a(k)z^k \qquad z \in \mathbb{C}\backslash\{0\}$$

and $\|\widetilde{a}(z)\|_p := \|a\|_p$. The subdivision operator associated with a is

$$S_a\lambda(i) := \sum_{k\in\mathbb{Z}} \lambda(k)a(i - 2k), \qquad i \in \mathbb{Z}, \lambda \in (\ell_0(\mathbb{Z}))^{r\times r}.$$

Observe that $\widetilde{S_a\lambda}(z) = \widetilde{\lambda}(z)\widetilde{a}(z^2)$. Let $\delta_I := \text{diag}(\delta, \cdots, \delta)$ where δ is the sequence on $\mathbb{Z}$ given by $\delta(0) = 1$ and $\delta(k) = 0$ for $k \in \mathbb{Z}\backslash\{0\}$. By induction,

$$Q_a^n\phi_0 = \sum_{k\in\mathbb{Z}} S_a^n\delta_I(k)\phi_0(2^n \cdot - k). \tag{11}$$

Based on the results in [11] we have the following:

Lemma 2.1. *Let* $a \in (\ell([-N, N]))^{r \times r}$ *for some positive integer* N. *Suppose* a *satisfies condition* (2) *and the pyramid scheme associated with* a *converges in the* L_p *norm. Then there exist* $\eta > 0, \nu > 0$, *and* $C > 0$ *such that for any* $b \in (\ell([-N, N]))^{r \times r}$, *if* b *satisfies* (8) *and* $\|a - b\|_1 < \eta$, *then the pyramid scheme associated with* b *converges in the* L_p *norm and*

$$\|\widetilde{b}_n(z) \operatorname{diag}(1 - z, 1, \cdots, 1)\|_p \le C 2^{(1/p - \nu)n} \qquad \forall \, n \in \mathbb{N}, \qquad (12)$$

where $\widetilde{b}_n(z) := \widetilde{S_b^n \delta_I}(z) = \widetilde{b}(z^{2^{n-1}}) \widetilde{b}(z^{2^{n-2}}) \cdots \widetilde{b}(z^{2^0})$.

Proof: Since $\rho_p(\{A_0|_{U_N}, A_1|_{U_N}\}) < 2^{1/p}$, there exist $\nu > 0$ and a positive integer m such that $\|\{A_0|_{U_N}, A_1|_{U_N}\}^m\|_p^{1/m} < 2^{1/p - 2\nu}$. So there is a positive number η such that for any element b in $(\ell([-N, N]))^{r \times r}$ satisfying (8) and $\|a - b\|_1 < \eta$,

$$\|\{B_0|_{U_N}, B_1|_{U_N}\}^m\|_p^{1/m} < 2^{1/p - \nu}. \qquad (13)$$

Here, B_0 and B_1 are defined by $B_\varepsilon \lambda(i) := \sum_{k \in \mathbb{Z}} b(\varepsilon + 2i - k)\lambda(k)$ for $i \in \mathbb{Z}$, $\lambda \in (\ell_0(\mathbb{Z}))^r$. So, the pyramid scheme associated with b converges in the L_p norm. Based on [11, Lemma 4.3], inequality (12) comes directly from the above inequality (13). See [7, Lemma 2.1] for a detailed proof of this fact. □

Similarly to Han [7, Lemma 2.2], we have the following result:

Proposition 2.2. *Let* a *be an element in* $(\ell([-N, N]))^{r \times r}$ *for some positive integer* N. *Let* q *satisfy* $1/p + 1/q = 1$. *Then for any* b *in* $(\ell_0(\mathbb{Z}))^{r \times r}$,

$$\|\widetilde{b}(z^{2^k}) \widetilde{a}(z^{2^{k-1}}) \cdots \widetilde{a}(z^{2^0})\|_p = \|\widetilde{b}(z^{2^k}) \widetilde{S_a^k \delta_I}(z)\|_p \le (2N + 1)^{\frac{1}{q}} \|b\|_p \cdot \|S_a^k \delta_I\|_p.$$

Proof: Note that $S_a^k \delta_I$ is supported in $[-(2^k - 1)N, (2^k - 1)N]$. Thus we can decompose $S_a^k \delta_I$ as $S_a^k \delta_I = \sum_{j=-N}^{N} a_j$ where each element a_j is supported in $(2^k j + [0, 2^k)) \cap \mathbb{Z}$. Hence

$$\|\widetilde{b}(z^{2^k}) \widetilde{S_a^k \delta_I}(z)\|_p \le \sum_{j=-N}^{N} \|\widetilde{b}(z^{2^k}) \widetilde{a}_j(z)\|_p \le \sum_{j=-N}^{N} \|b\|_p \cdot \|a_j\|_p.$$

Now, by Hölder's inequality, we have

$$\sum_{j=-N}^{N} \|a_j\|_p \le (2N + 1)^{\frac{1}{q}} \left(\sum_{j=-N}^{N} \|a_j\|_p^p \right)^{\frac{1}{p}} = (2N + 1)^{1/q} \|S_a^k \delta_I\|_p,$$

where $1/p + 1/q = 1$. This completes the proof. □

The next proposition's straightforward proof has been omitted for brevity.

Proposition 2.3. *Let* a *and* b *be elements in* $(\ell([-N, N]))^{r \times r}$ *for some positive integer* N. *If* a *(resp.* b*) satisfies* (6) *(resp.* (8)*), then there exists a unique sequence* c *in* $(\ell([-N, N]))^{r \times r}$ *such that*

$$\widetilde{a}(z) - \widetilde{b}(z) = \operatorname{diag}(1 - z^2, 1, \cdots, 1)\widetilde{c}(z).$$

Moreover $\|c\|_1 \le C_N \|a - b\|_1$ *where* C_N *is a constant depending only on* N.

Finally, we will need one more auxiliary result.

Lemma 2.4. *Let a be a sequence of $r \times r$ matrices on $\mathbb{Z}$. If the pyramid scheme associated with a converges in the L_p norm, then there exists a positive constant C such that*

$$\|S_a^n \delta_I\|_p \leq C 2^{n/p} \qquad \forall\, n \in \mathbb{N}.$$

Proof: From Equation (11) and since the shifts of ϕ_0 are stable, there exists a positive constant C_1 such that $\|S_a^n \delta_I\|_p \leq C_1 2^{n/p} \|Q_a^n \phi_0\|_p$. Since the sequence $(Q_a^n \phi_0,\ n \in \mathbb{N})$ converges in the L_p norm, there exists a constant C_2 such that $\|Q_a^n \phi_0\|_p \leq C_2$ for all $n \in \mathbb{N}$. $\square$

§3. Proof and Sharpness of the Main Theorem

Proof: Take η as in Lemma 2.1. Let b be any element in $(\ell([-N, N]))^{r \times r}$ satisfying Equations (8) and $\|a - b\|_1 < \eta$. From (11),

$$Q_a^n \phi_0 - Q_b^n \phi_0 = \sum_{k \in \mathbb{Z}} (S_a^n \delta_I(k) - S_b^n \delta_I(k)) \phi_0(2^n \cdot - k).$$

Since $\phi_0 \in L_p(\mathbb{R})^r$ is compactly supported, there exists a constant $C_1 > 0$ such that

$$\|Q_a^n \phi_0 - Q_b^n \phi_0\|_p \leq C_1 2^{-n/p} \|S_a^n \delta_I - S_b^n \delta_I\|_p. \tag{14}$$

Write $S_a^n \delta_I - S_b^n \delta_I$ in the following form

$$S_a^n \delta_I - S_b^n \delta_I = \sum_{k=1}^{n} S_a^{k-1} (S_a - S_b) S_b^{n-k} \delta_I = \sum_{k=1}^{n} S_a^{k-1} S_{a-b} S_b^{n-k} \delta_I. \tag{15}$$

Since a and b satisfy Equations (6) and (8), Proposition 2.3 guarantees the existence of a sequence c in $(\ell([-N, N]))^{r \times r}$ and a constant C_N for which

$$\widetilde{a}(z) - \widetilde{b}(z) = \mathrm{diag}(1 - z^2, 1, \cdots, 1) \widetilde{c}(z) \quad \text{with} \quad \|c\|_1 \leq C_N \|a - b\|_1.$$

Let $G_k(z)$ denote the symbol of $S_a^{k-1} S_{a-b} S_b^{n-k} \delta_I$. Then it follows from the above equality that

$$G_k(z) = \widetilde{b}(z^{2^{n-1}}) \cdots \widetilde{b}(z^{2^k})(\widetilde{a}(z^{2^{k-1}}) - \widetilde{b}(z^{2^{k-1}})) \widetilde{a}(z^{2^{k-2}}) \cdots \widetilde{a}(z^{2^1}) \widetilde{a}(z^{2^0})$$

$$= \widetilde{b}(z^{2^{n-1}}) \cdots \widetilde{b}(z^{2^k}) \widetilde{v}_0(z^{2^k}) \widetilde{c}(z^{2^{k-1}}) \widetilde{a}(z^{2^{k-2}}) \cdots \widetilde{a}(z^{2^1}) \widetilde{a}(z^{2^0}),$$

where $\widetilde{v}_0(z) := \mathrm{diag}(1 - z, 1, \cdots, 1)$. Since a is supported in $[-N, N] \cap \mathbb{Z}$, by Proposition 2.2, we have

$$\|G_k(z)\|_p \leq (2N + 1)^{1/q} \|\widetilde{b}(z^{2^{n-1}}) \cdots \widetilde{b}(z^{2^k}) \widetilde{v}_0(z^{2^k}) \widetilde{c}(z^{2^{k-1}})\|_p \cdot \|S_a^{k-1} \delta_I\|_p,$$

where $1/p + 1/q = 1$. On the other hand, by Young's inequality, we have

$$\|\widetilde{b}(z^{2^{n-1}}) \cdots \widetilde{b}(z^{2^k}) \widetilde{v}_0(z^{2^k}) \widetilde{c}(z^{2^{k-1}})\|_p \leq \|c\|_1 \cdot \|\widetilde{b}(z^{2^{n-1}}) \cdots \widetilde{b}(z^{2^k}) \widetilde{v}_0(z^{2^k})\|_p$$

$$= \|c\|_1 \cdot \|\widetilde{b}(z^{2^{n-k-1}}) \cdots \widetilde{b}(z^{2^0}) \widetilde{v}_0(z^{2^0})\|_p.$$

Hence, by inequality (12) in Lemma 2.1, there exists a constant C_2 such that

$$\|\widetilde{b}(z^{2^{n-1}})\cdots\widetilde{b}(z^{2^k})\widetilde{v}_0(z^{2^{k-1}})\widetilde{c}(z^{2^{k-1}})\|_p \leq C_2 2^{(1/p-\nu)(n-k)}\|c\|_1 \qquad \forall\, n \in \mathbb{N}.$$

Combining all the above inequalities with Lemma 2.4, we conclude that

$$\|G_k(z)\|_p \leq (2N+1)^{1/q}C_2C_3C_N\|a-b\|_1 2^{(1/p-\nu)(n-k)}2^{(k-1)/p}$$
$$= C_4\|a-b\|_1 2^{n/p}2^{-\nu(n-k)},$$

where C_3 is the constant of Lemma 2.4 and $C_4 := (2N+1)^{1/q}C_2C_3C_N 2^{-1/p}$. It follows from equality (15) that

$$\|S_a^n\delta_I - S_b^n\delta_I\|_p \leq \sum_{k=1}^{n}\|S_a^{k-1}S_{a-b}S_b^{n-k}\delta_I\|_p \leq C_4\|a-b\|_1 2^{n/p}\sum_{k=1}^{n}2^{-\nu(n-k)}$$
$$\leq C_4\|a-b\|_1 2^{n/p}\sum_{k=0}^{\infty}2^{-\nu k}.$$

Combining the above inequality with (14), we complete the proof. $\square$

Example. We modify an example from [7] to show that (9) is sharp by giving a mask a for which

$$\|Q_a^n\phi_0 - Q_b^n\phi_0\|_p \text{ is not } o(\|a-b\|_1) \text{ as } \|a-b\|_1 \to 0.$$

Define $a \in (\ell([0,2]))^{r\times r}$ by $\widetilde{a}(z) = \text{diag}((1+z)^2/2, 0, \cdots, 0)$. Then $Q_a^2\phi_0 = Q_a\phi_0 = (h, 0, \cdots, 0)^T$. So the pyramid scheme associated with a converges in the L_∞ norm to $(h, 0, \cdots, 0)^T$. Now define $b \in (\ell([0,2]))^{r\times r}$ by $\widetilde{b}(z) :=$ $\text{diag}((1+z)((1+\eta)+(1-\eta)z)/2, 0, \cdots, 0)$, where η is a positive number to be determined. Then $\|a-b\|_1 = \eta$. So, by the Main Theorem, the pyramid scheme associated with b converges for sufficiently small η. Fix such η. Now, by direct computation, $S_a^n\delta_I(2^{n-1}-1) = \text{diag}(1/2, 0, \cdots, 0)$ and $S_b^n\delta_I(2^{n-1}-1) = \text{diag}((1-\eta)/2, 0, \cdots, 0)$. Therefore,

$$\|S_a^n\delta_I - S_b^n\delta_I\|_\infty \geq \|S_a^n\delta_I(2^{n-1}-1) - S_b^n\delta_I(2^{n-1}-1)\|_\infty \geq \eta/2 = \|a-b\|_1/2.$$

By (11) and since the shifts of ϕ_0 are stable, there exists $C > 0$ such that

$$\|Q_a^n\phi_0 - Q_b^n\phi_0\|_\infty \geq 2C\|S_a^n\delta_I - S_b^n\delta_I\|_\infty \geq C\|a-b\|_1 \qquad \forall\, n \in \mathbb{N}.$$

Thus inequality (9) is sharp for $p = \infty$. It is also sharp for any $1 \leq p \leq \infty$. Lastly, our results generalize easily to the multivariate case.

Acknowledgments. Research of the first author was supported by Killam Trust under Isaak Walton Killam Memorial Scholarship.

References

1. Cavaretta, A. S., W. Dahmen, and C. A. Micchelli, *Stationary Subdivision*, Memoirs of Amer. Math. Soc. Volume **93**, 1991.

2. Cohen, A., I. Daubechies, and G. Plonka, Regularity of refinable function vectors, J. Fourier Anal. Appl. **3** (1997), 295–324.

3. Dahmen, W., B. Han, R. Q. Jia, and A. Kunoth, Biorthogonal multiwavelets on the interval: Cubic Hermite splines, 1998, preprint.

4. Dahmen, W. and C. A. Micchelli, Biorthogonal wavelet expansions, Constr. Approx. **13** (1997), 293–328.

5. Donovan, G., J. S. Geronimo, D. P. Hardin, and P. R. Massopust, Construction of orthogonal wavelets using fractal interpolation functions, SIAM J. Math. Anal. **27** (1996), 1158–1192.

6. Daubechies, I. and Y. Huang, How does truncation of the mask affect a refinable function? Constr. Approx. **11** (1995), 365–380.

7. Han, B., Error estimate of a subdivision scheme with a truncated refinement mask, 1997, preprint.

8. Heil, C. and D. Colella, Matrix refinement equations and subdivision schemes, J. Fourier Anal. Appl. **2** (1996), 363–377.

9. Heil, C., G. Strang, and V. Strela, Approximation by translates of refinable functions, Numer. Math. **73** (1996), 75–94.

10. Jia, R. Q., Subdivision schemes in L_p spaces, Advances in Comp. Math. **3** (1995), 309–341.

11. Jia, R. Q., S. Riemenschneider, and D. X. Zhou, Vector subdivision schemes and multiple wavelets, Math. Comp., to appear.

12. Rota, G.-C. and G. Strang, A note on the joint spectral radius, Indag. Math. **22** (1960), 379–381.

13. Wang, Y., Two-scale dilation equations and the mean spectral radius, Random & Computational Dynamics **4** (1996), 49–72.

Bin Han
Department of Mathematical Sciences
University of Alberta
Edmonton, Alberta
Canada T6G 2G1
bhan@math.ualberta.ca
http://www.math.ualberta.ca/~bhan

Thomas A. Hogan
Department of Mathematics
Vanderbilt University
Nashville, TN 37240
hogan@math.vanderbilt.edu
http://www.math.vanderbilt.edu/~hogan

Approximation to Small Deformation of Surfaces and Its Application

Matthew He and Chandra Kambhamettu

Abstract. Let S be a surface given by a parametric form

$$\mathbf{x} = \mathbf{x}(u,v) = (x(u,v), y(u,v), z(u,v)).$$

Consider a general small deformation of the S, in which the displacement $\mathbf{s}$ of each point is a small quantity. Suppose that the point, whose original position vector is $\mathbf{r}$, undergoes a small displacement $\mathbf{s}$. The position vector $\mathbf{r}^*$ of a corresponding point on the deformed surface S^* is given by

$$\mathbf{r}^* = \mathbf{r} + \mathbf{s}.$$

Explicit representations of first fundamental coefficients, discriminant and Gaussian curvature after general small deformation are presented. Based on these differential geometric changes during general small deformations, we are able to estimate the nonrigid motion and point correspondences. Polynomial functions are used locally to approximate for the nonrigid motion involved. Simulations and experiments on real data are performed to illustrate performance and accuracy of the derived algorithms.

§1. Introduction

In recent years, motion analysis has become an important and very active field in mathematics and computer applications. There has been extensive research in this area due to its importance in dynamic scene understanding. In most of the work, rigidity is assumed to estimate for the motion parameters. However, growing interest is seen in recent years towards research in nonrigid motion analysis due to such potential applications as medical imaging, model-based image compression, and atmospheric sciences.

Analysis of nonrigid motion is complicated by the fact that its varying structure cannot be defined by any specific set of parameters without certain restrictions on object's behavior. Differential geometry is one of the tools used in the analysis of nonrigid motion. Classical differential geometry provides a

Approximation Theory IX, Volume 2: Computational Aspects   105
Charles K. Chui and Larry L. Schumaker (eds.), pp. 105–112.
Copyright ℗ 1998 by Vanderbilt University Press, Nashville, TN.
ISBN 0-8265-1326-3.

complete local description of smooth surfaces. We use such surface descriptors and follow differential geometric approaches as presented in [1, 2, 3].

Previous work in this area includes [5], where the authors present linear stretching estimation in conformal motion using point correspondence information. As an extension to this, [4] presented linear stretching and point correspondence estimation in conformal motion without any a priori knowledge of point correspondence information. The problem of small nonrigid motion analysis is addressed in [3]. However, the authors ignore the higher order terms of deformation to estimate for the motion.

This paper introduces a general small nonrigid motion analysis problem, with all higher order terms of displacement. The motion parameter used here is a vector point function, also called displacement function. We characterize first fundamental coefficients, discriminant and Gaussian curvature under general small deformations. The paper is organized as follows: In Section 2, we introduce general small deformations and give explicit expressions for the first fundamental coefficients. The variations of discriminant and Gaussian curvature under general small deformation will be presented in Section 3. We give our simulation experiments on real data in Section 4.

§2. General Small Deformations

Let a surface S be given by parametric form

$$\mathbf{x} = \mathbf{x}(u, v) = (x(u, v), y(u, v), z(u, v)), \tag{1}$$

where the Cartesian coordinates x, y, z of a surface point are differentiable functions of the parameters u and v. In order to avoid potential problems with undefined normal vectors, we assume that

$$\mathbf{x}_u \times \mathbf{x}_v \neq \mathbf{0}. \tag{2}$$

Consider a general small deformation of the S, in which the displacement $\mathbf{s}$ of each point is a small quantity. Differential geometry of small displacement $\mathbf{s}$ with only first order was studied in [7]. Suppose that the point, whose original position vector is $\mathbf{r}$, undergoes a small displacement $\mathbf{s}$. The position vector $\mathbf{r}^*$ of a corresponding point on the deformed surface S^* is given by

$$\mathbf{r}^* = \mathbf{r} + \mathbf{s}. \tag{3}$$

Let the corresponding points of the surfaces S and S^* be characterized by the same values of the parameters. In terms of any convenient parameters u and v on the surface, the displacement $\mathbf{s}$ may be expressed as

$$\mathbf{s} = (x^*(u, v) - x(u, v), y^*(u, v) - y(u, v), z^*(u, v) - z(u, v)), \tag{4}$$

or

$$\mathbf{s} = (X(u, v), Y(u, v), Z(u, v)).$$

Let the corresponding points of the surfaces S, S^* be characterized by the same values of the parameters, and let the suffixes 1, 2 denote differentiations with respect to these parameters. The coefficients of first fundamental form of S can be denoted by E, F, G, and are given by

$$E = \mathbf{r}_1^2, \quad F = \mathbf{r}_1 \cdot \mathbf{r}_2, \quad G = \mathbf{r}_2^2.$$

Those for S^* after small motion have the values

$$E^* = \mathbf{r}_1^* \cdot \mathbf{r}_1^* = (\mathbf{r_1} + \mathbf{s_1})^2 = E + \theta^{11}, \tag{5}$$

$$F^* = \mathbf{r}_1^* \cdot \mathbf{r}_2^* = (\mathbf{r_1} + \mathbf{s_1}) \cdot (\mathbf{r_2} + \mathbf{s_2}) = F + \theta^{12}, \tag{6}$$

$$G^* = \mathbf{r}_2^* \cdot \mathbf{r}_2^* = (\mathbf{r_2} + \mathbf{s_2})^2 = G + \theta^{22}, \tag{7}$$

where

$$\theta^{11} = 2\mathbf{r_1} \cdot \mathbf{s_1} + \mathbf{s_1} \cdot \mathbf{s_1},$$

$$\theta^{12} = \mathbf{r_1} \cdot \mathbf{s_2} + \mathbf{r_2} \cdot \mathbf{s_1} + \mathbf{s_1} \cdot \mathbf{s_2},$$

$$\theta^{22} = 2\mathbf{r_2} \cdot \mathbf{s_2} + \mathbf{s_2} \cdot \mathbf{s_2}.$$

Discriminant and Gaussian curvature are the invariant differential geometric parameters that have been utilized for nonrigid motion analysis. We define the discriminant of a surface in terms of its first fundamental form

$$D^2 = EG - F^2. \tag{8}$$

For Gaussian curvature, denoted by K, we use the following representation [6],

$$K = \frac{1}{D^4}(A - B), \tag{9}$$

where

$$A = \begin{vmatrix} -\frac{1}{2}E_{22} + F_{12} - \frac{1}{2}G_{11} & \frac{1}{2}E_1 & F_1 - \frac{1}{2}E_2 \\ F_2 - \frac{1}{2}G_1 & E & F \\ \frac{1}{2}G_2 & F & G \end{vmatrix},$$

$$B = \begin{vmatrix} 0 & \frac{1}{2}E_2 & \frac{1}{2}G_1 \\ \frac{1}{2}E_2 & E & F \\ \frac{1}{2}G_1 & F & G \end{vmatrix}.$$

§3. Changes of Discriminant and Gaussian Curvature

In this section, we derive expressions for the discriminant D^* and Gaussian curvature K^* of the surface S^* after motion in terms of the discriminant D and Gaussian curvature K of the surface S before motion. The expressions are used to estimate the deformations of the surface.

Theorem 1. Let D and D^* be the discriminants of the surface S before and after motion. Then

$$D^* = D\theta, \tag{10}$$

where

$$\theta = \sqrt{1 + \frac{(E\theta^{22} - F\theta^{12}) + (G\theta^{11} - F\theta^{12}) + (\theta^{11}\theta^{22} - \theta^{12}\theta^{12})}{D^2}}. \tag{11}$$

Proof: Let $D^2 = EG - F^2$ be the discriminant for S defined as in (8). Then the discriminant on S^* can be derived as the following. Using (8), (5), (6), and (7), we have

$$
\begin{aligned}
(D^*)^2 &= E^*G^* - F^{*2} \\
&= (E + \theta^{11})(G + \theta^{22}) - (F + \theta^{12})^2 \\
&= EG - F^2 + (E\theta^{22} - F\theta^{12}) + (G\theta^{11} - F\theta^{12}) + (\theta^{11}\theta^{22} - \theta^{12}\theta^{12}) \\
&= D^2 + (E\theta^{22} - F\theta^{12}) + (G\theta^{11} - F\theta^{12}) + (\theta^{11}\theta^{22} - \theta^{12}\theta^{12}) \\
&= D^2 \left(1 + \frac{(E\theta^{22} - F\theta^{12}) + (G\theta^{11} - F\theta^{12}) + (\theta^{11}\theta^{22} - \theta^{12}\theta^{12})}{D^2} \right) \\
&= D^2\theta^2.
\end{aligned}
$$

Hence, we have $D^* = D\theta$. □

Note that if we assume orthogonal parameterization both before and after motion, then the following holds.

$$F = r_1 \cdot r_2 = 0, F_2^* = r_1^* \cdot r_2^* = 0.$$

$$
\begin{aligned}
D^{*2} &= E^*G^* \\
&= (E + \theta^{11})(G + \theta^{22}) \\
&= EG + E\theta^{22} + G\theta^{11} + \theta^{11}\theta^{22} \\
&= D^2 + E\theta^{22} + G\theta^{11} + \theta^{11}\theta^{22} \\
&= D^2 \left(1 + \frac{(E\theta^{22}) + (G\theta^{11}) + (\theta^{11}\theta^{22})}{D^2} \right) \\
&= D^2\theta^2,
\end{aligned}
$$

where

$$\theta = \sqrt{1 + \frac{(E\theta^{22}) + (G\theta^{11}) + (\theta^{11}\theta^{22})}{D^2}}. \tag{12}$$

Next we present an expression for the Gaussian curvature K^* of S^*.

Theorem 2. *Let K and K^* be the Gaussian curvatures of the surface before and after motion. Then*

$$K^* = \frac{1}{\theta^4}K + \frac{1}{D^4\theta^4}f(u,v), \tag{13}$$

where

$$f(u,v) = \frac{1}{(1+\theta)^4}(\delta_1 + \delta_2 + \delta_3 - \delta_4 - \delta_5 - \delta_6), \tag{14}$$

$$\delta_1 = \begin{vmatrix} -\frac{1}{2}\theta_{22}^{11} + \theta_{12}^{12} - \frac{1}{2}\theta_{11}^{22} & \frac{1}{2}(E_1 + \theta_1^{11}) & F_1 + \theta^{12} - \frac{1}{2}(E_2 + \theta_2^{11}) \\ \theta_2^{12} - \frac{1}{2}\theta_1^{22} & E + \theta^{11} & F + \theta^{12} \\ \frac{1}{2}\theta_2^{22} & F + \theta^{12} & G + \theta^{22} \end{vmatrix},$$

$$\delta_2 = \begin{vmatrix} -\frac{1}{2}E_{22} + F_{12} - \frac{1}{2}G_{11} & \frac{1}{2}\theta_1^{11} & F_1 + \theta^{12} - \frac{1}{2}(E_2 + \theta_2^{11}) \\ F_2 - \frac{1}{2}G_1 & \theta^{11} & F + \theta^{12} \\ \frac{1}{2}G_2 & \theta^{12} & G + \theta^{22} \end{vmatrix},$$

$$\delta_3 = \begin{vmatrix} -\frac{1}{2}E_{22} + F_{12} - \frac{1}{2}G_{11} & \frac{1}{2}E_1 & \theta^{12} - \frac{1}{2}\theta_2^{11} \\ F_2 - \frac{1}{2}G_1 & E & \theta^{12} \\ \frac{1}{2}G_2 & F & \theta^{22} \end{vmatrix},$$

$$\delta_4 = \begin{vmatrix} 0 & \frac{1}{2}(E_2 + \theta_2^{11}) & \frac{1}{2}(G_1 + \theta_1^{22}) \\ \frac{1}{2}\theta_2^{11} & E + \theta^{11} & F + \theta^{12} \\ \frac{1}{2}\theta_1^{22} & F + \theta^{12} & G + \theta^{22} \end{vmatrix},$$

$$\delta_5 = \begin{vmatrix} 0 & \frac{1}{2}\theta_2^{11} & \frac{1}{2}(G_1 + \theta_1^{22}) \\ \frac{1}{2}E_2 & \theta^{11} & F + \theta^{12} \\ \frac{1}{2}G_1 & \theta^{12} & G + \theta^{22} \end{vmatrix}, \quad \delta_6 = \begin{vmatrix} 0 & \frac{1}{2}E_2 & \frac{1}{2}\theta_1^{22} \\ \frac{1}{2}E_2 & E & \theta^{12} \\ \frac{1}{2}G_1 & F & \theta^{22} \end{vmatrix}.$$

Proof: Use (9), (5), (6), and (7), and denote by A^*, B^* the matrices that correspond to the matrices A, B after motion as defined in Section 2. Then

$$K^* = \frac{1}{(D^*)^4}(A^* - B^*)$$

$$= \frac{1}{\theta^4}K + \frac{1}{D^4\theta^4}f(u,v),$$

where we have applied certain properties of the discriminant. $\square$

We remark that when the parametric curves are orthogonal before and after motion, then $F^* = 0$ and $(D^*)^2 = E^*G^*$, so that the Gaussian curvature K^* takes the form

$$K^* = -\frac{1}{2D^*}\left[\left(\frac{G_1^*}{D^*}\right)_1 + \left(\frac{E_2^*}{D^*}\right)_2\right].$$

(See [6].) From Equations (5) and (7), we can rewrite K^* as

$$K^* = -\frac{1}{2D\theta}\left[\left(\frac{G_1 + \theta_1^{22}}{D\theta}\right)_1 + \left(\frac{E_2 + \theta_2^{11}}{D\theta}\right)_2\right]$$

$$= -\frac{1}{2D\theta}\left[\left(\frac{G_1}{D\theta}\right)_1 + \left(\frac{\theta_1^{22}}{D\theta}\right)_1 + \left(\frac{E_2}{D\theta}\right)_2 + \left(\frac{\theta_2^{11}}{D\theta}\right)_2\right] \qquad (15)$$

$$= \frac{1}{\theta^2}K + \frac{1}{D^2\theta^2}f(u, v),$$

where

$$f(u, v) = \frac{D_1\theta_1^{22} + D_2\theta_2^{11}}{D} + \frac{\theta_1(G_1 + \theta_1^{22}) + \theta_2(E_2 + \theta_2^{11})}{\theta} - (\theta_{11}^{22} + \theta_{22}^{11}).$$

§4. Experimental Results

It is known that point correspondences between a surface and its deformed counterpart is an important, yet difficult problem [4]. We propose to use the relations presented in this paper to find point matches between surfaces under deformation. We can derive various energy measures assuming general small deformations and thus using the equations presented in this work. We now present one such implementation.

Discriminant changes (with orthogonal parameterization) can be used to define an error measure as shown below:

$$ER_D = [D^{*2} - D^2\theta^2]^2.$$

Gaussian curvature changes (with orthogonal parameterization) can also be used to define an error measure such as

$$ER_K = \left[K^{*2} - \left(\frac{1}{\theta^2}K + \frac{1}{D^2\theta^2}f(u, v)\right)\right]^2.$$

Here, ER_D (resp. ER_K) represents least-square error of the discriminant (resp. of the Gaussian curvature), and is ideally zero if the surface follows assumed deformation. θ is dependent on the displacement function s (see previous sections for definition of s). The displacement s can be assumed as a first or second order polynomial so as to define the above error measure. Hypothesis point correspondences may be first defined, from which the best match can be picked by minimizing the least-square error defined above. Note that as we minimize the above error, we also solve for the displacement function, s. In our experiments, we assumed a first-order polynomial for s and applied the above relation over a region (called, *template window*) centered around each point of interest.

We have performed various experiments on the general deformation of surfaces using the presented motion model. The results have been very encouraging, showing potential applications for the algorithm in areas such as

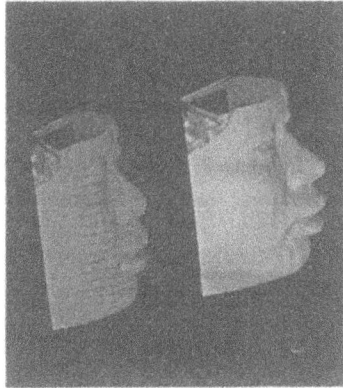

Fig. 1. Point correspondences for simulated motion.

Fig. 1. Point correspondences for real motion.

teleconferencing, biomedical image analysis and analysis of remote sensing images. We present our experiments in Figures 1 and 2 on certain face data and explain our algorithm's applicability.

We have used the data provided to us by Dr. Thomas S. Huang of the University of Illinois. The surface points of a face were obtained in a cylindrical fashion. The data consist of 512 slices, where each slice has 512 points distributed radially around a central axis. We have used the front view of the face, sampled into (100 by 100) points. The differential geometric parameters were calculated by fitting a quadratic surface over a 5 by 5 square window. We have first generated a simulated data set by pulling each point on the face as a function of its depth (i.e, $z = \delta \times z$). We have arbitrarily chosen δ to be 1.3, making sure not to violate the general deformation constraint. Figure 1 shows the face without pull, face with the pull, and point correspondences between them. All the estimated point correspondences have been found to be correct with no errors.

However, the output starts to fail slowly as the motion is increased (i.e, as δ is being increased). In such case, one may increase the region around each point which is used for error calculations (i.e, increase the *template* size) in order to obtain better results as the motion increases. Hence, the template size is inversely proportional to the underlying motion with an overhead of computations as the size is increased. Figure 2 shows three facial expressions of the same person with estimated point correspondences shown between them.

Acknowledgments. This research was supported in part by C. K. Wong Education Foundation, Hong Kong, and the Faculty Research Fund at NSU.

References

1. Goldgof, D., H. Lee, and T. Huang, Motion analysis of nonrigid surfaces, *Proceedings of the IEEE Computer Society Conference on Computer Vision and Pattern Recognition*, 1988.

2. He, M., D. Goldgof, and C. Kambhamettu, Variation of Gaussian curvature under conformal mapping and its application, Computers and Mathematics With Applications, **26** (1993), 63–74.

3. Kambhamettu, C., D. Goldgof, and M. He, Determination of motion parameters and estimation of point correspondences in small nonrigid deformations, *Proceedings of IEEE Conference on Computer Vision and Pattern Recognition*, 1994, pp 943–946.

4. Kambhamettu, C. and D. Goldgof, Point correspondence recovery in nonrigid motion, *Proceedings of the IEEE Computer Society Conference on Computer Vision and Pattern Recognition*, June, 1992.

5. Mishra, S., D. Goldgof, T. Huang, and C. Kambhamettu, Curvature-based non-rigid motion analysis from 3D point correspondences, *International Journal of Imaging Systems and Technology* **4** (1993), 214–225.

6. Kreyszig, E., *Differential Geometry*, The Univ. of Toronto Press,Toronto, 1959.

7. Weatherburn, C., *Differential Geometry in Three Dimensions*, Vol. II, Cambridge University Press, Cambridge, 1930.

Matthew He
Department of Mathematics
Nova Southeastern University
Ft. Lauderdale, FL 33314
hem@polaris.nova.edu

Chandra Kambhamettu
Dept. of Computer and Information Science
University of Delaware
Newark, DE 19716-2712
chandra@eecis.udel.edu

Biorthogonal Spline Approximation

and Its Wavelet Analysis

Tian Xiao He

Abstract. In this paper, we will discuss spline approximation by constructing and using compactly supported biorthogonal spline functions. For the spline functions constructed, the corresponding Daubechies-type wavelets will be used in the refinement of the approximation.

§1. Introduction

For application in many numerical analysis and approximation problems, we need orthogonal or biorthogonal smooth scaling and wavelet functions with compact support. Obviously, spline-type scaling functions and the corresponding wavelets meet all of these requirements except for orthogonality and biorthogonality. Some methods of constructing spline scaling functions and their duals that are not both compactly supported can be found in Battle and Lemarié [1], Chui [3], Chui and Wang [4], de Boor, Höllig, and Riemenschneider [2], and Riemenschneider and Shen [17].

In this paper, I construct biorthogonal scaling functions with compact support and derive certain approximation properties from biorthogonal spline functions by using Daubechies' approach. Thus, our first step is to construct biorthogonal splines. Let ϕ and $\tilde{\phi}$ be biorthogonal scaling functions that will be constructed from biorthogonal splines by using Daubechies' approach in Section 2. Denote $\{V_j\}$ and $\{\tilde{V}_j\}$ as the biorthogonal MRA generated by ϕ and $\tilde{\phi}$, respectively, and ψ and $\tilde{\psi}$ as the corresponding wavelets of ϕ and $\tilde{\phi}$, respectively. Therefore, for any function $f \in L_2$, we have an approximation $f_j = \sum_{jk} \langle f, \tilde{\phi}_{jk} \rangle \phi_{jk} \in V_j$, where $\phi_{jk} = 2^{j/2} \phi(2^j t - k)$ and $\tilde{\phi}_{jk} = 2^{j/2} \tilde{\phi}(2^j t - k)$. In addition, an improvement, f_{j+1}, of the above approximation can be obtained by using the following refinement: $f_{j+1} = f_j + g_j \in V_{j+1}$, where $g_j = \sum_{jk} \langle f, \tilde{\psi}_{jk} \rangle \psi_{jk} \in W_j$, $\psi_{jk} = 2^{j/2} \psi(2^j t - k)$, and $\tilde{\psi}_{jk} = 2^{j/2} \tilde{\psi}(2^j t - k)$. By using Ciesieski's results (see [6]) on dyadic spline approximation, we have

Approximation Theory IX, Volume 2: Computational Aspects
Charles K. Chui and Larry L. Schumaker (eds.), pp. 113–120.
Copyright © 1998 by Vanderbilt University Press, Nashville, TN.
ISBN 0-8265-1326-3.

$\|f - f_k\|_2 \leq C_r \omega_r(f, 2^{-k})_2$, $k = j, j + 1$, where $\omega_r(f, 2^{-k})_2$ is the rth modulus of smoothness of $f \in L_2$ with step size h, $0 < h \leq 2^{-k}$.

Let N_n be the cardinal B-spline of order n (i.e., degree $n - 1$) and S the linear space consisting of integer dilation and translations of N_n. In Section 1, we construct a C^{n-2} spline function $\tilde{N}_n$ of degree $n - 1$ from S such that N_n and $\tilde{N}_n$ are biorthogonal ($\langle \tilde{N}_n(t - \ell), N_n(t) \rangle = \delta_{0\ell}$) and the shift space of $\tilde{N}_n$ contains all polynomials of degree $n - 2$. In Section 2, we will discuss the biorthogonal MRA generated by the scaling functions that are constructed from N_n and $\tilde{N}_n$ by using Daubechies' approach. As an example, a type of biorthogonal scaling functions with the property of the partition of unit will be given.

§2. Construction of Biorthogonal Splines

In this section, we use the dilations and translations of B-splines to construct splines that are biorthogonal with the B-splines and have some basic approximation properties.

Theorem 1. *Let $N_n : R \to R$ be the B-spline of order n, and let*

$$\tilde{N}_n(t) = \sum_{j=0}^{n-1} \sum_{k=0}^{m-1} c_{jk} N_n(m(t - j) - k), \tag{1}$$

be a spline function of order n such that $\langle \tilde{N}_n(t - \ell), N_n(t) \rangle = \delta_{0\ell}$. Then the shift space of $\tilde{N}_n$ contains all polynomials of degree $n - 1$ if and only if

$$\sum_{j=o}^{n-1} j^\alpha c_{jk} = \left(-\frac{k}{m}\right)^\alpha \tag{2}$$

for $k = 0, 1, \cdots, m - 1$ and $\alpha = 0, 1, \cdots, n - 1$.

Proof: *Sufficiency.* We only need to prove that condition (2) implies $\widehat{\tilde{N}_n}(0) = 1$ and $\widehat{\tilde{N}_n}^{(\beta)}(2\ell\pi) = 0$ for $\ell \in Z \backslash \{0\}$ and $\beta = 0, 1, \cdots, n - 1$. In fact, from the biorthogonality of $\tilde{N}_n$ and N_n, we have

$$\widehat{\tilde{N}_n}(0) = \langle \tilde{N}_n(t), 1 \rangle$$
$$= \langle \tilde{N}_n(t), \sum_{\ell \in Z} N_n(t - \ell) \rangle$$
$$= \sum_{\ell \in Z} \langle \tilde{N}_n(t + \ell), N_n(t) \rangle$$
$$= \sum_{\ell \in Z} \delta_{0\ell}$$
$$= 1. \tag{3}$$

Taking a Fourier transform on both sides of expression (1), we obtain

$$\widehat{\tilde{N}_n}(\omega) = \frac{1}{m} \sum_{j=0}^{n-1} \sum_{k=0}^{m-1} c_{jk} e^{-i(j+k/m)\omega} \widehat{N_n}\left(\frac{\omega}{m}\right). \tag{4}$$

Thus

$$\widehat{\tilde{N}_n}^{(\beta)}(\omega)$$

$$= \frac{1}{m} \sum_{k=0}^{m-1} \sum_{j=0}^{n-1} c_{jk} \sum_{\gamma=0}^{\beta} \binom{\beta}{\gamma} (-i)^\gamma \left(j + \frac{k}{m}\right)^\gamma e^{-i(j+k/m)\omega} \widehat{N_n}^{(\beta-\gamma)}\left(\frac{\omega}{m}\right)$$

$$= \frac{1}{m} \sum_{k=0}^{m-1} \sum_{\gamma=0}^{\beta} \binom{\beta}{\gamma} (-i)^\gamma \sum_{\alpha=0}^{\gamma} \binom{\gamma}{\alpha} \sum_{j=0}^{n-1} c_{jk} j^\alpha \left(\frac{k}{m}\right)^{\gamma-\alpha}$$

$$\times e^{-i(j+k/m)\omega} \widehat{N_n}^{(\beta-\gamma)}\left(\frac{\omega}{m}\right).$$

Substituting $\omega = 2\ell\pi$ into the last equation and noting condition (2), we obtain

$$\widehat{\tilde{N}_n}^{(\beta)}(2\ell\pi)$$

$$= \frac{1}{m} \sum_{k=0}^{m-1} \sum_{\gamma=0}^{\beta} \binom{\beta}{\gamma} (-i)^\gamma \sum_{\alpha=0}^{\gamma} \binom{\gamma}{\alpha} (-1)^\alpha \left(\frac{k}{m}\right)^\gamma e^{-2i\ell k\pi/m} \widehat{N_n}^{(\beta-\gamma)}\left(\frac{2\ell\pi}{m}\right)$$

$$= \frac{1}{m} \sum_{k=0}^{m-1} e^{-2i\ell k\pi/m} \widehat{N_n}^{(\beta)}\left(\frac{2\ell\pi}{m}\right)$$

$$+ \frac{1}{m} \sum_{k=0}^{m-1} \sum_{\gamma=1}^{\beta} \binom{\beta}{\gamma} (-i)^\gamma (1-1)^\gamma \left(\frac{k}{m}\right)^\gamma e^{-2i\ell k\pi/m} \widehat{N_n}^{(\beta-\gamma)}\left(\frac{2\ell\pi}{m}\right)$$

$$= \frac{1}{m} \left(\sum_{k=0}^{m-1} e^{-2i\ell k\pi/m}\right) \widehat{N_n}^{(\beta)}\left(\frac{2\ell\pi}{m}\right)$$

$$= 0.$$

Necessity. From equation (4), we have

$$\widehat{\tilde{N}_n}(0) = \frac{1}{m} \sum_{j=0}^{n-1} \sum_{k=0}^{m-1} c_{jk}.$$

This equation and equation (3) give

$$\sum_{j=0}^{n-1} \sum_{k=0}^{m-1} c_{jk} = m. \tag{5}$$

If $\tilde{N}_n(t - \ell)$, $\ell \in Z$, is a partition of unit, then $\widehat{\tilde{N}_n}(2\ell\pi) = 0$; i.e.,

$$\sum_{k=0}^{m-1} \left(\sum_{j=0}^{n-1} c_{jk} \right) e^{-i\frac{2\ell k}{m}\pi} = 0, \tag{6}$$

$\ell = 1, 2, \cdots, m - 1$. It is easy to see that the system of equations made up by (5) and (6) has the unique solution $\sum_{j=0}^{n-1} c_{jk} = 1$, $k = 0, 1, \cdots, m - 1$. Hence, we have proved that condition (2) for $\alpha = 0$ is necessary for the shift space of $\tilde{N}_n$ to contain all constants. By using mathematical induction, we can prove the necessity of conditions of $\alpha = 1, \cdots, n - 1$ for the shift space of $\tilde{N}_n$ to contain all polynomials of degree $n - 1$. $\square$

Obviously, if we need $\mathrm{supp}(\tilde{N}_n) = [0, n]$ and biorthogonal condition:

$$\langle \tilde{N}_n(t), N_n(t - \ell) \rangle = \delta_{0\ell}, \tag{7}$$

then the shift space of $\tilde{N}_n$ contains all polynomials of degree of at most $n - 2$. Hence, we have the following result.

Theorem 2. *There exists a unique spline function $\tilde{N}_n$ defined by (1) with $m = 3n - 3$ and $\mathrm{supp}(\tilde{N}_n) = [0, n]$ that satisfies equation (7) and the shift space of $\tilde{N}_n$ contains all polynomials of degree $n - 2$.*

Proof: Noting the supports of N_n and $\tilde{N}_n$, we immediately have $\langle \tilde{N}_n(t - \ell), N_n(t) \rangle = 0$ for all $\ell \geq n$ or $\ell \leq -n$. Hence, for $-n + 1 \leq \ell \leq n - 1$, we have

$$\delta_{0\ell} = \langle \tilde{N}_n(t), N_n(t - \ell) \rangle$$

$$= \sum_{j=0}^{n-1} \sum_{k=0}^{m-1} c_{jk} \langle N_n(m(t - j) - k), N_n(t - \ell) \rangle$$

$$= \frac{1}{2m\pi} \sum_{j=0}^{n-1} \sum_{k=0}^{m-1} c_{jk} \int_{-\infty}^{\infty} \widehat{N_n}\left(\frac{\omega}{m}\right) e^{-i(j+k/m)\omega} \overline{\widehat{N_n}(\omega)} e^{-i\ell\omega} d\omega$$

$$= \frac{1}{2m\pi} \sum_{j=0}^{n-1} \sum_{k=0}^{m-1} c_{jk} \int_{-\infty}^{\infty} \widehat{N_n}\left(\frac{\omega}{m}\right) e^{-i(j+k/m)\omega} \widehat{N_n}(\omega) e^{i(\ell+n)\omega} d\omega$$

$$= \frac{1}{2m^{n+1}\pi} \sum_{j=0}^{n-1} \sum_{k=0}^{m-1} c_{jk} \int_{-\infty}^{\infty} \widehat{N_{2n}}\left(\frac{\omega}{m}\right) \left(1 + e^{-i\omega/m} + e^{-i2\omega/m} + \cdots \right.$$

$$\left. + e^{-i(m-1)\omega/m}\right)^n e^{i(\ell+n-j-k/m)\omega} d\omega$$

$$= \frac{1}{2m^{n+1}\pi} \sum_{j=0}^{n-1} \sum_{k=0}^{m-1} \sum_{|s|=n} c_{jk} \int_{-\infty}^{\infty} \widehat{N_{2n}}\left(\frac{\omega}{m}\right) \frac{n!}{s!} e^{-i\omega(s_1 + 2s_2 + \cdots + (m-1)s_{m-1})/m}$$

$$\times e^{i(\ell+n-j-k/m)\omega} d\omega$$

$$= \sum_{j=0}^{n-1} \sum_{k=0}^{m-1} a_{jk\ell} c_{jk},$$

where

$$a_{jk\ell} = \frac{1}{m^n} \left(\sum_{|s|=n} \frac{n!}{s!} N_{2n}(m(\ell + n - j) - k - \sum_{\alpha=1}^{m-1} \alpha s_\alpha) \right) \qquad (8)$$

and $s = (s_1, s_2, \cdots, s_{m-1})$, $|s| = s_1 + s_2 + \cdots + s_{m-1}$, and $s! = s_1! s_2! \cdots s_{m-1}!$. From the requirement of $\mathrm{supp}(\tilde{N}_n) = [0, n]$, we need to set

$$c_{n-1,m-k} = 0, \quad k = 1, 2, \cdots, n - 1. \qquad (9)$$

Therefore, the coefficients c_{jk} in expression (1) can be solved uniquely from the system consisting of equations (9), equations (2) for $k = 0, 1, \cdots, m - 1$ and $\alpha = 0, 1, \cdots, n - 2$, and equations $\sum_{j=0}^{n-1} \sum_{k=0}^{m-1} a_{jk\ell} c_{jk} = \delta_{0\ell}$ for $\ell = -n + 1, -n + 2, \cdots, n - 1$. In fact, the biorthogonal conditions in (7) imply equation (5). Thus, we omit equation $\sum_{j=0}^{n-1} c_{j,m-1} = 1$ from the system obtained, and the remaining system possesses a $nm \times nm$ coefficient matrix with full rank if $m = 3n - 3$. This completes the proof of Theorem 2. $\square$

As an example, we consider the case of $n = 2$. The corresponding $m = 3$, and

$$\tilde{N}_2(t) = -\frac{7}{6} N_2(3t) + \frac{13}{6} N_2(3t - 1) + N_2(3t - 2) + \frac{13}{6} N_2(3t - 3) - \frac{7}{6} N_2(3t - 4).$$

§3. Biorthogonal Scaling Functions Associated with $\tilde{N}_n$ and N_n

In this section, we will use $\tilde{N}_n$ and N_n to construct compactly supported scaling functions, $\tilde{\phi}$ and ϕ, that generate a biorthogonal MRA and possess partition of unit. (Biorthogonal MRA with other approximation properties will be discussed in another paper.) Obviously, we can make use of Daubechies' approach to get the following two-scale relation.

$$\tilde{\phi}(t) = \sum_{j \in \tilde{J}} \tilde{c}_j \tilde{\phi}(2t - j) \qquad (10)$$

and

$$\phi(t) = \sum_{j \in J} c_j \phi(2t - j), \qquad (11)$$

where $\tilde{J}$ and J are finite sets. Substituting the initial $\tilde{\phi}$ and ϕ on the right-hand side of (10) and (11) with $\tilde{N}_n$ and N_n, respectively, we obtain $\tilde{\phi}$ and ϕ from the iteration process.

Denote the two-scale symbols of $\{\tilde{c}_j\}$ and $\{c_j\}$ in equations (10) and (11), respectively, by $\tilde{m}_0 \left(\frac{\omega}{2} \right)$ and $m_0 \left(\frac{\omega}{2} \right)$, respectively. From Cohen, Daubechies, and Feauveau [8], we have the biorthogonality condition expressed in terms of $\tilde{m}_0$ and m_0.

$$m_0(\omega)\overline{\tilde{m}_0(\omega)} + m_0(\omega + \pi)\overline{\tilde{m}_0(\omega + \pi)} = 1, \qquad (12)$$

which, coupled with the requirement that $\tilde{m}_0(0) = 1$ and $m_0(0) = 1$, gives the four expressions

$$\sum_{j \in J} c_j = 2, \qquad \sum_{j \in \tilde{J}} \tilde{c}_j = 2, \qquad (13)$$

$$\sum_{j \in J} (-1)^j c_j = 0, \qquad \sum_{j \in \tilde{J}} (-1)^j \tilde{c}_j = 0. \qquad (14)$$

Next, we need the following biorthonormality of $\tilde{\phi}$ and ϕ,

$$\langle \tilde{\phi}(t), \phi(t) \rangle = 1,$$

which becomes

$$\sum_{j \in J \cap \tilde{J}} c_j \tilde{c}_j = 2. \qquad (15)$$

Finally, if we require some degree of smoothness for $\tilde{\phi}$ and ϕ, then some wavelet moments must be zero. These convert into some additional conditions of $\{\tilde{c}_j\}$ and $\{c_j\}$.

As an example, we consider the trivial case of $|J| = |\tilde{J}| = 2$ and set $\tilde{J} = \{0, 1\}$ and $J = \{0, 1\}$ in expressions (10) and (11), respectively. Thus, equations (13)–(15) have solutions $c_0 = c_1 = \tilde{c}_0 = \tilde{c}_1 = 1$.

Another example is for $|J| = |\tilde{J}| = 3$, $J = \{0, 1, 2\}$, and $\tilde{J} = \{1, 2, 3\}$. To determine the coefficients $\{\tilde{c}_j\}$ and $\{c_j\}$ in expressions (10) and (11), we need one of the following additional smoothness conditions. Hence,

$$\sum_{j \in J} (-1)^j j c_j = 0, \qquad (16)$$

and

$$\sum_{j \in \tilde{J}} (-1)^j j \tilde{c}_j = 0. \qquad (17)$$

Equations (16) and (17) are from the wavelet moments $\langle t, \psi(t) \rangle = 0$ and $\langle t, \tilde{\psi}(t) \rangle = 0$, respectively, where $\tilde{\psi}$ and ψ are the corresponding wavelets of $\tilde{\phi}$ and ϕ, respectively.

Solving equations (13)–(16), we obtain $c_0 = c_2 = 1/2$, $c_1 = 1$, $\tilde{c}_1 = 3/2$, $\tilde{c}_2 = 1$, and $\tilde{c}_3 = -1/2$. Equations (13)–(15) and (17) give solutions $c_0 = 3/2$, $c_1 = 1$, $c_2 = -1/2$, $\tilde{c}_1 = \tilde{c}_3 = 1/2$, and $\tilde{c}_2 = 1$. Hence, we have the following result.

Theorem 3. *There exist only one pair of functions $\phi(t)$, $t \in D$, and $\tilde{\phi}(t)$, $t \in \frac{1}{3}D$, that satisfy the following conditions, where $D = \cup_{j \in Z} D_j$ and $D_j = \{k/2^j : k \in Z\}$.*

(i) ϕ and $\tilde{\phi}$ satisfy the two-scaling relations

$$\phi(t) = c_0 \phi(2t) + \phi(2t - 1) + (1 - c_0)\phi(2t - 2), \quad t \in D,$$

$$\tilde{\phi}(t) = \tilde{c}_1 \tilde{\phi}(2t - 1) + \tilde{\phi}(2t - 2) + (1 - \tilde{c}_1)\tilde{\phi}(2t - 3), \quad t \in \frac{1}{3}D,$$

where $c_0 = 1/2$ *and* $\tilde{c}_1 = 3/2$ *or* $c_0 = 3/2$ *and* $\tilde{c}_1 = 1/2;$

(ii) $\sum_{\ell \in Z} \phi(t - \ell) = 1$ *and* $\sum_{\ell \in Z} \tilde{\phi}(t - \ell) = 1;$

(iii) $\operatorname{supp}(\phi) = [0, 2]$ *and* $\operatorname{supp}(\tilde{\phi}) = [\frac{1}{2}, 3];$

(iv) $\langle \tilde{\phi}(t - \ell), \phi(t) \rangle = \delta_{0\ell}.$

The proof of Theorem 3 is similar to the argument in Section 2 of Pollen [16].

If $c_0 = 1/2$, then $\phi(t)$ is the B-spline $N_2(t)$. It can be shown that the corresponding $\tilde{\phi}(t)$ is not in $L^2(\mathbb{R})$. It also can be shown that ϕ is both right- and left-differentiable but is not differentiable at any point in $D \cup [0, 2)$ and $\tilde{\phi}$ is both right- and left-differentiable but is not differentiable at any point in $\frac{1}{3}D \cup [\frac{1}{2}, 3)$.

Acknowledgments. Many thanks to the referee and Ingrid Daubechies for suggesting improvements.

References

1. Battle, G., A block spin construction of ondelettes Part I: Lemarié functions, Comm. Math. Phys. **110** (1987), 601–615.
2. de Boor, C., K. Höllig, and S. Riemenschneider, *Box Splines*, Springer-Verlag, New York, 1993.
3. Chui, C. K., *An Introduction to Wavelet*, Academic Press, Boston, 1992.
4. Chui, C. K. and J. Z. Wang, On compactly supported spline-wavelets and a duality principle, Trans. Amer. Math. Soc. **330** (1992), 903–915.
5. Chui, C. K., J. Stöckler, and J. D. Ward, Compactly supported box spline wavelets, Approx. Theory Appl. **8** (1992), 77–100.
6. Ciesielski, Z., Constructive function theory and spline systems, Studia Math. **52** (1973), 277–302.
7. Cohen, A. and I. Daubechies, A stability criterion for biorthogonal wavelet bases and their related subband coding scheme, Duke Math. J. **68** (1992), 313–335.
8. Cohen, A., I. Daubechies, and J. C. Feauveau, Biorthogonal bases of compactly supported wavelets, Comm. Pure Appl. Math. **45** (1992), 485–500.
9. Daubechies, I., *Ten Lectures on Wavelets*, CBMS-NSF Series in Appl. Math., SIAM Publ., Philadelphia, 1992.
10. He, T. X., Spline interpolation and its wavelet analysis, in *Approximation Theory VIII*, C. K. Chui and L. L. Schumaker (eds.), World Scientific Publishing Co., Inc., New Jersey, 1995, pp. 143–150.
11. He, T. X., Construction of boundary quadrature formulas using wavelets, in *Wavelet Applications in Signal and Image Processing III*, A. F. Laine and M. A. Unser (eds.), SPIE-The International Society for Optical Engineering, 1995, pp. 825–836.
12. He, T. X., Spline wavelet transforms, in *Proceedings of International Conference on Scientific Computing & Modeling*, Charleston Illinois, S. K. Dey and J. Ziebarth (eds.), 1996, pp. 139–142.

13. He, T. X., Spline wavelet transforms II, in *Wavelet Applications in Signal and Image Processing IV*, M. A. Unser, A. Aldroubi, and A. F. Laine (eds.), SPIE-The International Society for Optical Engineering, 1996, pp. 488–491.

14. He, T. X., Short time Fourier transform, integral wavelet transform, and wavelet functions associated with splines, to appear.

15. Meyer, Y., *Ondelettes et Opérateurs*, Herman, Paris, 1990.

16. Pollen, D., Daubechies' scaling function on [0,3], in *Wavelets: A Tutorial in Theory and Applications*, C. K. Chui (ed.), Academic Press, New York, 1992, pp. 3–13.

17. Riemenschneider, D. and Z. Shen, Box splines, cardinal series, and wavelets, in *Approximation Theory and Functional Analysis*, C. K. Chui (ed.), Academic Press, New York, 1991, pp. 133–149.

18. Walter, G. G., *Wavelets and Other Orthogonal Systems with Applications*, CRC Press, Ann Arbor, 1994.

Tian Xiao He
Department of Mathematics
Illinois Wesleyan University
Bloomington, IL 61702-2900
the@sun.iwu.edu

A New Sufficient Condition for the Orthonormality of Refinable Functions

Wenjie He and Ming-Jun Lai

Abstract. A sufficient condition for the orthonormality of refinable function vectors with multiplicity $r = 2$ is given. We shall apply this sufficient condition to check the orthonormality of several well-known examples of multiscaling functions.

§1. Introduction

In recent years, multiscaling functions and multiwavelets have been studied extensively. In [7], a characterization of multiscaling functions and multiwavelets with semi-orthogonality was established. Examples of symmetric and/or antisymmetric compactly supported multiscaling functions were constructed in [2] and [6]

One of the main difficulties in the construction of compactly supported multiscaling functions is how to ensure orthonormality. Currently, there are some criteria about orthonormality for refinable functions. The basic ideas mainly come from the uniscale case, but more complicated situation needs to be considered due to the complexity of matrix products. In ([5, 9, 10] and [11]), orthonormality conditions based on the so-called *transfer operator* were studied for univariate or multivariate setting. The transfer operator condition can be converted to a spectral condition to the representing matrix, that is a generalization of Lawton's condition for scaling functions (see [4]). Several versions of necessary and sufficient orthonormality or stability conditions can be found in [8, 10, 11] and [12]. However, those conditions are not easily applicable.

In this paper, we shall give another sufficient orthonormality condition for refinable functions. This sufficient condition is based on the structure of the zeros of the determinant of the symbol which generalizes Cohen's condition in [3] for the univariate and uniscale situation. Our sufficient condition is different from the one in [10]. We shall demonstrate the feasibility of our

Approximation Theory IX, Volume 2: Computational Aspects
Charles K. Chui and Larry L. Schumaker (eds.), pp. 121–128.
Copyright ℗ 1998 by Vanderbilt University Press, Nashville, TN.
ISBN 0-8265-1326-3.

sufficient condition by checking orthonormality of some well-known examples of Chui and Lian's refinable functions with support on $[0, 2]$ and $[0, 3]$ (see [2]) and Geronimo, Hardin, and Massopust's multiscaling function (see [6]).

§2. A New Orthonormality Condition

Let P_k be a real valued matrix of size $r \times r$, and let $\phi := (\phi_1, ..., \phi_r)^T$ with $\phi_1, ..., \phi_r \in L_2(\mathbf{R})$ satisfy the refinement equation

$$\phi(x) = \sum_{k=0}^{n} P_k \phi(2x - k), \quad x \in \mathbf{R}.$$

Such a ϕ is called a refinable function (vector). Its Fourier transform is

$$\hat{\phi}(\omega) = \mathcal{P}(\omega/2)\hat{\phi}(\omega/2) \tag{1}$$

with $\mathcal{P}(\omega) = \frac{1}{2}\sum_{k=0}^{n} P_k e^{-ik\omega}$. The symbol $\mathcal{P}(\omega)$ is an $r \times r$ matrix with trigonometric polynomial entries. By repeated applications of (1), we have

$$\hat{\phi}(\omega) = \left(\prod_{j=1}^{\infty} \mathcal{P}(\omega/2^j) \right) \hat{\phi}(0).$$

If the infinite product above converges, the $\hat{\phi}$ is well-defined and we will say ϕ is generated by $\mathcal{P}$. The following lemma ensures the convergence of the above infinite product. (See [1] for a proof).

Lemma 1. *The infinite matrix product* $\prod_{j=1}^{\infty} \mathcal{P}(\omega/2^j)$ *converges uniformly on compact sets to a continuous matrix-valued function if and only if* $\mathcal{P}(0)$ *has eigenvalues* $\lambda_1 = \cdots = \lambda_s = 1$ *and* $|\lambda_{s+1}|, ..., |\lambda_r| < 1$, *with the eigenvalue 1 non-degenerate for* $s \geq 1$.

For a continuous orthonormal refinable function $\phi(x)$, the eigenvalue 1 must be simple (see [12]). That is, we have

Lemma 2. *Suppose that a refinable function* ϕ *generated by* $\mathcal{P}$ *is continuous and its integer translates form an orthogonal set. Then 1 must be a simple eigenvalue of* $\mathcal{P}(0)$, *and all other eigenvalues* λ *of* $\mathcal{P}(0)$ *must have* $|\lambda| < 1$.

To study the orthonormality, we define the Gramian matrix

$$\mathbf{\Lambda}(\omega) := \left[\sum_{n \in Z} \hat{\phi}_l(\omega + 2\pi n)\overline{\hat{\phi}_m(\omega + 2\pi n)} \right]_{1 \leq l, m \leq r}. \tag{2}$$

The following lemmas are well-known.

Lemma 3. $\{\phi_1(x), ..., \phi_r(x)\}$ *is an orthonormal refinable function if and only if* $\Lambda(\omega)$ *as in (2) is an identity matrix* $\mathbf{I}_r$.

Lemma 4. *If* $\phi(x)$ *is an orthonormal refinable function, then* $\mathcal{P}(\omega)$ *satisfies the following identity:*

$$\mathcal{P}\mathcal{P}^*(\omega) + \mathcal{P}\mathcal{P}^*(\omega + \pi) = \mathbf{I}_r. \tag{3}$$

We thus define the transfer operator $\mathbf{P_0}$ on $r \times r$ matrix $\mathbf{L}(\omega)$ as

$$(\mathbf{P_0}\mathbf{L})(\omega) := \mathcal{P}\mathbf{L}\mathcal{P}^*(\omega/2) + \mathcal{P}\mathbf{L}\mathcal{P}^*(\omega/2 + \pi). \tag{4}$$

It is easy to see that $\Lambda(\omega)$ is a fixed point of the operator $\mathbf{P_0}$. Define

$$\Theta(\omega) := \Lambda(\omega) - c\mathbf{I}_2 \tag{5}$$

for a constant $c \geq 0$ such that $\Theta(\omega)$ is non-negative definite and $\det(\Theta(\omega))$ has at least one zero. One can see that $\Theta(\omega)$ is also a fixed point of the operator $\mathbf{P_0}$. The main result in this paper is as follows. We refer to [5] for a proof.

Theorem 5. *Let* $\mathcal{P}(\omega)$ *be a symbol of a refinable function* ϕ *with multiplicity* $r = 2$ *and satisfying the condition in Lemma 2. Suppose that* $\mathcal{P}$ *satisfies (3). Then* ϕ *is an orthonormal refinable function if the symbol* $\mathcal{P}$ *satisfies the following conditions:*

1° $\det(\mathcal{P}(\omega + \pi))$ *does not have zeros which are in a non-trivial finite cycle invariant under the operator* τ, *where* $\tau\omega = 2\omega \bmod 2\pi$;

2° *The two operators* $\mathbf{V}_i, i = 1, 2$ *defined by*

$$\left(\mathbf{V}_0 \begin{bmatrix} f_1 \\ f_2 \end{bmatrix}\right)(\omega) := \frac{1}{2}\left(\left(\mathcal{P}\begin{bmatrix} f_1 \\ f_2 \end{bmatrix}\right)(\omega/2) + \left(\mathcal{P}\begin{bmatrix} f_1 \\ f_2 \end{bmatrix}\right)(\omega/2 + \pi)\right)$$

$$\left(\mathbf{V}_1 \begin{bmatrix} f_1 \\ f_2 \end{bmatrix}\right)(\omega) := \frac{e^{i\omega/2}}{2}\left(\left(\mathcal{P}\begin{bmatrix} f_1 \\ f_2 \end{bmatrix}\right)(\omega/2) - \left(\mathcal{P}\begin{bmatrix} f_1 \\ f_2 \end{bmatrix}\right)(\omega/2 + \pi)\right)$$

have no common eigenvectors.

Remark 6. When $r = 2$, under the conditions and assumption of Theorem 5, $\det(\Lambda(\omega))$ cannot have a unique zero at $\omega = 0$ (see [5]). □

Remark 7. Condition 2° of Theorem 5 actually ensures that $\det(\Theta(\omega)) \not\equiv 0$ (see [5]), where $\Theta(\omega)$ is defined as in (5). Therefore the integer translates of the refinable function are algebraically independent (see [10]). □

§3. Verifications of the Orthonormality of Some Refinable Functions

We shall verify that the integer translates of Chui and Lian's refinable functions [2] and those of Geronimo, Hardin, and Massopust's refinable function [6] are orthonormal.

Example 8. Chui and Lian's refinable function with support $[0, 2]$. The symbol $\mathcal{P}(\omega)$ is given by

$$\mathcal{P}(\omega) = \frac{1}{2} \begin{bmatrix} \frac{1}{2} + z + \frac{1}{2}z^2 & \frac{1}{2} - \frac{1}{2}z^2 \\ -\frac{\sqrt{7}}{4} + \frac{\sqrt{7}}{4}z^2 & -\frac{\sqrt{7}}{4} + \frac{1}{2}z - \frac{\sqrt{7}}{4}z^2 \end{bmatrix} = \frac{1}{2}(P_0 + P_1 z + P_2 z^2)$$

where $z = e^{-i\omega}$, $P_0 = \begin{bmatrix} \frac{1}{2} & \frac{1}{2} \\ -\frac{\sqrt{7}}{4} & -\frac{\sqrt{7}}{4} \end{bmatrix}$, $P_1 = \begin{bmatrix} 1 & 0 \\ 0 & \frac{1}{2} \end{bmatrix}$, and $P_2 = \begin{bmatrix} \frac{1}{2} & -\frac{1}{2} \\ \frac{\sqrt{7}}{4} & -\frac{\sqrt{7}}{4} \end{bmatrix}$.

Note that $\mathcal{P}(0) = \begin{bmatrix} 1 & 0 \\ 0 & \frac{1-\sqrt{7}}{4} \end{bmatrix}$ satisfies the condition in Lemmas 1 and 2. And $\det(\mathcal{P}(\omega)) = \frac{1-\sqrt{7}}{8}z(1 + z)^2$. The only zero of $\det(\mathcal{P}(\omega + \pi))$ is $\omega = 0$; thus, it satisfies $1°$ of Theorem 5.

Now we look at the operators $\mathbf{V}_0$ and $\mathbf{V}_1$ in Theorem 5. They act on trigonometric polynomial vectors $[f_1, f_2]^T(\omega) = \mathbf{x}_0 + \mathbf{x}_1 z + \mathbf{x}_2 z^2 \neq \mathbf{0}$ with $z = e^{-i\omega}$ and

$$\mathbf{x}_i = [a_i, b_i]^T \in \mathbf{R}^2 \qquad \text{for} \quad i = 0, 1, 2.$$

Let the eigenvalues for $\mathbf{V}_0$ and $\mathbf{V}_1$ be $C(\omega) = c_1 + c_2 z + c_3 z^2$ and $D(\omega) = d_1 + d_2 z + d_3 z^2$, respectively. The polynomial vectors

$$\mathbf{V}_0\left(\begin{bmatrix} f_1 \\ f_2 \end{bmatrix}\right), \quad C(\omega)\begin{bmatrix} f_1 \\ f_2 \end{bmatrix}, \quad \mathbf{V}_1\left(\begin{bmatrix} f_1 \\ f_2 \end{bmatrix}\right), \quad D(\omega)\begin{bmatrix} f_1 \\ f_2 \end{bmatrix}$$

can be expressed in a form of real vectors in the order of the coefficients of $1, z, z^2, \ldots$. That is

$$\begin{bmatrix} P_0 & 0 & 0 \\ P_2 & P_1 & P_0 \\ 0 & 0 & P_2 \end{bmatrix} \begin{bmatrix} \mathbf{x}_0 \\ \mathbf{x}_1 \\ \mathbf{x}_2 \end{bmatrix}, \quad c_1 \begin{bmatrix} \mathbf{x}_0 \\ \mathbf{x}_1 \\ \mathbf{x}_2 \\ 0 \\ 0 \end{bmatrix} + c_2 \begin{bmatrix} 0 \\ \mathbf{x}_0 \\ \mathbf{x}_1 \\ \mathbf{x}_2 \\ 0 \end{bmatrix} + c_3 \begin{bmatrix} 0 \\ 0 \\ \mathbf{x}_0 \\ \mathbf{x}_1 \\ \mathbf{x}_2 \end{bmatrix},$$

$$\begin{bmatrix} P_1 & P_0 & 0 \\ 0 & P_2 & 0 \\ 0 & 0 & 0 \end{bmatrix} \begin{bmatrix} \mathbf{x}_0 \\ \mathbf{x}_1 \\ \mathbf{x}_2 \end{bmatrix}, \quad d_1 \begin{bmatrix} \mathbf{x}_0 \\ \mathbf{x}_1 \\ \mathbf{x}_2 \\ 0 \\ 0 \end{bmatrix} + d_2 \begin{bmatrix} 0 \\ \mathbf{x}_0 \\ \mathbf{x}_1 \\ \mathbf{x}_2 \\ 0 \end{bmatrix} + d_3 \begin{bmatrix} 0 \\ 0 \\ \mathbf{x}_0 \\ \mathbf{x}_1 \\ \mathbf{x}_2 \end{bmatrix}.$$

For simplicity, denote the two block matrices with entries P_i's by V_0 and V_1.

I : Suppose $\mathbf{x}_2 \neq 0$. We must have $c_2 = c_3 = d_2 = d_3 = 0$. Then

$$V_0 \begin{bmatrix} \mathbf{x}_0 \\ \mathbf{x}_1 \\ \mathbf{x}_2 \end{bmatrix} = c_1 \begin{bmatrix} \mathbf{x}_0 \\ \mathbf{x}_1 \\ \mathbf{x}_2 \end{bmatrix}, \quad V_1 \begin{bmatrix} \mathbf{x}_0 \\ \mathbf{x}_1 \\ \mathbf{x}_2 \end{bmatrix} = d_1 \begin{bmatrix} \mathbf{x}_0 \\ \mathbf{x}_1 \\ \mathbf{x}_2 \end{bmatrix}.$$

However, we can immediately see $d_1 = 0$ which is a contradiction.

II : Suppose $x_2 = 0$, but $x_1 \neq 0$. We must have $c_3 = d_3 = 0$. Then

$$V_0 \begin{bmatrix} x_0 \\ x_1 \\ 0 \end{bmatrix} = c_1 \begin{bmatrix} x_0 \\ x_1 \\ 0 \end{bmatrix} + c_2 \begin{bmatrix} 0 \\ x_0 \\ x_1 \end{bmatrix},$$

$$V_1 \begin{bmatrix} x_0 \\ x_1 \\ 0 \end{bmatrix} = d_1 \begin{bmatrix} x_0 \\ x_1 \\ 0 \end{bmatrix} + d_2 \begin{bmatrix} 0 \\ x_0 \\ x_1 \end{bmatrix}.$$

By $x_1 \neq 0$, we can easily see that $c_2 = d_2 = 0$. Also

$$P_0 x_0 = c_1 x_0, \quad P_2 x_1 = d_1 x_1 \quad \text{and} \quad P_1 x_0 + P_0 x_1 = d_1 x_0.$$

From the third equation, we can see that $x_0 \neq 0$. Since $c_1 \neq 0$, from the first equation, we get

$$x_0 = [2, -\sqrt{7}], \quad c_1 = 1/2 - \sqrt{7}/4.$$

Since $d_1 \neq 0$, from the second equation, we get

$$x_1 = [2, \sqrt{7}], \quad d_1 = 1/2 - \sqrt{7}/4.$$

One can check that the third equation is not satisfied.
III : Suppose $x_2 = x_1 = 0$, but $x_0 \neq 0$. Then we have

$$V_0 \begin{bmatrix} x_0 \\ 0 \\ 0 \end{bmatrix} = \begin{bmatrix} c_1 x_0 \\ c_2 x_0 \\ c_3 x_0 \end{bmatrix}, \quad V_1 \begin{bmatrix} x_0 \\ 0 \\ 0 \end{bmatrix} = \begin{bmatrix} d_1 x_0 \\ d_2 x_0 \\ d_3 x_0 \end{bmatrix}.$$

We can see $c_3 = d_2 = d_3 = 0$ and x_0 is a common eigenvector of matrices P_0, P_1, and P_2, which is impossible.

Therefore, $\mathbf{V}_0$ and $\mathbf{V}_1$ do not have any common eigenvectors and the condition $2°$ of Theorem 5 holds. Hence, this refinable function is orthonormal. □

Example 9. The Chui and Lian's refinable function with support [0,3]. Recall the symbol associated with Chui and Lian's refinable function with support $[0, 3]$ is $\mathcal{P}(\omega) = \begin{bmatrix} P_{11}(z) & P_{12}(z) \\ P_{21}(z) & P_{22}(z) \end{bmatrix}$, where

$$P_{11}(z) = \frac{10 - 3\sqrt{10}}{80}(1 + z)(1 + (38 + 12\sqrt{10})z + z^2)$$

$$P_{12}(z) = \frac{5\sqrt{6} - 2\sqrt{15}}{80}(1 - z)(1 + z)^2$$

$$P_{21}(z) = \frac{5\sqrt{6} - 3\sqrt{15}}{80}(1 - z)(1 - 10(3 + \sqrt{10})z + z^2)$$

$$P_{22}(z) = \frac{5 - 3\sqrt{10}}{1040}(1 + z)(13 - (10 + 6\sqrt{10})z + 13z^2),$$

and

$$\det(\mathcal{P}(z)) = -\frac{1}{640}(1+z)^2(-158 + 51\sqrt{10} + 2(-22 + 39\sqrt{10})z$$
$$+ 6(-106 + 9\sqrt{10})z^2 + 2(-22 + 39\sqrt{10})z^3 + (-158 + 51\sqrt{10})z^4).$$

Thus, $\det(\mathcal{P}(\omega + \pi))$ has only one zero at $\omega = 0$, which satisfies $1°$ of Theorem 5. We can also write $\mathcal{P}(\omega) = \frac{1}{2}(P_0 + P_1 z + P_2 z^2 + P_3 z^3)$ where

$$P_0 = \begin{bmatrix} \frac{10-3\sqrt{10}}{40} & \frac{5\sqrt{6}-2\sqrt{15}}{40} \\ \frac{5\sqrt{6}-3\sqrt{15}}{40} & \frac{5-3\sqrt{10}}{40} \end{bmatrix}, P_1 = \begin{bmatrix} \frac{30+3\sqrt{10}}{40} & \frac{5\sqrt{6}-2\sqrt{15}}{40} \\ -\frac{5\sqrt{6}+7\sqrt{15}}{40} & \frac{15-3\sqrt{10}}{40} \end{bmatrix}$$

$$P_2 = \begin{bmatrix} \frac{30+3\sqrt{10}}{40} & -\frac{5\sqrt{6}-2\sqrt{15}}{40} \\ \frac{5\sqrt{6}+7\sqrt{15}}{40} & \frac{15-3\sqrt{10}}{40} \end{bmatrix}, P_3 = \begin{bmatrix} \frac{10-3\sqrt{10}}{40} & -\frac{5\sqrt{6}-2\sqrt{15}}{40} \\ -\frac{5\sqrt{6}-3\sqrt{15}}{40} & \frac{5-3\sqrt{10}}{40} \end{bmatrix}.$$

Let $[f_1, f_2]^T(\omega) = \mathbf{x}_0 + \mathbf{x}_1 z + \mathbf{x}_2 z^2 + \mathbf{x}_3 z^3 \neq \mathbf{0}$ and $\mathbf{x}_i = [a_i, b_i]^T \in \mathbf{R}^2$ for $i = 0, 1, 2, 3$. The operators $\mathbf{V}_0$ and $\mathbf{V}_1$ can be expressed as follows.

$$\mathbf{V}_0\left(\begin{bmatrix} f_1 \\ f_2 \end{bmatrix}\right): V_0 \begin{bmatrix} \mathbf{x}_0 \\ \mathbf{x}_1 \\ \mathbf{x}_2 \\ \mathbf{x}_3 \end{bmatrix} \qquad \mathbf{V}_1\left(\begin{bmatrix} f_1 \\ f_2 \end{bmatrix}\right): V_1 \begin{bmatrix} \mathbf{x}_0 \\ \mathbf{x}_1 \\ \mathbf{x}_2 \\ \mathbf{x}_3 \end{bmatrix},$$

where $V_0 = \begin{bmatrix} P_0 & 0 & 0 & 0 \\ P_2 & P_1 & P_0 & 0 \\ 0 & P_3 & P_2 & P_1 \\ 0 & 0 & 0 & P_3 \end{bmatrix}$ and $V_1 = \begin{bmatrix} P_1 & P_0 & 0 & 0 \\ P_3 & P_2 & P_1 & P_0 \\ 0 & 0 & P_3 & P_2 \\ 0 & 0 & 0 & 0 \end{bmatrix}$. We consider

four cases:

$\quad$ I : $\mathbf{x}_3 \neq 0$.
$\quad$ II : $\mathbf{x}_3 = 0, \mathbf{x}_2 \neq 0$.
$\quad$ III : $\mathbf{x}_3 = \mathbf{x}_2 = 0, \mathbf{x}_1 \neq 0$.
$\quad$ IV : $\mathbf{x}_3 = \mathbf{x}_2 = \mathbf{x}_1 = 0, \mathbf{x}_0 \neq 0$.
Similar to Example 8, we can show that this refinable function is also orthonormal. $\square$

Example 10. The Geronimo, Hardin, and Massopust refinable function. The symbol associated with the Geronimo, Hardin, and Massopust's refinable function is

$$\mathcal{P}(\omega) = \frac{1}{40}\begin{bmatrix} 12(1+z) & 16\sqrt{2} \\ -\sqrt{2}(z+1)(z^2 - 10z + 1) & -6 + 20z - 6z^2 \end{bmatrix}$$
$$= \frac{1}{2}(P_0 + P_1 z + P_2 z^2 + P_3 z^3)$$

where

$$P_0 = \begin{bmatrix} \frac{3}{5} & \frac{4\sqrt{2}}{5} \\ -\frac{\sqrt{2}}{20} & -\frac{3}{10} \end{bmatrix}, \quad P_1 = \begin{bmatrix} \frac{3}{5} & 0 \\ \frac{9\sqrt{2}}{20} & 1 \end{bmatrix},$$

$$P_2 = \begin{bmatrix} 0 & 0 \\ \frac{9\sqrt{2}}{20} & -\frac{3}{10} \end{bmatrix}, \quad P_3 = \begin{bmatrix} 0 & 0 \\ -\frac{\sqrt{2}}{20} & 0 \end{bmatrix}.$$

Note that $\det(\mathcal{P}(\omega)) = -\frac{1}{40}(1+z)^3$. The only zero of $\det(\mathcal{P}(\omega + \pi))$ is $\omega = 0$. 1° of Theorem 5 is satisfied. Suppose a common eigenvector of $\mathbf{V}_0$ and $\mathbf{V}_1$ is $[f_1 \quad f_2]^T(\omega) = \mathbf{x}_0 + \mathbf{x}_1 z + \mathbf{x}_2 z^2$. As in Example 7, letting

$$V_0 = \begin{bmatrix} P_0 & 0 & 0 \\ P_2 & P_1 & P_0 \\ 0 & P_3 & P_2 \end{bmatrix} \quad \text{and} \quad V_1 = \begin{bmatrix} P_1 & P_0 & 0 \\ P_3 & P_2 & P_1 \\ 0 & 0 & P_3 \end{bmatrix},$$

we have

$$\mathbf{V}_0\left(\begin{bmatrix} f_1 \\ f_2 \end{bmatrix}\right) : V_0 \begin{bmatrix} \mathbf{x}_0 \\ \mathbf{x}_1 \\ \mathbf{x}_2 \end{bmatrix}, \quad \text{and} \quad \mathbf{V}_1\left(\begin{bmatrix} f_1 \\ f_2 \end{bmatrix}\right) : V_1 \begin{bmatrix} \mathbf{x}_0 \\ \mathbf{x}_1 \\ \mathbf{x}_2 \end{bmatrix}.$$

I : Suppose $\mathbf{x}_2 \neq 0$. We have

$$V_0 \begin{bmatrix} \mathbf{x}_0 \\ \mathbf{x}_1 \\ \mathbf{x}_2 \end{bmatrix} = c_1 \begin{bmatrix} \mathbf{x}_0 \\ \mathbf{x}_1 \\ \mathbf{x}_2 \end{bmatrix} \quad V_1 \begin{bmatrix} \mathbf{x}_0 \\ \mathbf{x}_1 \\ \mathbf{x}_2 \end{bmatrix} = d_1 \begin{bmatrix} \mathbf{x}_0 \\ \mathbf{x}_1 \\ \mathbf{x}_2 \end{bmatrix}.$$

It follows from $P_3\mathbf{x}_2 = d_1\mathbf{x}_2$ that d_1 must be zero which is a contradiction.

II : Suppose $\mathbf{x}_2 = 0, \mathbf{x}_1 \neq 0$. We have

$$V_0 \begin{bmatrix} \mathbf{x}_0 \\ \mathbf{x}_1 \\ 0 \end{bmatrix} = c_1 \begin{bmatrix} \mathbf{x}_0 \\ \mathbf{x}_1 \\ 0 \end{bmatrix} + c_2 \begin{bmatrix} 0 \\ \mathbf{x}_0 \\ \mathbf{x}_1 \end{bmatrix},$$

$$V_1 \begin{bmatrix} \mathbf{x}_0 \\ \mathbf{x}_1 \\ 0 \end{bmatrix} = d_1 \begin{bmatrix} \mathbf{x}_0 \\ \mathbf{x}_1 \\ 0 \end{bmatrix} + d_2 \begin{bmatrix} 0 \\ \mathbf{x}_0 \\ \mathbf{x}_1 \end{bmatrix}.$$

We can see $d_2 = 0$, and $c_2 = 0$ from $P_3\mathbf{x}_1 = c_2\mathbf{x}_1$. We find that when

$$\mathbf{x}_0 = [\sqrt{2}/2, -1/2]^T, \mathbf{x}_1 = [0, -1/2]^T \quad \text{and} \quad c_1 = d_1 = -1/5,$$

the above two equations hold. This implies that the operators $\mathbf{V}_0, \mathbf{V}_1$ have a common eigenvector, which is $[\sqrt{2}/2, -1/2(1+z)]^T$. In this situation, we need to consider $\Theta(\omega)$ defined as in (5). We have

$$\Theta(2\omega) = \mathcal{P}\Theta\mathcal{P}^*(\omega) + \mathcal{P}\Theta\mathcal{P}^*(\omega + \pi).$$

If $\det(\Theta(\omega)) \equiv 0$, then up to a constant multiple,

$$\Theta(\omega) = \begin{bmatrix} \sqrt{2}/2 \\ -1/2(1 + e^{-i\omega}) \end{bmatrix} [\sqrt{2}/2, -1/2(1 + e^{i\omega})]$$

(see [5]). Moreover,

$$\mathcal{P}(\omega) \begin{bmatrix} \sqrt{2}/2 \\ -1/2(1+e^{-i\omega}) \end{bmatrix} = -\left(\frac{1+e^{-i\omega}}{10} \right) \begin{bmatrix} \sqrt{2}/2 \\ -1/2(1+e^{-2i\omega}) \end{bmatrix}.$$

We get a contradiction since

$$\mathcal{P}\Theta\mathcal{P}^*(\omega) + \mathcal{P}\Theta\mathcal{P}^*(\omega+\pi) = \frac{1}{25}\Theta(2\omega) \neq \Theta(2\omega).$$

Therefore, $\det(\Theta(\omega)) \not\equiv 0$ in this case.

III : Suppose $\mathbf{x}_2 = \mathbf{x}_1 = 0$, but $\mathbf{x}_0 \neq 0$. We have

$$V_0 \begin{bmatrix} \mathbf{x}_0 \\ 0 \\ 0 \end{bmatrix} = \begin{bmatrix} c_1\mathbf{x}_0 \\ c_2\mathbf{x}_0 \\ c_3\mathbf{x}_0 \end{bmatrix}, \quad V_1 \begin{bmatrix} \mathbf{x}_0 \\ 0 \\ 0 \end{bmatrix} = \begin{bmatrix} d_1\mathbf{x}_0 \\ d_2\mathbf{x}_0 \\ d_3\mathbf{x}_0 \end{bmatrix}.$$

It follows that $c_3 = d_3 = 0$. Also we can see that $\mathbf{x}_0$ is a common eigenvector of P_0, P_1, P_2, and P_3, but P_0 and P_1 do not have any common eigenvectors. Thus, condition $2°$ of Theorem 5 holds. Therefore, this refinable function is orthonormal. $\square$

References

1. Cabrelli, C., C. Heil, and U. Molter, Accuracy of lattice translates of several multidimensional refinable functions, J. Approx. Theory 1998, to appear.

2. Chui, C. K. and J. Lian, A study of orthonormal multiwavelets, Applied Numerical Mathematics **20** (1996), 273–298.

3. Cohen, A., *Ondelettes, Analyses Multiresolutions et Traitement Numérique du Signal*, Ph.D. Thesis, Université Paris, Dauphine, 1990.

4. Daubechies, I., *Ten Lectures on Wavelets*, SIAM, Philadelphia, 1992.

5. He, W., *Compactly Supported Multivariate Multiwavelets: Theory and Construction*, Ph.D. Thesis, University of Georgia, Athens, Georgia, 1998.

6. Geronimo, J. S., D. P. Hardin, and P. R. Massopust, Fractal functions and wavelet expansions based on several scaling functions, J. Approx. Theory **78** (1994), 373–401.

7. Goodman, T. N. T., S. L. Lee, and W. S. Tang, Wavelets in wandering subspaces, Trans. Amer. Math. Soc. **338** (1993), 639–654.

8. Hogan, T. A., Stability and independence of shifts of finitely many refinable functions, J. Fourier Anal. Appl. **3** (1997), 757–774.

9. Lian, J., Orthogonality criteria for multiscaling functions, 1996, preprint.

10. Plonka, G., Necessary and sufficient conditions for orthonormality of scaling vectors, in *Multivariate Approximation and Splines*, G. Nürnberger, J. W. Schmidt, and G. Walz (eds.), Birkhäuser, Basel, 1997, pp. 205–218.

11. Shen, Z., Refinable function vectors, SIAM J. Math. Anal. **29** (1998), 235–250.

12. Wang, Y., On orthogonal refinable function vectors, 1997, preprint.

Optimal Triangulations and Smoothness Conditions for Bivariate Splines

Don Hong, Huan-Wen Liu, and Ram Mohapatra

Abstract. In this paper, we survey the results on optimal triangulations for bivariate splines. Here the optimality means the full order of approximation and is achieved from bivariate spline spaces over such triangulations. We focus on the C^1 quartic and C^1 cubic splines. An example from a MatLab package is displayed for optimal triangulation using an edge swapping algorithm for C^1 quartic splines. Also, we reformulate some smoothness conditions and conformality conditions for these splines via interpolation data at a set of B-net domain points.

§1. Introduction

In general, for a given set V of data sites, there are many different triangulations with vertex set V. On a triangulation Δ of a polygonal domain $\Omega \subset \mathbb{R}^2$ with vertex set V, one of the most important problems in application is to represent scattered data defined on V by C^r smooth piecewise polynomial (*pp* or *spline*) functions. Of course, one usually wants to find an optimal triangulation of the given sample sites. Here, we are concerned with optimal order of approximation with respect to the given order r of smoothness and degree k of the polynomial pieces of the smooth spline functions. In the study of spline functions on a triangulation Δ, the notation $S_k^r(\Delta)$ is used to denote the subspace of $C^r(\Omega)$ of all *pp* functions with total degree $\leq k$ and with grid lines given by the edges of Δ.

We use the term optimal triangulation of a given set V of data sites to mean that (i) the set V of sample sites is the same as the set of vertices of the triangulation, and (ii) the space of *pp* functions with degree k and smoothness order r on this triangulation achieves the full order of approximation. The full order of approximation from $S_k^r(\Delta)$, of course, is $k + 1$.

Approximation Theory IX, Volume 2: Computational Aspects
Charles K. Chui and Larry L. Schumaker (eds.), pp. 129–136.
Copyright © 1998 by Vanderbilt University Press, Nashville, TN.
ISBN 0-8265-1326-3.

Definition 1. *For a given set V of data sites, of degree k and smoothness order r, any triangulation Δ with vertex set V is called optimal (of type (k, r)) if the spline space $S_k^r(\Delta)$ admits full approximation order $k + 1$.*

We survey the recent study on optimal triangulations for bivariate splines in the next section. In real world applications, only the spline elements with small degree versus smoothness order are widely used. One obvious reason is that the low-degree bivariate spline functions can achieve the efficiency (because of the simple calculation property) and also preserve the accuracy (because of the smoothness conditions) required by computer aided geometric design and by numerical approximation. Therefore, we will only consider C^1 quartic and C^1 cubic spline approximation. An example from a MatLab package is displayed for optimal triangulation using an edge swapping algorithm for C^1 quartic splines in the next section. Since most results on the optimal triangulations are obtained by carefully studying the smoothness conditions and conformality conditions of bivariate splines, we also present some reformulations of smoothness conditions and conformality conditions for C^1 quartic and cubic splines, via interpolation data on a set of B-net domain points, in the final section.

§2. Optimal Triangulations

It is well known that any triangulation Δ is optimal for spline space $S_k^r(\Delta)$ when $k \geq 3r + 2$. This result was shown by de Boor and Hölling [2] using an abstract analysis method and by Chui, Hong and Jia [6] using a constructive method. A three-directional mesh $\Delta^{(1)}$, also called a type-1 triangulation, is a rectangular partition plus north-east diagonals. A four-directional mesh $\Delta^{(2)}$, or a type-2 triangulation, is a rectangular partition plus all diagonals. de Boor and Jia [3] proved that the approximation order of the spline space $S_k^r(\Delta^{(1)})$ is at most k provided that $k \leq 3r + 1$. This shows that $\Delta^{(1)}$ cannot be an optimal triangulation for $S_k^r(\Delta)$ when $k \leq 3r + 1$. Therefore, it is important to find optimal triangulations for bivariate splines when $k \leq 3r+1$. A local Clough-Tocher refinement procedure is introduced in [4] to construct an optimal triangulation based on an arbitrary given triangulation for C^1 quartic splines. On the other hand, Jia [15] proved that $S_k^r(\Delta^{(2)})$ has full order of approximation if $k \geq 2r + 1$. Therefore, $\Delta^{(2)}$ is an optimal triangulation for $S_k^r(\Delta)$ when $k \geq 2r + 1$. However, one can see that a four-directional mesh uses more data sites (intersections of diagonals) than a three-directional mesh. In [13], we constructed a mixed three-directional mesh (a rectangular partition plus only one diagonal in each rectangle) and proved that such a mixed three-directional mesh is an optimal triangulation for C^1 quartic splines. Therefore, the mixed three-directional mesh is better than a three-directional mesh in the sense that the corresponding spline space has a higher order of approximation, and also the mixed three-directional mesh is better than local refinements or a four-directional mesh in the sense that the C^1 quartic spline space achieves the optimal approximation order by using a smaller number of data sites in the interpolation.

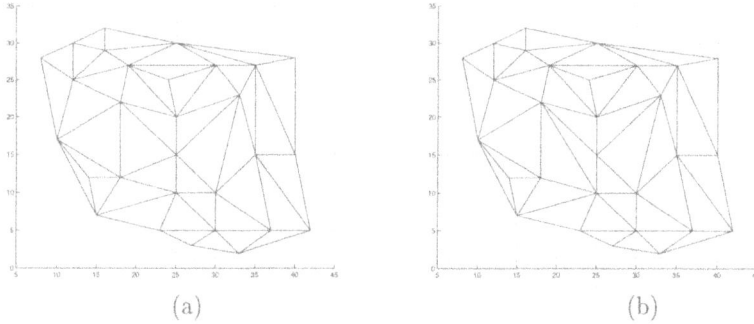

(a) (b)

Fig. 1. Triangulation and optimal triangulation for a set of data sites.

To represent scattered data over an arbitrary set of sites, we recently developed an algorithm to construct an optimal triangulation with these sample sites as the only vertices for C^1 quartic splines (see [5]). A so-called type-O triangulation is then introduced and we found that any type-O triangulation is an optimal triangulation for C^1 quartic splines (see [10]). Starting with an arbitrary triangulation, we proved, by providing a so-called edge swapping algorithm, that an optimal triangulation can be obtained by swapping edges from the given triangulation for C^1 quartic splines. Therefore, for given data on arbitrarily scattered sites, it is always possible to find an interpolation scheme such that the C^1 quartic spline space achieves the full order of approximation. In [8], we developed a MatLab package to implement this algorithm. The MatLab package includes a main function `swap.m` as well as subfunctions `consecv.m`, `findrow.m`, `delrow.m`, `findtri.m`, and `trimesh2.m`, a modification of the MatLab 5.0 function `trimesh.m`. One can see that the vertex located at (25, 15) in Fig. 1 (a) is not a type-O vertex as defined in [10]. The program `swap.m` then makes an edge swap to form an optimal triangulation in Figure 1 (b).

Jia proved using a box spline technique in [15] that a four-directional mesh $\Delta^{(2)}$ is an optimal triangulation for spline space $S_k^r(\Delta^{(2)})$ as long as $k \geq 2r + 1$. Thus, $\Delta^{(2)}$ is an optimal triangulation for C^1 cubic splines. Lai [16] reobtained this specific result by using a B-net analysis. Together with Schumaker [17], he found that a so-called quadrangulation is also an optimal triangulation for C^1 cubic splines. It would be very interesting to find some more general triangulations which are optimal for C^1 cubic splines.

§3. Smoothness Conditions and Conformality Conditions

Bernstein-Bezier representations (B-net for short) play an important role in the study of multivariate spline functions. The general description of a B-net for multivariate splines can be found in [1] (see also [14]). Some recent studies of multivariate splines are surveyed in [11]. In this section we present some reformulations of smoothness conditions and conformality conditions for bivariate C^1 cubic and C^1 quartic splines.

3.1. C^1 Cubic Splines

Let $\delta = [v_0, v_1, v_2]$ be a triangle with vertices $v_i = (x_i, y_i) \in \mathbb{R}^2$, $i = 0, 1, 2$, and $\xi = (\xi_0, \xi_1, \xi_2)$ the barycentric coordinate of $x \in \mathbb{R}^2$ with respect to δ_1. We introduce

$$\lambda_i := y_j - y_k, \quad \mu_i := -(x_j - x_k), \quad \nu_i := x_j y_k - y_j x_k,$$

where (i, j, k) is a permutation of the subscripts in cyclic order $0 \to 1 \to 2 \to 0$. Then the barycentric coordinate can be expressed as

$$\begin{pmatrix} \xi_0 \\ \xi_1 \\ \xi_2 \end{pmatrix} = \frac{1}{A^{(1)}} \begin{pmatrix} \nu_0 & \lambda_0 & \mu_0 \\ \nu_1 & \lambda_1 & \mu_1 \\ \nu_2 & \lambda_2 & \mu_2 \end{pmatrix} \begin{pmatrix} 1 \\ x \\ y \end{pmatrix},$$

where $A^{(1)}$ is the oriented area of the simplex δ.

For a bivariate polynomial p with total degree k and B-net representation b_α, i.e.,

$$p = \sum_{|\alpha|=k} b_\alpha B_{\alpha,\delta} = \sum_{|\alpha|=k} b_{\alpha_0 \alpha_1 \alpha_2} B_{\alpha,\delta},$$

where $B_\alpha(x) = \binom{|\alpha|}{\alpha} \xi^\alpha$, $\alpha \in \mathbb{Z}_+^3$, and $\xi^\alpha = \xi_0^{\alpha_0} \xi_1^{\alpha_1} \xi_2^{\alpha_2}$, we have the following equations:

$$\begin{pmatrix} \frac{\partial p}{\partial x} \\ \frac{\partial p}{\partial y} \end{pmatrix} = \begin{pmatrix} \frac{\partial \xi_0}{\partial x} & \frac{\partial \xi_1}{\partial x} \\ \frac{\partial \xi_0}{\partial y} & \frac{\partial \xi_1}{\partial y} \end{pmatrix} \begin{pmatrix} \frac{\partial p}{\partial \xi_0} \\ \frac{\partial p}{\partial \xi_1} \end{pmatrix} = \frac{1}{A^{(1)}} \begin{pmatrix} \lambda_0 & \lambda_1 \\ \mu_0 & \mu_1 \end{pmatrix} \begin{pmatrix} \frac{\partial p}{\partial \xi_0} \\ \frac{\partial p}{\partial \xi_1} \end{pmatrix}$$

$$= \sum_{|\alpha|=k-1} \frac{k}{A^{(1)}} \binom{|\alpha|}{\alpha} \begin{pmatrix} \lambda_0 & \lambda_1 \\ \mu_0 & \mu_1 \end{pmatrix} \begin{pmatrix} \triangle_{13} b_\alpha \\ \triangle_{23} b_\alpha \end{pmatrix} \xi_0^{\alpha_0} \xi_1^{\alpha_1} \xi_2^{\alpha_2},$$

where

$$\triangle_{1\,3} b_\alpha = b_{\alpha_0+1\,\alpha_1\,\alpha_2} - b_{\alpha_0\,\alpha_1\,\alpha_2+1} \quad \text{and}$$

$$\triangle_{2\,3} b_\alpha = b_{\alpha_0\,\alpha_1+1\,\alpha_2} - b_{\alpha_0\,\alpha_1\,\alpha_2+1}.$$

Let $f \in C^1(\Omega)$. We choose interpolation data on a set of B-net points of δ as follows. Chui and Lai gave a more general discussion on interpolation conditions to determine a polynomial on a simplex in [7].

$$p(v_i) = f(v_i) =: f_i, \quad p(\frac{v_0 + v_1 + v_2}{3}) = f(\frac{v_0 + v_1 + v_2}{3}) =: f_3,$$

$$\frac{\partial p}{\partial x}\big|_{v_i} = \frac{\partial f}{\partial x}\big|_{v_i} =: D_x f_i, \quad \text{and} \quad \frac{\partial p}{\partial y}\big|_{v_i} = \frac{\partial f}{\partial y}\big|_{v_i} =: D_y f_i, \quad i = 0, 1, 2. \tag{1}$$

The interpolation data can be expressed in terms of a B-net representation b_α and vice versa. We can use the interpolation data set to simplify the smoothness conditions and conformality conditions of cubic splines.

Let $\delta_1 = [v_0, v_1, v_2]$ and $\delta_2 = [v_0, v_1, v_3]$ be two adjacent triangles with a common edge $[v_0, v_1]$ and $v_3 = (x_3, y_3)$. For convenience, let $A^{(1)}$ denote the area of the triangle δ_1 and

$$A_0^{(1)} = \text{area}[v_3, v_1, v_2], \quad A_1^{(1)} = \text{area}[v_0, v_3, v_2], \quad A_2^{(1)} = \text{area}[v_0, v_1, v_3].$$

Suppose that the cubic spline $F(x, y)$ is defined on $\delta_1 \cup \delta_2$ by

$$F|_{\delta_1} = \sum_{|\alpha|=3} b_\alpha^{(1)} B_{\alpha, \delta_1} := f, \text{ and } F|_{\delta_2} = \sum_{|\alpha|=3} b_\alpha^{(2)} B_{\alpha, \delta_2} := g. \quad (2)$$

Then, the smoothness conditions given in [14] can be written as follows. For $\alpha = (\alpha_0, \alpha_1, 0) \in \mathbb{Z}+^3$, $b_\alpha^{(2)} = b_\alpha^{(1)}$, for $|\alpha| = 3$, and

$$b_{\alpha+e_3}^{(2)} = \sum_{\beta \in \mathbb{Z}+^3, |\beta|=1} b_{\alpha+\beta}^{(1)} \frac{(A_0^{(1)})^{\beta_0} (A_1^{(1)})^{\beta_1} (A_2^{(1)})^{\beta_2}}{A^{(1)}}, \text{ for } |\alpha| = 2.$$

Using the interpolation data (1), we arrive at the following theorem (see [12]).

Theorem 1. *Suppose that the cubic spline $F(x, y)$ is defined on $\delta_1 \cup \delta_2$ by (2). Let $b_\alpha^{(1)}$, $b_\alpha^{(2)}$ be the B-net representations of F on δ_1 and δ_2, respectively. Then $F(x, y) \in C^1(\delta_1 \cup \delta_2)$ if and only if*

$$f_i = g_i, \ D_x f_i = D_x g_i, \ D_y f_i = D_y g_i, \text{ for } i = 0, 1, \text{ and}$$
$$b_{111}^{(2)} = \frac{1}{A^{(1)}} (b_{210}^{(1)} A_0^{(1)} + b_{120}^{(1)} A_1^{(1)} + b_{111}^{(1)} A_2^{(1)}), \quad (3)$$

where

$$b_{210}^{(1)} = f_0 + \frac{1}{3} \mu_2 D_x f_0 - \frac{1}{3} \lambda_2 D_y f_0, \text{ and } b_{120}^{(1)} = f_1 - \frac{1}{3} \mu_2 D_x f_1 + \frac{1}{3} \lambda_2 D_y f_1. \quad (4)$$

The number of the edges emanating from a vertex v is called the *degree* of v and denoted by $deg(v)$. The union of all the triangles with a common interior vertex v of Δ is called a *standard cell* and denoted by Δ_v. The boundary vertices of Δ_v, in the counter clockwise direction, are denoted by $v_j, j = 1, 2, ..., d$. For a standard cell Δ_v with interior vertex v, we define $v_0 = v, e_j = [v_0, v_j]$, $A^{(j)} = \text{area}[v_0, v_j, v_{j+1}]$, $A_0^{(j)} = \text{area}[v_{j+2}, v_j, v_{j+1}]$, $A_1^{(j)} = \text{area}[v_0, v_{j+2}, v_{j+1}]$, and $A_2^{(j)} = \text{area}[v_0, v_j, v_{j+2}]$, for $j = 1, ..., d$, $j+1 \bmod(d)$ and $j+2 \bmod(d)$. The conditions (or linear equations) that a spline $s \in S_k^r(\Delta)$ satisfies around the vertex v are called conformality conditions. The subspace of super splines of smoothness order r and degree $\leq k$ with enhanced smoothness order $\theta \geq r$ is defined as

$$S_k^{r,\theta}(\Delta) = \{s \in S_k^r(\Delta) : \ s \in C^\theta \text{ at each vertex of } \Delta\}.$$

Now we consider the conformality conditions for bivariate splines based on the cubic super spline space $S_3^{0,1}(\Delta)$. We obtained the following result on conformality conditions of super cubic splines in [12].

Theorem 2. *Suppose* $s(x,y) \in S_3^{0,1}(\Delta_v)$ *is a bivariate super cubic spline defined on a standard cell* Δ_v *with an interior vertex* v. *Then a necessary and sufficient condition for* $s(x,y) \in S_3^1(\Delta_v)$ *is that*

  i) *if* d *is an even number* $d = 2N$, *then*

$$\sum_{j=1}^{2N}(-1)^j \frac{1}{A^{(j)}A^{(j+1)}}\left[b_{102}^{(j)} A_2^{(j)} + b_{201}^{(j)} A_0^{(j)} \right] = 0, \qquad (5)$$

  ii) *if* d *is an odd number* $d = 2N + 1$, *then*

$$b_{111}^{(1)} = \frac{1}{2}\sum_{j=1}^{2N+1}(-1)^{j+1}\frac{A^{(1)}}{A^{(j)}A^{(j+1)}}\left[b_{102}^{(j)} A_2^{(j)} + b_{201}^{(j)} A_0^{(j)} \right], \qquad (6)$$

where

$$\begin{aligned} b_{102}^{(j)} &= s_{j+1} - \frac{1}{3}(x_{j+1} - x_0)\, D_x s_{j+1} - \frac{1}{3}(y_{j+1} - y_0)\, D_y s_{j+1}, \\ b_{201}^{(j)} &= s_0 + \frac{1}{3}(x_{j+1} - x_0)\, D_x s_0 + \frac{1}{3}(y_{j+1} - y_0)\, D_y s_0. \end{aligned} \qquad (7)$$

3.2. C^1 Quartic Splines

For quartic spline on $\delta_1 \cup \delta_2$, we choose the interpolation data from the function values and values of partial derivatives on vertices, and the function value on the mid-point of each edge. Then, we give a similar discussion for C^1 quartic splines to C^1 cubic splines as in [12] and simplify the smoothness conditions as follows.

Theorem 3. *Suppose that the quartic spline* $F(x,y)$ *is defined on* $\delta_1 \cup \delta_2$ *and* $F|_{\delta_1} = f$, $F|_{\delta_2} = g$. *Let* $b_\alpha^{(1)}$, $b_\alpha^{(2)}$ *be the B-net representation of* F *on* δ_1 *and* δ_2, *respectively. Then* $F(x,y) \in C^1(\delta_1 \cup \delta_2)$ *if and only if*

$$\begin{aligned} f_i = g_i, \;\; i = 0, 1, \qquad f(\frac{v_0 + v_1}{2}) = g(\frac{v_0 + v_1}{2}), \\ D_x f_i = D_x g_i, \text{ and } D_y f_i = D_y g_i, \;\; i = 0, 1 \end{aligned} \qquad (8)$$

and

$$\begin{aligned} b_{211}^{(2)} &= \frac{1}{A^{(1)}}\left[b_{310}^{(1)}A_0^{(1)} + b_{220}^{(1)}A_1^{(1)} + b_{211}^{(1)}A_2^{(1)} \right], \\ b_{121}^{(2)} &= \frac{1}{A^{(1)}}\left[b_{220}^{(1)}A_0^{(1)} + b_{130}^{(1)}A_1^{(1)} + b_{121}^{(1)}A_2^{(1)} \right], \end{aligned} \qquad (9)$$

where

$$b_{310}^{(1)} = f_0 + \frac{1}{4}(\mu_2 D_x f_0 - \lambda_2 D_y f_0), \;\; b_{130}^{(1)} = f_1 - \frac{1}{4}(\mu_2 D_x f_1 - \lambda_2 D_y f_1), (10)$$

$$\begin{aligned} b_{220}^{(1)} = \frac{1}{6}(16f(\frac{v_0 + v_1}{2}) - 5f_0 - 5f_1 - \mu_2 D_x f_0 + \lambda_2 D_y f_0 \\ + \mu_2 D_x f_1 - \lambda_2 D_y f_1). \end{aligned} \qquad (11)$$

The following theorem gives conformality conditions of bivariate quartic splines.

Theorem 4. *Suppose* $s(x, y) \in S_4^{0,1}(\Delta_v)$ *is a bivariate super quartic spline defined on a standard cell* Δ_v *with an interior vertex* v, *then the conformality condition for* $s(x, y) \in S_4^1(\Delta_v)$ *is that*
  i) *if* $d = 2N$ *(d is even), then*

$$\sum_{j=1}^{2N} (-1)^j \frac{1}{A^{(j)} A^{(j+1)}} \left[b_{202}^{(j)} A_2^{(j)} + b_{301}^{(j)} A_0^{(j)} \right] = 0, \tag{12}$$

  ii) *if* $d = 2N + 1$ *(d is odd), then*

$$b_{211}^{(1)} = \frac{1}{2} \sum_{j=1}^{2N+1} (-1)^{j+1} \frac{A^{(1)}}{A^{(j)} A^{(j+1)}} \left[b_{202}^{(j)} A_2^{(j)} + b_{301}^{(j)} A_0^{(j)} \right], \tag{13}$$

where

$$b_{301}^{(j)} = s_0 + \frac{1}{4}(x_{j+1} - x_0) D_x s_0 + \frac{1}{4}(y_{j+1} - y_0) D_y s_0, \tag{14}$$

$$b_{202}^{(j)} = \frac{1}{6} \left[16 s(\frac{v_0 + v_{j+1}}{2}) - 5s_0 - 5s_{j+1} + (x_{j+1} - x_0)(D_x s_0 - D_x s_{j+1}) \right.$$
$$\left. + (y_{j+1} - y_0)(D_y s_0 - D_y s_{j+1}) \right]. \tag{15}$$

Acknowledgments. The first author is supported by a research development grant (RDC 2-25311) from East Tennessee State University. The authors are grateful to the referees for their comments on this paper.

References

1. de Boor, C., B-form basis, in *Geometric Modeling*, G. Farin (ed.), SIAM, Philadelphia, 1987, pp. 21–28.
2. de Boor, C. and K. Höllig, Approximation power of smooth bivariate *pp* functions, Math. Z. **197** (1988), 343–363.
3. de Boor, C. and R. Q. Jia, A sharp upper bound on the approximation order of smooth bivariate *pp* functions, J. Approx. Theory **72** (1993), 24–33.
4. Chui, C. K. and D. Hong, Construction of local C^1 quartic spline elements for optimal-order approximation, Math. Comp. **65** (1996), 85-98. MR **96d:65023**.
5. Chui, C. K. and D. Hong, Swapping edges of arbitrary triangulations to achieve the optimal order of approximation, SIAM J. Numer. Anal. **34** (1997), 1472–1482.
6. Chui, C. K., D. Hong, and R. Q. Jia, Stability of optimal-order approximation by splines over arbitrary triangulations, Trans. of Amer. Math. Soc. **374** (1995), 3301–3318. MR **96d:41012**.

7. Chui, C. K. and M. J. Lai, On bivariate super vertex splines, Constr. Approx. **6** (1990), 399–419.

8. Dyer, B. and Don Hong, A MatLab package for constructing optimal triangulations for bivariate C^1 quartic splines, manuscript (1998).

9. Hong, D., A new formulation of Bernstein-Bézier based smoothness conditions for pp functions, Approx. Theory and Appl. **11** (1995), 67–75.

10. Hong, D., Optimal triangulations for the best C^1 quartic spline approximation, in *Approximation Theory VIII, Vol. 1: Approximation and Interpolation*, Charles K. Chui and Larry L. Schumaker (eds.), World Scientific Publishing Co., Inc., Singapore, 1995, pp. 249–256.

11. Hong, D., Recent progress on multivariate splines, in *Approximation Theory: In Memory of A. K. Varma*, N. K. Govil et al. (eds.), Marcel Dekker, Inc,. New York, NY, 1998, pp. 265–291.

12. Hong, D. and H.-W. Liu, Some new formulations of smoothness conditions and conformality conditions for bivariate splines, Computers and Mathematics with Applications, to appear.

13. Hong D. and R. N. Mohapatra, Optimal-order approximation by mixed three-directional spline elements, Computers and Mathematics with Applications, to appear.

14. Jia, R. Q., B-net Representation of Multivariate Splines, Ke Xue Tong Bao (A Monthly Journal of Science) **11** (1987), 804–807.

15. Jia, R. Q., *Lecture Notes on Multivariate Splines*, Department of Mathematics, University of Alberta, Edmonton, Canada, 1990.

16. Lai, M. J., Approximation order from bivariate C^1 cubic on a four directional mesh is full, Comput. Aided Geom. Design **11** (1994), 215–223.

17. Lai, M. J. and L. L. Schumaker, On the approximation power of splines on triangulated quadrangulations, SIAM J. Numer. Anal., to appear.

Don Hong
Department of Mathematics
East Tennessee State University
Johnson City, TN 37614-0663
hong@etsu.edu www.etsu.edu/math/hong

Huan-Wen Liu
Department of Mathematics
University of Wollongong
Wollongong, NSW 2522, Australia
Huanwen_Liu@uow.edu.au

Ram Mohapatra
Department of Mathematics
University of Central Florida
Orlando, FL 32816
ramm@pegasus.cc.ucf.edu

Norming Sets and Scattered Data Approximation on Spheres

Kurt Jetter, Joachim Stöckler, and Joseph D. Ward

Abstract. This short note deals with approximation order of spaces spanned by $\kappa(x, \cdot)$, $x \in X$, with κ a positive definite kernel on a sphere and X a given set of nodes. We estimate both the L_2- and the L_∞-error, if the function to be approximated is assumed to be rather smooth, thus deviating from the usual assumption that the function stems from the 'native space' of the kernel. It is of interest that, based on our former results on norming sets, we can bound p-norms of right inverses of collocation matrices which originate from evaluating the orthonormal basis of spherical harmonics up to a given degree L at X.

§1. Introduction and Notations

The notion of norming sets proved to be an essential tool for our investigation of scattered data interpolation on spheres $S^{n-1} \subset \mathbb{R}^n$; see [2]. There, our study was based on the so-called native spaces; these are certain Hilbert spaces of continuous functions on S^{n-1} which possess a positive definite reproducing kernel $\kappa(x, y)$, $x, y \in S^{n-1}$. In such a setting, the same kernel serves two different purposes: firstly, it defines the Hilbert space H_κ of functions to be interpolated (see Section 3), and secondly, it also generates the space of interpolants,

$$V_X := \operatorname{span} \{\kappa(x, \cdot); \ x \in X\}, \tag{1}$$

where $X \subset S^{n-1}$ is the given (finite) set of interpolation nodes. Then the Golomb-Weinberger theory of optimal interpolation of linear functionals can be employed in order to find error estimates for the interpolation. Norming sets have entered our estimation technique, and they can be favourably used when the knots are scattered.

In the present paper we wish to establish further results of a similar character, but in a more general setting. In fact, we use the same space V_X as before, generated by a positive definite kernel κ. But we investigate upper bounds for the best approximation (rather than interpolation) of functions

Approximation Theory IX, Volume 2: Computational Aspects
Charles K. Chui and Larry L. Schumaker (eds.), pp. 137–144.
Copyright © 1998 by Vanderbilt University Press, Nashville, TN.
ISBN 0-8265-1326-3.

that may lie in a Banach space $\mathcal{B} \subset C(S^{n-1})$ which is different from the native space H_κ; in other words, we estimate

$$\text{dist}(f, V_X) = \inf_{v \in V_X} \|f - v\|, \qquad f \in \mathcal{B}. \tag{2}$$

Here, the norm in which the error is measured is either the max-norm $\|\cdot\|_\infty$, or the L_2-norm $\|\cdot\|_2$. The motivation for this study originates from the fact, that the space V_X can often yield better approximation rates than predicted by the Golomb-Weinberger theory, when we work with a subset $\mathcal{B} \subset H_\kappa$ of very smooth functions. Endowing $\mathcal{B}$ with a different norm (e.g., higher derivatives involved than for the native space norm) leads to a gain on the achievable approximation order.

For this purpose we shall first develop some stronger consequences of the results in [2] concerning norming sets for spherical harmonics; this will be the contents of Section 2. An application of these results to finding bounds for the approximation error will be given in Section 3.

Let us list some of the necessary notations. Every function $f \in L_2(S^{n-1})$ has a (Fourier) series expansion

$$f = \sum_{\ell=0}^{\infty} \sum_{k=1}^{N(n,\ell)} \widehat{f}_{\ell,k} Y_{\ell,k} \tag{3}$$

with respect to the orthonormal basis of spherical harmonics $Y_{\ell,k}$. Here, the index ℓ refers to the degree of the spherical harmonics, and $N(n, \ell)$ is the dimension of the space of spherical harmonics of degree ℓ. The above series converges in $L_2(S^{n-1})$, i.e., in the norm induced by the (normalized) inner product

$$\langle f, g \rangle = \frac{1}{\omega_{n-1}} \int_{S^{n-1}} f(x)\overline{g(x)}\, dx, \qquad f, g \in L_2(S^{n-1}), \tag{4}$$

with ω_{n-1} the surface area of the unit sphere $S^{n-1} \subset \mathbb{R}^n$. For more details we refer to the books [4, 6], the recent survey [1], or [2]. Classical Hilbert spaces, such as Sobolev spaces $H^s(S^{n-1})$ of real order $s \geq 0$, can be defined by introducing a weight in the summability condition on the Fourier coefficients, e.g., see [5].

For two points $x, y \in S^{n-1}$ and a set $X \subset S^{n-1}$ we denote the spherical distance of the two points by $d(x, y) = \arccos(x \cdot y)$, the distance from y to X by $d(y, X) = \inf_{x \in X} d(y, x)$, and the mesh norm of X by

$$h(X) = \sup_{y \in S^{n-1}} d(y, X). \tag{5}$$

The point evaluation functional at x is denoted by δ_x, and its restriction to a subspace W of $C(S^{n-1})$ by $\delta_x|_W$. We note further that the norm of the functional $\sum_{x \in X} a_x \delta_x$, considered on the Banach space $C(S^{n-1})$, is given by $\sum_{x \in X} |a_x|$.

§2. Norming Sets and Least Squares Matrices

For the reader's orientation we recall the notion of a norming set.

Definition 1. *Let V be a normed linear space with (continuous) dual V^*. Given two subspaces $W \subset V$ and $Z \subset V^*$, the set Z is called a norming set of W if there exists some $c > 0$, the norming constant, so that*

$$\sup_{z \in Z, \|z\|=1} |z(w)| \geq c\|w\| \qquad \text{for all } w \in W.$$

A first application in view of the sampling of finite dimensional subspaces of $C(S^{n-1})$ was given in [2].

Proposition 2. *Let W be a finite dimensional subspace of $C(S^{n-1})$, and let $Z = \text{span}\{\delta_x;\ x \in X\}$, where $X \subset S^{n-1}$ is a finite knot set. Assume that Z is a norming set of W with norming constant $c > 0$. Then for any $w^* \in W^*$ with norm $\|w^*\| = 1$, one can find complex numbers a_x, $x \in X$, so that*

$$w^* = \sum_{x \in X} a_x \delta_x|_W \qquad \text{and} \qquad \sum_{x \in X} |a_x| \leq \frac{1}{c}.$$

Note that the above formulations do not refer to a specific basis of W. For the technique required in Section 3, however, it is more appropriate to look at a least-squares matrix which results from sampling a special type of basis of W at the points X. Such matrices also appear in the study of spherical cubature formulas as investigated in [3].

In the following we let $X = \{x_1, \ldots, x_M\}$ for convenience.

Theorem 3. *With the assumptions of Proposition 2, let*

$$A = \begin{pmatrix} w_1(x_1) & w_1(x_2) & \cdots & w_1(x_M) \\ w_2(x_1) & w_2(x_2) & \cdots & w_2(x_M) \\ \vdots & \vdots & \cdots & \vdots \\ w_N(x_1) & w_N(x_2) & \cdots & w_N(x_M) \end{pmatrix} \in \mathbf{C}^{N \times M},$$

where $\{w_1, \ldots, w_N\}$ is an orthonormal basis of the space W with respect to the inner product in (4). Then there exist matrices $B_1, B_2 \in \mathbf{C}^{M \times N}$ so that

$$AB_1 = AB_2 = I_N \quad \text{and} \quad \|B_1\|_1 \leq \frac{1}{c}, \quad \|B_2\|_2 \leq \frac{1}{c}.$$

Here I_N denotes the identity matrix of size N, and the norms are the operator p-norms of the matrices. B_1 and B_2 are right-inverses (or generalized inverses) of the matrix A. The least possible upper bound for the 2-norm is attained if B_2 is the Moore-Penrose pseudoinverse of A. The statement of the theorem necessarily implies that the dimension N of the space W is less than or equal to the number M of points in X.

Proof: We first consider the assertion concerning B_1. Every linear functional

$$\lambda_i(w) := \langle w, w_i \rangle, \qquad w \in W, \qquad i = 1, \ldots, N,$$

(considered as an element in W^*, where W carries the maximum norm) has norm ≤ 1. This follows from (4), since

$$|\langle w, w_i \rangle| \leq \|w\|_\infty \cdot \frac{1}{\omega_{n-1}} \int_{S^{n-1}} |w_i(x)| \, dx \leq \|w\|_\infty$$

by an apparent application of the Cauchy-Schwarz inequality. By Proposition 2, for each index i there exist constants $b_{i,j}$, $1 \leq j \leq M$, so that

$$\sum_{j=1}^{M} b_{i,j} w_k(x_j) = \lambda_i(w_k) = \begin{cases} 1, & \text{if } i = k \\ 0, & \text{otherwise} \end{cases} \qquad \text{and} \qquad \sum_{j=1}^{M} |b_{i,j}| \leq \frac{1}{c}.$$

Taking $(b_{i,1}, \ldots, b_{i,M})'$ as the i-th column of the matrix B_1 yields the desired result.

As far as the matrix B_2 is concerned, it suffices to show that

$$\inf\{\|A^* d\|_2; \; d \in \mathbb{C}^N \text{ and } \|d\|_2 = 1\} \geq c.$$

Then the smallest singular value of A is bounded below by the norming constant c, and the Moore-Penrose pseudoinverse of A can be chosen as a searched solution for B_2. The above relation can be shown as follows. For such a vector d we let $p_d = \sum_{i=1}^{N} d_i w_i \in W$. Clearly,

$$\|p_d\|_\infty \geq \langle p_d, p_d \rangle^{1/2} = \|d\|_2 = 1.$$

Since $A^* d = p_d|_X$ holds by the definition of A, the assertion follows from

$$\|A^* d\|_2 \geq \max_{x \in X} |p_d(x)| \geq c \cdot \max_{y \in S^{n-1}} |p_d(y)| = c\|p_d\|_\infty,$$

where the latter estimate is an immediate consequence of Proposition 2 (we choose $w^* = \delta_y$ at a point y where $|p_d(y)|$ is maximal). $\square$

Remark. The result of Theorem 3 can be viewed as a stability result for the sampling of an orthogonal basis at scattered data sites. The price to pay for placing the knots at arbitrary positions is that we have to take more nodes in X than given by the dimension of the space W, in general. However, this type of oversampling may not be needed all the time. The simplest case for this occurs for the circle S^1, if the nodes are equidistributed. Here, W is the space of trigonometric polynomials of degree ℓ and A is the square matrix of the discrete Fourier transform, $A = (\exp(ijk/(2\ell + 1)))_{j,k=-\ell,\ell}$. Its (unique) inverse $B = \frac{1}{2\ell+1} A^*$ has norms $\|B\|_1 = 1$ and $\|B\|_2 = 1/\sqrt{2\ell + 1}$. This simple example demonstrates, that the asymptotic behavior of the 1-norm is precisely described by the result of Theorem 3, but the bound for the 2-norm may be too pessimistic in certain cases.

Let us add in closing this section how the mesh norm of X can be associated with a space of spherical harmonics

$$W_L = \text{span} \{Y_{\ell,k}; \; 0 \leq \ell \leq L, \; 1 \leq k \leq N(n, \ell)\}$$

of maximal degree L. The following result [3, Prop. 2] is a direct consequence of Bernstein's inequality.

Proposition 4. *Let $X \subset S^{n-1}$ be a finite knot set with mesh norm $h(X) \leq \frac{1-c}{L}$, for some $0 < c < 1$ and $L \in \mathbb{N}$. Then the set $Z_X = \text{span}\{\delta_x; \ x \in X\}$ is a norming set of W_L with norming constant c.*

§3. Approximation Order by Positive Definite Kernels

A similar strategy as in [2] is employed in order to find bounds for the approximation order of spaces generated by the evaluation of a positive definite kernel on a finite knot set.

Let us begin with the definition of a positive definite kernel

$$\kappa(x, y) = \sum_{\ell=0}^{\infty} \alpha_\ell \sum_{k=1}^{N(n,\ell)} Y_{\ell,k}(x) Y_{\ell,k}(y), \qquad x, y \in S^{n-1},$$

with all coefficients $\alpha_\ell > 0$. We further assume that $\kappa \in C(S^{n-1} \times S^{n-1})$. The considered space V_X of approximants is defined as in (1). Just for sake of comparison to our previous results in [2], we mention that the interpolation methods employ the function space

$$\mathcal{H}_\kappa := \{f \in L_2(S^{n-1}); \ \|f\|_\kappa^2 = \sum_{\ell,k} \frac{|\widehat{f_{\ell k}}|^2}{\alpha_\ell} < \infty\},$$

which is called native space in a context of radial basis functions (adapted to the sphere).

Instead of introducing new function spaces and norms, we prefer to state our estimates in terms of the Fourier coefficients of the function f in (3). These can be further specialized according to the needs of the given situation. The subsequent corollaries provide an easy access for possible applications. The estimates become meaningless when either of the involved series diverges. We first deal with uniform error bounds, and then turn to L_2-error estimates.

3.1 Uniform Error Bounds

Suppose the function $f \in C(S^{n-1})$ is given, and assume that its Fourier expansion (3) converges uniformly. For a set X of nodes we choose $L = L(X) \in \mathbb{N}$ so that

$$\frac{1}{2(L+1)} < h(X) \leq \frac{1}{2L}. \tag{6}$$

In view of Proposition 4, this corresponds to $c = 1/2$ as a norming constant of the space Z_X with respect to W_L.

Theorem 5. *Under the above assumptions,*

$$\min_{v \in V_X} \|f - v\|_\infty \leq \sum_{\ell > L(X)} \sum_{k=1}^{N(n,\ell)} |\widehat{f_{\ell k}}| \sqrt{N(n,\ell)} + 2\|f\|_{L(X),1} \sum_{\ell > L(X)} \alpha_\ell N(n,\ell) \tag{7}$$

with

$$\|f\|_{L(X),1} = \sum_{\ell=0}^{L(X)} \sum_{k=1}^{N(n,\ell)} \frac{|\widehat{f_{\ell k}}|}{\alpha_\ell}.$$

Proof: Instead of minimizing over the space V_X, we choose a linear approximation process as follows:

$$f \mapsto w_d = \sum_{j=1}^{M} d_j \kappa(x_j, \cdot) \quad \text{with} \quad d = B_1 \left(\frac{\widehat{f_{\ell k}}}{\alpha_\ell} \right)_{0 \le \ell \le L, \; 1 \le k \le N(n,\ell)}, \qquad (8)$$

where B_1 is a matrix as in Theorem 3 (with the basis $\{w_1, \dots, w_N\}$ of $W = W_L$ given by $\{Y_{\ell,k}; 0 \le \ell \le L, 1 \le k \le N(n, \ell)\}$, ordered lexicographically with respect to (ℓ, k)). Then it is clear that

$$\sum_{j=1}^{M} d_j Y_{\ell,k}(x_j) = \frac{\widehat{f_{\ell k}}}{\alpha_\ell}, \qquad 0 \le \ell \le L, \; 1 \le k \le N(n, \ell),$$

i.e., the approximation process reproduces the polynomial part in W_L of the Fourier series of f, and the error reduces to

$$f(x) - w_d(x) = \sum_{\ell > L(X)} \sum_{k=1}^{N(n,\ell)} \left[\widehat{f_{\ell k}} - \alpha_\ell \sum_{j=1}^{M} d_j Y_{\ell,k}(x_j) \right] Y_{\ell,k}(x). \qquad (9)$$

Now we simply use the triangle inequality to obtain the estimate in (7). Concerning the first term we use the fact $\|Y_{\ell,k}\|_\infty \le \sqrt{N(n,\ell)}$. For the second term we apply the addition theorem of spherical harmonics (see [4, Thm. 2]), the expression of the surface area disappears according to our normalization in (4)) to yield

$$\left| \sum_{\ell > L(X)} \sum_{k=1}^{N(n,\ell)} \alpha_\ell \sum_{j=1}^{M} d_j Y_{\ell,k}(x_j) Y_{\ell,k}(x) \right|$$

$$\le \sum_{\ell > L(X)} \alpha_\ell \sum_{j=1}^{M} |d_j| \left| \sum_{k=1}^{N(n,\ell)} Y_{\ell,k}(x_j) Y_{\ell,k}(x) \right|$$

$$= \sum_{\ell > L(X)} \alpha_\ell N(n, \ell) \sum_{j=1}^{M} |d_j| \left| C_\ell^{(n-2)/2}(x_j \cdot x) \right|$$

$$\le \|d\|_1 \sum_{\ell > L(X)} \alpha_\ell N(n, \ell).$$

The notation C_ℓ^s refers to the corresponding Gegenbauer polynomial of degree ℓ normalized to $C_\ell^s(1) = 1$. Estimating $\|d\|_1$ according to the choice in (8), using Theorem 3 with $c = 1/2$ gives the final bound. $\square$

At first sight it seems as if the expression $\|f\|_{L(X),1}$ must stay bounded in order that a reasonable error estimate can be derived from Theorem 5. But imposing such a condition on f is not necessary. Indeed, the following result is a special case of Theorem 5, and its proof is quite elementary.

Corollary 6. *Let $(\beta_\ell)_{\ell \geq 0}$ be a positive and summable sequence so that the ratio β_ℓ / α_ℓ is monotonically increasing. Then the error estimate*

$$\min_{v \in V_X} \|f - v\|_\infty \leq 3\|f\|_{\beta,1} \sum_{\ell > L(X)} \beta_\ell N(n, \ell) \tag{10}$$

holds for any function f with $\|f\|_{\beta,1} := \sum_{\ell=0}^{\infty} \sum_{k=1}^{N(n,\ell)} \frac{|\widehat{f_{\ell k}}|}{\beta_\ell} < \infty$.

Proof: For the second expression in (7) we use

$$\|f\|_{L(X),1} \sum_{\ell > L(X)} \alpha_\ell N(n, \ell) = \sum_{\lambda=0}^{L(X)} \sum_{k=1}^{N(n,\lambda)} \frac{|\widehat{f_{\lambda k}}|}{\beta_\lambda} \frac{\beta_\lambda}{\alpha_\lambda} \sum_{\ell > L(X)} \frac{\alpha_\ell}{\beta_\ell} \beta_\ell N(n, \ell).$$

Now the assumptions on the sequence (β_ℓ) give

$$\frac{\beta_\lambda}{\alpha_\lambda} \frac{\alpha_\ell}{\beta_\ell} \leq 1 \quad \text{for all} \quad \lambda \leq \ell,$$

and this leads to the right-hand side of (10). The first expression in (7) is treated in a similar way. $\square$

The previous result shows that the general form of Theorem 5 leaves much freedom for obtaining error estimates for a whole range of function spaces. Without dwelling on this we only note that the case $\beta_\ell = \alpha_\ell$ requires f to be much smoother than a generic function in the corresponding native space (where $\beta_\ell \approx \sqrt{\alpha_\ell}$). On the other hand, the error bound in (10) is roughly the square of the bound stemming from the Golomb-Weinberger theory of interpolation (see [2, Theorem 2]).

3.2 L_2-error Estimates

A similar technique based on the orthogonality of the Fourier expansion yields the following. We modify the approximation process in the latter proof by replacing the matrix B_1 in (8) by the matrix B_2 of Theorem 3. Then Equation (9) persists to hold with the given vector d, hence

$$\|f - w_d\|_2^2 = \sum_{\ell > L(X)} \sum_{k=1}^{N(n,\ell)} \left| \widehat{f_{\ell k}} - \alpha_\ell \sum_{j=1}^{M} d_j Y_{\ell,k}(x_j) \right|^2. \tag{11}$$

A slightly more involved analysis than before leads to the following result.

Theorem 7. *Given $L(X)$ in (6) and $M = \#X$, for any $f \in L_2(S^{n-1})$, we have*

$$\min_{v \in V_X} \|f - v\|_2 \leq \left\{ 2 \sum_{\ell > L(X)} \sum_{k=1}^{N(n,\ell)} |\widehat{f_{\ell k}}|^2 + 8M\|f\|_{L(X),2}^2 \sum_{\ell > L(X)} \alpha_\ell^2 N(n, \ell) \right\}^{1/2} \tag{12}$$

with

$$\|f\|_{L(X),2} = \left\{ \sum_{\ell=0}^{L(X)} \sum_{k=1}^{N(n,\ell)} |\frac{\widehat{f_{\ell k}}}{\alpha_\ell}|^2 \right\}^{1/2}.$$

Here the number of nodes enters in (12) as a result of estimating the eigenvalues of an $M \times M$-matrix whose entries are certain values of the Gegenbauer polynomials. A related technique was used in our previous work [2, Theorem 2].

By an analogy with Corollary 6 we can derive the following.

Corollary 8. *Let* $(\beta_\ell)_{\ell \geq 0}$ *be a positive and summable sequence so that the ratio* β_ℓ/α_ℓ^2 *is monotonically increasing. Then the error estimate*

$$\min_{v \in V_X} \|f - v\|_2 \leq \|f\|_{\beta,2} \left\{ (8M + 2) \sum_{\ell > L(X)} \beta_\ell N(n,\ell) \right\}^{1/2}$$

holds for any function f *with* $\|f\|_{\beta,2} := \left\{ \sum_{\ell=0}^{\infty} \sum_{k=1}^{N(n,\ell)} \frac{|\widehat{f_{\ell k}}|^2}{\beta_\ell} \right\}^{1/2} < \infty.$

Acknowledgments. The second author received travel support from the Deutsche Forschungsgemeinschaft. The work of the third author was partially supported by Air Force Grant #F49620-98-1-0204.

References

1. Freeden, W., M. Schreiner, and R. Franke, A survey on spherical spline approximation, Surv. Math. Ind. **7** (1997), 29–85.

2. Jetter, K., J. Stöckler, and J. D. Ward, Scattered data interpolation on spheres, Math. Comp., to appear.

3. Jetter, K., J. Stöckler, and J. D. Ward, Norming sets and spherical cubature formulas, preprint.

4. Müller, C., *Spherical Harmonics*, Lecture Notes in Mathematics, vol. **17**, Springer-Verlag, Berlin-Heidelberg, 1966.

5. Narcowich, F. J., Generalized Hermite interpolation and positive definite kernels on a Riemannian manifold, J. Math. Analysis and Applications **190** (1995), 165–193.

6. Reimer, M., *Constructive Theory of Multivariate Functions*, B. I. Wissenschaftsverlag, Mannheim, 1990.

K. Jetter and J. Stöckler
Institut für Angewandte
Mathematik und Statistik
Universität Hohenheim
D-70593 Stuttgart, Germany
kjetter@uni-hohenheim.de
stockler@uni-hohenheim.de

J. D. Ward
Department of Mathematics
Texas A&M University
College Station, TX 77843
jward@math.tamu.edu

Radial Quasi Interpolation on S^2

A. K. Kushpel and J. Levesley

Abstract. In this paper we consider a simple method of radial quasi-interpolation by polynomials on S^2 and present rates of convergence for this method on a wide range of smooth functions.

§1. Harmonic Analysis and Sets of Smooth Functions on S^2

Let $\langle x, y \rangle = x_1 y_1 + x_2 y_2 + x_3 y_3$ be the usual scalar product in Euclidean 3–space $\mathbb{R}^3$, and S^2 be the 2-dimensional unit sphere in $\mathbb{R}^3$, i.e.,

$$S^2 = \{x \mid x \in \mathbb{R}^3, \ \langle x, x \rangle = 1\}.$$

Let $d\mu$ be the normalized rotation invariant measure on the sphere, and

$$\|\varphi\|_p = \begin{cases} (\int_{S^2} |\varphi(x)|^p d\mu(x))^{1/p}, & 1 \le p < \infty, \\ \operatorname{ess\,sup}\{|\varphi(x)| \mid x \in S^2\}, & p = \infty. \end{cases}$$

Let $L_p = \{\varphi \mid \|\varphi\|_p < \infty\}$, and $U_p = \{\varphi \mid \|\varphi\|_p \le 1\}$.

The space L_2 has the orthogonal decomposition

$$L_2 = \bigoplus_{k=0}^{\infty} H_k$$

where H_k is the space of spherical harmonic polynomials of degree k. It is known that H_k has dimension $2k+1$. Let $\{Y_{-k}^{(k)}, \ldots, Y_k^{(k)}\}$ be an orthonormal basis for H_k. For each $k \in N$, H_k is an eigenspace of the Laplace-Beltrami operator for the sphere, Δ, corresponding to the eigenvalue $\gamma_k = -k(k+1)$, i.e., $\Delta Y_i^{(k)} = \gamma_k Y_i^{(k)}$, $i = -k, \ldots, k$.

A function Z_η is zonal with respect to a pole $\eta \in S^2$ if it is invariant under the action of all rotations σ of S^2 which fix η, i.e., $Z_\eta(x) = Z_\eta(\sigma x)$ for all

Approximation Theory IX, Volume 2: Computational Aspects
Charles K. Chui and Larry L. Schumaker (eds.), pp. 145–152.

$x \in S^2$, and $\sigma \in SO(3)$ with $\sigma\eta = \eta$. Then $Z_\eta(x) = \tilde{Z}(\langle x, \eta \rangle)$ for some $\tilde{Z}$ defined on $[-1, 1]$.

Equipped with a zonal kernel we may define the convolution

$$(\varphi * Z_\eta)(x) = \int_{S^2} \varphi(y)\tilde{Z}(\langle x, y \rangle)d\mu(y) = \int_{S^2} \varphi(y)Z_\eta(\sigma^{-1}y)d\mu(\eta)$$

where $x = \sigma\eta$.

The real zonal polynomial

$$Z_\eta^{(k)}(x) = \sum_{m=-k}^{k} \overline{Y_m^{(k)}(\eta)}Y_m^{(k)}(x) \tag{1}$$

is a kernel for orthogonal projection onto H^k. Furthermore,

$$\|Z_\eta^{(k)}\|_2^2 = \sum_{m=-k}^{k} |Y_m^{(k)}(\eta)|^2 = 2k + 1, \ \forall \ \eta \in S^2.$$

The zonal harmonics have a simple expression in terms of the Legendre polynomials P_k, which can be defined in terms of generating function

$$\frac{1}{(1 - 2\rho t + \rho^2)^{1/2}} = \sum_{k=0}^{\infty} \rho^k P_k(t),$$

where $0 \le |\rho| < 1$ and $|t| \le 1$. It is known that

$$Z_\eta^{(k)}(x) = \tilde{Z}^{(k)}(\langle x, \eta \rangle) = (2k + 1)P_k(\cos\theta),$$

where $\cos\theta = \langle x, \eta \rangle$. In the sequel, where there is no possibility of confusion, we shall not make explicit reference to the pole η.

Each $\varphi \in L_1$ has a formal Fourier expansion in terms of complex spherical harmonics

$$\varphi \sim c_{0,1}(\varphi) + \sum_{k=1}^{\infty} \sum_{m=-k}^{k} c_{k,m}(\varphi)Y_m^{(k)}, \tag{2}$$

where the Fourier coefficients $c_{k,m}(\varphi)$ are given by

$$c_{k,m}(\varphi) = \int_{S^2} \varphi(z)\overline{Y_m^{(k)}(z)}d\mu(z).$$

A zonal function has a Fourier series in zonal polynomials

$$Z \sim c_0(Z) + \sum_{k=1}^{\infty} c_k(Z)Z^{(k)},$$

where

$$c_k(Z) = \frac{1}{2k+1} \int_{S^2} Z(z) Z^{(k)}(z) d\mu(z).$$

With the earlier definition of convolution we obtain the following familiar expression for the Fourier series of a convolution:

$$\varphi * Z \sim c_{0,1}(\varphi) c_0(Z) + \sum_{k=1}^{\infty} c_k(Z) \sum_{m=-k}^{k} c_{k,m}(\varphi) Y_m^{(k)}. \tag{3}$$

We also have Young's inequality

$$\|\varphi * Z\|_q \leq \|\varphi\|_p \|Z\|_r, \tag{4}$$

where $1 \leq p, q, r \leq \infty$, $1/q = 1/p + 1/r - 1$.

For more information on harmonic analysis on the sphere the reader should consult references [1, 2, 5, 7, 8, 10, 11, 12].

We shall introduce a wide range of smooth functions on a sphere in terms of multiplier operators, which, via (3), can often be realized as convolution operators. Given a sequence $\lambda = \{\lambda_k\}_{k \in \mathbb{N}}$, we shall say that the function f is in $\Lambda U_p \oplus \mathbb{R}$ if

$$f \sim c + \sum_{k=1}^{\infty} \lambda_k \sum_{m=-k}^{k} c_{k,m}(\varphi) Y_m^{(k)},$$

where $c \in \mathbb{R}$ and $\varphi \in U_p$. If the function $K \in L_1$ and

$$K \sim \sum_{k=1}^{\infty} \lambda_k Z^{(k)}$$

then the convolution $K * \varphi(x)$ is well defined and the function $f(x)$ can be written in the form

$$f(x) = K * \varphi(x) + c,$$

where $c \in \mathbb{R}$ is some constant. In this case we shall say $\phi \in K * U_p \oplus \mathbb{R}$.

We remark that the sets $\Lambda U_p \oplus \mathbb{R}$ are shift-invariant, i.e., for any $f \in \Lambda U_p \oplus \mathbb{R}$ and any $\sigma \in SO(3)$, we have

$$f_\sigma(x) = f(\sigma^{-1} x) \in \Lambda U_p \oplus \mathbb{R}.$$

§2. Error Estimates

In this section the letter C will denote a constant which will not necessarily have the same value at each occurrence.

As usual, we parametrize the points z on S^2 by their spherical coordinates $x = (\theta, \phi) \in [0, \pi] \times [0, 2\pi)$. In what follows we shall be concerned with approximation based on the equiangular grid points

$$\theta_s = \frac{\pi s}{2b}, \quad \phi_t = \frac{\pi t}{b}, \quad 0 \leq s \leq 2b - 1, \quad 0 \leq t \leq 2b - 1.$$

Let $x_{s,t} = (\theta_s, \phi_t)$.

For a continuous function f we consider the sequence of polynomials

$$T_b(f, x) = \sum_{k=0}^{b} \sum_{j=-k}^{k} \alpha_{k,j}(f) Y_j^{(k)}(x),$$

where

$$\alpha_{k,j}(f) = \frac{\sqrt{2\pi}}{2b} \sum_{s=0}^{2b-1} a_s^{(b)} \sum_{t=0}^{2b-1} \overline{Y_j^{(k)}(\theta_s, \phi_t)} f(\theta_s, \phi_t),$$

and

$$a_s^{(b)} := \frac{2^{3/2}}{b} \sin\left(\frac{\pi s}{b}\right) \sum_{l=0}^{b-1} \frac{1}{2l+1} \sin\left((2l+1)\frac{\pi s}{b}\right).$$

This is the form of the discrete Fourier series suggested by Driscoll and Healy [3].

We may rewrite T_b as a quasi-interpolant

$$T_b(f, z) = \sum_{s=0}^{2b-1} \sum_{t=0}^{2b-1} f(\theta_s, \phi_t) L_{s,t}^{(b)}(z),$$

where

$$L_{s,t}^{(b)}(z) = \frac{\sqrt{2\pi} a_s^{(b)}}{b} \sum_{k=0}^{b} \sum_{j=-k}^{k} \overline{Y_j^{(k)}(\theta_s, \phi_t)} Y_j^{(k)}(z)$$

$$= \frac{\sqrt{2\pi} a_s^{(b)}}{b} \sum_{k=0}^{b} Z_{x_{s,t}}(z)$$

by (1). Thus, the quasi-Lagrange function for $x_{s,t}$, $L_{s,t}$, is zonal about $x_{s,t}$. Using Szegö [10, 4.5.3] we have the following very simple form for the quasi-Lagrange function:

$$L_{s,t}^{(b)}(z) = \frac{\sqrt{\pi/2} a_s^{(b)}(b+1)}{b} P_b^{(1,0)}(\langle z, x_{s,t} \rangle), \tag{5}$$

where $P_b^{(1,0)}$ is the Jacobi polynomial with weight $(1-x)$.

We shall say that the multiplier sequence $\lambda = \{\lambda_k\}_{k \in \mathbf{N}} \in A_p$ if

$$\begin{aligned}
\lim_{k\to\infty} |\lambda_k| k^{1/2} &= 0, & p &< 4/3, \\
\lim_{k\to\infty} |\lambda_k| k^{1/2} (\ln n)^{3/4} &= 0, & p &= 4/3, \\
\lim_{k\to\infty} |\lambda_k| k^{2(1-1/p)} &= 0, & p &> 4/3.
\end{aligned}$$

If

$$K \sim \sum_{k=1}^{\infty} \lambda_k Z^{(k)}$$

for $\lambda \in A_p$, we shall (with a slight abuse of notation) say $K \in A_p$.

The main result of this article is

Theorem 1. *Let* $K(z)$ *be an integrable radial (zonal) function on* S^2 *with Fourier series*

$$K(z) \sim \sum_{k=1}^{\infty} \lambda_k Z^{(k)}(z),$$

where $\{\lambda_k\}_{k \in N} \in A_p$. *Then,*

$$\sup_{f \in K * U_p \oplus R} \|f(z) - T_{b-1}(f, z)\|_{\infty} \leq C b^{1/2} \sum_{k=b}^{\infty} |\Delta^2 \lambda_k| k^{1+2/p}, \quad b \to \infty.$$

Remark 1. By a proper choice of of the multiplier sequence $\{\lambda_k\}_{k \in \mathbf{N}}$ we obtain the following concrete examples:

1) Let

$$\lambda_k = (k(k+1))^{-\alpha/2}, \ \alpha > 0, \ k \in N,$$

then the resulting convolution class $K * U_p \oplus R$ is the Sobolev's class W_p^{α}; see [1, 4]. In this case we have

$$\sup_{f \in W_p^{\alpha}} \|f(z) - T_b(f, z)\|_{\infty} \leq C b^{-\alpha + 2/p + 1/2}.$$

2) The function $f(x)$, $x \in R^3$ is analytic in $x_0 \in R^3$ if there is a series

$$\sum_{\alpha} a_{\alpha} (x - x_0)^{\alpha}$$

which converges to $f(x)$ in some closed ball $|x - x_0| \leq \rho$, $\rho > 0$. It is known (see [9, p. 494]) that if $f(x)$ is analytic on the sphere S^2 then $|Y_k(f, x)| \leq K e^{-\nu k}$, where $Y_k(f, x) = Z^{(k)} * f(x)$, $k \in N$ is the spherical harmonic of f of degree k, $K > 0$ and $\nu > 0$ are some fixed constants. So, if we put

$$\lambda_k = e^{-\alpha k}, \ \alpha > 0, \ k \in N,$$

then the resulting convolution class $K * U_p \oplus R$ will be the set of analytic functions on S^2. In this situation we have

$$\sup_{f \in K * U_p \oplus R} \|f(z) - T_b(f, z)\|_{\infty} \leq C e^{-\alpha b} b^{3/2 + 2/p}$$

3) Let

$$\lambda_k = e^{-\alpha k^r}, \ \alpha > 0, \ r > 0, \ k \in \mathbf{N}.$$

Then the set $K * U_p \oplus R$ is the set of entire (if $r > 1$) or infinitely differentiable (if $0 < r < 1$) functions on S^2 and

$$\sup_{f \in K * U_p \oplus R} \|f(z) - T_b(f, z)\|_{\infty} \leq C e^{-\alpha b} b^{\min\{1, r\} + 1/2 + 2/p}.$$

In order to prove the main result of this paper we require several lemmas. The first which we state without proof bounds the discrete 1 norm by the continuous norm. Interested readers may find a proof of this result in [6].

Lemma 1. *For any zonal (radial) polynomial $P_b(z)$ of degree $\leq b$ on S^2, we have*

$$\sum_{0\leq s,t\leq 2b-1} \sin\theta_s |P_b(\theta_s,\phi_t)| < \frac{b(\pi(b+3)+2b+1)}{2\pi^3} \int_{S^2} d\mu(z)|P_b(z)|.$$

Let us define the Lebesgue function L_b for T_b by

$$L_b(\theta,\phi) := \sum_{0\leq s,t\leq 2b-1} \left| L_{s,t}^{(b)}(\theta,\phi) \right|,$$

The next statement gives us the order of growth of Lebesgue constants.

Lemma 2. *The Lebesgue constants of quasi interpolation satisfy*

$$\max_{z\in S^2} L_b(z) \leq Cb^{1/2}, \quad b\to\infty.$$

Proof: From (5) we have

$$L_b((z) = \sum_{1\leq s,t\leq 2b-1} \frac{\sqrt{\pi/2}a_s^{(b)}(b+1)}{b} |P_b^{(1,0)}(\langle z, x_{s,t}\rangle)|.$$

Noting that for all $0\leq s,t\leq 2b-1$, we have $b^{-1}a_s^{(b)} \leq Cb^{-2}\sin\theta_s$, $b\to\infty$, we see that

$$L_b((z) \leq C(b+1) \sum_{1\leq s,t\leq 2b-1} b^{-2}\sin\theta_s |P_b^{(1,0)}(\langle z, x_{s,t}\rangle)|.$$

Now, from Lemma 1 and the last inequality, we see that

$$\max_{z\in S^2} L_b(z) \leq Cb\|P_b^{(1,0)}(\langle\cdot,\eta\rangle)\|_1 \leq Cb^{1/2},$$

for any fixed $\eta \in S^2$, since $\|P_b^{(1,0)}(\langle\cdot,\eta\rangle)\|_1 \asymp b^{-1/2}$ (see [4]).

Lemma 3. *Let $K \in A_p$., Then there is a radial (zonal) polynomial $t_{b-1}(K,\cdot)$ such that*

$$\|K(\cdot) - t_{b-1}(K,\cdot)\|_p \leq C\sum_{k=b}^{\infty} |\Delta^2\lambda_k|k^{3-2/p}, \quad b\to\infty.$$

Proof: We will need some facts about Cesàro means. For the generating function $(1-\rho)^{-m-1}$ we have

$$\frac{1}{(1-\rho)^{m+1}} = \sum_{s=0}^{\infty} A_s^m \rho^n = \sum_{s=0}^{\infty} \frac{\Gamma(s+m+1)}{\Gamma(m+1)\Gamma(s+1)}\rho^s,$$

$$A_s^m \le C s^m, \quad s \to \infty.$$

Let

$$C_n^\delta = \frac{1}{A_n^\delta} \sum_{k=0}^{n} A_{n-k}^\delta Z^{(k)}$$

be the Cesàro kernel of order n and index δ. It is known (see [5]) that

$$\|C_n^\delta\|_p \le C \begin{cases} n^{1/2-\delta}, & p < 4/(3+2\delta), \\ n^{1/2-\delta}(\ln n)^{3/4}, & p = 4/(3+2\delta), \\ n^{2(1-1/p)}, & p > 4/(3+2\delta). \end{cases} \tag{11}$$

Applying the Abel transform we are get the following representation which is valid, in L_p, for any $K \in A_p$:

$$K = \sum_{k=1}^{\infty} \Delta^2 \lambda_k k C_k^1,$$

where $\Delta^2 \lambda_k := \lambda_k - 2\lambda_{k+1} + \lambda_{k+2}$.

As in [2, 4] let us consider the sequence of polynomials

$$t_b = \sum_{k=1}^{b} \Delta^2 \lambda_k k C_k^1.$$

Using the asymptotic estimates for Cesàro kernels (11) we can complete the proof, since

$$\|K(z) - \sum_{k=1}^{b-1} \Delta^2 \lambda_k k C_k^1(z)\|_p \le C$$

$$\sum_{k=b}^{\infty} |\Delta^2 \lambda_k| k^{3-2/p}. \quad \square$$

Corollary 1. *Let* $K \in A_r$ *where* $r = 1/(1 - (1/p - 1/q)_+)$. *Then there is a sequence of polynomials* $t_b(f)$ *such that*

$$\sup_{f \in \Lambda U_p \oplus R} \|f - t_{b-1}(f)\|_q \le C \sum_{k=b}^{\infty} |\Delta^2 \lambda_k| k^{3-2/r}, \quad b \to \infty.$$

Corollary 1 follows from the Young inequality (4) and Lemma 2.

Proof of Theorem 1.

Observing that the operator $T_b(f, \cdot)$ is a linear projector onto the space of polynomials of degree $\le b - 1$ (see [3]), we can apply Lebesgue inequality

$$|f(z) - T_b(f)(z)| \le E_{b-1}(f)(1 + L_b(z)), \quad z \in S^2, \tag{12}$$

where $f \in C(S^2)$, $E_{b-1}(f) = \inf\{\|f - t_{b-1}\|_\infty|\ t_{b-1} \in \mathcal{T}_{b-1}\}$ and $\mathcal{T}_{b-1}$ is the subspace of polynomials of degree $\le b-1$. The theorem follows via Corollary 1, Lemma 1 and (12). $\square$

References

1. Askey R. and S. Wainger, On the behavior of special classes of ultraspherical expansions–I, II, J. Anal. Math. **15** (1965), 193–244.

2. Bonami A., and J. Leclerk, Sommes de Cesàro et multiplicateurs des developments en harmoniques spheriques, Trans. Amer. Math. Soc. **183** (1973), 223–263.

3. Driscoll, J. R. and D. M. Healy, Computing Fourier transforms and convolutions on the 2-sphere, Adv. Appl. Math. **15** (1994), 202–250.

4. Kamzolov, A. I., On the best approximation of sets of function $W_p^\alpha(S^d)$ by polynomials, Mat. Zametki **32** (1982), 285–293 (in Russian).

5. Kogbetliantz, E., Recherches sur la sommabilite des series ultraspheriques par la methode des moyennes arithmetiques, J. Math. **3** (1924), 107–187.

6. Kushpel A. K. and J. Levesley, Quasi interpolation of the 2-sphere using zonal polynomials, Research Report 1998/1, Department of Mathematics and Computer Science, University of Leicester, Leicester LE1 7RH, UK.

7. Müller C., *Spherical Harmonics*, Springer-Verlag, Berlin, 1966.

8. Murnaghan, F. D., *The Unitary and Rotation Groups*, Spartan Books, Washington, D. C, 1962.

9. Sobolev, S. L., *Introduction to the Theory of Cubature Formulas*, Nauka, Moscow, 1974 (in Russian).

10. Stein, E. M. and G. Weiss, *Introduction to Fourier Analysis on Euclidean Spaces*, Princeton Univ. Press, 1971.

11. Szegö G., *Orthogonal Polynomials*, AMS, New York, 1939.

12. Vilenkin, N. J., *Special Functions and the Theory of Representation of Groups*, Nauka, Moscow, 1965 (in Russian).

13. Zygmund, A., *Trigonometric Series*, v.2, Cambridge University Press, 1959.

A. K. Kushpel
IMECC-UNICAMP
CAIXA Postal 6065
1381-970, Campinas SP
Brazil.
ak99@mcs.le.ac.uk

J. Levesley
Department of Mathematics and Computer Science
University of Leicester
Leicester LE1 7RH
UK
jl1@mcs.le.ac.uk

Bivariate Spline Method for
Navier-Stokes Equations:
Domain Decomposition Technique

Ming-Jun Lai and Paul Wenston

Abstract. We have used the bivariate spline method to numerically solve the steady state Navier-Stokes equations in the stream function formulation. Galerkin's method is applied to the resulting nonlinear fourth order equation, and Newton's iterative method is then used to solve the resulting nonlinear system. The steady state and time dependent Navier-Stokes equations were treated in [6] and [7], respectively. In this paper, we present a domain decomposition technique for solving this nonlinear system which has been implemented in MATLAB. Our numerical experiments show that the numerical solutions from the domain decomposition technique converge to the exact solution.

§1. Introduction

We are interested in solving the Navier-Stokes equations numerically. The well-known Navier-Stokes equations are as follows:

$$
\begin{cases}
-\nu\Delta\mathbf{u} + (\mathbf{u}\cdot\nabla)\mathbf{u} + \nabla p = \mathbf{f}, & (x,y)\in\Omega, \\
\operatorname{div}\mathbf{u} = 0, & (x,y)\in\Omega, \\
\mathbf{u} = \mathbf{g}, & (x,y)\in\partial\Omega,
\end{cases}
\tag{1.1}
$$

where Δ denotes the usual Laplacian operator and ∇ the gradient operator. We shall express the equations in the stream function formulation. Let φ be such that $\mathbf{u} = \operatorname{curl}\varphi$, i.e., $u_1 = \frac{\partial\varphi}{\partial y}$, $u_2 = -\frac{\partial\varphi}{\partial x}$. φ is called a stream function associated with the flow. By a simple calculation, the Navier-Stokes equations become the following fourth order nonlinear equation (cf. [3])

$$
\begin{cases}
\nu\Delta^2\varphi - \dfrac{\partial}{\partial y}\left(\dfrac{\partial\varphi}{\partial y}\dfrac{\partial^2\varphi}{\partial x\partial y} - \dfrac{\partial\varphi}{\partial x}\dfrac{\partial^2\varphi}{\partial y^2}\right) \\
\qquad - \dfrac{\partial}{\partial x}\left(\dfrac{\partial\varphi}{\partial y}\dfrac{\partial^2\varphi}{\partial x^2} - \dfrac{\partial\varphi}{\partial x}\dfrac{\partial^2\varphi}{\partial x\partial y}\right) = h, & \text{in } \Omega, \\
\dfrac{\partial\varphi}{\partial x} = -g_2, & \text{on } \partial\Omega, \\
\dfrac{\partial\varphi}{\partial y} = g_1, & \text{on } \partial\Omega.
\end{cases}
\tag{1.2}
$$

Approximation Theory IX, Volume 2: Computational Aspects
Charles K. Chui and Larry L. Schumaker (eds.), pp. 153–160.
Copyright © 1998 by Vanderbilt University Press, Nashville, TN.
ISBN 0-8265-1326-3.

If the nonlinear term above is omitted, we get the Stokes equations which is a biharmonic equation:

$$\begin{cases} \nu \Delta^2 \varphi = h, & \text{in } \Omega, \\ \frac{\partial \varphi}{\partial x} = -g_2, & \text{on } \partial\Omega, \\ \frac{\partial \varphi}{\partial y} = g_1, & \text{on } \partial\Omega. \end{cases} \tag{1.3}$$

Let $H^2(\Omega)$ be the usual Sobolev space of order 2 and $H_0^2(\Omega)$ be the subspace of $H^2(\Omega)$ of functions whose derivatives of order less than or equal to one, all vanish on the boundary $\partial\Omega$. Define the bilinear form $a_2(\varphi, \psi)$ and trilinear form $b(\theta, \varphi, \psi)$ by

$$a_2(\varphi, \psi) = \int_\Omega \Delta\varphi(x, y) \Delta\psi(x, y) dxdy$$

$$b(\theta, \varphi, \psi) = \int_\Omega \Delta\theta(x, y) \left(\frac{\partial \varphi}{\partial x} \frac{\partial \psi}{\partial y} - \frac{\partial \varphi}{\partial y} \frac{\partial \psi}{\partial x} \right) dxdy$$

and denote the $L_2(\Omega)$ inner product by

$$\langle h, \psi \rangle = \int_\Omega h(x, y) \psi(x, y) dxdy.$$

We say $\varphi \in H^2(\Omega)$ is a **weak solution** of the Stokes equations (1.3) if φ satisfies the following

$$\begin{cases} \nu a_2(\varphi, \psi) = \langle h, \psi \rangle, & \forall \psi \in H_0^2(\Omega), \\ \frac{\partial \varphi}{\partial x} = -g_2, & \text{on } \partial\Omega, \\ \frac{\partial \varphi}{\partial y} = g_1, & \text{on } \partial\Omega. \end{cases} \tag{1.4}$$

Similarly, a function $\varphi \in H^2(\Omega)$ is a weak solution of the Navier-Stokes equations (1.2) if φ satisfies

$$\begin{cases} \nu a_2(\varphi, \psi) + b(\varphi, \varphi, \psi) = \langle h, \psi \rangle, & \forall \psi \in H_0^2(\Omega), \\ \frac{\partial \varphi}{\partial x} = -g_2, & \text{on } \partial\Omega, \\ \frac{\partial \varphi}{\partial y} = g_1, & \text{on } \partial\Omega. \end{cases} \tag{1.5}$$

Such weak formulations are referred as the stream function formulation of the Stokes and Navier-Stokes equations, respectively. The reason for using the stream function formulation is that it leads to a smaller size nonlinear system than the traditional velocity-pressure formulation and vorticity-stream function formulation. (See, e.g., [3] and [6].)

We shall use the C^1 cubic spline space to solve these equations. Let $\Diamond$ be a quadrangulation of Ω which consists of non-degenerate convex quadrilaterals and $\oplus$ be the special triangulation obtained from $\Diamond$ by adding the two diagonals of each quadrilateral in $\Diamond$. See, e.g., Figure 1 for an example of a quadrangulation of an L-shape domain and the derived triangulation. Let

$$S_3^1(\oplus) = \{ s \in C^1(\Omega) : s \mid_t \in \mathbf{P}_3, \ \forall \, t \in \oplus \}$$

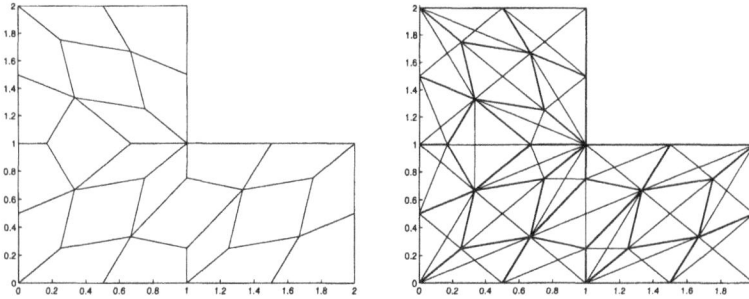

Fig. 1. Quadrangulation and Derived triangulation.

be the C^1 spline space over the triangulation $\diamondsuit$, where $\mathbf{P}_3$ is the space of all polynomials of total degree ≤ 3 and t denotes any triangle in $\diamondsuit$. Although this spline space is not new (it is the space of Fraeijs de Veubeke and Sander's finite elements which were introduced in the mid 60s (cf. [2] and [9])), this spline space has the smallest dimension among all C^1 finite element spaces and C^1 spline spaces. See [7] for a comparison of the dimensions of various C^1 finite element or spline spaces having the same approximation power and the same size of triangulations. For more accurate approximations, we should use the spline space $S_{3r}^r(\diamondsuit)$ for general r. The approximation properties of this space was studied in [4].

We remark here that the special triangulation $\diamondsuit$ is general enough to partition any given polygonal domains.

We have used the spline space to solve both the steady state and time dependent Navier-Stokes equations in the stream function formulation numerically. (See [6] and [7].) In these papers, we demonstrate that the bivariate spline method is very efficient and effective for numerically solving the Navier-Stokes equations.

When the domain is large or the refinement level is high, the system of equations may be too large to solve. In this case, a domain decomposition technique may be useful. In this paper, we shall present our analysis and numerical experiments with the domain decomposition technique for the numerical solution of the Navier-Stokes equations.

§2. Domain Decomposition Technique and Numerical Experiments

Let us first describe the domain decomposition technique for Stokes equations and for the Navier-Stokes equations. We first subdivide Ω into N subdomains Ω_1, Ω_2, ..., Ω_N such that $\Omega = \cup_{i=1}^{N}\Omega_i$. We assume that all subdomains are necessary in the sense that for each Ω_i, $\Omega_i \setminus \cup_{j \neq i} \Omega_j \neq \emptyset$ for $i = 1, \ldots, N$. We also need to assume that each Ω_i is the union of quadrilaterals of $\Diamond$. In addition, letting $\xi_i = \Omega_i \setminus \cup_{j \neq i} \Omega_j$ and defining $\gamma_i = \operatorname{dist}(\xi_i, \Omega \setminus \Omega_i)$, we assume that the overlapping size

$$\gamma = \min\{\gamma_i, i = 1, \cdots, N\} > 0. \tag{1}$$

This implies that each boundary segments of Ω_i, which is not part of $\partial\Omega$ must lie in the interior of another subdomain Ω_j.

Let $V = S_{3r}^r(\diamondsuit) \cap H_0^2(\Omega)$ and

$$V_i = \{s \in S_{3r}^r(\diamondsuit \cap \Omega_i) : D^\alpha s(v) = 0, |\alpha| \leq 1, \forall v \in \partial\Omega_i\}.$$

We may extend each element of $s \in V_i$ naturally to the whole domain Ω. We note that $V = \sum_{i=1}^{N} V_i$.

Schwarz's Domain Decomposition Method for the Navier-Stokes Equations: Starting with any initial guess $\varphi^{(0)}$ in $S_{3r}^r(\diamondsuit)$ which approximately satisfies the boundary condition, we construct a domain decomposition sequence $\{\varphi^{(n)}, n = 0, 1, \ldots,\}$ as follows: for $n = 0, 1, 2, \ldots$, and for $i = 0, \ldots, N - 1$,

$$\begin{cases} \nu a_2(\varphi^{(n+(i+1)/N)}, \psi) + b(\varphi^{(n+(i+1)/N)}, \varphi^{(n+(i+1)/N)}, \psi) \\ = \int_{\Omega_i} h\psi dxdy, \quad \forall \psi \in V_i, \\ D^\alpha(\varphi^{(n+(i+1)/N)} - \varphi^{(n+i/N)})|_{\partial\Omega_i} = 0, \quad |\alpha| \leq 1 \\ \varphi^{(n+(i+1)/N)}(x,y) = \varphi^{(n+i/N)}(x,y), \quad (x,y) \in \Omega\backslash\Omega_i. \end{cases} \tag{2}$$

When the nonlinear term above is omitted, we obtain the Schwarz domain decomposition method for Stokes equations. We now present several numerical experiments with the domain decomposition technique (2).

Example 1. Find $\varphi \in H^2(\Omega)$ such that

$$\begin{cases} \dfrac{\partial^4\varphi}{\partial x^4} + 2\dfrac{\partial^4\varphi}{\partial x^2\partial y^2} + \dfrac{\partial^4\varphi}{\partial y^4} = f, \quad (x,y) \in \Omega \\ \varphi = g_0, \dfrac{\partial\varphi}{\partial x} = g_1, \dfrac{\partial\varphi}{\partial y} = g_2, \quad (x,y) \in \partial\Omega, \end{cases}$$

where $f = 48$ and $\varphi(x,y) = x^4 + y^4$ on the boundary. Here, the domain Ω is the L-shape domain given in Figure 1. We divide Ω into two subdomains. See Figure 2 where the first subdomain consists of the quadrilaterals with integer 1 while the second subdomain consists of the quadrilaterals with integer 2.

The technique (2) without the nonlinear term is implemented. The maximum norm of the numerical solutions against the exact solutions is given below. The maximum norm is computed over 100×100 grid points over Ω.

Matrix Sizes	150×150	527×527	1971×1971
Errors	5.70×10^{-3}	4.89×10^{-4}	3.03×10^{-5}
Matrix Sizes	7619×7619	29955×29955	
Errors	1.815×10^{-6}	1.177×10^{-7}	

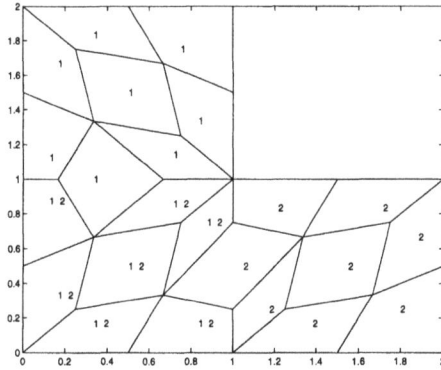

Fig. 2. Two Subdomains of the L-shape Domain.

Example 2. With L-shape domain Ω being divided as before, find $\varphi \in H^2(\Omega)$ such that

$$\begin{cases} \dfrac{\partial^4 \varphi}{\partial x^4} + 2\dfrac{\partial^4 \varphi}{\partial x^2 \partial y^2} + \dfrac{\partial^4 \varphi}{\partial y^4} = 1, & (x,y) \in \Omega \\ \varphi = 0, \dfrac{\partial \varphi}{\partial x} = 0, \dfrac{\partial \varphi}{\partial y} = 0, & (x,y) \in \partial\Omega. \end{cases}$$

We use both the direct method and the technique (2) to solve the above Stokes equations and compare their numerical solutions in the maximum norm. The maximum norm error is given below.

Matrix Sizes	150×150	527×527	1971×1971
Errors	7.58×10^{-7}	8.12×10^{-7}	1.23×10^{-6}
Matrix Sizes	7619×7619	29955×29955	
Errors	1.83×10^{-6}	2.44×10^{-6}	

Example 3. Find $\varphi \in H^2(\Omega)$ such that

$$\begin{cases} \nu \Delta^2 \varphi - \dfrac{\partial}{\partial y}\left(\dfrac{\partial \varphi}{\partial y}\dfrac{\partial^2 \varphi}{\partial x \partial y} - \dfrac{\partial \varphi}{\partial x}\dfrac{\partial^2 \varphi}{\partial y^2} \right) \\ \quad - \dfrac{\partial}{\partial x}\left(\dfrac{\partial \varphi}{\partial y}\dfrac{\partial^2 \varphi}{\partial x^2} - \dfrac{\partial \varphi}{\partial x}\dfrac{\partial^2 \varphi}{\partial x \partial y} \right) = f, & (x,y) \in \Omega \\ \varphi = g_0, \dfrac{\partial \varphi}{\partial x} = g_1, \dfrac{\partial \varphi}{\partial y} = g_2, & (x,y) \in \partial\Omega, \end{cases}$$

where f, g_0, g_1, g_2 are computed based on an artificial stream function $\varphi = x^4 + y^4$. Again Ω is the L-shape domain and is divided into two subdomains. We solve this problem by (2) and compare the numerical solution against the exact solution. The maximum norm errors are listed in the following table.

Matrix Sizes	150×150	527×527	1971×1971
Errors	5.3×10^{-3}	4.38×10^{-4}	3.08×10^{-5}

§3. Convergence Results

For the Stokes equations, we have the following preliminary results regarding the convergences of the technique (2).

Theorem 1. *Suppose that the size $|\Phi|$ is small enough and the subdomains satisfy* (1). *Then there exist constants K_1 and K_2 independent of the number N of subdomains and the size $|\Phi|$ such that for any $v \in V$, there exists a partition $v_i \in V_i, i = 1, \ldots, N$ with $v = v_1 + \ldots + v_N$ such that*

$$\sum_{i=1}^{N} a_2(v_i, v_i) \leq K_1 a_2(v, v).$$

Moreover, for any partition $v = v_1 + \cdots + v_N, v_i \in V_i$, we have

$$a_2(v, v) \leq K_2 \sum_{i=1}^{N} a_2(v_i, v_i).$$

We first note here that as illustrated by Figure 2, convergence may take place without the assumption (1).

Let $Q(f) = \sum_i \lambda_i(f)\phi_i$ be the quasi-interpolation operator given in [4]. Let $Q_0(f) = \sum_{\phi_i \in V} \lambda_i(f)\phi_i$ be an operator mapping $f \in H_0^r(\Omega)$ to V. It can be shown that $Q_0(f)$ has the same approximation power as $Q(f)$ (cf. [5]). When $|\Phi|$ is sufficiently small, we can find overlapping subdomains $\omega_i \subset \Omega_i$, $i = 1, \cdots, N$ such that $\cup_i \omega_i = \Omega$, their overlapping size is at least a half of γ and for each $f \in C_0^r(\omega_i), Q_0(f) \in V_i$ for all i's.

Let $\chi_i \in C_0^r(\omega_i), i = 1, \ldots, N$ be a partition of unity subordinate to ω_i satisfying

$$0 \leq \chi_i \leq 1, \sum_{k=1}^{N} \chi_k = 1, \text{ and } \|D_x^i D_y^j \chi_k\|_{L^\infty(\Omega)} \leq \frac{K_3}{(\gamma)^{i+j}}, \forall i + j \leq r.$$

For any $v \in V$, since $v = \sum_{k=1}^{N} \chi_k v$ and $v = Q_0(v)$, we let

$$v_k = Q_0(\chi_k v), \quad k = 1, \ldots, N.$$

Then we have $v_k \in V_k$ and $v = v_1 + \cdots + v_N$. Note that

$$\|D_x^i D_y^j Q_0(\chi_k v)\|_{L^2(\Omega)} = \|D_x^i D_y^j (Q_0(\chi_k v) - \chi_k v) + D_x^i D_y^j(\chi_k v)\|_{L^2(\Omega_k)}$$
$$\leq K|\Delta|^{r-i-j}|\chi_k v|_{r,\Omega_k} + \|D_x^i D_y^j(\chi_k v)\|_{L^2(\Omega_k)}.$$

By the well-known Leibniz differentiation formula, we can show that

$$\|v_k\|_{H^r(\Omega)}^2 = \sum_{\ell=0}^{r} \sum_{i+j=\ell} \|D_x^i D_y^j Q_0(\chi_k v)\|_{L^2(\Omega)}^2$$

$$\leq \sum_{\ell=0}^{r} 2 \sum_{i+j=\ell} \left(\frac{K_{i,j}}{(\gamma)^\ell}\right)^2 \|v\|_{H^r(\Omega_k)}^2 \leq \frac{K}{(\gamma)^{2r}} \|v\|_{H^r(\Omega_k)}^2$$

for a constant K independent of $|\Phi|$. Thus, we have

$$a(v_1, v_1) + \ldots + a(v_N, v_N) \leq K_3 \left[\|v_1\|_{H^r(\Omega)}^2 + \ldots + \|v_N\|_{H^r(\Omega)}^2\right]$$

$$\leq \frac{K_3 K}{(\gamma)^{2r}} \left[\|v\|_{H^r(\Omega_1)}^2 + \ldots + \|v\|_{H^r(\Omega_N)}^2\right] \leq \frac{\frac{K_3}{K_4} K}{(\gamma)^{2r}} N a(v, v).$$

This completes the proof with $K_1 = K_3 K N/(K_4 \gamma^{2r})$ for the first inequality. The second inequality may be proved easily using Cauchy-Schwarz's inequality. This completes the proof of Theorem 1.

With Theorem 1, we are able to prove the following convergence result by using the arguments in [8, 10], and [1]. We omit the details here.

Theorem 2. *Let $\varphi \in S_{3r}^r(\Phi)$ be the spline solution of Stokes' equations and $\varphi^{(n)}$ be the numerical solution obtained from (2) without the nonlinear term. Let $\mu = 1 - \dfrac{1}{K_1(1 + (N_0 K_2)^{1/2})^2}$. Then*

$$\|\varphi - \varphi^{(n)}\|_{H^2(\Omega)}^2 \leq C\mu^n \|\varphi - \varphi^{(0)}\|_{H^2(\Omega)}^2.$$

Here, N_0 denotes the maximum cardinality of the N index sets $\{j : \Omega_i \cap \Omega_j \neq \emptyset, 1 \leq j \leq N\}, i = 1, \cdots, N$.

Similar results for the Navier-Stokes equations are under further investigation. We now give numerical evidence of the convergence of the technique (2).

Example 3 (continued.) In Example 3, we computed twenty iterative solutions and used the last one as the approximate solution to compare with the exact solution. Let us list the maximum errors of each iterative solution (including the initial guess) against the exact solution. We also give the ratio of the ith error and $(i+1)^{\text{th}}$ error for $i = 0, \cdots, 12$.

Note from the Table below that after the ratios become 1, the maximum errors do not improve. That is, it reaches the maximum error of the bivariate spline solution obtained by a direct method over the given triangulation.

Iterative No.	527 × 527	ratios	1971 × 1971	ratios
Initial	16.4163		18.0035750	
1st	0.374027	43.890	0.3440805	52.32371
2nd	0.150826	2.4798	0.1567153	2.19557
3rd	0.058396	2.5828	0.0697791	2.24587
4th	0.022098	2.6425	0.0297737	2.34364
5th	0.008234	2.6838	0.0124528	2.39091
6th	0.002984	2.7588	0.0051630	2.41193

Iterative No.	527×527	ratios	1971×1971	ratios
7th	0.001053	2.8337	0.0021286	2.42552
8th	0.000438	2.4003	0.0008721	2.44073
9th	0.000438	1.0001	0.0003555	2.45317
10th			0.0001432	2.48101
11th			0.0000562	2.54689
12th			0.0000308	1.82347
13th			0.0000308	1.00006

References

1. Chan, T. F. and T. Mathew, Domain decomposition algorithm, Acta Numerica (1994), 61–143.

2. Fraeijs de Veubeke, B., Bending and stretching of plates, *Proc. Conf. Matrix Methods in Structure Mechanics*, Air Force Institute of Technology, Dayton, Ohio, 1965.

3. Girault, V. and P. A. Raviart, *Finite Element Methods for Navier-Stokes Equations*, Springer-Verlag, 1986.

4. Lai, M. J. and L. L. Schumaker, On the approximation power of splines on triangulated quadrangulations, SIAM Numer. Anal., to appear.

5. Lai, M. J. and P. Wenston, On Schwarz's domain decomposition methods for elliptic boundary value problems, submitted.

6. Lai, M. J. and P. Wenston, Bivariate spline method for numerical solution of steady state Navier-Stokes equations over polygons in stream function formulation, submitted.

7. Lai, M. J., C. Liu, and P. Wenston, Bivariate spline method for numerical solution of time evolution Navier-Stokes equations over polygons in stream function formulation, submitted.

8. Lions, P. L., On the Schwarz alternating method, I., in *Domain Decomposition Methods for Partial Differential Equations*, Glowinski, Golub, Meurant, and Periaux (eds.), SIAM Publications, 1988, pp. 1–42.

9. Sander, G., Bornes supérieures et inférieures dans l'analyse matricielle des plaques en flexion-torsion, Bull. Soc. r. Sci. Liege **33** (1964), 456–494.

10. Xu, J., Iterative methods by space decomposition and subspace corrections, SIAM Review **34** (1992), 581–613.

Ming-Jun Lai and Paul Wenston
Department of Mathematics
University of Georgia
Athens, GA 30602
mjlai@math.uga.edu, paul@math.uga.edu

Divergence-Free Multiwavelets

Joseph D. Lakey, Peter R. Massopust, and Maria C. Pereyra

Abstract. In this paper we construct $\mathbb{R}^n$-valued biorthogonal, compactly supported multiwavelet families such that one of the biorthogonal pairs consists of divergence-free vector wavelets. The construction is based largely on Lemarié's idea of multiresolution analyses intertwined by differentiation. We show that this technique extends nontrivially to multiwavelets via Strela's two-scale transform. An example based on the Donovan-Geronimo-Hardin-Massopust (DGHM) multiwavelets is given.

§1. Introduction

The study of divergence-free wavelets originated in the early 1990s with two completely different constructions: one due to Battle and Federbush [1], the other due to Lemarié [7]. The Battle-Federbush machine built interscale orthogonal wavelets. Later refinements gave orthogonal wavelets in dimensions 1, 2, 4, 8. None of these wavelet families have compact support. The Lemarié wavelets, on the other hand, are biorthogonal and compactly supported. Only one of the two biorthogonal families is divergence-free. His construction hinged on an important realization that multiresolution analyses (MRA's) could be related to one-another by differentiation, which is why biorthogonality is required (cf. [6]).

Federbush [3] used divergence-free wavelets to study uniqueness for the Navier-Stokes equations (NSE). Numerical analysis of NSE using variations of Lemarié wavelets has been implemented by K. Urban, et al. [9]. Other wavelet-Galerkin approaches to NSE do not use divergence-free wavelets, cf. Glowinski [5].

It is our thesis that, if wavelets are to have an impact on the numerical analysis of NSE, the wavelets should be divergence-free; in addition, they should have the best possible tradeoff between localization and approximation. This is a hypothesis at this stage: testing it will be the subject of future efforts. The burden of the present paper is to show how to construct adequate wavelets. The procedure is simple enough. We follow Lemarié's method for the most part, but we do so using multiwavelets. The tool for relating biorthogonal

Approximation Theory IX, Volume 2: Computational Aspects   161
Charles K. Chui and Larry L. Schumaker (eds.), pp. 161–168.
Copyright © 1998 by Vanderbilt University Press, Nashville, TN.
ISBN 0-8265-1326-3.

multi-MRA's (MMRA's) by differentiation is a result in Strela's thesis [8]. We build in the divergence-free property starting from the biorthogonal MMRA's. The biorthogonal filters associated to the DGHM [2] wavelets are recorded.

§2. MMRA's and Biorthogonality

First we need some handy, though abusive, notation. We will think of a real-valued multi-scaling function $\Phi(x)$ as a vector $[\phi^1(x)\ldots\phi^m(x)] : \mathbb{R} \to \mathbb{R}^m$. Translation and dilation are performed componentwise. Thus $\Phi_{jk}(x) = [\phi^1_{jk},\ldots,\phi^m_{jk}]$, $\phi^i_{jk} = 2^{j/2}\phi^i(2^j x - k)$. When we write a wavelet or multiresolution expansion, for example, when we write the projection of $f : \mathbb{R} \to \mathbb{R}$ onto V_0, the l^2-closed linear span of the translates $\Phi_k(x) = \Phi(x - k)$ of Φ, as

$$P_{V_0}f(x) = \sum_{k\in\mathbb{Z}}\langle f,\Phi_k\rangle\Phi_k(x), \tag{1}$$

we really mean the integral of $f(y)$ against the kernel

$$K_0(x,y) = \sum_{k\in\mathbb{Z}}\Phi_k(x)\cdot\Phi_k(y), \tag{2}$$

where the dot product is taken in $\mathbb{R}^m$. In particular, we will always interpret quantities like the right side of (1) as being scalar-valued.

Since much of our analysis is done at the filter level, we recall here the basic properties of scaling filters. The fact that Φ is an $\mathbb{R}^m$-valued scaling function means that there is a sequence $\{C_k\} \subset \mathcal{M}_{m\times m}(\mathbb{R})$ such that

$$\Phi(x) = \sum_{k\in\mathbb{Z}}C_k\Phi(2x - k) \tag{3}$$

which, upon taking Fourier transforms of both sides, can be rewritten

$$\hat{\Phi}(2\xi) = \frac{1}{2}\sum_k C_k e^{-2\pi ik\xi}\hat{\Phi}(\xi) = H(\xi)\hat{\Phi}(\xi). \tag{4}$$

It will be convenient to make the substitution $z = e^{-2\pi i\xi}$ so that the low pass filter $H(z) = H(\xi)$ is a matrix-valued Laurent series, or Laurent polynomial in the compactly supported case. If $H^*(\xi)$ denotes the conjugate transpose of $H(\xi)$ then, in terms of z, we have $H^*(z) = \sum_k C_k^* z^{-k}$ where C_k^* is the transpose of C_k. We will denote by the high pass filter $F(\xi)$ a matrix valued trigonometric series such that $\hat{\Psi}(2\xi) = F(\xi)\hat{\Phi}(\xi)$ defines a multiwavelet Ψ. For the sake of brevity we will not be pedantic about what conditions on H are needed to guarantee convergence/regularity of Φ, nor about issues in the design of F in terms of H; there is ample description in the literature. In any case, we shall assume throughout that Φ has enough regularity that convergence of both sides of any identity can be justified. Instead, we will focus on the biorthogonality conditions. In what follows, we will consider biorthogonal MMRA's having scaling functions Φ_+, Φ_- that come from smoothing or roughening the scaling function Φ from a fixed orthogonal MMRA. One could equally well derive new biorthogonal MMRA's starting from fixed biorthogonal MMRA's. The *conditions of biorthogonality* for the derived MMRA's are then (cf. [8])

$$
\begin{aligned}
H_+(z)H_-^*(z) + H_+(-z)H_-^*(-z) &= I \\
F_+(z)F_-^*(z) + F_+(-z)F_-^*(-z) &= I \\
H_+(z)F_-^*(z) + H_+(-z)F_-^*(-z) &= 0 \\
H_-(z)F_+^*(z) + H_-(-z)F_+^*(-z) &= 0.
\end{aligned}
\tag{5}
$$

The Fourier transforms of the biorthogonal scaling functions and wavelets, respectively, are then given by

$$
\hat{\Phi}_\pm(2\xi) = H_\pm(\xi)\hat{\Phi}_\pm(\xi); \qquad \hat{\Psi}_\pm(2\xi) = F_\pm(\xi)\hat{\Phi}_\pm(\xi).
\tag{6}
$$

In the case $H_+ = H_-$ and $F_+ = F_-$ these are the conditions of orthogonality.

§3. Two-Scale Transform and Differentiation

To relate a new scaling filter to an old one, Strela invented the two-scale transform. This means that one starts with a transition matrix $M(z) = M(\xi)$, $z = e^{-2\pi i\xi}$ and defines

$$
\begin{aligned}
H(z) &= \frac{1}{2}M(z^2)H_-(z)M^{-1}(z) \\
H_+(z) &= \frac{1}{2}M^*(z^2)H(z)M^{*-1}(z)
\end{aligned}
\tag{7}
$$

provided M^{-1} exists. If H satisfies the orthogonality condition $HH^*(z) + HH^*(-z) = I$, then since

$$
\begin{aligned}
H_+H_-^*(z) &= \left[\frac{1}{2}M^*(z^2)H(z)M^{*-1}(z)\right][2M^{-1}(z^2)H(z)M(z)]^* \\
&= M^*(z^2)HH^*(z)M^{*-1}(z^2),
\end{aligned}
$$

it follows that

$$
H_+H_-^*(z) + H_+H_-^*(-z) = I.
$$

Therefore, H_+ and H_- will give rise to a new pair of biorthogonal MMRA's under suitable convergence conditions. These and the role of differentiation are provided by the following proposition due to Strela [8]:

Proposition 1. *Let Φ be a scaling vector with filter H providing approximation order at least one. If $M(\xi)$ is invertible for all $\xi \neq 0$ ($z \neq 1$), $M(0)$ has left eigenvector $\hat{\Phi}(0)$ corresponding to the simple eigenvalue $\lambda(0) = 0$, and $\frac{d}{dz}\det M(z)|_{z=1} \neq 0$ then*

$$
\hat{\Phi}_+(\xi) = \frac{1}{2\pi i\xi}M^*(\xi)\hat{\Phi}(\xi)
$$

solves the scaling equation

$$\hat{\Phi}_+(2\xi) = H_+(\xi)\hat{\Phi}_+(\xi)$$

with H_+ defined by (7). Similarly, $2\pi i \xi \hat{\Phi}(\xi) = M(\xi)\hat{\Phi}_-(\xi)$.

Strictly speaking, some technical regularity conditions, which are met in the example of Section 6, should be imposed to insure convergence of $\Phi_\pm$. Alternatively, one can take convergence in the sense of distributions. Since $M^*(\xi)$ is a matrix trigonometric series, its inverse Fourier transform is a matrix T^* of generalized translation operators. The equations relating $\hat{\Phi}_\pm$ to $\hat{\Phi}$ become $D\Phi_+ = -T^*\Phi$ and $D\Phi = -T\Phi_-$ where $D = d/dx$ is taken componentwise. Then $D\Phi_+ \in V_0$ provided the coefficients of T^* are in $l^2(\mathbb{Z})$.

Given these scaling functions, how should the wavelets be designed? Typically one can choose the wavelets to satisfy as many desirable properties, such as symmetry or short support, as possible. In the present setting, the simplest high-pass filters are

$$F(z) = \frac{1}{2}F_-(z)M^{-1}(z)$$

$$F_+(z) = \frac{1}{2}F(z)M^{*-1}(z). \tag{8}$$

Lemma 2. *If the filters H, F satisfy the conditions of orthogonality then the filters $H_\pm, F_\pm$ satisfy the conditions (5) of biorthogonality.*

This is easy to see. For example,

$$F_+(z)F_-^*(z) = \left[\frac{1}{2}F(z)M^{*-1}(z)\right][2F(z)M(z)]^* = F(z)F^*(z)$$

so that, if $FF^*(z) + FF^*(-z) = I$ then $F_+F_-^*(z) + F_+F_-^*(-z) = I$ as well. The corresponding multiwavelets are related by differentiation in the simplest way possible.

Lemma 3. *The multiwavelets $\Psi_\pm$ satisfy*

$$D\Psi_+ = -\Psi$$

$$D\Psi = -\Psi_-.$$

Proof: In terms of Fourier transforms we have:

$$\hat{\Psi}_+(2\xi) = F_+(\xi)\hat{\Phi}_+(\xi) = \frac{1}{2\pi i\xi}F_+(\xi)M^*(\xi)\hat{\Phi}(\xi)$$

$$= \frac{1}{2\pi i\xi}\frac{1}{2}F(\xi)M^{*-1}(\xi)M^*(\xi)F^{-1}(\xi)\hat{\Psi}(2\xi) = \frac{1}{2\pi i\xi}\frac{1}{2}\hat{\Psi}(2\xi).$$

Substituting ξ for 2ξ and inverse Fourier transforming shows that $D\Psi_+ = -\Psi$. The proof that $D\Psi = -\Psi_-$ is similar. $\square$

§4. Differentiation and Multiresolution Expansions

To check that differentiation intertwines wavelet expansions let $\Phi, \Phi_\pm$ and $\Psi, \Psi_\pm$ be related by (7) and (8). Let V_0 and W_0 be the multiresolution and wavelet spaces spanned by the translates of Φ and Ψ, and let $V_{0,+}$ denote the span of translates of Φ_+, and $W_{0,+}$ the span of $\Psi_{+,k}$ in the sense of (1). The MRA properties imply that $V_{1,+} = V_{0,+} + W_{0,+}$, respecting biorthogonality with $V_{1,-} = V_{0,-} + W_{0,-}$, and replicable at any scale. In one sense, it is obvious that the MRA expansions commute with differentiation.

Lemma 4. *If $f \in V_{0,+}$ and $g \in W_{0,+}$ then $Df \in V_0$, $Dg \in W_0$, and*

$$Df = \sum_{k \in \mathbb{Z}} \langle f, \Phi_{-,k} \rangle D\Phi_{+,k} = \sum_{k \in \mathbb{Z}} \langle Df, \Phi_k \rangle \Phi_k$$

$$Dg = \sum_{k \in \mathbb{Z}} \langle g, \Psi_{-,k} \rangle D\Psi_{+,k} = \sum_{k \in \mathbb{Z}} \langle Dg, \Psi_k \rangle \Psi_k.$$

As always, we assume that convergence is no problem. For example, if Φ has compact support and any Hölder regularity, then both sides of both identities will converge absolutely. Though the identities are obvious, the fact that they are related through summation by parts will be important below. The W_0 case is trivial, so we demonstrate the V_0 case. Since $D\Phi_+ = -T^*\Phi$, and the same holds for the translates Φ_k, we have

$$D \sum_k \langle f, \Phi_{-,k} \rangle \Phi_{+,k} = \sum_k \langle f, \Phi_{-,k} \rangle D\Phi_{+,k} = -\sum_k \langle f, \Phi_{-,k} \rangle T^*\Phi_k$$

$$= \sum_k \langle f, T\Phi_{-,k} \rangle \Phi_k = \sum_k \langle Df, \Phi_k \rangle \Phi_k.$$

The third equation is nothing but summation by parts and change of variables in the integrals defining the scaling coefficients.

§5. Form-Valued Wavelets

For multivariate expansions we need to take tensor-products of univariate wavelets. We follow the construction in [4] (cf. [9]) now to create vector-valued biorthogonal wavelet families. Then we will build in the divergence-free condition. Even if one is solely interested in $\mathbb{R}^n$-valued wavelets, the project is carried out most naturally at the level of r-forms.

To do this we need some notation. As usual, $\epsilon \in E^*$ means that $\epsilon = (\epsilon_1, \ldots, \epsilon_n) \in \{0,1\}^n \setminus (0,0)$. Set $i_\epsilon = \min\{i : \epsilon_i = 1\}$. Functions $f = \sum_{\alpha \in \mathcal{P}_r} f_\alpha e_\alpha$ where e_α is a basis for the r-th exterior product of $\mathbb{R}^n$ indexed by subsets $\alpha \in \mathcal{P}_r = \{1 \leq i_1 < i_2 < \ldots < i_r \leq n\}$ are r *forms*. If $j \in \alpha$ then $\alpha \setminus j \in \mathcal{P}_{r-1}$ is obtained by deleting j from α. For any fixed $\epsilon \in E^*$, set

$$\mathcal{T}_r = \mathcal{T}_r^\epsilon = \{\alpha \in \mathcal{P}_r : i_\epsilon \notin \alpha\}; \qquad \mathcal{N}_r = \mathcal{P}_r \setminus \mathcal{T}_r.$$

The decomposition $f = \sum_{\alpha \in \mathcal{T}_r} f_\alpha e_\alpha + \sum_{\alpha \in \mathcal{N}_r} f_\alpha e_\alpha = f_\mathcal{T} + f_\mathcal{N}$ is analogous to the Fourier decomposition of any form f into its 'tangential' and 'normal' components f_T and f_N: there f_T is the 'divergence-free' component; for us it will be a modification of f_T.

To get biorthogonal bases for $L^2(\mathbb{R}^n)$ we use tensor products. In each component we use either the original basis generated by Ψ or the derived bases generated by $\Psi_\pm$, depending on the *signature* (ϵ, α). Precisely, define

$$\Gamma_{\epsilon,\alpha}(x) = \prod_{j \in \alpha}(\epsilon_j \Psi_+(x_j) + (1 - \epsilon_j)\Phi_+(x_j)) \prod_{j \notin \alpha}(\epsilon_j \Psi(x_j) + (1 - \epsilon_j)\Phi(x_j))$$

$$\Theta_{\epsilon,\alpha}(x) = \prod_{j \in \alpha}(\epsilon_j \Psi_-(x_j) + (1 - \epsilon_j)\Phi_-(x_j)) \prod_{j \notin \alpha}(\epsilon_j \Psi(x_j) + (1 - \epsilon_j)\Phi(x_j)).$$

$$(9)$$

These functions, along with their translates and dilates, are biorthogonal to one another. Furthermore, for each fixed $\alpha \in \mathcal{P}_r$, the families $\{\Gamma_{\epsilon,\alpha}, \Theta_{\epsilon,\alpha}\}_{\epsilon \in E^*}$ give rise to biorthogonal bases for $L^2(\mathbb{R}^n)$.

To build bases for $L^2(\mathbb{R}^n, \Lambda_r(\mathbb{R}^n))$ we simply set

$$\Gamma^{\epsilon,\alpha} = \Gamma_{\epsilon,\alpha} e_\alpha; \qquad \Theta^{\epsilon,\alpha} = \Theta_{\epsilon,\alpha} e_\alpha.$$

Since $\{e_\alpha\}_{\alpha \in \mathcal{P}_r}$ forms an orthonormal basis for the Hilbert space $\Lambda_r(\mathbb{R}^n)$, the translates and dilates of $\{\Gamma^{\epsilon,\alpha}, \Theta^{\epsilon,\alpha}\}_{\epsilon \in E^*, \alpha \in \mathcal{P}_r}$ form biorthogonal bases for $L^2(\mathbb{R}^n, \Lambda_r(\mathbb{R}^n))$. It remains to construct a basis for the Hodge projection onto the 'divergence-free' r-forms. This is where $\mathcal{T}_r$ and $\mathcal{N}_r$ come into play.

For each fixed ϵ and $\alpha \in \mathcal{T}_r^\epsilon = \mathcal{T}_r$, set $a = \alpha \cup i_\epsilon$ and set $\Psi^{\epsilon,\alpha} = -\text{sgn}(i_\epsilon, a)d^*\Gamma^{\epsilon,a}$. The deRham operator $d^* = \sum_{i=1}^n \mu_i^* \partial/\partial x_i$. On vector fields $(r = 1)$ d^* is the divergence operator. Here, μ_i^* is the interior multiplication operator $\mu_i^* e_\alpha = \text{sgn}(i, \alpha)e_{\alpha \setminus i}$ where $\text{sgn}(i, \alpha)$ is 0 if $i \notin \alpha$ and ± 1 depending on whether i marks an odd/even place when the elements of α are listed in increasing order. One has

$$\Psi^{\epsilon,\alpha} = \Gamma^{\epsilon,\alpha} + \text{sgn}(i_\epsilon, a)\sum_{j \in \alpha}\text{sgn}(j, a)\Gamma^{\epsilon, a \setminus j}. \qquad (10)$$

The pairs $\{(\Psi^{\epsilon,\alpha}, \Theta^{\epsilon',\alpha'})\}_{\alpha \in \mathcal{T}_r}$ are thus biorthogonal because $(\Gamma^{\epsilon,\alpha}, \Theta^{\epsilon',\alpha'})$ are and if $\alpha \in \mathcal{T}_r$ and $j \in \alpha$ then $a \setminus j \in \mathcal{N}_r$.

Since the distributional image of d^* on $r + 1$-forms equals the distributional kernel of d^* on r-forms, the $\Psi^{\epsilon,\alpha}$ ($\alpha \in \mathcal{T}_r$) are divergence-free in the sense that $d^*\Psi^{\epsilon,\alpha} = 0$. Therefore, the $\Psi^{\epsilon,\alpha}$ will form a basis for the divergence-free subspace of $L^2(\mathbb{R}^n, \Lambda_r)$ provided they are complete for this subspace. In turn, it suffices that the expansion of any divergence-free f in terms of $\{(\Psi^{\epsilon,\alpha}, \Theta^{\epsilon,\alpha})\}_{\alpha \in \mathcal{T}_r}$ agrees with its expansion in terms of $\{(\Gamma^{\epsilon,\beta}, \Theta^{\epsilon,\beta})\}_{\beta \in \mathcal{P}_r}$. By rescaling, this is true if both expansions agree at unit scale when f is in the kernel of d^*. First, because of Lemma 4,

$$\mu_{i_\epsilon}^* \frac{\partial}{\partial x_{i_\epsilon}}\left(\sum_{\beta \in \mathcal{P}_r}\sum_{k \in \mathbb{Z}^n}\langle f, \Theta_k^{\epsilon,\beta}\rangle\Gamma_k^{\epsilon,\beta}\right) = \sum_{\delta \in \mathcal{T}_{r-1}}\sum_k\left\langle\mu_{i_\epsilon}^*\frac{\partial}{\partial x_{i_\epsilon}}f, \Theta_k^{\epsilon,\delta}\right\rangle\Gamma_k^{\epsilon,\delta}. \qquad (11)$$

On the other hand, if $\alpha \in \mathcal{T}_r$ then

$$\sum_k \langle f, \Theta_k^{\epsilon,\alpha}\rangle \Psi_k^{\epsilon,\alpha} = \sum_k \langle f, \Theta_k^{\epsilon,\alpha}\rangle \Gamma_k^{\epsilon,\alpha} \pm \sum_{j\in\alpha} \mathrm{sgn}(j,a) \sum_k \langle \mu_j^* \frac{\partial}{\partial x_j} f, \Theta_k^{\epsilon,\alpha\backslash j}\rangle \Gamma_k^{\epsilon,a\backslash j}.$$

Consequently we obtain

$$\mu_{i_\epsilon}^* \frac{\partial}{\partial x_{i_\epsilon}} \left(\sum_{k\in\mathbb{Z}^n} \langle f, \Theta_k^{\epsilon,\alpha}\rangle \Psi_k^{\epsilon,\alpha} \right) = -\sum_{j\in\alpha} \sum_k \langle \mu_j^* \frac{\partial}{\partial x_j} f, \Theta_k^{\epsilon,\alpha\backslash j}\rangle \Gamma_k^{\epsilon,a\backslash j}. \tag{12}$$

Summing (12) over all α and comparing to (11) proves

Theorem 5. *If* $f \in L^2(\mathbb{R}^n, \Lambda_r(\mathbb{R}^n))$ *satisfies* $d^* f = 0$ *then, for* $\epsilon \in E^*$,

$$\sum_{\beta\in\mathcal{P}_r} \sum_{k\in\mathbb{Z}^n} \langle f, \Theta_k^{\epsilon,\beta}\rangle \Gamma_k^{\epsilon,\beta} = \sum_{\alpha\in\mathcal{T}_r} \sum_{k\in\mathbb{Z}^n} \langle f, \Theta_k^{\epsilon,\alpha}\rangle \Psi_k^{\epsilon,\alpha}.$$

Similar identities hold for every scale. Thus the translates and dilates of $\Psi^{\epsilon,\alpha}$
($\epsilon \in E^*, \alpha \in \mathcal{T}_r$) form a basis for the divergence free subspace.

§6. The Case of DGHM Multiwavelets

The scaling filter $H(z)$ in (4) for the DGHM orthogonal multiwavelets is

$$H(z) = \frac{1}{20} \begin{bmatrix} 6(1+z) & 8\sqrt{2} \\ (-1+9z+9z^2-z^3)/\sqrt{2} & -3+10z-3z^2 \end{bmatrix}.$$

It is easily checked that, together with the high pass filter

$$F(z) = \frac{1}{20} \begin{bmatrix} (-1+9z+9z^2-z^3)/\sqrt{2} & -3-10z-3z^2 \\ (-1+9z-9z^2+z^3) & 6(z^2-1)/\sqrt{2} \end{bmatrix},$$

the conditions of orthogonality hold. The wavelets give approximation order two.

Now it is a simple matter [8] to check that the matrix

$$M(z) = \begin{bmatrix} 0 & \sqrt{2} \\ 1-z & -1-z \end{bmatrix}$$

satisfies the hypotheses of Proposition 1 for H. Then by (7), $H_+(z)$ is

$$\frac{1}{80z^2} \begin{bmatrix} z-1+40z^2+40z^3+z^4-z^5 & (1-z)(1+z)^2(1-10z+z^2) \\ (z-1)(1-26z^2+z^4) & 1-9z+4z^2+4z^3-9z^4+z^5 \end{bmatrix},$$

whereas

$$H_-(z) = \frac{1}{10} \begin{bmatrix} 5(1+z) & 0 \\ 8(1-z) & -2(1+z) \end{bmatrix}.$$

The corresponding high pass filters are

$$F_+(z) = \frac{1}{80} \begin{bmatrix} 1-z+z^2-z^3 & -1+9z+9z^2-z^3 \\ \sqrt{2}(1-z-z^2+z^3) & \sqrt{2}(-1+9z-9z^2+z^3) \end{bmatrix},$$

whereas

$$F_-(z) = \frac{1}{10} \begin{bmatrix} (-1+z)(3+10z+3z^2) & 2(1+z)(1+10z+z^2) \\ -3(1-z^2)(1-z) & 2\sqrt{2}(1-z)(1+7z+z^2) \end{bmatrix}.$$

It is simple to check that the conditions of biorthogonality are met for these filters, so that the corresponding wavelets form biorthogonal bases that are related by differentiation.

References

1. Battle, G., P. Federbush, and P. Uhlig, Wavelets for quantum gravity and divergence-free wavelets, Appl. Comput. Harmonic Anal. **1** (1993), 295–297.

2. Donovan, G., J. Geronimo, D. Hardin, and P. Massopust, Construction of orthogonal wavelets using fractal interpolation functions, SIAM J. Math. Anal. **27** (1996), 1791–1815.

3. Federbush, P., Navier and Stokes meet the wavelet, Comm. Math. Phys. **155** (1993), 219–248.

4. Gilbert, J., J. Hogan, and J. Lakey, Characterization of Hardy spaces by singular integrals and divergence-free wavelets, preprint.

5. Glowinski, R., Ensuring well-posedness by analogy; Stokes problem and boundary control for the wave equation, J. Comput. Phys. **103** (1992), 189–221.

6. Lemarié-Rieusset, P. G., Un théorème d'inexistence pour les ondelettes vecteurs á divergence nulle, C. R. Acad. Sci. Paris Sér. I Math. **319** (1994), 811–813.

7. Lemarié-Rieusset, P. G., Analyses multi-résolutions non-orthogonales,- commutation entre projecteurs et dérivation et ondellettes vecteurs á divergence nulle, Rev. Mat. Iberoamericana **8** (1992), 221–237.

8. Strela, V., *Multiwavelets: Theory and Applications*, Ph.D. Thesis, MIT, 1996.

9. Urban, K, On divergence-free wavelets, Adv. Comput. Math. **4** (1995), 51–81.

Joseph D. Lakey
Department of Mathematical Sciences
New Mexico State University
Las Cruces, NM 88003-8001
jlakey@nmsu.edu

Peter R. Massopust
Department of Mathematics
Sam Houston State University
Huntsville, TX 77341
mth_prm@shsu.edu

Maria Cristina Pereyra
Department of Mathematics
University of New Mexico
Albuquerque, NM 87131
crisp@math.unm.edu

On Intra- and Inter-orthogonal Scaling and Wavelet Vectors

Jian-ao Lian

Abstract. Compactly supported intra- and inter-orthogonal scaling and semi-orthogonal wavelet vectors are investigated. For a scaling vector $\phi = [\phi_1, \cdots, \phi_r]^{\mathrm{T}}$, its *intra-* and *inter-*orthogonality are defined by

$$\langle \phi_\ell(\cdot), \, \phi_\ell(\cdot - k) \rangle = \delta_{k,0}, \quad 1 \le \ell \le r, \quad \text{and}$$

$$\langle \phi_\ell(\cdot), \, \phi_{\ell'}(\cdot - k) \rangle = 0, \quad \ell \ne \ell'; \; 1 \le \ell, \ell' \le r; \quad k \in \mathbb{Z},$$

respectively. Furthermore, compactly supported dual scaling and semi-orthogonal wavelet vectors are introduced. Examples with $r = 2$ and small supports, namely, $\operatorname{supp} \phi_1 = [0, 1]$ and $\operatorname{supp} \phi_2 = [0, 2]$, are demonstrated.

§1. Introduction

Let $\phi := [\phi_1, \cdots, \phi_r]^{\mathrm{T}}$ be a compactly supported scaling vector and $\psi := [\psi_1, \cdots, \psi_r]^{\mathrm{T}}$ be its corresponding compactly supported wavelet vector with multiplicity r (see, e.g., [1, 2, 3, 8, 10].) That is, for $M, N \in \mathbb{Z}_+$, ϕ and ψ satisfy the following two-scale relations

$$\phi(x) = \sum_{j=0}^{M} P_j \phi(2x - j) = \sum_{j \in \mathbb{Z}} P_j \phi(2x - j), \quad \text{with} \qquad (1.1)$$

$$P_0, P_M \ne 0; \; P_j := 0, \; j < 0, j > M; \quad \text{and} \qquad (1.2)$$

$$\psi(x) = \sum_{k=0}^{N} Q_k \phi(2x - k) = \sum_{k \in \mathbb{Z}} Q_k \phi(2x - k), \quad \text{with} \qquad (1.3)$$

$$Q_0, Q_N \ne 0; \; Q_k := 0, \; k < 0, k > N. \qquad (1.4)$$

The matrix sequences $\{P_j\}_{j \in \mathbb{Z}}, \{Q_k\}_{k \in \mathbb{Z}} \subset \mathbb{R}^{r \times r}$, are called two-scale sequences of ϕ and ψ. Moreover, these ϕ's are ℓ^2-stable, i.e., they generate MRAs $\{V_j\}_{j \in \mathbb{Z}}$ with their complementary (wavelet) subspaces $\{W_j\}_{j \in \mathbb{Z}}$ generated by ψ's, i.e., $L^2 = \bigcup_{j \in \mathbb{Z}} V_j$, where

$$V_j := \operatorname{clos}_{L^2(\mathbb{R})} \langle 2^{j/2} \phi_\ell(2^j \cdot -k_\ell) : \; k_\ell \in \mathbb{Z}; \; 1 \le \ell \le r \rangle,$$

$$W_j := \operatorname{clos}_{L^2(\mathbb{R})} \langle 2^{j/2} \psi_\ell(2^j \cdot -k_\ell) : \; k_\ell \in \mathbb{Z}; \; 1 \le \ell \le r \rangle, \quad j \in \mathbb{Z},$$

Approximation Theory IX, Volume 2: Computational Aspects
Charles K. Chui and Larry L. Schumaker (eds.), pp. 169–178.
Copyright ⊜ 1998 by Vanderbilt University Press, Nashville, TN.
ISBN 0-8265-1326-3.

and
$$V_{j+1} = V_j \bigoplus W_j, \quad j \in \mathbb{Z}.$$
The Fourier transformation of (1.1)–(1.2) is
$$\widehat{\phi}(\omega) = P(z)\widehat{\phi}\left(\frac{\omega}{2}\right), \quad \text{and} \tag{1.5}$$
$$\widehat{\psi}(\omega) = Q(z)\widehat{\phi}\left(\frac{\omega}{2}\right), \tag{1.6}$$
where $z = e^{-i\omega/2}$, and
$$P(z) := \frac{1}{2}\sum_{j=0}^{M} P_j z^j = \frac{1}{2}\sum_{j \in \mathbb{Z}} P_j z^j, \tag{1.7}$$
$$Q(z) := \frac{1}{2}\sum_{k=0}^{N} Q_k z^k = \frac{1}{2}\sum_{k \in \mathbb{Z}} Q_k z^k, \tag{1.8}$$
are called two-scale symbols of ϕ and ψ, respectively. In short, with a slight abuse of notations, we will call (ϕ, ψ) a scaling-wavelet (s-w) vector with two-scale symbol (P, Q).

One of the objectives of this paper is to introduce *intra-* and *inter-*orthogonal scaling and semi-orthogonal wavelet vectors. The other objective is to introduce *compactly supported* dual scaling and semi-orthogonal wavelet vectors. To this end, we use the same demonstrative example throughout the paper. We will begin with some preliminaries in Section 2 and establishing the scaling vectors ϕ with *small supports* in Section 3. Intra- and inter-orthogonal s-w vectors (ϕ, ψ) will be considered in Section 4. Compactly supported dual s-w vectors $(\widetilde{\phi}, \widetilde{\psi})$ will be investigated in Section 5.

§2. Preliminaries

In the sequel, we will define autocorrelation symbol F relative to a vector-valued function $f = [f_1, \cdots, f_r]$ as the Gramian matrix
$$F(z) := \sum_{n \in \mathbb{Z}} \left(\langle f_j(\cdot), f_k(\cdot - n)\rangle\right)_{1 \le j, k \le r} z^n,$$
and use Φ and Ψ to represent the autocorrelation symbols relative to scaling and wavelet vectors ϕ and ψ, respectively.

We will need the following.

Lemma 1. [Cf., e.g., 3, 6] *The two vector-valued functions ϕ and ψ in (1.1)–(1.4) are scaling and wavelet vectors if and only if both $\Phi(z)$ and $\Psi(z)$ are positive definite on $|z| = 1$, i.e., both Φ and Ψ are invertible on $|z| = 1$. Furthermore, ϕ and ψ are orthonormal (o.n.) if and only if $\Phi(z) = \Psi(z) = I_r$, $|z| = 1$, where I_r denotes the identity matrix of order r.*

For two polynomial matrices U and V in $\mathbb{R}^{r \times r}$, we define the $2r \times 2r$ polynomial matrix $M_{U,V}$ by
$$M_{U,V}(z) := \begin{bmatrix} U(z) & U(-z) \\ V(z) & V(-z) \end{bmatrix}. \tag{2.1}$$

With the notation $M_{U,V}$ in (2.1), a s-w vector (ϕ, ψ) with two-scale symbol (P, Q) satisfies

$$M_{P,Q}(z) \begin{bmatrix} \Phi(z) & 0 \\ 0 & \Phi(-z) \end{bmatrix} M_{P,Q}(z)^* = \begin{bmatrix} \Phi(z^2) & 0 \\ 0 & \Psi(z^2) \end{bmatrix}, \quad |z| = 1, \quad (2.2)$$

where * represents the conjugate transposition. An o.n. s-w vector (ϕ, ψ) is self-dual. If either a scaling vector ϕ or its corresponding wavelet vector ψ is not o.n., we have their *dual* s-w vector $(\tilde{\phi}, \tilde{\psi})$ defined, via their Fourier transforms, by

$$\widetilde{\hat\phi}(\omega) := \Phi\left(e^{-i\omega}\right)^{-1} \hat\phi(\omega), \quad \text{and}$$
$$\widetilde{\hat\psi}(\omega) := \Psi\left(e^{-i\omega}\right)^{-1} \hat\psi(\omega). \quad (2.3)$$

Indeed, $\tilde{\phi}$ and $\tilde{\psi}$ are scaling and wavelet vectors with two-scale symbols

$$\widetilde{P}(z) := \Phi(z^2)^{-1} P(z) \Phi(z), \quad \text{and}$$
$$\widetilde{Q}(z) := \Psi(z^2)^{-1} Q(z) \Phi(z), \quad (2.4)$$

respectively. It is then clear from (2.1) that (2.2) becomes

$$M_{P,Q}(z) M_{\widetilde{P},\widetilde{Q}}(z)^* = I_{2r}, \quad |z| = 1. \quad (2.5)$$

We end this section by giving the following.

Lemma 2. [Cf., e.g., 4, 7, 8, 10] *A compactly supported scaling vector ϕ has approximation order ≥ 2 if and only if there are row vectors $a^{0,0} \neq 0$ and $a^{0,1}$, in $\mathbb{R}_r$, satisfying*

$$a^{0,0} P(1) = a^{0,0},$$
$$a^{0,0} P(-1) = 0; \quad \text{and} \quad (2.6)$$

$$a^{0,1} P(1) - \frac{1}{2} a^{0,0} P'(1) = \frac{1}{2} a^{0,1},$$
$$a^{0,1} P(-1) + \frac{1}{2} a^{0,0} P'(-1) = 0. \quad (2.7)$$

Furthermore, ϕ has approximation order ≥ 3 if and only if there are row vectors $a^{0,0} \neq 0$, $a^{0,1}$, and $a^{0,2}$, in $\mathbb{R}_r$, satisfying both (2.6)–(2.7) and

$$a^{0,2} P(1) - a^{0,1} P'(1) + \frac{1}{4} a^{0,0}(P''(1) + P'(1)) = \frac{1}{4} a^{0,2},$$
$$a^{0,2} P(-1) + a^{0,1} P'(-1) + \frac{1}{4} a^{0,0}(P''(-1) - P'(-1)) = 0. \quad (2.8)$$

§3. Scaling Vectors with Small Supports

From a practical point of view, we will study s-w vectors with small supports. We will also focus on $r = 2$, rewrite the two-scale symbol P of ϕ as

$$P(z) = \begin{bmatrix} P_{1,1}(z) & P_{1,2}(z) \\ P_{2,1}(z) & P_{2,2}(z) \end{bmatrix}, \quad (3.1)$$

and consider ϕ's with the following properties:

Property A: supp $\phi_1 = [0, 1]$, supp $\phi_2 = [0, 2]$;
Property B: Both ϕ_1 and ϕ_2 are symmetric [1], i.e,

$$\phi_1(x) = \phi_1(1 - x), \quad \phi_2(x) = \phi_2(2 - x); \tag{3.2}$$

Property C: Neither ϕ_1 nor ϕ_2 is self-refinable; and
Property D: ϕ provides approximation order at least 2.
Then, by applying (2.6)–(2.7) in Lemma 2, we have the following.

Theorem 1. *The scaling function ϕ, denoted by $^2\phi$, with two-scale symbol P in (3.1) and satisfying Properties A–D in (3.2), must have the following polynomial entries*

$$P_{1,1}(z) = \frac{1+z}{2}\mu, \qquad P_{1,2}(z) = \tau,$$

$$P_{2,1}(z) = -\frac{1+z}{4}\frac{\nu}{\tau}(2\mu - 1 - 2z + (2\mu - 1)z^2), \tag{3.3}$$

$$P_{2,2}(z) = \frac{1}{4} - \nu + \frac{1}{2}z + \left(\frac{1}{4} - \nu\right)z^2,$$

where $\mu, \tau, \nu \in \mathbb{R}$, satisfying

$$\nu\tau \neq 0 \quad \text{and} \quad \mu - 1 < 2\nu < \mu + 1. \tag{3.4}$$

The two row vectors $a^{0,0}$ and $a^{0,1}$ in (2.6)–(2.7) are given by

$$a^{0,0} = [2\nu, \tau] \quad \text{and} \quad a^{0,1} = [\nu, \tau]. \tag{3.5}$$

Furthermore, under the assumption $\|\phi_1\|_2 = \|\phi_2\|_2 = 1$, the autocorrelation symbol Φ relative to ϕ is given by

$$\Phi(z) = \begin{bmatrix} 1 & \alpha\left(\dfrac{1}{z} + 1\right) \\ \alpha(1+z) & \dfrac{\beta}{z} + 1 + \beta z \end{bmatrix}, \quad \text{where} \tag{3.6}$$

$$\alpha = -\frac{1}{4}\frac{4\nu^2 - 8(\mu - 1)\nu + 3(\mu - 1)(\mu + 1)}{2\mu\nu^2 - (\mu - 1)(3\mu - 1)\nu + (\mu - 1)^2(\mu + 1)}\tau, \tag{3.7}$$

$$\beta = \frac{1}{4}\frac{8\nu^3 - 4(2\mu - 1)\nu^2 + 4(\mu - 1)\nu + (\mu - 1)^2(\mu + 1)}{2\mu\nu^2 - (\mu - 1)(3\mu - 1)\nu + (\mu - 1)^2(\mu + 1)}, \quad \text{and} \tag{3.8}$$

$$\tau^2 = \frac{2\mu\nu^2 - (\mu - 1)(3\mu - 1)\nu + (\mu - 1)^2(\mu + 1)}{\mu - 2\nu + 2}, \tag{3.9}$$

so that

$$\det\left(\Phi(z)\right)$$

$$= \frac{1}{16}\frac{(\mu - 2\nu + 1)(\mu - 2\nu - 1)^2}{(\mu - 2\nu + 2)(2\mu\nu^2 - (\mu - 1)(3\mu - 1)\nu + (\mu - 1)^2(\mu + 1))}$$

$$\times \left(\frac{10\nu - 5\mu - 1}{z} + 2(2\nu - \mu + 7) + (10\nu - 5\mu - 1)z\right). \tag{3.10}$$

Proof: Property A and Property B require $P_{1,1}$ and $P_{1,2}$ have the form in (3.3), and $P_{2,1}$ and $P_{2,2}$ have the form

$$P_{2,1}(z) = \frac{1+z}{2}(\xi_1 + \eta_1 z + \xi_1 z^2), \quad \text{and} \quad P_{2,2}(z) = \xi_2 + \eta_2 z + \xi_2 z^2,$$

for some appropriate constants $\xi_1, \ldots, \eta_2$. By assuming $a^{0,0} = (x_0, y_0)$ and $a^{0,1} = (x_1, y_1)$, Property D leads to, from (2.6)–(2.7),

$$\xi_1 = -\frac{1}{2}\eta_1(2\mu - 1), \quad \xi_2 = \frac{1}{4} - \tau\eta_1, \quad \eta_2 = \frac{1}{2}, \quad \text{and} \qquad (3.11)$$

$$a^{0,0} = (2\eta_1, 1)y_0, \quad a^{0,1} = (\eta_1, 1)y_0, \quad y_0 \neq 0. \qquad (3.12)$$

Property C is equivalent to $\tau\eta_1 \neq 0$. Hence, the new parameter $\nu := \tau\eta_1$ yields $P_{2,1}$ and $P_{2,2}$ in (3.3), with $\nu\tau \neq 0$, the first inequality in (3.4), and (3.5) with the choice $y_0 = \tau$ in (3.12). Here, since ϕ is a scaling vector, i.e., ℓ^2-stable, $a^{0,0}$ is unique up to a constant multiple. Next, under the assumption $\|\phi_1\|_1 = \|\phi_2\|_2 = 1$, Φ can be *evaluated*, as given by (3.6)–(3.9), by using an identity from (2.2), namely,

$$P(z)\Phi(z)P(z)^* + P(-z)\Phi(-z)P(-z)^* = \Phi(z^2), \quad |z| = 1. \qquad (3.13)$$

Observe that, with P given, (3.13) is a *linear* matrix equation of Φ, in the sense that it is equivalent to a *linear* matrix system by comparing all the coefficient matrices. Or, on the other hand, by using the *Kronecker product* [5, p. 243] of two matrices, (3.13) is also equivalent to the *linear* system [9]

$$\left(\overline{P(z)} \otimes P(z)\right) \text{vec}\left(\Phi(z)\right)$$
$$+ \left(\overline{P(-z)} \otimes P(-z)\right) \text{vec}\left(\Phi(-z)\right) = \text{vec}\left(\Phi(z^2)\right), \qquad (3.14)$$

where for each matrix $A = (a_{jk}) \in C^{m \times n}$,

$$\text{vec}\, A := (a_{11}, \cdots, a_{m1}, a_{12}, \cdots, a_{m2}, \cdots, a_{1n}, \cdots, a_{mn})^{\mathrm{T}},$$

i.e., vec A is the stacked columns of A; and, with $B \in C^{p \times q}$, $A \otimes B := (a_{jk}B) \in C^{mp \times nq}$. In fact, the ℓ^2-stability of ϕ is equivalent to the fact that (3.13) has a *unique* solution for Φ up to a constant multiple. The orthonormality of ϕ is equivalent to $\Phi(z) = I_r, |z| = 1$, i.e., (3.13) or (3.14) has a unique solution I_r for Φ [9]. Finally, (3.10) follows from (3.6)–(3.9); the positivity of Φ on $|z| = 1$, the requirement for the ℓ^2-stability of ϕ or ϕ being a scaling vector, is equivalent to the second inequality in (3.4). Indeed, from (3.10), we introduce a new parameter $t := 2\nu - \mu$, and replace ν by $(\mu + t)/2$ to reach

$$\det\left(\Phi(z)\right) = \frac{(1-t)(1+t)^2}{8(2-t)((2-t)\mu^2 + (t^2 + 4t - 3)\mu + (2-t))}$$
$$\times \left(\frac{5t-1}{z} + 2(t+7) + (5t-1)z\right).$$

Then it takes a little effort of algebra, which is omitted here, to show that $\det\left(\Phi(z)\right) > 0, |z| = 1$, iff $|t| < 1$, i.e., $\mu - 1 < 2\nu < \mu + 1$, the second inequality in (3.4). □

Similarly, let ϕ be a scaling vector satisfying (3.2), but with Property D replaced by

Property D' : ϕ provides approximation order at least 3.   (3.15)

Then, by continuing applying (2.8) in Lemma 2 to P in (3.3) and $a^{0,0}, a^{0,1}$ in (3.12), the two-scale symbol P of a scaling vector ϕ, we have the following.

Theorem 2. *The scaling vector ϕ, denoted by $^3\phi$, with two-scale symbol P in (3.1) and satisfying Properties A–C in (3.2) and Property D' in (3.15), must have the polynomial entries as in (3.3), with μ, ν, and τ satisfying both (3.4) and $\nu = (4\mu - 1)/8$. The row vectors $a^{0,0}, a^{0,1}$, and $a^{0,2}$ in (2.6)–(2.8) are*

$$a^{0,0} = [4\mu - 1, 4\tau], \quad a^{0,1} = [(4\mu - 1)/2, 4\tau], \quad \text{and} \quad a^{0,2} = [2\mu - 1, 4\tau]. \quad (3.16)$$

In addition, Φ is given by (3.6), with

$$\alpha = -\frac{8\mu - 7}{16\tau} \qquad \beta = \frac{16\mu^2 - 28\mu + 11}{64\tau^2}$$
$$\tau^2 = (4\mu^2 - 7\mu + 4)/8, \quad \text{and} \qquad\qquad (3.17)$$
$$\det(\Phi(z)) = \frac{5}{256\tau^2}\left(-\frac{1}{z} + 6 - z\right), \qquad \mu \in \mathbb{R}.$$

Furthermore, the two components ϕ_1 and ϕ_2 of such a scaling vector, i.e., $^3\phi = [\phi_1, \phi_2]^T$, are continuous piecewise quadratic polynomial splines that are explicitly given by

$$\phi_1(x) = a\,x(1-x)\chi_{[0,1)}(x),$$
$$\phi_2(x) = x(b + cx)\chi_{[0,1)}(x) + (2-x)(b + c(2-x))\chi_{[1,2)}(x), \quad (3.18)$$
$$\text{where} \quad a^2 = 30, \quad b = -\frac{2\mu - 1}{4\tau}a, \quad c = \frac{4\mu - 1}{8\tau}a.$$

Here the standard notation for the characteristic function χ_A of a set A has been used.

Proof: Indeed, by using (3.11) and $a^{0,0}$ and $a^{0,1}$ in (3.12), with η_1 replaced by ν/τ, the matrix system (2.8) in Lemma 2 leads to

$$\nu = \frac{4\mu - 1}{8}, \quad \text{and} \quad a^{0,2} = \left(\frac{2\mu - 1}{4\tau}, 1\right)y_0.$$

Hence, (3.16) follows from the choice $y_0 = 4\tau$, and (3.17) from (3.7)–(3.10) with ν being replaced by $(4\mu - 1)/8$. Finally, it is no surprising that ϕ_1 and ϕ_2 are piecewise quadratic polynomials since the *saturated* approximation order 3. A rigorous proof can be given by verifying directly that ϕ in (3.18) satisfies the same two-scale relation with two-scale symbol P determined by (3.1) and (3.3)–(3.4), and applying the fact that, since ϕ is a (ℓ^2-stable) scaling vector, it is *uniquely* determined from its two-scale symbol.   □

Turn to a corresponding wavelet vector ψ (with minimal support), its two-scale symbol $Q(z) = 1/2 \sum_{j \in \mathbb{Z}} Q_j z^j$ can be determined, from (2.2), by

$$Q(z)\Phi(z)P(z)^\star + Q(-z)\Phi(-z)P(-z)^\star = 0, \quad \text{and} \tag{3.19}$$

$$Q(z)\Phi(z)Q(z)^\star + Q(-z)\Phi(-z)Q(-z)^\star = \Psi(z^2), \quad |z| = 1, \tag{3.20}$$

where the autocorrelation symbol Ψ relative to ψ is positive definite on $|z| = 1$. Due to the limitation of space, we will omit any expression of Q here.

§4. Intra- and Inter-orthogonal Scaling Vectors

For a scaling vector ϕ, its intra- and inter-orthogonality are defined by

$$\begin{aligned}
\langle \phi_\ell(\cdot), \phi_\ell(\cdot - k) \rangle &= \delta_{k,0}, \quad 1 \le \ell \le r, \quad \text{and} \\
\langle \phi_\ell(\cdot), \phi_{\ell'}(\cdot - k) \rangle &= 0, \quad \ell \ne \ell'; \, 1 \le \ell, \ell' \le r; \quad k \in \mathbb{Z},
\end{aligned} \tag{4.1}$$

respectively. The definition of the intra- and inter-orthogonality of a wavelet vector ψ is similar. By using the similar notation of P in (3.1) for Φ, namely, $\Phi(z) = (\Phi_{j,k}(z))_{j,k=1}^r$, we have the following.

Theorem 3. *A scaling vector ϕ is intra-orthogonal if and only if the main diagonal entries of the autocorrelation symbol $\Phi(z)$ relative to ϕ are 1's, i.e.,*

$$\Phi_{j,j}(z) = 1, \quad 1 \le j \le r, \quad |z| = 1, \tag{4.2}$$

and ϕ is inter-orthogonal if and only if $\Phi(z)$ is diagonal, i.e.,

$$\Phi_{j,k}(z) = 0, \quad |z| = 1; \quad j < k, \quad 1 \le j, k \le r. \tag{4.3}$$

Moreover, ϕ is o.n. if and only if ϕ is both intra- and inter-orthogonal.

Let $^2\phi$ be a scaling function with two-scale symbol P determined by (3.1) and (3.3)–(3.4). Then it follows from (3.6)–(3.9) in Theorem 1 and (4.2) that $^2\phi$ is intra-orthogonal, iff $\beta = 0$, i.e.,

$$8\nu^3 - 4(2\mu - 1)\nu^2 + 4(\mu - 1)\nu + (\mu - 1)^2(\mu + 1) = 0, \quad |2\nu - \mu| < 1. \tag{4.4}$$

Observe that (4.4), together with (3.9), has the parametric form

$$\mu = \frac{2 + 26\sin\theta - 3\sin^2\theta + 2\sqrt{5}(1 + 4\sin\theta)\cos\theta}{5\sin\theta(2 - 7\sin\theta)},$$

$$\nu = \frac{3 + 9\sin\theta - 12\sin^2\theta + \sqrt{5}(1 + 4\sin\theta)\cos\theta}{5\sin\theta(2 - 7\sin\theta)}, \tag{4.5}$$

$$\tau^2 = \frac{4(1 - \sin\theta)(1 + 4\sin\theta)^2}{25\sin^2\theta(2 - 7\sin\theta)},$$

$$\text{where} \quad \theta \in \left(\pi + \sin^{-1}\left(\frac{1}{4}\right), 2\pi - \sin^{-1}\left(\frac{1}{4}\right) \right).$$

That is, we have constructed a family of intra-orthogonal scaling vectors $^2\phi$ with two-scale symbols P determined by (3.1), (3.3), and (4.5). See Fig. 1 (a) for a typical $^2\phi$ where $\theta = 4$. Analogously, it follows from (3.6)–(3.9) and (4.3) that $^2\phi$ is inter-orthogonal, iff $\alpha = 0$, i.e.,

$$4\nu^2 - 8(\mu - 1)\nu + 3(\mu - 1)(\mu + 1) = 0, \quad |2\nu - \mu| < 1. \tag{4.6}$$

The equation in (4.6), together with (3.9), has the parametric form

$$
\begin{aligned}
\mu &= 4 + 3\csc\vartheta, \\
\nu &= 3 + 3\csc\vartheta + \frac{3}{2}\cot\vartheta, \\
\tau^2 &= -\frac{3}{2}(1 + \csc\vartheta)(5 + 3\csc\vartheta), \quad \vartheta \in \left(\pi + \sin^{-1}\frac{3}{5}, \frac{3\pi}{2}\right).
\end{aligned}
\tag{4.7}
$$

In other words, we have constructed a family of inter-orthogonal scaling vectors $^2\phi$ with two-scale symbols P determined by (3.1), (3.3), and (4.7). See Figure 1 (b) for a typical $^2\phi$ where $\vartheta = 4$.

As another example, let $^3\phi = [\phi_1, \phi_2]^T$ be a scaling vector in Theorem 2, i.e., its two-scale symbol P is in (3.1) and (3.3), with μ and ν satisfying $2\nu - \mu = -1/4$. That is, ϕ_1 and ϕ_2 are continuous quadratic piecewise polynomial splines in (3.18). Then it follows from (3.6) and (3.17) that $^3\phi$ is intra-orthogonal iff $\mu = (7 \pm \sqrt{5})/8$. It is inter-orthogonal iff $\mu = 7/8$.

§5. Compactly Supported Dual S-W Vectors $(\widetilde{\phi}, \widetilde{\psi})$

It is known that non-orthonormal *uni*-scaling functions, i.e., scaling vectors with multiplicity $r = 1$, have *no* compactly supported duals. However, we will see that it is possible to construct scaling or wavelet vectors, with multiplicity $r \geq 2$, that have *compactly supported* dual scaling or wavelet vectors.

First, we have the following.

Theorem 4. *A scaling vector ϕ (or wavelet vector ψ) has compactly supported dual $\widetilde{\phi}$ (or $\widetilde{\psi}$) in (2.3) if and only if*

$$\det(\Phi(z)) = \text{constant} \quad (\text{or} \quad \det(\Psi(z)) = \text{constant}), \quad |z| = 1. \tag{5.1}$$

Let $^2\phi$ be a scaling function with two-scale symbol P determined by (3.1) and (3.3)–(3.4). Then it follows from (3.10) and (3.4) that (5.1) leads to $2\nu - \mu = 1/5$. That is, from (3.3)–(3.4) and (3.6)–(3.9), we have constructed a family of scaling vectors $^2\phi$ with compactly supported duals $^2\widetilde{\phi}$. These scaling vectors $^2\phi$ have two-scale symbols P in (3.3) with parameters μ, ν and τ satisfying $2\nu - \mu = 1/5$ and (3.9). Indeed, (3.7)–(3.10) become to

$$
\alpha = -\frac{5\mu - 3}{10\tau}, \quad \beta = \frac{(5\mu - 3)^2}{100\tau^2}, \quad \det(\Phi(z)) = \frac{8}{25\tau^2},
\tag{5.2}
$$
$$
\text{where} \quad \tau^2 = (5\mu^2 - 6\mu + 5)/10, \quad \mu \in \mathbb{R}.
$$

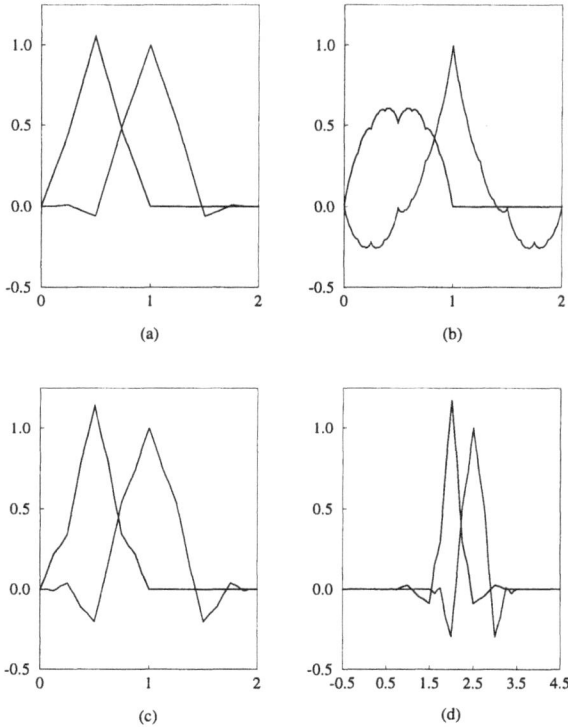

Fig. 1. (a) Intra-orthogonal $^2\phi$ with $\theta = 4$ in (4.5); (b) inter-orthogonal $^2\phi$ with $\vartheta = 4$ in (4.7); (c)–(d) $^2\phi$ and $^2\tilde{\phi}$ with $\mu = 1/2, \nu = 7/20$, and $\tau = \sqrt{130}/20$ in (3.3). For each graph, the one from left is ϕ_1 (or $\tilde{\phi}_1$ in (d)) and the one from right is ϕ_2 (or $\tilde{\phi}_2$ in (d)).

Observe from (5.2) that, among these scaling vectors, with two-scale symbols P determined by (3.1) and (3.3) and α, β, τ in (5.2), there is a *unique* (up to constant multiples) intra- and inter-orthogonal, i.e., o.n. scaling vector, when $\nu = 2/5$ (so that $\mu = 3/5$ and $\tau^2 = 8/25$). This ϕ is exactly the GHM o.n. scaling vector [2].

Furthermore, it follows from the definition of $\widetilde{P}$ in (2.4) that, with similar notation of P in (3.1), the polynomial entries of $z^3\widetilde{P}(z)$ are given by

$$z^3\widetilde{P}_{1,1}(z) = -\frac{(1+z)z}{160}\big[(5\mu - 3) - 10(5\mu - 3)z + 2(5\mu - 27)z^2$$
$$- 10(5\mu - 3)z^3 + (5\mu - 3)z^4\big],$$

$$z^3\widetilde{P}_{1,2}(z) = \frac{\sqrt{10}}{1600\sqrt{5\mu^2 - 6\mu + 5}}\big[(5\mu - 3)^2 - 40\mu(5\mu - 3)z$$
$$- (5\mu - 3)(85\mu - 83)z^2 - 80(5\mu^2 + 3\mu - 10)z^3$$

$$- (5\mu - 3)(85\mu - 83)z^4 - 40\mu(5\mu - 3)z^5 + (5\mu - 3)^2 z^6 \big],$$

$$z^3 \widetilde{P}_{2,1}(z) = -\frac{\sqrt{5\mu^2 - 6\mu + 5}}{16\sqrt{10}} z^3 (1 + z)(1 - 10z + z^2),$$

$$z^3 \widetilde{P}_{2,2}(z) = \frac{1}{160} z^2 \big[(5\mu - 3) - 40\mu z - (90\mu - 134)z^2$$
$$- 40\mu z^3 + (5\mu - 3)z^4 \big].$$

See Figure 1 (c)–(d) for $^2\phi$ and $^2\widetilde{\phi}(x)$ with $\mu = 1/2, \nu = 7/20, \tau = \sqrt{130}/20$. Observe that supp $\phi_1 \subset [-1/2, 9/2]$ and supp $\widetilde{\phi}_2 \subset [1/2, 9/2]$.

However, for any $^3\phi = [\phi_1, \phi_2]^\mathrm{T}$, i.e., ϕ_1 and ϕ_2 being given by (3.18), it follows from (3.17) that $^3\phi$ *cannot* have compactly supported dual scaling vector for any choice of the parameter $\mu \in \mathbb{R}$.

Acknowledgments. This work was partially supported by Texas Higher Education Coordinating Board Grant 000015-109.

References

1. Chui, C. K. and J.-A. Lian, A study of orthonormal multi-wavelets, J. Appl. Numer. Math. **20** (1996), 272–298.
2. Geronimo, J. S., D. P. Hardin, and P. R. Massopust, Fractal functions and wavelet expansions based on several scaling functions, J. Approx. Theory **78** (1994), 373–401.
3. Goodman, T. N. T. and S. L. Lee, Wavelets of multiplicity r, Trans. Amer. Math. Soc. **342** (1994), 307–324.
4. Heil, C., G. Strang, and V. Strela, Approximation by translates of refinable functions, Numerische Math. **73** (1996), 75–94.
5. Horn, R. A. and C. R. Johnson, *Topics in Matrix Analysis*, Cambridge, 1991.
6. Jia, R. Q. and C. A. Micchelli, Using the refinement equations for the construction of pre-wavelets II: Powers of two, in *Curves and Surfaces*, P. J. Laurent, A. Lé Méhauté, and L. L. Schumaker (eds.), Academic Press, New York, 1991, pp. 209–246.
7. Jia, R. Q., S. D. Riemenschneider, and D. X. Zhou, Approximation by multiple refinable functions, Canadian J. Math. **49** (1997), 944–962.
8. Lian, J.-A., On the order of polynomial reproduction for multi-scaling functions, Appl. Comput. Harmonic Anal. **3** (1996), 358–365.
9. Lian, J.-A., Orthogonality criteria for multi-scaling functions, Appl. Comput. Harmonic Anal. **5** (1998), 277–311.
10. Plonka, G., Approximation order provided by refinable function vectors, Constr. Approx. **13** (1997), 221–244.

Jian-ao Lian
Department of Mathematics
Prairie View A&M University
Prairie View, Texas 77446
jian-ao_lian@pvamu.edu

General Moments Evaluation of Scaling Functions and Wavelets

Jian-ao Lian and George A. Roberts

Abstract. The general inner products (GIP) of two compactly supported functions f and g in $L^2(\mathbb{R})$ are defined by

$$\langle f, g(\cdot - \ell)\rangle_{[k]} := \int_{-\infty}^{\infty} t^k f(t)\overline{g(t-\ell)}\,dt, \quad k, \ell \in \mathbb{Z}_+.$$

For a compactly supported function $f \in L^2(\mathbb{R})$, its general moments (GM) are defined by $\langle f, f\rangle_k$, $k \in \mathbb{Z}$. We investigate the evaluation of the GIP of the integer shifts of various real-valued scaling functions ϕ_1 and their corresponding wavelets ψ_1. Examples will be given.

§1. Introduction

For two compactly supported functions f and g in $L^2(\mathbb{R})$, their general inner products (GIP) are defined as

$$\langle f, g(\cdot - \ell)\rangle_{[k]} := \int_{-\infty}^{\infty} t^k f(t)\overline{g(t-\ell)}\,dt, \quad k, \ell \in \mathbb{Z}_+. \tag{1.1}$$

For a single compactly supported function $f \in L^2(\mathbb{R})$, its general moments (GM) are defined as $\langle f, f\rangle_k$, i.e.,

$$\langle f, f\rangle_{[k]} = \int_{-\infty}^{\infty} t^k |f(t)|^2\,dt, \quad k \in \mathbb{Z}_+. \tag{1.2}$$

Let ϕ be a given compactly supported scaling function, with two-scale sequence $\{p_j\}_{j\in\mathbb{Z}}$ of finitely many nonzero entries, and ψ_1 be its corresponding compactly supported wavelets, with two-scale sequence $\{q_k\}_{k\in\mathbb{Z}}$ of finitely

Approximation Theory IX, Volume 2: Computational Aspects
Charles K. Chui and Larry L. Schumaker (eds.), pp. 179–187.
Copyright © 1998 by Vanderbilt University Press, Nashville, TN.
ISBN 0-8265-1326-3.

many nonzero entries. To be precise, let ϕ_1 and ψ_1 satisfy the real-valued two-scale relations

$$\phi_1(x) = \sum_{j \in \mathbb{Z}} p_j \phi_1(2x - j), \tag{1.3}$$

$$\psi_1(x) = \sum_{k \in \mathbb{Z}} q_k \phi_1(2x - k). \tag{1.4}$$

As usual, let

$$P_{\phi_1}(z) = \frac{1}{2} \sum_{j \in \mathbb{Z}} p_j z^j, \tag{1.5}$$

$$Q_{\psi_1}(z) = \frac{1}{2} \sum_{k \in \mathbb{Z}} q_k z^k, \tag{1.6}$$

be the two-scale symbols relative to ϕ_1 and ψ_1, respectively. Then, by taking the Fourier transforms of (1.3)–(1.4), we have, with $z := e^{-i\omega/2}$,

$$\widehat{\phi}_1(\omega) = P_{\phi_1}(z)\widehat{\phi}_1\left(\frac{\omega}{2}\right), \tag{1.7}$$

$$\widehat{\psi}_1(\omega) = Q_{\psi_1}(z)\widehat{\phi}_1\left(\frac{\omega}{2}\right). \tag{1.8}$$

Here, ϕ_1 and ψ_1 can be either semiorthogonal (s.o.) or orthonormal (o.n.)

One of the main objectives of this paper is to evaluate, for various *given* ϕ_1 and ψ_1 and fixed $r \in \mathbb{Z}_+$, the first $2r - 1$ GM of ϕ_1 and ψ_1, namely,

$$\langle \phi_1, \phi_1 \rangle_{[k]} = \int_{-\infty}^{\infty} t^k \left|\phi_1(t)\right|^2 dt, \tag{1.9}$$

$$\langle \psi_1, \psi_1 \rangle_{[k]} = \int_{-\infty}^{\infty} t^k \left|\psi_1(t)\right|^2 dt, \qquad k = 0, \cdots, 2r - 2. \tag{1.10}$$

To this end, we will give some more notations and preliminaries in Section 2. In order to get GM of ϕ_1 and ψ_1 in (1.9)–(1.10), algorithms for evaluating the GIP of the integer shifts of ϕ_1 or ψ_1 will be established in Section 3. Examples of how to apply the algorithms to various compactly supported scaling functions and wavelets will be demonstrated in Section 4.

§2. More Notations and Preliminaries

Let r be a positive integer, and $\phi = [\phi_1, \cdots, \phi_r]^{\mathrm{T}}$ be a vector-valued function with compactly supported components $\phi_\ell \in L^2(\mathbb{R})$, $\ell = 1, \cdots, r$. Throughout this paper, we will not require ϕ to be a scaling vector [cf., e.g., 2, 5, 6], i.e., $\{\phi_\ell(\cdot - k_\ell) : k_\ell \in \mathbb{Z}; \ell = 1, \cdots, r\}$ does not necessarily generate a Riesz basis of a subspace of $L^2(\mathbb{R})$. In other words, ϕ does not necessarily generate a multiresolution analysis (MRA.)

The autocorrelation symbol Φ relative to ϕ is defined by

$$\Phi(z) := \sum_{n \in \mathbb{Z}} \left(\langle \phi_j(\cdot), \phi_k(\cdot - n) \rangle \right)_{1 \le j,k \le r} z^n. \tag{2.1}$$

Here, the standard notation $\langle \cdot, \cdot \rangle$ of the inner product on L^2 has been used. Observe from (1.2) that $\langle \cdot, \cdot \rangle = \langle \cdot, \cdot \rangle_{[0]}$. Since ϕ_ℓ, $\ell = 1, \cdots, r$, are compactly supported, the semidefinite matrix Φ in (2.1) is a $r \times r$ matrix with Laurent polynomial entries. Moreover, Poisson summation formula leads to

$$\Phi\left(e^{-i\omega/2}\right) = \sum_{n \in \mathbb{Z}} \widehat{\phi}\left(\frac{\omega}{2} + 2\pi n\right) \widehat{\phi}\left(\frac{\omega}{2} + 2\pi n\right)^\star, \quad \omega \in \mathbb{R}, \tag{2.2}$$

where $\star$ denotes the transpose of complex conjugation.

Assume, in addition, that there are $M+1$ $r \times r$ matrices $P_j, j = 0, \cdots, M$, with P_0, $P_M \ne 0$, so that ϕ satisfies a two-scale equation

$$\phi(x) = \sum_{j=0}^{M} P_j \phi(2x - j). \tag{2.3}$$

Then the Fourier transformation of (2.3) yields

$$\widehat{\phi}(\omega) = P\left(e^{-i\omega/2}\right) \widehat{\phi}\left(\frac{\omega}{2}\right), \tag{2.4}$$

where P is the two-scale symbol of ϕ, defined by

$$P(z) := \frac{1}{2} \sum_{j=0}^{M} P_j z^j, \tag{2.5}$$

and $\{P_j\}_{0 \le j \le M}$ is called the two-scale sequence of ϕ. Furthermore, with Φ in (2.1) or (2.2) as the autocorrelation symbol relative to ϕ, we have

$$P(z)\Phi(z)P(z)^\star + P(-z)\Phi(-z)P(-z)^\star = \Phi(z^2), \quad |z| = 1. \tag{2.6}$$

Similarly, let $\psi = [\psi_1, \cdots, \psi_r]^{\mathrm{T}}$ be a vector-valued function satisfying

$$\psi(x) = \sum_{k=0}^{N} Q_k \phi(2x - k). \tag{2.7}$$

Then

$$\widehat{\psi}(\omega) = Q\left(e^{-i\omega/2}\right) \widehat{\phi}\left(\frac{\omega}{2}\right), \tag{2.8}$$

$$\text{where} \quad Q(z) := \frac{1}{2} \sum_{k=0}^{N} Q_k z^k. \tag{2.9}$$

The polynomial Q in (2.9) is called the two-scale symbol of ψ.

Let Ψ be the autocorrelation symbol relative to ψ. Then Q and Ψ satisfy the matrix identity

$$Q(z)\Phi(z)Q(z)^* + Q(-z)\Phi(-z)Q(-z)^* = \Psi(z^2), \quad |z| = 1. \qquad (2.10)$$

However, since we have *no* requirement on ψ, the three matrices P, Φ, and Q do not, in general, satisfy

$$P(z)\Phi(z)Q(z)^* + P(-z)\Phi(-z)Q(-z)^* = 0, \quad |z| = 1. \qquad (2.11)$$

Back to the scaling function ϕ_1 and wavelet ψ_1 in (1.3)–(1.4), with two-scale symbols P_{ϕ_1} and Q_{ψ_1} in (1.5)–(1.8), we denote by Φ_1 and Ψ_1 the Euler-Frobenius polynomials of ϕ_1 and ψ_1, respectively, i.e.,

$$\Phi_1(z) = \sum_{n\in\mathbb{Z}} \langle \phi_1(\cdot), \phi_1(\cdot - n)\rangle z^n, \qquad (2.12)$$

$$\Psi_1(z) = \sum_{n\in\mathbb{Z}} \langle \psi_1(\cdot), \psi_1(\cdot - n)\rangle z^n. \qquad (2.13)$$

Again, Poisson summation formula gives

$$\Phi_1(e^{-i\omega/2}) = \sum_{k\in\mathbb{Z}} \left|\hat{\phi}_1\left(\frac{\omega}{2} + 2\pi k\right)\right|^2, \qquad (2.14)$$

$$\Psi_1(e^{-i\omega/2}) = \sum_{k\in\mathbb{Z}} \left|\hat{\psi}_1\left(\frac{\omega}{2} + 2\pi k\right)\right|^2. \qquad (2.15)$$

Then it follows from (1.5)–(1.8) that (2.6), (2.10), and (2.11), which is satisfied by P_{ϕ_1}, Φ_1, and Q_{ψ_1}, yielding the following polynomial identities

$$|P_{\phi_1}(z)|^2 \Phi_1(z) + |P_{\phi_1}(-z)|^2 \Phi_1(-z) = \Phi_1(z^2), \qquad (2.16)$$

$$|Q_{\psi_1}(z)|^2 \Phi_1(z) + |Q_{\psi_1}(-z)|^2 \Phi_1(-z) = \Psi_1(z^2), \qquad (2.17)$$

$$P_{\phi_1}(z)\Phi_1(z)\overline{Q_{\psi_1}}\left(\frac{1}{z}\right) + P_{\phi_1}(-z)\Phi_1(-z)\overline{Q_{\psi_1}}\left(-\frac{1}{z}\right) = 0, \quad |z| = 1. \,(2.18)$$

§3. Algorithms for Evaluating GM

For convenience, we will first focus on evaluating GM of scaling functions ϕ_1 in (1.3). The evaluation of GM of a wavelet ψ_1 is similar. Rewrite (1.3) as

$$\phi_1(x) = \sum_{j=0}^{M} p_j \phi_1(2x - j), \quad p_0 p_M \neq 0. \qquad (3.1)$$

Observe that to get the GM $\langle \phi_1, \phi_1 \rangle_{[k]}$ in (1.9), it is necessary to evaluate a large group of GIP as follows,

$$u_{k;\ell} := \int_{\mathbf{R}} t^k \phi_1(t) \phi_1(t - \ell)\, dt, \quad |\ell| = 0, \cdots, M - 1. \tag{3.2}$$

for each $k = 0, \cdots, 2r - 2$, so that

$$\langle \phi_1, \phi_1 \rangle_{[k]} = u_{k;0}, \quad k = 0, \cdots, 2r - 2. \tag{3.3}$$

Introduce

$$\phi_j(x) := x^{j-1}\phi_1(x), \quad 2 \le j \le r. \tag{3.4}$$

Then, with the convention

$$p_j := 0, \quad j < 0 \quad \text{or} \quad j > M, \tag{3.5}$$

it follows from

$$x^{j-1} = \frac{1}{2^{j-1}} \left(n + (2x - n) \right)^{j-1}$$

$$= \frac{1}{2^{j-1}} \sum_{k=1}^{j} \binom{j-1}{k-1} n^{j-k}(2x - n)^{k-1},$$

that, for $1 \le j \le r$,

$$\phi_j(x) = x^{j-1}\phi_1(x) = x^{j-1} \sum_{n \in \mathbb{Z}} p_n \phi_1(2x - n)$$

$$= \sum_{n \in \mathbb{Z}} \left[\frac{1}{2^{j-1}} \sum_{k=1}^{j} \binom{j-1}{k-1} n^{j-k} p_n \right] \phi_k(2x - n). \tag{3.6}$$

Hence, by letting

$$P_0 := \text{diag}\left(p_0, \frac{1}{2}p_0, \cdots, \frac{1}{2^{r-1}}p_0 \right),$$

$$P_n := \left(\frac{1}{2^{j-1}} \binom{j-1}{k-1} n^{j-k} p_n \right)_{j,k=1}^{r}, \quad 1 \le n \le M, \tag{3.7}$$

which are lower triangular $r \times r$ matrices, we have

$$\phi(x) = \sum_{n=0}^{M} P_n \phi(2x - n), \tag{3.8}$$

$$\phi(x) := [\phi_1(x), \cdots, \phi_r(x)]^{\mathrm{T}}. \tag{3.9}$$

As we indicated earlier, ϕ in (3.8)–(3.9) is not necessarily a scaling vector, i.e., Φ is not necessarily invertible on $|z| = 1$ [2, 5, 6].

However, since each component of ϕ in (3.9) is compactly supported, the GIP in (3.2) of ϕ_1 can be obtained from the autocorrelation symbol Φ in (2.1) relative to ϕ through the identity (2.6). Precisely, let

$$\Phi(z) = [\Phi_{j,k}(z)]_{j,k=1}^r . \tag{3.10}$$

Then $\Phi_{1,1}(z) = \Phi_1(z)$, and that, from (2.1) and (3.4),

$$\Phi_{j,k}(z) = \sum_{n \in \mathbb{Z}} \langle \phi_j(\cdot), \phi_k(\cdot - n) \rangle z^n$$

$$= \sum_{n \in \mathbb{Z}} z^n \int_{-\infty}^{\infty} t^{j-1}(t-n)^{k-1} \phi_1(t) \phi_1(t-n)\, dt, \tag{3.11}$$

i.e., the coefficient matrix of z^0 of $\Phi(z)$ will be the GM $\langle \phi_1, \phi_1 \rangle_{[j+k-2]}$, $j, k = 1, \cdots, r$. All the GIP can be also obtained from Φ. For example,

$$\Phi_{k,1}(z) = \sum_{n \in \mathbb{Z}} u_{k-1;n} z^n. \tag{3.12}$$

To summarize, we give the following.

Algorithm 3.1: Evaluating GM $\langle \phi_1, \phi_1 \rangle_{[k]}$ of ϕ_1 in (1.9) and (3.1), with P_{ϕ_1} in (1.5).

1° Evaluate Φ_1 relative to ϕ_1 via (2.16).
2° Define ϕ_j, $2 \le j \le r$, as in (3.4), and $M+1$ $r \times r$ lower triangular matrices or two-scale sequence of ϕ in (3.9), i.e., P_n, $0 \le n \le M$, as in (3.7).
3° Evaluate the autocorrelation symbol Φ relative to ϕ as in (2.1)–(2.2), (3.10), and (3.8), via (2.6).
4° If $\Phi(z) = \sum_{j=-(M-1)}^{M-1} \Phi^j z^j$, and $\Phi^0 = [\Phi_{j,k}^0]_{j,k=1}^r$, then

$$\langle \phi_1, \phi_1 \rangle_{[j+k-2]} = \Phi_{j,k}^0, \quad j, k = 1, \cdots, r.$$

Algorithm 3.2: Evaluating GM $\langle \psi_1, \psi_1 \rangle_{[k]}$ of ψ_1 in (1.10) and (2.7), with $\psi_1(x) = \sum_{j=0}^N q_j \phi_1(2x - j)$, $q_0 q_N \ne 0$, and Q_{ψ_1} in (1.6).

1° Evaluate Ψ_1 relative to ψ_1 via (2.17).
2° Define

$$\psi_j(x) := x^{j-1}\psi_1(x), \quad 2 \le j \le r, \quad \text{and}$$

$$Q_n := \left(\frac{1}{2^{j-1}} \binom{j-1}{k-1} n^{j-k} q_n \right)_{j,k=1}^r, \quad 0 \le n \le N.$$

3° Evaluate the autocorrelation symbol Ψ relative to $\psi = [\psi_1, \cdots, \psi_r]^{\mathrm{T}}$ in (2.7)–(2.9), via (2.10).
4° If $\Psi(z) = \sum_{j=-(N-1)}^{N-1} \Psi^j z^j$, and $\Psi^0 = [\Psi_{j,k}^0]_{j,k=1}^r$, then

$$\langle \psi_1, \psi_1 \rangle_{[j+k-2]} = \Psi_{j,k}^0, \quad j, k = 1, \cdots, r.$$

Remark. The algorithms can be applied recursively. Take ϕ_1 as an example. If all the $2r - 1$ GIP of ϕ_1 are obtained through the vector-valued function ϕ, then the $(2r)^{\text{th}}$ and $(2r + 1)^{\text{st}}$ GIP of ϕ_1 can be obtained by introducing 1 more component $\phi_{r+1}(x) := x^r \phi_1(x)$, to $\phi = [\phi_1, \cdots, \phi_r]^T$, and reapplying Algorithm 3.1 to the new block vector-valued function $[\phi^T, \phi_{r+1}]^T$.

§4. Demonstrative Examples

We now demonstrate how to apply our algorithms in Section 3 to various compactly supported scaling functions ϕ_1 and their corresponding compactly supported wavelets ψ_1. For simplicity, we consider $r = 2$.

Example 1. $\phi_1 = {}^D\phi_3$, Daubechies' orthonormal (o.n.) scaling function [3, 4] with approximation order 2 and support $[0, 3]$. By applying Algorithm 3.1 and with $\Phi_1(z) \equiv 1$ and $\phi_2(x) := x\phi_1(x)$, we have the autocorrelation symbol Φ relative to $\phi = [\phi_1, \phi_2]^T$ given by $\Phi(z) = [\Phi_{j,k}(z)]_{j,k=1}^2$, where

$$\Phi_{1,1}(z) = \Phi_1(z) = 1,$$

$$\Phi_{1,2}(z) = \Phi_{2,1}(z) = \frac{\sqrt{3}}{784}\left(\frac{1}{z^2} + z^2\right) - \frac{2\sqrt{3}}{49}\left(\frac{1}{z} + z\right) + \frac{3}{2} - \frac{165\sqrt{3}}{392},$$

$$\Phi_{2,2}(z) = \left(\frac{3\sqrt{3}}{784} - \frac{127}{117600}\right)\left(\frac{1}{z^2} + z^2\right)$$

$$\hspace{3cm} (4.1)$$

$$- \left(\frac{6\sqrt{3}}{49} - \frac{1807}{29400}\right)\left(\frac{1}{z} + z\right) + \frac{56433}{19600} - \frac{495\sqrt{3}}{392}.$$

Hence, we have the two GM of ${}^D\phi_3$ as follows,

$$\langle {}^D\phi_3, {}^D\phi_3 \rangle_{[1]} = \int_{-\infty}^{\infty} t \left| {}^D\phi_3(t) \right|^2 dt = \frac{3}{2} - \frac{165\sqrt{3}}{392},$$

$$\langle {}^D\phi_3, {}^D\phi_3 \rangle_{[2]} = \int_{-\infty}^{\infty} t^2 \left| {}^D\phi_3(t) \right|^2 dt = \frac{56433}{19600} - \frac{495\sqrt{3}}{392},$$

and, moreover, all the nonzero GIP of ${}^D\phi_3$ are given by

$$\langle {}^D\phi_3, {}^D\phi_3(\cdot - 1) \rangle_{[1]} = -\frac{2\sqrt{3}}{49} = \langle \phi_1, \phi_1(\cdot + 1) \rangle_{[1]},$$

$$\langle {}^D\phi_3, {}^D\phi_3(\cdot - 2) \rangle_{[1]} = \frac{\sqrt{3}}{784} = \langle {}^D\phi_3, {}^D\phi_3(\cdot + 2) \rangle_{[1]};$$

$$\langle {}^D\phi_3, {}^D\phi_3(\cdot - 1) \rangle_{[2]} = -\left(\frac{8\sqrt{3}}{49} - \frac{1807}{29400}\right),$$

$$\langle {}^D\phi_3, {}^D\phi_3(\cdot + 1) \rangle_{[2]} = -\left(\frac{4\sqrt{3}}{49} - \frac{1807}{29400}\right),$$

$$\langle {}^D\phi_3, {}^D\phi_3(\cdot - 2) \rangle_{[2]} = \frac{5\sqrt{3}}{784} - \frac{127}{117600},$$

$$\langle {}^D\phi_3, {}^D\phi_3(\cdot + 2) \rangle_{[2]} = \frac{\sqrt{3}}{784} - \frac{127}{117600}.$$

Example 2. $\psi_1 = {}^D\!\psi_3$, Daubechies' o.n. wavelet [3, 4] corresponding to ${}^D\!\phi_3$ with 2 vanishing moments and support $[0,3]$. By applying Algorithm 3.2 and with $\Psi_1(z) \equiv 1$ and $\psi_2(x) := x\psi_1(x)$, we have the autocorrelation symbol $\boldsymbol{\Psi}$ relative to $\psi = [\psi_1, \psi_2]^{\mathrm{T}}$ given by $\boldsymbol{\Psi}(z) = [\Psi_{j,k}(z)]_{j,k=1}^2$, where

$$\Psi_{1,1}(z) = 1,$$

$$\Psi_{1,2}(z) = \Psi_{2,1}(z) = -\frac{\sqrt{3}}{784}\left(\frac{1}{z^2} + z^2\right) + \frac{33\sqrt{3}}{784}\left(\frac{1}{z} + z\right) + \frac{3}{2},$$

$$\Phi_{2,2}(z) = -\left(\frac{3\sqrt{3}}{784} - \frac{127}{117600}\right)\left(\frac{1}{z^2} + z^2\right) \qquad (4.2)$$

$$+ \left(\frac{99\sqrt{3}}{784} - \frac{2461}{78400}\right)\left(\frac{1}{z} + z\right) + \frac{13281}{5600}.$$

Hence, the two GM of ${}^D\!\psi_3$ are given by

$$\langle {}^D\!\psi_3, {}^D\!\psi_3 \rangle_{[1]} = \int_{-\infty}^{\infty} t\left| {}^D\!\psi_3(t)\right|^2 dt = \frac{3}{2},$$

$$\langle {}^D\!\psi_3, {}^D\!\psi_3 \rangle_{[2]} = \int_{-\infty}^{\infty} t^2\left| {}^D\!\psi_3(t)\right|^2 dt = \frac{13281}{5600},$$

and all the nonzero GIP of ${}^D\!\psi_3$ can also be obtained from (4.2).

Example 3. $\phi_1 = N_3(x)$, the 3rd-order cardinal B-spline [1]. Without using the explicit expression of N_3, by Algorithm 3.1 and with $\phi_2(x) := x\phi_1(x)$, we have the autocorrelation symbol $\boldsymbol{\Phi}$ relative to $\phi = [\phi_1, \phi_2]^{\mathrm{T}}$ given by

$$\Phi_{1,1}(z) = \Phi_1(z) = \frac{1}{120z^2}(1 + 26z + 66z^2 + 26z^3 + z^4),$$

$$\Phi_{1,2}(z) = \Phi_{2,1}\left(\frac{1}{z}\right) = \frac{1}{120z^2}(5 + 104z + 198z^2 + 52z^3 + z^4),$$

$$\Phi_{2,2}(z) = \frac{1}{280z^2}(3 + 128z + 368z^2 + 128z^3 + 3z^4),$$

so that

$$\langle N_3, N_3 \rangle_{[1]} = \int_{-\infty}^{\infty} t\,|N_3(t)|^2\, dt = \frac{33}{40},$$

$$\langle N_3, N_3 \rangle_{[2]} = \int_{-\infty}^{\infty} t^2\,|N_3(t)|^2\, dt = \frac{46}{35}.$$

In addition, with the two-scale symbol of the corresponding spline wavelet ψ_1 given by

$$Q_{\psi_1}(z) = \frac{1}{960}(1-z)^3(1 - 26z + 66z^2 - 26z^3 + z^4),$$

and $\psi_2(x) := x\psi_1(x)$, we can also reach, by applying Algorithm 3.2,

$$\langle \psi_1, \psi_1 \rangle_{[0]} = \int_{-\infty}^{\infty} |\psi_1(t)|^2 \, dt = \frac{17011}{172800},$$

$$\langle \psi_1, \psi_1 \rangle_{[1]} = \int_{-\infty}^{\infty} t \, |\psi_1(t)|^2 \, dt = \frac{17011}{69120},$$

$$\langle \phi_1, \phi_1 \rangle_{[2]} = \int_{-\infty}^{\infty} t^2 \, |\psi_1(t)|^2 \, dt = \frac{1541701}{2419200}.$$

Acknowledgments. This work was partially supported by Texas Higher Education Coordinating Board Grant #000015-109.

References

1. Chui, C. K., *An Introduction to Wavelets*, Academic Press, Boston, 1992.
2. Chui, C. K. and J.-A. Lian, A study of orthonormal multi-wavelets, J. Appl. Numer. Math. **20** (1996), 272–298.
3. Daubechies, I., Orthonormal basis of compactly supported wavelets, Comm. Pure and Applied Math. **41** (1988), 909–996.
4. Daubechies, I., *Ten Lectures on Wavelets*, CBMS-NSF Series in Applied Math #61, SIAM Publ., Philadelphia, 1992.
5. Goodman, T. N. T. and S. L. Lee, Wavelets of multiplicity r, Trans. Amer. Math. Soc. **342** (1994), 307–324.
6. Jia, R. Q. and C. A. Micchelli, Using the refinement equations for the construction of pre-wavelets II: Powers of two, in *Curves and Surfaces*, P. J. Laurent, A. Lé Méhauté, and L. L. Schumaker (eds.), Academic Press, New York, 1991, pp. 209–246.

Jian-ao Lian and George A. Roberts
Department of Mathematics
Prairie View A&M University
Prairie View, Texas 77446
jian-ao_lian@pvamu.edu
groberts@zeno.math.pvamu.edu

Properly Posed Set of Nodes for Bivariate Lagrange Interpolation

Xue-Zhang Liang and Chun-Mei Lu

Abstract. This paper develops a theorem based on a problem proposed by the first author in 1965, and proves some theorems concerning the geometrical structure of certain properly posed set of nodes for bivariate Lagrange interpolation.

§1. Introduction

In this paper we discuss the bivariate Lagrange interpolation problem in the two-dimensional complex plane $\mathbf{C}^2$ (or in the two-dimensional real plane $\mathbf{R}^2$). Let n be a nonnegative integer. Π_n denotes the space of all bivariate polynomials of total degree $\leq n$

$$\Pi_n = \left\{ \sum_{0 \leq i+j \leq n} a_{ij} x^i y^j \,|\, a_{ij} \in \mathbf{C} \right\}.$$

Let $q_i = (x_i, y_i) \in \mathbf{C}^2, i = 1, 2, \ldots, k$, be k distinct points, where $k = \frac{1}{2}(n+1)(n+2)$. We consider the following Lagrange interpolation problem:

Problem 1. *Given any set of complex numbers*

$$\{ f_i \in \mathbf{C} \,|\, i = 1, 2, \ldots, k \}.$$

We seek a polynomial $P(x, y) \in \Pi_n$ satisfying

$$P(q_i) = f_i, \quad i = 1, 2, \ldots, k. \tag{1}$$

We call the set $\Re = \{q_i\}_{i=1}^k$ a properly posed set of nodes (or PPSN, for short) for Π_n if there exists a unique solution for equations (1).

It is clear that the condition

$$k = \dim \Pi_n = \frac{1}{2}(n+1)(n+2)$$

is necessary for $\Re$ being a PPSN for Π_n. But the condition is not sufficient. This situation is quite different from the univariate case. As well known for univariate polynomial interpolation, we have the following trivial observation

Charles K. Chui and Larry L. Schumaker (eds.), pp. 189–196.
Copyright © 1998 by Vanderbilt University Press, Nashville, TN.
ISBN 0-8265-1326-3.

Observation. Let $x_1, x_2, \ldots, x_{n+1}$ be $n+1$ distinct points in $\mathbf{C}^1$. Then given $n+1$ complex numbers $f_1, f_2, \ldots, f_{n+1}$, there exists a unique polynomial $P(x)$ of degree $\leq n$ such that

$$P(x_i) = f_i, \quad i = 1, 2, \ldots, n+1.$$

This observation is fundamental to the numerical analyst. But it cannot be simply generalized to the multivariate case. Therefore, the study of properly posed sets of nodes (namely PPSN) constitutes the first problem of multivariate polynomial interpolation. There are two approaches to this research:

(1) to find the properly posed sets of interpolation conditions for a given space of interpolating polynomials;

(2) to find the properly posed space of interpolating polynomials for a given set of interpolation conditions.

C. de Boor and A. Ron have obtained some results (see [3]) on the second approach. Our research work (see [6, 7, 8]) was focused on (1).

In 1965, X. Z. Liang gave the following lemma and theorem:

Lemma 1. $\{q_i\}_{i=1}^k$ $(k = \frac{1}{2}(n+1)(n+2))$ is a PPSN for Π_n, if and only if $\{q_i\}_{i=1}^k$ is not contained in any curve in Π_n. (We call $P(x, y) = 0$ a curve in Π_n if $P(x, y) \in \Pi_n$ and if $P(x, y)$ is not identically zero.)

Theorem 1. If $\{q_i\}_{i=1}^k$ is a PPSN for Π_n, and if none of these points lie on an irreducible curve $Q(x, y) = 0$ of degree l ($l = 1$ or $l = 2$; $l = 1$ means a straight line; $l = 2$ means a conic), then $\{q_i\}_{i=1}^k$ with the $(n+3)l - 1$ points being distinct and selected freely on the irreducible curve must constitute a PPSN for Π_{n+l}.

The purpose of this paper is to generalize the above theorem to the case $l \geq 3$, and finally deduce the geometrical structure of PPSN. To this end, we must introduce a few new concepts.

§2. Lagrange Interpolation along an Algebraic Curve without Multiple Factors

First we introduce the following

Definition 1. Let n, l be two natural numbers,

$$k = \begin{cases} \binom{n+2}{2} - \binom{n+2-l}{2}, & \text{if } n \geq l, \\ \binom{n+2}{2}, & \text{if } n \leq l-1. \end{cases}$$

Suppose that $Q(x, y) = 0$ is an algebraic curve without multiple factors (or ACWMF) of degree l and $\{q_i\}_{i=1}^k$ are distinct points on $Q(x, y) = 0$. If the assumptions

$$P(x, y) \in \Pi_n \quad \text{and} \quad P(q_i) = 0, \quad i = 1, 2, \ldots, k,$$

imply

$$P(x,y) \equiv 0$$

on the curve $Q(x,y) = 0$, *then we call* $\{q_i\}_{i=1}^k$ *a properly posed set of nodes for the interpolation of degree* n *along the ACWMF* $Q(x,y) = 0$ *of degree* l, *and write*

$$\{q_i\}_{i=1}^k \in I_n(Q).$$

The following theorem gives a general method to produce a PPSN for the interpolation of degree n along an ACWMF. It also shows the existence of PPSN for the interpolation.

Theorem 2. *Suppose* $Q(x,y) = 0$ *is an ACWMF of degree* l, *and* $Q(x,y)$ *is factorized as*

$$Q(x,y) = Q_1(x,y)\ldots Q_m(x,y),$$

where $Q_i(x,y)(i = 1, 2, \ldots, m)$ *are distinct irreducible polynomials of degree* l_i, *respectively, and* $l_1 + l_2 + \ldots + l_m = l$. *Let* n *be a nonnegative integer,*

$$r = \begin{cases} nl + m - (n+1)(n+2)/2, & \text{if } n \leq l-1, \\ m + l(l-3)/2, & \text{if } n \geq l. \end{cases}$$

Suppose that $nl_i + 1$ *distinct points are freely chosen on each factor curve* $Q_i(i = 1, 2, \ldots, m)$. *Then we can properly delete altogether* r *points from their aggregate, such that the remaining points constitute a PPSN for the interpolation of degree* n *along curve* Q.

Proof: The proof of this theorem for the case $n \leq l-1$ is simple. Hence, we only give a proof for the case $n \geq l$. Let $\mathcal{T} = \{q_i\}_{i=1}^k$ be the points chosen on the curve Q, where $k = nl + m$. After deleting some r points from them, the number of remaining points is exactly equal to the number of points in a PPSN for the interpolation of degree n along Q.

If $P(x,y) \in \Pi_n$ is the solution of system of equations

$$P(q_i) = 0, \quad i = 1, 2, \ldots, k, \tag{3}$$

then the nth algebraic curve P and l_ith algebraic curve Q_i intersect at least at $nl_i + 1$ distinct points $(i = 1, 2, \ldots, m)$. Due to Bezout's Theorem (see [9]), there exists $R(x,y) \in \Pi_{n-l}$ such that

$$P(x,y) = Q_1(x,y)\ldots Q_m(x,y)R(x,y) = Q(x,y)R(x,y). \tag{4}$$

Conversely, any $P(x,y) \in \Pi_n$ expressed by Equation (4) is a solution of Equation (3).

Let $\{T_1(x,y), \ldots, T_{s_1}(x,y)\}$, $\{R_1(x,y), \ldots, R_{s_2}(x,y)\}$ be a basis for Π_n, Π_{n-l}, respectively, $s_1 = \frac{1}{2}(n+1)(n+2)$, $s_2 = \frac{1}{2}(n-l+1)(n-l+2)$. From Equation (4), there exists in the solution space only s_2 linealy independent polynomials of degree n, say

$$P_i(x,y) = Q(x,y)R_i(x,y), \quad i = 1, 2, \ldots, s_2,$$

which pass through T. This shows the rank of the coefficient matrix $[T_j(q_i)]$, $(i = 1, 2, \ldots, k; j = 1, 2, \ldots, s_1)$ of equation (3) must be

$$s_1 - s_2 = nl - \frac{1}{2}(l^2 - 3l).$$

So there exists a nonsingular submatrix of order $s_1 - s_2$ and the $s_1 - s_2$ points related to the submatrix belong to $I_n(Q)$. This completes the proof. $\square$

In particular, we have

Theorem 3. *Suppose $Q(x, y) = 0$ is a lth $(l \geq 1)$ irreducible algebraic curve in $\mathbf{C}^2$. And suppose that $nl + 1$ distinct points are freely chosen in the curve Q. Then we can properly delete $r = \frac{1}{2}(l - 1)(l - 2)$ points from them, such that the remaining points belong to $I_n(Q)$.*

Let us recall (see [4]) the following

Cramer Strange Proposition. *Suppose the curve $P(x, y) = 0$ of degree n and the curve $Q(x, y) = 0$ of degree n meet exactly at n^2 distinct points $\{q_i\}_{i=1}^{n^2}$. Then any curve P_0 of degree n which passes through $\frac{1}{2}n(n + 3) - 1$ of these points must pass through some of the other $\frac{1}{2}(n - 1)(n - 2)$ points of intersection.*

From this proposition and Theorem 2 we can get the following

Theorem 4. *Suppose the curve $P(x, y) = 0$ of degree n and the curve $Q(x, y) = 0$ of degree n meet exactly at n^2 distinct points $\{q_i\}_{i=1}^{n^2}$. On the curve $Q(x, y) = 0$ we choose a point q_0 beyond $\{q_i\}_{i=1}^{n^2}$. Then q_0 with the $\frac{1}{2}n(n + 3) - 1$ points being distinct and selected freely in $\{q_i\}_{i=1}^{n^2}$ must constitute a PPSN for the interpolation of degree n along curve Q.*

Proof: We let $\{q_i\}_{i=1}^{k}$ $(k = \frac{1}{2}n(n + 3) - 1)$ be the points taken from $\{q_i\}_{i=1}^{n^2}$. Then the number of points in $\{q_i\}_{i=0}^{k}$ is $\frac{1}{2}n(n + 3)$, which is exactly equal to the number of points in a PPSN for the interpolation of degree n along curve Q.

Next we show $\{q_i\}_{i=0}^{k} \in I_n(Q)$. In fact, if

$$P(q_i) = 0, \quad i = 1, \ldots, k,$$

then by the Cramer Strange Proposition, the curve P must pass through the other $\frac{1}{2}(n - 1)(n - 2)$ points of intersection, so that

$$P(q_i) = 0, \quad i = 1, \ldots, n^2.$$

Noticing that $P(q_0) = 0$, then the nth algebraic curve P and the irreducible curve Q intersect at $n^2 + 1$ points. Since $Q(x, y)$ is irreducible, it follows from Bezout's Theorem, that there exists a constant c such that

$$P(x, y) = cQ(x, y)$$

This proves the theorem. $\square$

Furthermore, we prove the following

Theorem 5. *Suppose the curve $Q(x, y) = 0$ of degree k and the straight line $P(x, y) = 0$ meet exactly at k distinct points $\{q_i\}_{i=1}^k$, and suppose $\mathcal{T}_n \in I_n(Q)$ $(n \geq k - 2)$ and $\mathcal{T}_n \cap \{q_i\}_{i=1}^k = \emptyset$. Then we have*

$$\mathcal{T}_n \cup \{q_i\}_{i=1}^k \in I_{n+1}(Q).$$

Proof: We suppose that

$$P(x, y) = ax - y + c$$

and

$$q_i = (x_i, y_i), \quad i = 1, \ldots, k$$

with $x_1, x_2, \ldots, x_k$ are pairwise distinct.(Otherwise, we can make a coordinate rotation transform). Then we have

$$Q(x, y) = (ax - y + c)Q_1(x, y) + Q(x, ax + c),$$

with $\deg Q_1 \leq k - 1$, $\deg Q(x, ax + c) \leq k$ and

$$Q(x_i, ax_i + c) = 0, \quad i = 1, 2, \ldots, k.$$

So there must exist a constant $c_1(\neq 0)$ such that

$$Q(x, ax + c) = c_1(x - x_1)(x - x_2)\ldots(x - x_k).$$

On the other hand, for any given bivariate polynomial $F(x, y)$ with $\deg F$ at most $n + 1$ and satisfying

$$F(q) = 0 \text{ for all } q \text{ in } \mathcal{T}_n; \quad F(q_i) = 0, \quad i = 1, 2, \ldots, k,$$

we have

$$F(x, y) = (ax - y + c)F_1(x, y) + F(x, ax + c),$$

where $\deg F_1 \leq n$, $\deg F(x, ax + c) \leq n + 1$ and

$$F(x_i, ax_i + c) = F(x_i, y_i) - (ax_i - y_i + c)F_1(x_i, y_i) = 0, i = 1, 2, \ldots, k.$$

It is clear that there exists a univariate polynomial $G(x)$ of degree $\leq n - k + 1$ such that

$$F(x, ax + c) = G(x)(x - x_1)(x - x_2)\ldots(x - x_k).$$

Therefore, we have

$$\begin{aligned} F(x, y) &= (ax - y + c)F_1(x, y) + G(x)[Q(x, y) - (ax - y + c)Q_1(x, y)]/c_1 \\ &= (ax - y + c)[F_1(x, y) - G(x)Q_1(x, y)/c_1] + G(x)Q(x, y)/c_1 \\ &= (ax - y + c)H(x, y) + G(x)Q(x, y)/c_1 \end{aligned}$$

with $\deg H \leq n$. Since $F(q) = 0$ for all q in $\mathcal{T}_n$, we have $H(q) = 0$ for all q in $\mathcal{T}_n$. Since $\mathcal{T}_n \in I_n(Q)$, we have

$$H(x, y) = R(x, y)Q(x, y),$$

$$F(x, y) = ((ax - y + c)R(x, y) + G(x)/c_1)Q(x, y)$$

which completes the proof. $\square$

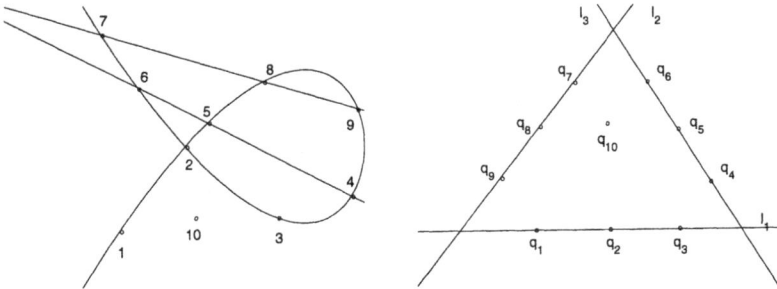

Fig. 1. Cubic superposition process.

§3. The Lagrange Interpolation in $\mathbf{C}^2$ (or $\mathbb{R}^2$)

In this section we study the structure of PPSN for Lagrange interpolation in $\mathbf{C}^2$ (or $\mathbb{R}^2$). The following theorem relates the interpolation problem in $\mathbf{C}^2$ to the problem of interpolation along an ACWMF.

Theorem 6. (Recursive Construction Theorem). *Let* $\Re$ *be a PPSN for* Π_n, *and the number of points in* $\Re$ *be* $\frac{1}{2}(n+1)(n+2)$. *If none of these points lie on an ACWMF* $Q(x,y) = 0$ *of degree* l, *then for any* $\mathcal{T} \in I_{n+l}(Q)$, $\mathcal{T} \cup \Re$ *must be a PPSN for* Π_{n+l}.

Proof: The number of points in $\mathcal{T} \cup \Re$ is

$$\frac{1}{2}(n+1)(n+2) + (n+l)l - \frac{1}{2}(l^2 - 3l) = \frac{1}{2}(n+l+1)(n+l+2),$$

which is exactly equal to the dimension of $\Pi_{n+l}(\mathbf{C}^2)$.

If $\mathcal{T} \cup \Re$ is not a PPSN for Π_{n+l}, then there must exist a curve $P(x,y) \in \Pi_{n+l}$ such that P passes through all of these points in $\mathcal{T} \cup \Re$. Since $\mathcal{T} \in I_{n+l}(Q)$, we have

$$P(x,y) = Q(x,y)R(x,y),$$

where $R(x,y) \in \Pi_n$. Since the curve P passes through all the points in set $\Re$, which are not on the curve Q, the nth algebraic curve $R(x,y) = 0$ passes through $\Re$, which contradicts the assumption that $\Re$ is a PPSN for Π_n. Therefore, $\mathcal{T} \cup \Re$ must be a PPSN for Π_{n+l}. $\square$

Example 1. Cubic Superposition Process. Let $Q(x,y) : x^2 - cy^2 + y^3 = 0 (c \geq 0)$ be a cubic curve (see the figure on the left in Figure 1).

$\{1,2,3\} \in I_1(Q)$ by Definition 1.

$\{1,\ldots,6\} \in I_2(Q)$ by Theorem 5.

$\{1,\ldots,9\} \in I_3(Q)$ by Theorem 5.

$\{1,\ldots,10\}$ is a PPSN for Π_3 by Theorem 6.

Example 2. Three intersecting lines l_1, l_2, l_3 make up a triangle in $\mathbb{R}^2$. First we divide each side of the triangle to four equal sections. Then select from each side the three equally distant points beyond the three vertices, and get a point set $\{q_i\}_{i=1}^9$. We can prove $\{q_i\}_{i=1}^9$ is a PPSN for the interpolation of degree 3 along the 3rd ACWMF $l_1 l_2 l_3$. Then $\{q_i\}_{i=1}^9$ with a point q_{10} selected freely inside the triangle must constitute a PPSN for Π_3 (see the figure on the right in Figure 1).

The following theorem can be regarded as the converse proposition to Theorem 6 and finally it gives the geometrical structure of PPSN.

Theorem 7. (Structure of PPSN). *Let $\mathfrak{R}$ be a PPSN for Π_n. Then $\mathfrak{R}$ can be generated by the Recursive Construction Theorem.*

Proof: Suppose $k = \frac{1}{2}(n+1)(n+2)$, $\mathfrak{R} = \{q_i\}_{i=1}^k$, and $L_1(x,y), \ldots, L_k(x,y)$ are the fundamental Lagrange interpolation polynomials which satisfy

$$L_i(q_j) = \begin{cases} 1, & \text{if } i = j, \\ 0, & \text{if } i \neq j, \end{cases}$$

where $i, j = 1, \ldots, k$. We can assert that every $L_i(x,y)$ is a nth ACWMF $(i = 1, 2, \ldots, k)$.

In what follows we prove that if we select a point in $\mathfrak{R}$, it might as well be the point q_1. Then the remaining points $\{q_i\}_{i=2}^k$ must be a PPSN for the interpolation of degree n along the nth ACWMF $L_1(x,y) = 0$. In fact, suppose that

$$P(x,y) \in \Pi_n, \quad P(x,y) \neq 0, \quad P(q_i) = 0, \quad i = 2, \ldots, k.$$

Since $\{q_i\}_{i=1}^k$ is a PPSN for Π_n, we have

$$P(q_1) \neq 0.$$

(Otherwise, it contradicts the basic Lemma in introduction.) We set

$$\widetilde{P}(q_i) = P(x,y)/P(q_1).$$

Then

$$\widetilde{P}(q_1) = 1, \quad \widetilde{P}(q_i) = 0, \quad i = 2, \ldots, k.$$

So we have

$$\widetilde{P}(x,y) = L_1(x,y)$$

and

$$P(x,y) = P(q_1)L_1(x,y).$$

Therefore $\{q_i\}_{i=2}^k$ is a PPSN for the interpolation of degree n along algebraic curve L_1. The theorem thus holds. $\square$

Acknowledgments. Supported by the National Natural Science Foundation of China.

References

1. Chui, C. K. and M. J. Lai, Vandermonde determinant and Lagrange interpolation in R^s, in *Nonlinear and Convex Analysis*, B. L. Lin (ed), Marcel Dekker, New York, 1987, pp. 23–36.

2. Chung, K. C. and T. H. Yao, On lattices admitting unique Lagrange interpolation, SIAM J. Numer. Anal. **14** (1977), 735–743.

3. de Boor, C. and A. Ron, The least solution for the polynomial interpolation problem, Math. Z. **210** (1992), 347–378.

4. Fang, D. Z. and Y. P. Chen, *Projective Geometry*, Advanced Education Press, Beijing, 1983.

5. Gasca, M. and J. I. Maeztu, On Lagrange and Hermite interpolation in R^k, Numer. Math. **39** (1982), 1–14.

6. Liang, X. Z., *On the Interpolation and Approximation in Several Variables*, Postgraduate Thesis, Jilin Univ., Changchun, 1965.

7. Liang, X. Z., Lagrange representation of multivariate interpolation, Science in China (Ser. A) **4** (1989), 385–396.

8. Liang, X. Z. and L. Q. Li, On bivariate osculatory interpolation, J. Comput. Appl. Math. V **38** (1991), 271–282.

9. Walker, R. J., *Algebraic Curves*, Princeton Univ. Press, NJ, 1950.

Xue-Zhang Liang
Institute of Mathematics
Jilin University
Changchun, 130023
P. R. China
xzliang@mail.jlu.edu.cn

Chun-Mei Lu
Department of Mathematics
University of South Carolina
Columbia, SC 29208
USA
clu@math.sc.edu

On the Closure of the Union of Nested Subspaces of $L^2(\mathrm{R}^d)$

R. A. Lorentz and W. R. Madych

Abstract. We characterize the closure of the union of a nested sequence $\{V_j\}_{j \in \mathbb{Z}}$ of closed subspaces of $L^2(\mathbb{R}^d)$ under various assumptions on the subspaces. A typical case is that V_0 is invariant under shifts from $\mathbb{Z}^d$ and that V_j consists of the A^j-dilates of V_0 for some "expanding" $d \times d$-matrix A. In the case that the closure is not dense, we use the characterization to show that there exists another nested sequence of spaces with the same translation and dilation properties, orthogonal to the original sequence, such that together they span all of $L^2(\mathbb{R}^d)$.

§1. Introduction

We investigate the closure of a nested sequence of closed subspaces of $L^2(\mathbb{R}^d)$, each of which is invariant under certain shifts. Our objective is to characterize this closure, which includes giving necessary and sufficient conditions that it be all of $L^2(\mathbb{R}^d)$, and if it is not, to capture the missing part with another similarly nested sequence. Situations of this type are multiresolution analyses (MRAs) without the density property and the shift invariant spaces investigated in [1]. In the previous paper [5], we looked at univariate MRAs with this goal in mind. We show that almost all of the results achieved in [5] hold for multivariate MRAs with arbitrary dilation matrices. The main difficulties in this extension are the necessity of a version of Lebesgue theorem valid for families of nonnested sets (Lemma 2) and the construction of a kind of pseudo-norm for which the application of the dilation matrix is the same as multiplication by a constant (in the proof of Theorem 1).

We use the following terminology. $\mathcal{V} = \{V_j\}_{j \in \mathbb{Z}}$ is a nested sequence of closed subspaces of $L^2(\mathbb{R}^d)$; $V_j \subset V_{j+1}$ for $j \in \mathbb{Z}$. By V_∞, we denote the L^2-closure of the union of the V_j. We say that a closed subspace U of $L^2(\mathbb{R}^d)$ is shift invariant if for any $u \in U$, $u(\cdot - k) \in U$ for any $k \in \mathbb{Z}^d$. If A is any $d \times d$-matrix, then the subspace U is said to be A-shift invariant if for any

Approximation Theory IX, Volume 2: Computational Aspects
Charles K. Chui and Larry L. Schumaker (eds.), pp. 197–204.
Copyright © 1998 by Vanderbilt University Press, Nashville, TN.
ISBN 0-8265-1326-3.

$u \in U$, $u(\cdot - Ak) \in U$ for any $k \in \mathbb{Z}^d$. Finally, the sequence $\mathcal{V}$ is said to be A-dilation invariant if for each $j \in \mathbb{Z}$, $V_{j+1} = \{v(A\cdot) \mid v \in V_j\}$.

The most general situation we consider is a $\mathcal{V}$ such that the V_j are A^{-j}-shift invariant for some invertible matrix A. The case $A = 2I$, where I is the identity on $\mathbb{R}^d$, was investigated in [1, 2] in regard to density, i.e., as to when $V_\infty = L^2(\mathbb{R}^d)$ and with regard to the approximation power of the spaces V_j.

We call a $d \times d$-matrix a dilation if all of the eigenvalues of A are in modulus strictly larger than 1. Note that we do not assume that A maps $\mathbb{Z}^d$ to $\mathbb{Z}^d$ as would be usual for a MRA. For a dilation matrix A, we define B to be the adjoint of A: $B = A^*$. The assumption that A is a dilation already suffices to show that for any nested A^{-j}-shift invariant nested sequence $\mathcal{V}$, V_∞ is translation invariant, i.e., if $v \in V_\infty$, then $v(\cdot - y) \in V_\infty$ for any $y \in \mathbb{R}^d$ (see Lemma 1). In Corollary 1, a characterization for V_∞ is given. This characterization leads to a necessary and sufficient condition that $V_\infty = L^2(\mathbb{R}^d)$ (see Corollary 2). These conditions coincide with those in [1] for the case that each V_j is generated by one function, $V_j = V(\phi_j)$, and with $A = 2I$. We say that $\phi \in L^2(\mathbb{R}^d)$ generates the subspace V of $L^2(\mathbb{R}^d)$ if V is the L^2-closure of the space of all finite linear combinations of the functions $\phi(\cdot - k)$, $k \in \mathbb{Z}^d$.

Suppose now that $\mathcal{V}$ is A-dilation invariant in addition, so that $V_j = \{v(A^j\cdot) \mid v \in V_0\}$ for all $j \in \mathbb{Z}$. In Theorem 1, we show that there is then another nested sequence $\mathcal{U} = \{U_j\}_{j \in \mathbb{Z}}$ with the same translation and dilation properties as $\mathcal{V}$ such that

$$V_\infty \oplus U_\infty = L^2(\mathbb{R}^d) .$$

Moreover, U_0 has a Riesz basis consisting of the integer shifts of some function. This situation is very close to that of a multiresolution analysis.

We say that a collection $\mathcal{V} = \{V_j\}_{j \in \mathbb{Z}}$ of closed linear subspaces of $L^2(\mathbb{R}^d)$ is a multiresolution analysis if it satisfies

(A1) $V_j \subset V_{j+1}$ $j \in \mathbb{Z}$;
(A2) $f(x) \in V_j$ if and only if $f(Ax) \in V_{j+1}$;
(A3)   There is a φ in V_0 such that $\{\varphi(x - k)\}_{k \in \mathbb{Z}^d}$ is a Riesz basis for V_0.

Here A, which we call the MRA-dilation matrix, is a scaling matrix which satisfies $A\mathbb{Z}^d \subset \mathbb{Z}^d$. The function φ in (A3) is referred to as the scaling function. If in addition

(A4) $\bigcup_{j \in \mathbb{Z}} V_j$ is dense in $L^2(\mathbb{R}^d)$,

or equivalently, the scaling function φ satisfies

(A4)$_1$ $\lim_{k \to \infty} \frac{1}{|B^{-k}Q|} \int_{B^{-k}Q} \frac{|\hat{\varphi}(\xi)|^2}{\sum_{k \in \mathbb{Z}^d} |\hat{\varphi}(\xi - 2\pi k)|^2} d\xi = 1$

for all cubes Q of finite diameter, then we say that $\mathcal{V}$ is a multiresolution analysis for $L^2(\mathbb{R}^d)$ and note that this definition is equivalent to the usual one, see [6]. Here, $B = A^*$, $|U|$ is the measure of the measurable set A and $\hat{f} = \mathcal{F}f$ is the Fourier transform of f defined by $\hat{f}(\xi) = \mathcal{F}f(\xi) = \int_{\mathbb{R}^d} f(x)e^{-ix\xi} dx$ for $f \in L^1(\mathbb{R}^d)$ and distributionally otherwise.

When discussing MRAs in the following, we don't really need dilations which map $\mathbb{Z}^d$ to $\mathbb{Z}^d$, we just need the properties demanded of the eigenvalues. Moreover, we use the Riesz basis property only for Theorem 3 and property $(A4)_3$ of Theorem 4.

Theorem 2 shows that for each incomplete MRA $\mathcal{V}$, there is another MRA $\mathcal{U}$ with the same dilation matrix such that $V_\infty \oplus U_\infty = L^2(\mathbb{R}^d)$. In Theorem 4, we give two other necessary and sufficient conditions that a MRA be complete.

§2. The Main Results

A subset Γ of $\mathbb{R}^d$ is called absorbing with respect to a $d \times d$ invertible matrix B, or B-absorbing, if, for almost all $\xi \in \mathbb{R}^d$, there is an integer k_0, which may depend on ξ, such that $B^{-k}\xi \in \Gamma$ for all integers k with $k \geq k_0$.

A subset Γ of $\mathbb{R}^d$ is called B-dilation invariant with respect to a $d \times d$ invertible matrix B if $\xi \in \Gamma$ implies that $B^{-k}\xi \in \Gamma$ for all $k \in \mathbb{Z}$.

In general, formulas are valid only up to sets of measure zero even if not explicitly so stated.

Lemma 1. *Let A be a $d \times d$-matrix. Let $\mathcal{V} = \{V_j\}_{j\in\mathbb{Z}}$ be a nested sequence of closed subspaces of $L^2(\mathbb{R}^d)$ such that V_j is A^{-j}-shift invariant. Then V_∞ is translation invariant if $\cup_{j\in\mathbb{Z}} A^{-j}\mathbb{Z}^d$ is dense in $\mathbb{R}^d$.*

Using the notation $\mathcal{F}^{-1}V = \{f \mid \hat{f} \in V\}$ for subsets V of $L^2(\mathbb{R}^d)$ and $L^2(A) = \{f \in L^2(\mathbb{R}^d) \mid f(x) = 0 \text{ whenever } x \in \mathbb{R}^d \setminus A\}$, the translation invariance of V_∞ is equivalent to $V_\infty = \mathcal{F}^{-1}\{L^2(\Omega)\}$ for some measurable subset Ω of $\mathbb{R}^d$, see [3]. Thus we have

Corollary 1. *Under the assumptions of Lemma 1, assume in addition that A is a dilation matrix. Then V_∞ is translation invariant and $V_\infty = \mathcal{F}^{-1}\{L^2(\Omega)\}$ for the measurable set*

$$\Omega = \cup_{j\in\mathbb{Z}} \cup_{\ell\in\mathbb{Z}} \{\xi \mid \hat{f}_{j,\ell}(\xi) \neq 0\}, \tag{1}$$

where, for each $j \in \mathbb{Z}$, $\{f_{j,\ell}\}_{\ell\in\mathbb{Z}}$ is a countable dense subset of V_j.

From Corollary 1, we immediately have

Corollary 2. *Under the assumptions of Corollary 1, $V_\infty = L^2(\mathbb{R}^d)$ if and only if*

$$\cup_{j\in\mathbb{Z}} \cup_{\ell\in\mathbb{Z}} \{\xi \mid \hat{f}_{j,\ell}(\xi) \neq 0\} = \mathbb{R}^d,$$

for some choice of $f_{f,\ell}$ such that for each $j \in \mathbb{Z}$, $\{f_{j,\ell}\}_{\ell\in\mathbb{Z}}$ is a countable dense subset of V_j

The situation becomes simpler if the spaces are dilates of each other.

Corollary 3. *In addition to the assumptions of Lemma 1 and Corollary 1, let, for each $j \in \mathbb{Z}$, V_j be the A^j-dilate of V_0. Then the Ω of Corollary 1 is B-dilation invariant and is of the form*

$$\Omega = \cup_{j\in\mathbb{Z}} \cup_{\ell\in\mathbb{Z}} \{\xi \mid \hat{f}_\ell(B^{-j}\xi) \neq 0\}, \tag{2}$$

where $\{f_\ell\}_{\ell \in \mathbb{Z}}$ is a countable dense subset of V_0 and $B = A^*$.

Theorem 1. *Let Ω be a measurable subset of $\mathbb{R}^d$ of positive measure which is B-dilation invariant, with $B = A^*$ for a dilation matrix A. Then there is nested sequence $\mathcal{U}$ of A-dilation invariant spaces U_j such that $U_\infty = \mathcal{F}^{-1}\{L^2(\Omega)\}$. Moreover, U_0 has a Riesz basis consisting of the integer shifts of one function. Thus if A is a MRA-dilation, $\mathcal{U}$ forms a MRA.*

Combining Corollary 3 and Theorem 1, we obtain

Corollary 4. *Under the assumptions of Corollary 3, there is a nested sequence $\mathcal{U}$ of A-dilation invariant closed spaces U_j such that*

$$V_\infty \oplus U_\infty = L^2(\mathbb{R}^d) . \tag{3}$$

Let $V_0 = V(\phi)$, i.e., let V_0 be generated by ϕ. Using the characterization

$$v \in V(\phi) \Leftrightarrow \hat{v} = h\hat{\phi}$$

for some 2π-periodic function h (see, e.g., [2] or [6]), we have the following corollary.

Corollary 5. *In addition to the assumptions of Corollary 4, assume that ϕ generates V_0. Then the Ω of Corollary 1 has the form*

$$\Omega = \{\xi \mid \hat{\phi}(B^{-j}\xi) \neq 0 \text{ for some } j \in \mathbb{Z}\} . \tag{4}$$

Of course, the rest of Corollary 4 holds.

Finally, we come to the case of an MRA, which follows directly from Theorem 1 and Corollaries 4 and 5.

Theorem 2. *Let $\mathcal{V}$ be a multiresolution analysis with scaling function φ and MRA-scaling matrix A and $B = A^*$. Then $V_\infty = \mathcal{F}^{-1}\{L^2(\Omega)\}$ for*

$$\Omega = \{\xi \mid \hat{\phi}(B^{-j}\xi) \neq 0 \text{ for some } j \in \mathbb{Z}\} , \tag{5}$$

and there is another multiresolution analysis $\mathcal{U} = \{U_j\}_{j \in \mathbb{Z}}$ which complements $\mathcal{V}$ in the sense that we have as an orthogonal sum:

$$V_\infty \oplus U_\infty = L^2(\mathbb{R}).$$

In [5], the following theorem was proved in the univariate case. The proof for a multivariate MRA is the same.

Theorem 3. *Let $\mathcal{V}$ be a multiresolution analysis with scaling function φ and MRA-scaling matrix A. Then the set Ω of Theorem 2 is given by*

$$\Omega = \{\xi \mid \lim_{j \to \infty} \frac{|\hat{\varphi}(B^{-j}\xi)|}{\{\sum_{k \in \mathbb{Z}^d} |\hat{\varphi}(B^{-j}\xi - 2\pi k)|^2\}^{1/2}} = 1\}, \tag{6}$$

where $B = A^*$.

Theorems 2 and 3 can be used to derive alternative equivalent formulations of the density condition (A4) or (A4)$_1$ for MRAs.

Theorem 4. *A multiresolution analysis $\mathcal{V}$ with scaling function φ and scaling matrix A, with $B = A^*$, is a multiresolution analysis of $L^2(\mathbb{R}^d)$ if and only if one of the following equivalent conditions is satisfied*
(A4)$_2$ *The set $\{\xi : |\hat{\varphi}(\xi)| > 0\}$ is B–absorbing;*
(A4)$_3$ $\lim_{j \to \infty} |\hat{\varphi}(B^{-j}\xi)| > 0$ *for almost all $\xi \in \mathbb{R}^d$.*

§3. Details

We use the notation $[f]$ to denote the function

$$[f](\xi) = \sum_{k \in \mathbb{Z}^d} |f(x - 2\pi k)|^2$$

which is well defined whenever f is in $L^2(\mathbb{R}^d)$. For any $\varphi \in L^2(\mathbb{R}^d)$, the fact that its integer translates form a Riesz basis is equivalent to

$$0 < A_1 \leq [\hat{\varphi}](\xi) \leq A_2 < \infty \tag{7}$$

for some positive constants A_1 and A_2. Without loss of generality, we assume that this inequality holds for all ξ in $\mathbb{R}^d$.

Proof of Lemma 1: Since any $v \in \cup_{j \in \mathbb{Z}}V_j$ belongs to some V_J, $v(\cdot - A^{-m}k) \in \cup_{j \in \mathbb{Z}}V_j$ for any $m \geq J$ and $k \in \mathbb{Z}^d$. Since the translation operator $T_y v(\cdot) := v(\cdot - y)$ on $L^2(\mathbb{R}^d)$ depends continuously on y and since, by assumption, $\cup_{j \in \mathbb{Z}}A^{-j}\mathbb{Z}^d$ is dense in $\mathbb{R}^d$, $v(\cdot - y) \in V_\infty$ for any $y \in \mathbb{R}^d$. Now let $f \in V_\infty$. Then there is a $v \in \cup_{j \in \mathbb{Z}}V_j$ arbitrarily close to f, and so $v(\cdot - y)$ is arbitrarily close to $f(\cdot - y)$ for any $y \in \mathbb{R}^d$. Since V_∞ is closed, $f \in V_\infty$. □

Proof of Corollary 1: We first show that V_∞ is translation invariant by showing that $\cup_{j \in \mathbb{Z}}A^{-j}\mathbb{Z}^d$ is dense in $\mathbb{R}^d$. Let $y \in \mathbb{R}^d$ and $\varepsilon > 0$ be given. Since $\sqrt[j]{\|A^{-j}\|}$ converges to the spectral norm of A^{-1} which is less than 1, there is a $J > 0$ for which $\|A^{-j}\| < \varepsilon/\sqrt{d}$. Letting z be the point in $\mathbb{Z}^d$ closest to y, we have $|A^J y - z| \leq d^{1/2}$ and so

$$|y - A^{-J}z| \leq \|A^{-J}\|d^{1/2} < \varepsilon ,$$

which proves the density and , by Lemma 1, that V_∞ is translation invariant. Thus, $V_\infty = \mathcal{F}^{-1}\{L^2(\Omega)\}$ for a measurable subset Ω of $\mathbb{R}^d$.
In [3], the formula given for Ω is $\Omega = \cup_{n=1}^{\infty}\{\xi \mid \hat{f}_n(\xi) \neq 0\}$ for any countable dense subset $\{f_n\}$ of V_∞. The set $\cup_{j \in \mathbb{Z}} \cup_{\ell \in \mathbb{Z}} \hat{f}_{j,\ell}$, where $\{f_{j,\ell}\}_{\ell \in \mathbb{Z}}$ is a countable dense subset of V_j, is such a subset. □

Proof of Corollary 3: Corollary 3 follows from Corollary 1 by noting that if $\{f_\ell\}_{\ell \in \mathbb{Z}}$ is a countable dense subset of V_0, then $\{f_\ell(A^j \cdot)\}_{\ell \in \mathbb{Z}}$ is a countable dense subset of V_j and that $\mathcal{F}f(A \cdot)(\xi) = |\det A|^{-1}\hat{f}(B^{-1}\xi)$. □

Theorem 1 is much more difficult to prove. We need two lemmas for this, the first being a special type of Lebesgue theorem. We could not find the particular form we need in the literature, but Stöckler [9], has pointed out, that the standard version, e.g., in [4], can be modified. The essential fact needed for its proof, which we omit, is that B^{-k} is a contraction for some k large enough.

Lemma 2. *Let* $f \in L^1(\mathbb{R}^d)$, *B be the transpose of a dilation matrix and $\mathcal{Q}$ be the half-open cube* $[-1, 1)^d$. *For $k \in \mathbb{Z}$, we define a function F_k as follows. For each $x \in \mathbb{R}^d$, there is a unique $\ell(x) \in \mathbb{Z}^d$ for which $x \in B^{-k}(\mathcal{Q} + \ell(x))$. Let*

$$F_k(x) = |B^{-k}\mathcal{Q}|^{-1} \int_{B^{-k}(\mathcal{Q}+\ell(x))} f(y)dy. \tag{8}$$

Then F_k converges to f in $L^1(\mathbb{R}^d)$ as $k \to \infty$.

Lemma 3. *Let Ω be a measurable subset of $\mathbb{R}^d$ of positive measure. Let A be a scaling matrix and $B = A^*$ and $\Omega_T := \cup_{\ell \in \mathbb{Z}^d}(\Omega + 2\pi\ell)$. If Ω is B-dilation invariant, then*

$$\Omega_T = \mathbb{R}^d$$

up to a set of measure zero.

Proof: Fix $x \in \Omega$ and let $U = \{y \in \mathbb{R}^d \mid |y| \leq 2|x| + 1\}$. Then $\chi_{\Omega \cap U}$ belongs to $L^1(\mathbb{R}^d)$, χ_A being the characteristic function of the set A. Thus, given an $\varepsilon > 0$, we can use Lemma 2 for the function $f = \chi_{\Omega \cap U}$ to conclude that there are k and $\ell(x)$ with $|\chi_{\Omega \cap U}(x) - F_k(x)| < \varepsilon$ where F_k is from (8). So,

$$\frac{|B^{-k}(\mathcal{Q} + \ell(x)) \cap \Omega \cap U|}{|B^{-k}(\mathcal{Q} + \ell(x))|} \geq 1 - \varepsilon.$$

By taking k larger if necessary, we can guarantee that $B^{-k}(\mathcal{Q} + \ell(x)) \subset U$ so that

$$|B^{-k}(\mathcal{Q} + \ell(x)) \cap \Omega| \geq (1 - \varepsilon)|A^{-k}(\mathcal{Q} + \ell(x))|.$$

Thus we also have

$$|B^{-k}(\mathcal{Q} + \ell(x)) \cap \Omega| \geq (1 - \varepsilon)|\mathcal{Q} + \ell(x)| = (1 - \varepsilon)(2\pi)^d$$

since Ω is B-dilation invariant. From the definition of Ω_T, $|\mathcal{Q} \cap \Omega_T| \geq (1 - \varepsilon)(2\pi)^d$ for any $\varepsilon > 0$ and so $|\mathcal{Q} \cap \Omega_T| = (2\pi)^d$. Now the lemma follows directly from the definition of Ω_T. $\square$

Continuation of the proof of Theorem 1. To construct the function ϕ mentioned in Theorem 1, we need a nonnegative function $F : \mathbb{R}^d \to \mathbb{R}_+$ with the following properties:
a) F is measurable,
b) $F(\xi) > 0$ for $|\xi| > 0$,
c) $F(B\xi) = \lambda F(\xi)$ for some $\lambda > 0$,

d) $F(\xi) \leq C|\xi|^{-1}$ for some $C > 0$ and all ξ large enough.

Here $|x|$ is the Euclidean norm of $x \in \mathbb{R}^d$.

Let $\ell > 0$ be such that $|B^\ell \xi| > 2|\xi|$ for any $\xi \in \mathbb{R}^d$. For this ℓ, let $g = \chi_{\{\xi \mid 1 \leq |\xi| \leq \|B^\ell\|\}}$. With $\lambda = \|B^\ell\|^{-1/2}$, so that $\lambda < 1$, we define F by

$$F(\xi) = \sum_{j \in \mathbb{Z}} \lambda^j g(|B^j \xi|).$$

and claim that F satisfies a)–d).

Noticing that for each $\xi \in \mathbb{R}^d$, there is a j such that $1 \leq |B^{-j}\xi| \leq \|B^\ell\|$, we have b). Since B^ℓ is expanding, the sum defining F has, for each ξ, at most a finite number r (which we use below), independent of ξ, of nonzero summands. Thus F is a step function, which shows a). Property c) follows directly from the definition of F.

To show d), let $\xi \in \mathbb{R}^d$ and k be such that $1 \leq |B^{-k}\xi| \leq \|B^\ell\|$. Let $k = m\ell + s$ with $0 \leq s \leq \ell - 1$, then

$$\|B^\ell\| \geq |B^{-k}\xi| \geq \|B^\ell\|^{-m}\|B^s\|^{-1}|\xi| \geq C_1\|B^\ell\|^{-k/\ell}|\xi| = C_1\lambda^k|\xi|$$

for some $C_1 > 0$ independent of ξ and k. Thus, for this ξ, $\lambda^k \leq C_2|\xi|^{-1}$ and so,

$$F(\xi) \leq \sum_{j=k-r}^{k+r} \lambda^j \leq (2r+1)\lambda^{k-r} \leq C_3|\xi|^{-1}$$

where r is the constant mentioned above.

Now let Ω be a subset of $\mathbb{R}^d$ of positive measure which is B-dilation invariant. Let

$$h(\xi) = \chi_\Omega(\xi)F(\xi)^s \tag{9}$$

for some $s > d/2$ and let φ be defined via

$$\hat{\varphi}(\xi) = \frac{h(\xi)}{\{[h](\xi)\}^{1/2}}.$$

Note that $h \in L^2(\mathbb{R}^d)$ and that by Lemma 3, $[h] > 0$ a.e. so that $\hat{\varphi}$ is well defined. Also $0 \leq \hat{\varphi} \leq 1$ and $\hat{\varphi} \in L^2(\mathbb{R}^d)$. Moreover, $[\hat{\varphi}] = 1$, so that, by (7), the integer translates of φ form a Riesz basis the space they generate. Since $\chi_\Omega(B^{-1}\xi) = \chi_\Omega(\xi)$, $\hat{\varphi}$ satisfies a scaling equation with

$$S(\xi) = \lambda^s \left\{ \frac{[h](\xi)}{[h](B(\xi))} \right\}^{1/2}.$$

Thus, if $U = U(\phi)$, then U is shift invariant, the A-dilates U_j of U form a nested sequence and $U_\infty = \mathcal{F}^{-1}\{L^2(\Omega)\}$. Moreover, if A is a MRA-dilation then $\mathcal{U}$ forms a MRA. $\square$

Proof of Corollary 4: By Corollary 3, $V_\infty = \mathcal{F}^{-1}\{L^2(\Omega)\}$. By assumption, $\Omega^c = \mathbb{R}^d \setminus \Omega$ is a measurable set of positive measure. Since Ω is B-dilation invariant, so is Ω^c. Thus, by Theorem 1, there is a an A-invariant nested sequence $\mathcal{U}$ with $U_\infty = \mathcal{F}^{-1}\{L^2(\Omega^c)\}$. Thus $V_\infty \perp U_\infty$. Since $\Omega \cup \Omega^c = \mathbb{R}^d$,

$$L^2(\mathbb{R}) = \mathcal{F}^{-1}(L^2(\mathbb{R})) = \mathcal{F}^{-1}(L^2(\Omega) \oplus L^2(\Omega^c)) = V_\infty \oplus U_\infty. \quad \square$$

Proof of Theorem 4: Condition $(A4)_2$ is just a reformulation of the formula (5). For $(A4)_3$, we observe that since $\{\phi(\cdot - k)\}$ is a Riesz basis, by (6), $\{\sum_{k\in\mathbb{Z}^d} |\hat{\varphi}(B^{-j}\xi - 2\pi k)|^2\}^{1/2} \leq A_2$, for some $A_2 > 0$. This shows that $\Omega = L^2(\mathbb{R}^d)$ is equivalent to $\lim_{j\to\infty} |\hat{\phi}(B^{-j}\xi)| \geq A_2^{-1}$. $\quad \square$

References

1. de Boor, C., R. A. DeVore, and A. Ron, On the construction of multivariate (pre)wavelets, Constr. Approx. **9** (1993), 123–126.

2. de Boor, C., R. A. DeVore, and A. Ron, Approximation from shift-invariant subspaces of $L_2(\mathbb{R}^d)$, Trans. Amer. Math. Soc. **341** (1994), 787–806.

3. Dym, H. and H. P. McKean, *Fourier Series and Integrals*, Academic Press, New York, 1972.

4. Edwards, R. E. and G. I. Gaudry, *Littlewood-Paley Multiplier Theorems*, Springer, Berlin, 1977.

5. Lorentz, R. A. and W. R. Madych, Translation and dilation invariant subspaces of $L^2(\mathbb{R})$ and multiresolution analyses, Appl. Comp. Harmonic Anal., to appear.

6. Madych, W. R., Some elementary properties of multiresolution analyses of $L^2(\mathbb{R}^n)$, in *Wavelets - A Tutorial in Theory and Applications*, C. K. Chui (ed.), Academic Press, Boston, 1992, pp. 259–294.

7. Mallat, S., Multiresolution approximations and wavelet orthonormal bases of $L^2(\mathbb{R})$, Trans. Amer. Math. Soc. **315** (1989), 69–87.

8. Meyer, Y., *Wavelets and Operators*, Cambridge University Press, Cambridge, 1992.

9. Stöckler, J., *Multivariate Affine Frames*, Habilitation, University of Duisburg, 1995.

R. A. Lorentz
GMD
Schloss Birlinghoven
53757 St. Augustin, Germany
lorentz@gmd.de

W. R. Madych
Department of Mathematics, U-9
University of Connecticut
Storrs, CT 06269, USA
madych@uconnvm.uconn.edu

Calculating Joint Spectral Radius of Matrices

and Hölder Exponent of Wavelets

Mohsen Maesumi

Abstract. The joint spectral radius (jsr) of a bounded collection of matrices $\mathcal{M}$ is the smallest positive number r such that $\mathcal{M}/r$ generates a norm-bounded semi-group. This quantity can be used to determine Hölder regularity of compactly supported wavelets. The finiteness Conjecture of Lagarias and Daubechies states that if $\mathcal{M}$ is finite then a certain optimal product P of n elements of $\mathcal{M}$ attains the maximal growth rate, $\rho(P) = \mathrm{jsr}(\mathcal{M})^n$. We describe an algorithm which is conjectured to terminate iff P is an optimal product. Experimental results of the algorithm relating to the Hölder exponent of four-coefficient multiresolution analyses are presented.

§1. Introduction

The concept of iteration is a cornerstone of many mathematical disciplines. It is a favorite tool in approximation, cascade algorithm for wavelets of compact support [9–11], refinement algorithm for computer aided design [6, 23], and image analysis techniques [2]. It is also the central theme in dynamical systems [22], Markov Chains [25], asynchronous processes in control theory [26, 27], fractal generation [1, 2, 22], and certain functional equations [17]. There are several general dividing lines. Iteration can be performed on a single function or a set of functions. The functions, in turn, can be linear or nonlinear. The style of iteration can be discrete or continuous. Each set of choices leads to a well established discipline. In any iterative system the measurement of the maximal rate of growth of the iterates plays an important role. This article focuses on the measurement of this rate in a linear iterative system and its application to Hölder regularity analysis of wavelets of compact support.

The concept of spectral radius for linear iterative systems has been defined through natural generalizations of the concept of spectral radius of a matrix to a bounded sets of matrices $\mathcal{M}$. There are three equivalent definitions for $\rho(\mathcal{M})$. Common spectral radius, $\mathrm{csr}(\mathcal{M})$, is the infimum of all matrix

Approximation Theory IX, Volume 2: Computational Aspects
Charles K. Chui and Larry L. Schumaker (eds.), pp. 205–212.
Copyright © 1998 by Vanderbilt University Press, Nashville, TN.
ISBN 0-8265-1326-3.

norms of $\mathcal{M}$. Joint spectral radius, jsr($\mathcal{M}$), is the infimum of positive numbers r for which $\mathcal{M}/r$ generates a norm-bounded semigroup. Generalized spectral radius, gsr($\mathcal{M}$), is the infimum of positive numbers r for which $\mathcal{M}/r$ generates an eigenvalue-bounded semigroup. A central question is the complexity of algorithms aimed at measuring the radius. In the positive direction there is the finiteness conjecture of Lagarias and Daubechies which states that if $\mathcal{M}$ is finite then there is an optimal product P of n elements of $\mathcal{M}$ satisfying $\rho(P) = \rho(\mathcal{M})^n$. We conjecture that the optimality of a given product P can be ascertained in a finite number of matrix calculations. An exact geometrical algorithm is described for this purpose. The algorithm has successfully mapped all essential information regarding the joint spectral radius of pairs of matrices associated with four-coefficient multiresolution analyses and wavelets. These findings have resolved two conjectures of Heil and Colella [7, 14] and coincided with the recent results of Bröker and Zhou [5]. They use a completely different method which focuses on properties of a broad class of pairs of 2×2 matrices. Their classification simplifies the search for the optimal product, however it is only applicable to pairs of 2×2 matrices. Our approach does not have a dimensionality restriction.

Some researchers believed or state that joint spectral radius cannot be effectively computed. An early paper [26] (subsequently withdrawn due to an error) attempted to show that the complexity is not a computable function of the problem size (that is, the problem is beyond NP-hard). A more recent paper [27] reports new but less negative results indicating that if an algorithm exists, then its cost is beyond polynomial in problem size. Our approach, however, suggests a possible improvement on the branch-and-bound method of Lagarias-Daubechies-Gripenberg [10, 11, 13], which is based on jsr and gsr definitions, by adaptive modification of the norm and the utilization of the csr definition. Moreover, at each stage of the branch-and-bound method we can check the optimality of the products exhibiting the fastest growth up to that stage by using the conjectured algorithm presented here.

§2. The Spectral Radius of a Set of Matrices

Consider a collection a of square complex matrices of the same size $\mathcal{M} = \{M_a \mid a \in \mathcal{A}\}$, where $\mathcal{A}$ is a nonempty index set. For each index sequence $\alpha = (a_1, a_2, \cdots)$ where $a_i \in \mathcal{A}$ we define the product of length k at α of $\mathcal{M}$ as $C_k(\alpha) = C_k(\alpha, \mathcal{M}) = M_{a_1} M_{a_2} \cdots M_{a_k}$, and write $C(\alpha) = \lim_{k \to \infty} C_k(\alpha)$ if the limit exists. $\mathcal{M}$ is said to be real definable if the index set $\mathcal{A} = \{0, \cdots, m-1\}$, $m > 1$, $C(\alpha)$ exists for each α and $M_i M_{m-1}^\infty = M_{i+1} M_0^\infty$ for $0 \le i < m - 1$. In this case α may be considered as a real number in $[0, 1]$, given by its base-m expansion $(a_1, a_2, \cdots)$, and the numbers which have non-unique expansions in base m will result in the same product of matrices.

We are interested in finding the optimal sequence α which produces the fastest growth of $C(\alpha)$. When $\mathcal{M}$ consists of just one matrix, then $\rho(\mathcal{M})$ determines the growth rate. When $\mathcal{M}$ is a bounded collection of matrices, then the spectral radius of $\mathcal{M}$ can be determined through three equivalent definitions. Each notion generalizes one of the definitions of the spectral radius of a matrix to a collection of matrices. Rota and Strang [24] defined joint spectral radius (JSR) in terms of norms of long products. Their definition, in above notation, reads as

$$\mathrm{jsr}(\mathcal{M}) = \limsup_{k \to \infty} \sup_{\alpha} ||C_k(\alpha, \mathcal{M})||^{1/k}. \tag{1}$$

Here $|| \cdot ||$ is an arbitrary norm; and lim sup can be replaced by either lim or inf [11, 24]. They proved JSR is equivalent to what we refer to as common spectral radius (CSR) and define as

$$\mathrm{csr}(\mathcal{M}) = \inf_{||\cdot||} ||\mathcal{M}|| = \inf_{||\cdot||} \sup_{f \in \mathcal{M}} ||f||, \tag{2}$$

where the infimum is over all submultiplicative norms $||fg|| \leq ||f|| ||g||$.

Daubechies and Lagarias [11] defined generalized spectral radius (GSR) as

$$\mathrm{gsr}(\mathcal{M}) = \limsup_{k \to \infty} \sup_{\alpha} \rho(C_k(\alpha, \mathcal{M}))^{1/k}, \tag{3}$$

where $\rho(A)$ is the usual spectral radius of the matrix A; and lim sup can be replaced by sup. Their conjecture, that JSR and GSR are equal, was demonstrated by Berger and Wang [4] and more recently by Elsner [12]. Therefore we may speak of *the* spectral radius of $\mathcal{M}$ and denote it as $\rho(\mathcal{M})$.

§3. Verifying the Optimality of a Product

In connection with the finiteness conjecture, we propose the following.

Conjecture 1. *Suppose $\mathcal{M}$ is finite and a certain product P is claimed to be the optimal product. Then the optimality of P can be verified in a finite number of matrix operations (described in Algorithm 1 below).*

The verification process uses an exact geometric algorithm which constructs the unit ball (of a certain norm) with respect to which the radius is achieved. First, we need a technical condition to assure us that such a ball exists. (The existence is not always guaranteed, not even when $\mathcal{M}$ consists of only one matrix A. There exists a norm $|| \cdot ||$ for which $\rho(A) = ||A||$ iff each eigenvalue λ of A with $|\lambda| = \rho(A)$ has the same algebraic and geometric multiplicity, *i.e.*, the corresponding Jordan block is a diagonal matrix.) For a set of matrices we have the following.

Definition 1. *A bounded set $\mathcal{M}$ is called ρ-diagonal if $\sup_{\alpha} ||C_k(\alpha, \mathcal{M})|| = \mathcal{O}(\rho(\mathcal{M})^k)$ as k tends to infinity.*

It is known [24] that there is a matrix norm $|| \cdot ||$ with respect to which $\rho(\mathcal{M}) = ||\mathcal{M}||$ iff $\mathcal{M}$ is ρ-diagonal. (This does not seem to give a practically verifiable condition similar to Jordan normal form of single matrix. It remains open to find such a condition.) Henceforth we will assume $\mathcal{M}$ is ρ-diagonal. As a result any product, including the optimal product, will also be ρ-diagonal. We also assume $\rho(\mathcal{M}) > 0$.

Definition 2. *Suppose M is a ρ-diagonal matrix with $\rho(M) = 1$. Then the maximal set Ω, of a finite positive radius in some norm, satisfying $M\Omega = \Omega$ is called an* invariant ball *of M. For a ρ-diagonal set $\mathcal{M}$ with a unit radius we define an* invariant ball *as a set Ω for which the convex hull of $\mathcal{M}\Omega$ is identical with Ω.*

Algorithm 1. *To verify the optimality of a product P:*
  1) *Scale all matrices so that the radius of the set is 1 (if P is indeed optimal), i.e., define $\mathcal{M}^* = \mathcal{M}/\rho(\mathcal{M})$ and $P^* = P/\rho(P)$.*
  2) *Find Ω_0, an invariant unit ball of P^*.*
  3) *For $q \geq 1$ define Ω_q as the convex hull of $\Omega_{q-1} \cup \mathcal{M}^*\Omega_{q-1}$.*
  4) *If at a certain stage q_c the convex hull does not grow, $\Omega_{q_c+1} = \Omega_{q_c}$, then P is an optimal product.*

If the algorithm terminates, then one easily shows P is an optimal product. Conjecture 1 states that the converse is also true. At the termination of the algorithm Ω_{q_c} can be considered as the unit ball of a norm $\| \cdot \|_c$ with respect to which $\mathcal{M}$ attains its radius $\rho(\mathcal{M}) = \|\mathcal{M}\|_c$. Moreover, Ω_{q_c} is also an invariant ball of $\mathcal{M}^*$. The value of q_c is defined as the critical index of the optimal product P, and Ω_{q_c} as the optimal ball of P.

We have applied a modified version of this algorithm to $\mathcal{M} = \{f_0, f_1\}$ where

$$f_0 = \begin{pmatrix} c_0 & 0 \\ -c_3 & 1 - c_0 - c_3 \end{pmatrix}, \quad f_1 = \begin{pmatrix} 1 - c_0 - c_3 & -c_0 \\ 0 & c_3 \end{pmatrix}, \qquad (4a)$$

$$(c_0 - 1/2)^2 + (c_3 - 1/2)^2 = 1/2. \qquad (4b)$$

These matrices are motivated by the Hölder regularity analysis of 4-coefficient multiresolution analyses and wavelets. In particular, Colella and Heil [7] conjectured that at $(c_0, c_3) = (0.6, -0.2)$ the radius of $\mathcal{M}$ attains its smallest value (*i.e.*, the corresponding wavelet is the smoothest) and the optimal product is $P = f_1 f_0^{12}$. We disproved the first statement and confirmed the second one. Our analysis showed [19, 20] that the optimal product at any point (c_0, c_3) is either $f_0 f_1^n$ or $f_1 f_0^n$. Furthermore, we obtained a very detailed picture of the structure of the optimal balls, dependence of n on (c_0, c_3), dependence of q_c on n, the smallest value of radius and the resulting smoothest wavelet, the critical arc on which $n > 0$, etc. Bröker and Zhou [5] also obtained most of these results without constructing the optimal balls.

The success of Algorithm 1 in determining the correct optimal products compels us to think that more efficient algorithms for calculating joint spectral radius can be designed. In particular the branch-and-bound method of Daubechies-Lagarias-Gripenberg [10, 11, 13] can perhaps be modified to provide much faster convergence. Their method relies on the JSR and GSR definitions of radius. We believe an adaptive branch-and-bound method which also utilizes the CSR definition deserves further study. This implies that we modify the vector norm and its induced matrix norm used in each step of the standard branch-and-bound method in a manner similar to the iterations of Ω_q in Algorithm 1.

§4. Numerical Results

We have applied a modified version of Algorithm 1 to determine the joint spectral radius of matrices (4) and the Hölder exponent of the associated wavelet. Here we summarize our results. (We emphasize that all matrix calculations here can be done in exact arithmetic, and even when they are performed in finite precision there is no significant roundoff error since the number of calculations is small.)

Consider the dilation equation $\phi(x) = c_0\phi(2x) + c_1\phi(2x - 1) + c_2\phi(2x - 2) + c_3\phi(2x - 3)$ and its circle of orthogonality $(c_0 - 1/2)^2 + (c_3 - 1/2)^2 = 1/2$ in the (c_0, c_3) plane. The Hölder exponent of the associated wavelet is $h = -\log_2(\rho(\mathcal{M}))$. To determine h for each wavelet, we will travel on the half-circle below $c_0 = c_3$, from $(0, 0)$ toward $(1, 1)$ in the counter clockwise direction. (The properties on the upper half can be described similarly.) First the optimal product is simply f_0 and the optimal ball is a quadrilateral. Then, starting at $(1/2, (1 - \sqrt{2})/2)$, there is a critical strip on which the optimal product, we conjecture, is of the form $f_1 f_0^n$ where n starts at infinity, descends to 11, and goes back to infinity. On an interval where n is constant there are typically three subintervals where the facial structure of the ball remains the same. However, anomalous intervals have been detected. At $n = 11$, there are five subintervals, as may be expected since 11 is the smallest value of n, but at $n = 16$ there are four subintervals. On the second stretch of the critical strip (when n goes from 11 to infinity) we pass through Heil-Colella point $(c_0, c_3) = (0.6, -0.2)$ which is on a subinterval where $n = 12$. The joint radius decreases throughout that interval and no minimum occurs. Next, there is a point on the border between $n = 22$ and $n = 23$ at which the smallest joint spectral radius and the smoothest multiresolution is realized. At this point the ball has 54 sides, $c_0 = 0.64319821225683$, $c_3 = -0.19245524910022$, $\rho(\mathcal{M}) = 0.64705462513820$, and the Hölder exponent of the resulting MRA is $h = 0.62804058345878$. As we leave the critical strip (at $c_3 = 1 - a^{1/3} - 1/3a^{-1/3}$ where $a = 1/4 + 33^{1/2}/36$, *i.e.*, $c_0 = 0.64779887126104$, and $c_3 = -0.19148788395312$) we enter an interval where once again the optimal product is of length one and the optimal ball is first a quadrilateral (Daubechies' D_4 is here) and then a hexagon. Finally we arrive at $(1, 1)$. At the two end points of the critical strip the length of the optimal product and the number of sides on the optimal ball go to infinity. One might suspect that this gives a counter-example to the Extremality Conjecture [18] of Lagarias and Wang which prescribes a piecewise-analytic ball with finite number of sides. However, there is no contradiction, since in the limit the ball with increasing number of sides approaches a quadrilateral.

Table 1 records sample values of $\rho(\mathcal{M})$ at different values of c_3 over the critical strip. Between two consecutively recorded values of c_3 the structure of optimal unit ball is determined. The quantity s is half of the number of vertices of the ball. The vector V, together with $-V$, represents the vertices of the ball and v is the eigenvector of the scaled optimal product associated with eigenvalue -1, $BA^n v = -v$, where $(A, B) = (f_0, f_1)/\rho(\{f_0, f_1\})$.

$c_3 = -0.20710678118655$
$s = \infty,\ BA^\infty v = -v$
$\cdots$

$c_3 = -0.20685451946438$
$s = 15,\ BA^{12}v = -v$
$c_3 = -0.20641657740770$
$s = 16,\ BA^{11}v = -v$
$c_3 = -0.20639313158185$
$s = 15,\ BA^{11}v = -v$
$c_3 = -0.20634605286404$
$s = 14,\ BA^{11}v = -v$
$c_3 = -0.20248452406185$
$s = 15,\ BA^{11}v = -v$
$c_3 = -0.20181564521458$
$s = 16,\ BA^{11}v = -v$
$c_3 = -0.20131323874003$
$s = 15,\ BA^{12}v = -v$
$c_3 = -0.19994273898044$
$s = 16,\ BA^{12}v = -v$
$c_3 = -0.19935467077442$
$s = 17,\ BA^{12}v = -v$
$c_3 = -0.19887220524860$
$\cdots$

anomalous interval starts
$c_3 = -0.19516075726816$
$s = 19,\ BA^{16}v = -v$
$c_3 = -0.19512218095930$
$s = 20,\ BA^{16}v = -v$
$c_3 = -0.19479024238150$
$s = 21,\ BA^{16}v = -v$
$c_3 = -0.19447589464925$
$s = 22,\ BA^{16}v = -v$
$c_3 = -0.19446922675618$
anomalous interval ends
$\cdots$

$c_3 = -0.19250565305303$
$s = 28,\ BA^{22}v = -v$
minimum value of ρ
$c_3 = -0.19245524910022$
$s = 27,\ BA^{23}v = -v$
$c_3 = -0.19240955523641$
$\cdots$

$s = \infty,\ BA^\infty v = -v$
$c_3 = -0.19148788395312$

$\rho = 0.70710678118655$
$V = [v \cdots A^\infty v, BA^\infty v]$
$\cdots$

$\rho = 0.69618860818864$
$V = [v \cdots A^{13}v, BA^{13}v]$
$\rho = 0.69004302279648$
$V = [v \cdots A^{13}v, BA^{13}v]$
$\rho = 0.68979383768344$
$V = [v \cdots A^{13}v, BA^{12}v]$
$\rho = 0.68930630911961$
$V = [v \cdots A^{12}v, BA^{12}v]$
$\rho = 0.66771960222144$
$V = [v \cdots A^{13}v, BA^{12}v]$
$\rho = 0.66530105883053$
$V = [v \cdots A^{13}v, BA^{13}v, BA^{12}v]$
$\rho = 0.66359100053031$
$V = [v \cdots A^{13}v, BA^{13}v]$
$\rho = 0.65951833373125$
$V = [v \cdots A^{14}v, BA^{13}v]$
$\rho = 0.65790899824005$
$V = [v \cdots A^{14}v, BA^{14}v, BA^{13}v]$
$\rho = 0.65663720229290$
$\cdots$

$\rho = 0.64928146835213$
$V = [v \cdots A^{17}v, BA^{17}v]$
$\rho = 0.64923016354484$
$V = [v \cdots A^{18}v, BA^{17}v]$
$\rho = 0.64879376983300$
$V = [v \cdots A^{18}v, BA^{18}v, BA^{17}v]$
$\rho = 0.64838816938558$
$V = [v \cdots A^{19}v, BA^{18}v, BA^{17}v]$
$\rho = 0.64837963968583$

$\cdots$

$\rho = 0.64705734026606$
$V = [v \cdots A^{25}v, BA^{24}v, BA^{23}v]$

$\rho = 0.64705462513820$
$V = [v \cdots A^{25}v, BA^{24}v]$
$\rho = 0.64705945464432$
$\cdots$

$V = [v \cdots A^\infty v, BA^\infty v]$
$\rho = 0.64779887126104$

Tab. 1. Sample values of $\rho(\mathcal{M})$.

Acknowledgments. This material is based in part upon work supported by the Texas Advanced Research Program under Grant No. 003581-005.

References

1. Barnsley, M. F., *Fractals Everywhere*, Academic Press, New York, 1988.

2. Barnsley, M. F., J. H. Elton, and D. P. Hardin, Recurrent Iterated Function Systems, Constr. Approx. **5** (1989), 3–31.

3. Belitskii, G. R. and Yu. I. Lyubich, *Matrix Norms and Their Applications*, Birkhäuser, Boston, 1988.

4. Berger, M. A. and Y. Wang, Bounded semi-groups of matrices, Linear Algebra Appl. **166** (1992), 21–27.

5. Bröker, M. and X. Zhou, Characterization of continuous four-coefficient scaling functions via matrix spectral radius, preprint.

6. Cavaretta, A. S. and C. A. Micchelli, The design of curves and surfaces by subdivision algorithms, in *Mathematical Methods in Computer Aided Geometric Design*, T. Lyche and L. Schumaker (eds.), Academic Press, Boston, 1989, pp. 113–153.

7. Colella, D. and C. Heil, The characterization of continuous, 4-coefficient scaling functions and wavelets, IEEE Trans. Inf. Th. **38** (1992), 876–881.

8. Daubechies, I., *Ten Lectures on Wavelets*, SIAM, Philadelphia, 1992.

9. Daubechies, I. and J. Lagarias, Two-scale difference equations, I. existence and global regularity of solutions, SIAM J. Math. Anal. **22** (1991), 1388–1410.

10. Daubechies, I. and J. Lagarias, Two-scale difference equations, II. Infinite matrix products, local regularity bounds and fractals, SIAM J. Math. Anal. **23** (1992), 1031–1079.

11. Daubechies, I. and J. Lagarias, Sets of matrices all infinite products of which converge, Linear Algebra Appl. **161** (1992), 227–263.

12. Elsner, L., The Generalized Spectral-Radius Theorem: An Analytic- Geometric Proof, Linear Algebra Appl. **220** (1995), 151–159.

13. Gripenberg, G., Computing the joint spectral radius, Linear Algebra and Applications **234** (1996), 43–60.

14. Heil, C. and D. Colella, Dilation equations and the smoothness of compactly supported wavelets, in *Wavelets: Mathematics and Applications*, J. J. Benedetto and M. W. Frazier (eds.), CRC Press, Boca Raton, FL, 1993, pp. 161–200.

15. Heil, C. and G. Strang, Continuity of the joint spectral radius: Application to wavelets, in *Linear Algebra for Signal Processing*, A. Bojanczyk and G. Cybenko (eds.), IMA Vol. Math. Appl., Springer, New York.

16. Horn, R. and C. R. Johnson, *Matrix Analysis*, Cambridge University Press, New York, 1988.

M. Maesumi

17. Kuczma, M., B. Choczewski, and R. Ger, *Iterative Functional Equations*, Cambridge University Press, 1990.

18. Lagarias, J. C. and Y. Wang, The finiteness conjecture for the generalized spectral radius of a set of matrices, Linear Algebra Appl. **214** (1995), 17–42.

19. Maesumi, M., Joint spectral radius and Hölder regularity of wavelets, Computer and Mathematics with Applications, submitted.

20. Maesumi, M., Optimum unit ball for joint spectral radius, an example from four-coefficient multiresolution analysis, in *Approximation Theory VIII, Vol 2: Wavelets and Multilevel Approximation*, C. K. Chui and L. L. Schumaker (eds.), World Scientific Publishing, Singapore, 1995, pp. 267–274.

21. Maesumi, M., An efficient lower bound for the generalized spectral radius of a set of matrices, Linear Algebra Appl. **240** (1996), 1–7.

22. Mandelbrot, B. B., *The Fractal Geometry of Nature*, Freeman, New York, 1982.

23. Micchelli, C. A. and H. Prautzsch, Uniform refinement of curves, Linear Algebra Appl. **114/115** (1989), 841–870.

24. Rota, G. C. and G. Strang, A note on the joint spectral radius, Kon. Nederl. Akad. Wet. Proc. A **63** (1960), 379–381.

25. Seneta, E., *Non-negative Matrices and Markov Chains*, 2nd Edition, Springer-Verlag, New York, 1981.

26. Toker, O., On the algorithmic unsolvability of asynchronous iterative processes, American Control Conference, 1995, preprint.

27. Tsitsiklis, T. N. and V. D. Blondel, The Lyapunov exponent and joint spectral radius of pairs of matrices are hard – when not impossible – to compute and to approximate, Math. of Control, Signals, and Systems **10** (1997), 31–40.

Mohsen Maesumi
Mathematics Department
Lamar University
Beaumont, TX 77710
maesumi@math.lamar.edu

On Interpolation by Hermite Tension
Splines of Arbitrary Order

Miljenko Marušić

Abstract. For a given partition $x_0 < x_1 < \cdots < x_n$, a Hermite tension spline of order $2k$ is a function that on each subinterval (x_i, x_{i+1}) satisfies the differential equation $D^{2k-2}(D^2 - p_i^2/h_i^2)u = 0$ ($h_i = x_{i+1} - x_i$ and p_i's are nonnegative real constants) and the interpolatory conditions $u^{(j)}(x_i) = f_i^j$, $j = 0, \ldots, k-1$, $i = 0, \ldots, n$ for prescribed real values f_i^j. For $p_i = 0$, a Hermite tension spline is a classical Hermite polynomial spline of order $2k$, whereas for $p_i \to \infty$ we obtain a Hermite polynomial spline of order $2k-2$. We discuss a behavior of such an interpolant, bounds for interpolation error and its behavior in the limit case when $p_i \to \infty$.

§1. Introduction

Let $x_0 < x_1 < \cdots < x_n$ be a partition of the interval $[0, 1]$, and let p_i, $i = 0, \ldots, n - 1$, be given nonnegative real numbers (called tension parameters). A tension spline of order $2k$ is a function that satisfies

$$D^{2k-2} \left(D^2 - \frac{p_i^2}{h_i^2} \right) u(x) = 0, \ \forall x \in (x_i, x_{i+1}), \ i = 0, \ldots, n - 1.$$

In general we denote the length of the interval $[x_i, x_{i+1}]$ by h_i, and D stands for the differentiation operator.

In the paper we study the asymptotic behavior of tension splines when $p_i \to \infty$. It will be convenient to use notation

$$\mathcal{L}_j u(x) := D^{j-2} \left(D^2 \frac{h_i^2}{p_i^2} - 1 \right) u(x) \text{ for } x \in [x_i, x_{i+1}].$$

Now we have

$$\lim_{p_i \to \infty} \mathcal{L}_{2k} f(x) = -f^{(2k-2)}(x)$$

for any function f that does not depend on p_i.

Approximation Theory IX, Volume 2: Computational Aspects   213
Charles K. Chui and Larry L. Schumaker (eds.), pp. 213–220.
Copyright © 1998 by Vanderbilt University Press, Nashville, TN.
ISBN 0-8265-1326-3.

Most papers on tension splines consider tension splines of order four. Applications of such splines are mostly shape-preserving approximations (starting with early work of Schweikert [5] and Späth [6]) and numerical solution of singularly perturbed two-point boundary-value problem for ODE [1, 4]. Published results on tension splines of higher order include only construction of B-splines basis, but there is still no paper about their usage. Application of tension splines of higher orders may be foreseen in a problem of approximation by functions with positive higher derivative (not only with positive first derivative in monotone approximation and second derivative in convex approximation), as well as in singularly perturbed boundary-value problems of higher degree (where we may obtain higher orders of convergence and raising the degree of tension splines).

In the paper we consider tension splines u of order $2k$ ($k \geq 2$) satisfying the interpolation conditions

$$u^{(j)}(x_i) = f_i^j, \text{ for } j = 0, \ldots, k-1, \ i = 0, \ldots, n,$$

where f_i^j are prescribed real values. Such splines are called Hermite tension splines. Existence and uniqueness of Hermite tension splines follow immediately from the fact that tension splines are Tchebycheffian splines [2]. If f_i^j are values of some underlaying function f and its derivatives at the knots x_i ($f_i^j = f^{(j)}(x_i)$), we say that a tension spline u interpolates function f. In this case, the interpolation error satisfies the following theorem.

Theorem 1. *Let $f \in C^{2k}[0,1]$, and let u be its Hermite tension spline of order $2k$ ($k \geq 2$). Then for all $x \in [x_i, x_{i+1}]$ the following estimates hold.*

$$\left| f^{(r)}(x) - u^{(r)}(x) \right| \leq h_i^{2k-2-r} C_{2k,r}(p_i) \| \mathcal{L}_{2k} f \|_i, \quad r = 0, 1, \ldots, 2k-1,$$

where the constants $C_{2k,r}(p)$ satisfy

$$0 < \lim_{p \to \infty} C_{2k,r}(p) < \infty, \quad r = 0, \ldots, k-1 \qquad (1)$$

$$0 < \lim_{p \to \infty} p^{k-1-r} C_{2k,r}(p) < \infty, \ r = k, \ldots, 2k-1. \qquad (2)$$

Let $\| \ \|_i$ denote the maximum norm on the interval $[x_i, x_{i+1}]$:

$$\|f\|_i := \max_{x \in [x_i, x_{i+1}]} |f(x)|.$$

Since tension splines belong to the null space of differential operator $\mathcal{L}_{2k}$, they are L-splines. Although the L-spline theory provides existence of coefficients $C_{2k,r}$ from the theorem, the theory does not explain the behavior of the coefficients described by (1) and (2).

The special case of the theorem when $k = 2$ is proved in [3]. Here we extend these results. The next corollary is a consequence of Theorem 1.

Corollary 2. Let $f \in C^{2k+m}[0,1]$, and let u be its Hermite tension spline of order $2k$ $(k \geq 2)$. For $x \in [x_i, x_{i+1}]$,

$$f(x) - u(x) = \sum_{j=0}^{m-1} \frac{h_i^{2k+j}}{j!} \mathcal{L}_{2k+j} f\left(x_i + \frac{h_i}{2}\right) F_{2k,j}\left(\frac{x - x_i}{h_i}\right) + R_{2k,m}(x), (3)$$

$$\left| R_{2k,m}^{(r)}(x) \right| \leq \frac{h_i^{2k-2+m-r}}{m! 2^m} C_{2k,r}(p_i) \| \mathcal{L}_{2k+m} f \|_i, \tag{4}$$

for $r = 0, 1, \ldots, 2k - 1$, where functions $F_{2k,j}$ are independent of f. The constants $C_{2k,r}(p_i)$ are defined in Theorem 1.

The theorem and its corollary are proved in the following sections. In the last section we describe a behavior of Hermite tension spline when $p_i \to \infty$.

§2. Interpolation Error

In this section we prove the assertion of Theorem 1. Our goal is to bound the error function $e := f - u$, and its derivatives. Since a Hermite spline is defined locally, we may consider only i-th subinterval, and to simplify notation we assume that $x_i = 0$ and $x_{i+1} = h$.

First, we note that the error function is a solution of the boundary-value problem

$$\mathcal{L}_{2k} e = \mathcal{L}_{2k} f, \tag{5}$$

with homogeneous boundary conditions

$$e^{(j)}(0) = e^{(j)}(h) = 0, \quad j = 0, \ldots, k - 1. \tag{6}$$

In [3] we proved the theorem for $k = 2$, by constructing Green's function G_4 for the problem (5) for $k = 2$, proving that the solution of the problem

$$\mathcal{L}_4 v = g$$

with homogeneous boundary conditions satisfies

$$|v^{(r)}(x)| \leq \int_0^h \left| \frac{\partial^r G_4(x,t)}{\partial x^r} \right| dt \|g\|_i \leq h^{2-r} C_{4,r}(p) \|g\|_i \tag{7}$$

for $r = 0, 1, 2, 3$. Coefficients $C_{4,r}(p)$ satisfy the properties (1)–(2). Here, we will avoid complicated construction of Green's function, but rather solve the equation (5) by finding a particular integral and a solution of the homogeneous problem.

To obtain a particular integral, we first consider the problem

$$\mathcal{L}_4 v = \mathcal{L}_{2k} f, \tag{8}$$

with homogeneous boundary conditions $v(0) = v'(0) = v(h) = v'(h) = 0$. From (7), it follows that solution v of problem (8) satisfies

$$\|v\|_\infty \leq h^2 C_{4,0}(p) \|\mathcal{L}_{2k} f\|_\infty.$$

Now, a particular integral is given by

$$e_I(x) := \int_0^x \int_0^{t_1} \cdots \int_0^{t_{2k-5}} v(t_{2k-4}) dt_{2k-4} dt_{2k-5} \cdots dt_1, \qquad (9)$$

and one proves easily that

$$\|e_I^{(r)}\|_\infty \le \frac{h^{2k-r-4}}{(2k-r-4)!}\|v\|_\infty \le \frac{h^{2k-2-r}}{(2k-r-4)!} C_{4,0}(p)\|\mathcal{L}_{2k}f\|_\infty, \qquad (10)$$

for $r = 0, 1, \ldots 2k - 4$ and

$$\|e_I^{(2k-3)}\|_\infty \le h^1 C_{4,1}(p)\|\mathcal{L}_{2k}f\|_\infty, \qquad (11)$$

$$\|e_I^{(2k-2)}\|_\infty \le h^0 C_{4,2}(p)\|\mathcal{L}_{2k}f\|_\infty, \qquad (12)$$

$$\|e_I^{(2k-1)}\|_\infty \le h^{-1} C_{4,3}(p)\|\mathcal{L}_{2k}f\|_\infty. \qquad (13)$$

In the next step, we consider a solution to the homogeneous problem

$$\mathcal{L}_{2k}e_0 = 0 \qquad (14)$$

with nonhomogeneous boundary conditions

$$e_0^{(j)}(0) = 0, \quad e_0^{(j)}(h) = -e_I^{(j)}(h), \quad j = 0, \ldots, k-1. \qquad (15)$$

Let us define the functions

$$P_i(t) := \frac{t^i}{i!},$$

$$\varphi_1(t) := \frac{\sinh pt}{\sinh p}$$

and

$$\varphi_{i+1}(t) := \int_0^t \varphi_i(\tau) d\tau, \quad i = 1, 2 \ldots .$$

It can be easily verified that φ_i are of the form

$$\varphi_{2j+1}(t) = \frac{\sinh pt - pt - \frac{(pt)^3}{3!} - \cdots - \frac{(pt)^{2j-1}}{(2j-1)!}}{p^{2j}\sinh p},$$

$$\varphi_{2j+2}(t) = \frac{\cosh pt - 1 - \frac{(pt)^2}{2!} - \cdots - \frac{(pt)^{2j}}{(2j)!}}{p^{2j+1}\sinh p}.$$

Furthermore, these functions satisfy

$$\varphi_i^{(j)}(x) = \varphi_{i-j}(x), \quad j = 0, 1, \ldots, i-1, \qquad (16)$$

$$\lim_{p \to 0} \varphi_i(x) = P_i(x) \text{ for } x \in [0, 1], \qquad (17)$$

$$\lim_{p \to \infty} p^k \varphi_i(x) = 0 \text{ for } x \in [0, 1), \ \forall k \text{ and} \qquad (18)$$

$$\lim_{p \to \infty} p^{i-1} \varphi_i(1) = 1. \qquad (19)$$

Now, with the abbreviation

$$t := \frac{x}{h}$$

the solution of the problem (14) can be written in the form

$$e_0(x) = \sum_{i=0}^{2k-3} \alpha_i P_i(t) + \alpha_{2k-2} p^{k-1} \varphi_{2k-1}(1-t) + \alpha_{2k-1} p^{k-1} \varphi_{2k-1}(t).$$

The coefficients α_j are determined by the boundary conditions (15), but written in the more convenient form:

$$h^j e_0^{(j)}(0) = 0, \text{ for } j = 0, \ldots, k-2,$$
$$h^j e_0^{(j)}(h) = -h^j e_I^{(j)}(h), \text{ for } j = 0, \ldots, k-2,$$
$$h^{k-1} e_0^{(k-1)}(0) = 0,$$
$$h^{k-1} e_0^{(k-1)}(h) = -h^{k-1} e_I^{(k-1)}(h).$$

This leads to the system of linear equations:

$$A_k(p)\alpha = b_k,$$

where

$$\alpha := [\alpha_0, \ldots, \alpha_{2k-1}]^T,$$
$$b_k := \left[0, \ldots, 0, -e_I(h), \ldots, -h^{k-2} e_I^{(k-2)}(h), 0, -h^{k-1} e_I^{(k-1)}(h)\right]^T.$$

The matrix A_k is of the form

$$A_k(p) := \begin{bmatrix} B_k & a_1(p) & a_2(p) \\ b_1^T & (-1)^{k-1} p^{k-1} \varphi_k(1) & 0 \\ b_2^T & 0 & p^{k-1} \varphi_k(1) \end{bmatrix}, \qquad (20)$$

with entries given by

$$(B_k)_{i,j} = P_{j-1}^{(i-1)}(0)$$
$$(B_k)_{k-1+i,j} = P_{j-1}^{(i-1)}(1)$$
$$(b_1)_j = P_{j-1}^{(k-1)}(0)$$
$$(b_2)_j = P_{j-1}^{(k-1)}(1), \quad j = 1, \ldots, 2k-2$$
$$(a_1(p))_i = (-1)^{i-1} p^{k-1} \varphi_{2k-1}^{(i-1)}(1),$$
$$(a_1(p))_{k-1+i} = (-1)^{i-1} p^{k-1} \varphi_{2k-1}^{(i-1)}(0) = 0,$$
$$(a_2(p))_i = p^{k-1} \varphi_{2k-1}^{(i-1)}(0) = 0,$$
$$(a_2(p))_{k-1+i} = p^{k-1} \varphi_{2k-1}^{(i-1)}(1), \quad i = 1, \ldots, k-1.$$

Since tension splines are Tchebycheffian, the matrix $\boldsymbol{A}_k(p)$ is regular. Moreover, since $\boldsymbol{B}_k$ is the matrix of interpolation conditions for Hermite polynomial spline of order $2k - 2$, it is also regular. Because of (19), we have

$$\lim_{p \to \infty} \boldsymbol{a}_1(p) = \boldsymbol{0} \quad \text{and} \quad \lim_{p \to \infty} \boldsymbol{a}_2(p) = \boldsymbol{0},$$

and the matrix

$$\lim_{p \to \infty} \boldsymbol{A}_k(p) = \begin{bmatrix} \boldsymbol{B}_k & 0 & 0 \\ \boldsymbol{b}_1^T & (-1)^{k-1} & 0 \\ \boldsymbol{b}_2^T & 0 & 1 \end{bmatrix}$$

is also regular. Therefore, for $p > 0$, matrix $\boldsymbol{A}_k^{-1}(p)$ exists, and $||\boldsymbol{A}_k^{-1}(p)||_\infty$ is bounded when $p \to \infty$. The bound for $\boldsymbol{\alpha}$ follows immediately:

$$||\boldsymbol{\alpha}||_\infty \leq ||\boldsymbol{A}_k^{-1}(p)||_\infty ||\boldsymbol{b}_k||_\infty \leq ||\boldsymbol{A}_k^{-1}(p)||_\infty \max_{0 \leq r \leq k-1} h^r |e_I^{(r)}(h)|$$

$$\leq ||\boldsymbol{A}_k^{-1}(p)||_\infty \max_{0 \leq r \leq k-1} \frac{h^{2k-2}}{(2k - r - 4)!} C_{4,0}(p) ||\mathcal{L}_{2k} f||_\infty$$

$$\leq ||\boldsymbol{A}_k^{-1}(p)||_\infty \frac{h^{2k-2}}{(k-3)!} C_{4,0}(p) ||\mathcal{L}_{2k} f||_\infty. \tag{21}$$

Since $P_i^{(j)} = P_{i-j}$ for $i \geq j$ and $P_i(t) \leq 1/i!$ for $t \in [0, 1]$, we have $\sum_{i=0}^{2k-1} |P_i^{(j)}(t)| \leq e$. Furthermore, $g_{k+2}(t) := \varphi_{k+2}(t) + \varphi_{k+2}(1 - t)$ satisfies $g_{k+2}(t) \geq 0$ and $g_{k+2}''(t) = g_k(t) \geq 0$. So it has maxima at the end-points of the interval $[0, 1]$ and $g_{k+2}(t) \leq g_{k+2}(1) = \varphi_{k+2}(1)$. Now,

$$||e_0^{(j)}||_\infty \leq ||\boldsymbol{\alpha}||_\infty \left[\sum_{i=0}^{2k-3} |P_i^{(j)}(t)| + p^{k-1}\left(\varphi_{2k-1-j}(t) + \varphi_{2k-1-j}(1 - t)\right) \right]$$

$$\leq ||\boldsymbol{\alpha}||_\infty \left[e + p^{k-1}\varphi_{2k-1-j}(1) \right]. \tag{22}$$

The error function e is given by $e = e_0 + e_I$, and by combining bounds (10), (11)–(13), (21), and (22), we obtain the coefficients $C_{2k,r}$ and prove the properties (1) and (2).

§3. Asymptotic Expansion of Interpolation Error

To prove Corollary 2 we again assume $x_i = 0$, $x_{i+1} = h$, $h_i = h$, and $p_i = p$. Since $f \in C^{2k+m}(0, h)$, we may expand $\mathcal{L}_{2k} f(x)$ as a Taylor series at $h/2$:

$$\mathcal{L}_{2k} f(x) = \sum_{j=0}^{m-1} \frac{1}{j!} \mathcal{L}_{2k+j} f\left(\frac{h}{2}\right) \left(x - \frac{h}{2}\right)^j + \mathcal{L}_{2k+m} f\left(\zeta(x)\right) \frac{\left(x - \frac{h}{2}\right)^m}{m!},$$

for some $\zeta(x) \in (0, h)$. Let us define the functions $F_{2k,j}$ and $R_{2k,m}$ as the solution of the problem

$$\mathcal{L}_{2k} F_{2k,j}(x) = \left(x - \frac{h}{2}\right)^j$$

and

$$\mathcal{L}_{2k} R_{2k,m}(x) = \mathcal{L}_{2k+m} f\left(\zeta(x)\right) \frac{\left(x - \frac{h}{2}\right)^m}{m!} \tag{23}$$

with homogeneous boundary conditions. The function

$$e(x) := \sum_{j=0}^{m-1} \frac{h_i^{2k+j}}{j!} \mathcal{L}_{2k+j} f\left(x_i + \frac{h_i}{2}\right) F_{2k,j}\left(\frac{x - x_i}{h_i}\right) + R_{2k,m}(x),$$

is the solution of the problem

$$\mathcal{L}_{2k} e = \mathcal{L}_{2k} f$$

with homogeneous boundary conditions, i.e., it is the error function. This proves (3).

From Theorem 1 and the previous section, it follows that $R_{2k,m}$, the solution of (23), satisfies

$$\left| R_{2k,m}^{(r)}(x) \right| \le h^{2k-2-r} C_{2k,r}(p_i) \left\| \frac{1}{m!} \mathcal{L}_{2k+m} f\left(\zeta(x)\right) \left(x - \frac{h}{2}\right)^m \right\|_i.$$

(We simply substitute $\mathcal{L}_{2k} f$ in (5) by the right-hand side of (23).) Taking into account that

$$\left| x - \frac{h}{2} \right| \le \frac{h}{2},$$

we prove (4), and with a final note that $F_{2k,j}$ does not depend on f, we have proved Corollary 2.

§4. Behavior of Hermite Tension Splines

In this section we consider the limit behavior of Hermite tension splines when $p \to \infty$. We again consider only the interval $[0, h]$. A Hermite tension spline u is of the form

$$u(x) = \sum_{i=0}^{2k-3} \alpha_i P_i(t) + \alpha_{2k-2} p^{k-1} \varphi_{2k-1}(1 - t) + \alpha_{2k-1} p^{k-1} \varphi_{2k+1}(t).$$

where $t := x/h$. The coefficients α_i are determined by the interpolation conditions

$$u^{(j)}(0) = f_0^j, \ u^{(j)}(h) = f_1^j, \ j = 0, \dots, k - 1. \tag{24}$$

Using the matrix $A_k(p)$ defined by (20), the interpolation conditions can be written in the matrix form

$$A_k(p)\boldsymbol{\alpha} = \boldsymbol{f},$$

where

$$\boldsymbol{f} = [f_0^0, \dots, h^{k-2} f_0^{k-2}, f_1^0, \dots, h^{k-2} f_1^{k-2}, h^{k-1} f_0^{k-1}, h^{k-1} f_1^{k-1}]^T.$$

Argumentation similar to that in Section 2 yields that $\bar{\alpha}_i := \lim_{p\to\infty} \alpha_i$ exists. Since

$$\lim_{p\to\infty} p^{k-1}\varphi_{2k-1} = 0$$

we have that

$$u_\infty := \lim_{p\to\infty} u$$

exists and u_∞ is a polynomial of degree $2k - 3$ on the interval $[0, h]$ with coefficients $\bar{\alpha}_0, \ldots, \bar{\alpha}_{2k-3}$. Using the notation

$$\bar{\alpha} := [\bar{\alpha}_0, \ldots, \bar{\alpha}_{2k-3}]^T \text{ and } \bar{f} = [f_0^0, \ldots, h^{k-2}f_0^{k-2}, f_1^0, \ldots, h^{k-2}f_1^{k-2}]^T,$$

one verifies easily that

$$B_k\bar{\alpha} = \bar{f}$$

is satisfied, i.e., that u_∞ satisfies interpolation conditions (24) for $i = 0, \ldots, k-2$. So, when $p_i \to \infty$ (for all i), the Hermite tension spline of order $2k$ tends to Hermite polynomial spline of order $2k - 2$.

Furthermore, because of (19), $\lim_{p\to\infty} u^{(k)}$ is discontinuous at 0 and h, while limits of higher derivatives do not exist at the end-points of the interval $[0, h]$:

$$u^{(k+i)}(0) \sim \frac{1}{p^i} \text{ and } u^{(k+i)}(h) \sim \frac{1}{p^i}.$$

Acknowledgments. Supported by Grant #037011 of the Croatian Ministry of Science and Technology.

References

1. Flaherty, J. E. and W. Mathon, Collocation with polynomial and tension splines for singularly-perturbed boundary value problems, SIAM J. Sci. Stat. Comp. **1** (1980), 260–289.

2. Marušić, M., Stable calculation by splines in tension, Grazer Math. Ber. **328** (1996), 65–76.

3. Marušić, M. and M. Rogina, Sharp error bounds for interpolating splines in tension, J. Comput. Appl. Math. **61** (1995), 205–223.

4. Marušić, M. and M. Rogina, A collocation method for singularly perturbed two-point boundary value problems with splines in tension, Adv. Comput. Math. **6** (1996), 65–76.

5. Schweikert, D.G., An interpolating curve using a spline in tension, J. Math. Physics **45** (1966), 312–317.

6. Späth, H., Exponential spline interpolation, Computing **4** (1969), 225–233.

Recent Developments in Approximation via Positive Definite Functions

Francis J. Narcowich

Abstract. Positive and conditionally positive definite functions, especially radial basis functions and similar functions for spheres, tori, and even Riemannian manifolds, are of interest because of the their well-known ability to *synthesize* a good surface fit from scattered data. More recently, positive definite basis functions have been employed to *analyze* scattered data. The methods used to do this involve constructing multiresolution analyses or multilevel approximations. This paper will discuss recent developments in the synthesis and analysis problems, point out new directions in their investigation, and remark on applications.

§1. Introduction

Positive definite and conditionally positive definite functions and kernels are used in areas that require fitting a surface to data taken at scattered points in Euclidean space or on some surface, a sphere or torus, say.

When the underlying space is Euclidean, radial basis functions (RBFs)—e.g., Gaussians, multiquadrics, and thin-plate splines—are employed. These are positive definite and conditionally positive definite kernels that are functions of only the distance between points. Because they retain their positivity no matter what the dimension of the space is, they are especially well suited for being activation functions in multi-input, feed-forward neural networks [5, 62]. They also produce pleasing, smooth surfaces [24] when used in fitting noisy data taken at two-dimensional sites. Some of the first RBF applications, which arose in geophysics [32, 33], were of this type. There are several good review articles on RBFs and related topics. Frequently cited ones are those written by Buhmann [9], Dyn [14], and Powell [63]. Cheney's article [11], although not specifically about RBFs, also includes a review of them. Finally, Schaback's article [71] contains a very nice review of many important properties of RBFs.

Positive definite functions on spheres, which Schoenberg [68] introduced, behave like RBFs, and have been called spherical (radial) basis functions (SBFs)

Approximation Theory IX, Volume 2: Computational Aspects
Charles K. Chui and Larry L. Schumaker (eds.), pp. 221–242.
Copyright © 1998 by Vanderbilt University Press, Nashville, TN.
ISBN 0-8265-1326-3.

or spherical splines; they have been applied to scattered data problems on the sphere by Freeden [25], Wahba [77, 78], and others [26]. Recent review articles dealing with positive definite functions on the sphere include Cheney's paper [11], a paper by Freeden *et al.* [26], and a comprehensive survey of approximation on the sphere by Fasshauer and Schumaker [21].

The general idea of positivity, including positive definiteness for kernels, is an old one. In [75], Stewart gives an historical review of it. The most recent review of positive definite kernels is in Cheney's article [11], which we mentioned above.

Fitting a surface to scattered data is the *synthesis* aspect of approximation with positive definite kernels. A second is the *analysis* of scattered data, both by wavelets [12, 13, 60, 61] and by a multi-level method introduced by Floater and Iske [22]; these methods make use of RBFs or similar positive definite kernels in constructing wavelets or fitting surfaces to residuals at various levels.

The paper is organized as follows. The notions of positive definite (PD) and strictly positive definite (SPD) functions and kernels are reviewed in Section 2. All of these are done from the point of view of distributions. A number of examples of such functions and kernels are also described there. In Section 3, we discuss how the SPD functions and kernels can be used in fitting surfaces to scattered data. The discussion includes variational principles, stability of approximations, "uncertainty relations," and rates of approximation.

The analysis of scattered data is discussed in Section 4. There we touch on wavelets developed for analyzing such data in Euclidean space and on spheres. In Section 5, we discuss applications to neural networks in radar, to fitting divergence-free vector fields such as those for magnetic fields and for incompressible fluids with neither sources nor sinks, to PDEs, and to positron emission tomography. In Section 6, we discuss possible avenues for new research.

The topics covered in this review reflect my own taste. For any gaps, omissions, or mistakes, I take responsibility—*mea culpa!*

§2. Positive Definite Functions and Kernels

2.1. Distributional Formulation

In this section, we want to define precisely the set of functions and kernels that we will be looking at. Throughout, M^m will be either $\mathbb{R}^m$ or a compact, C^∞, m-dimensional manifold, such as an m-sphere or a torus.

A continuous, complex-valued kernel $\kappa(\cdot, \cdot)$ is called *pointwise positive definite* on M^m if $\bar{\kappa}(q, p) = \kappa(p, q)$ and if for every finite set of points $\mathcal{C} = \{p_1, \ldots, p_n\}$ in M^m, the self-adjoint, $N \times N$ matrix with entries $\kappa(p_j, p_k)$ is positive semi-definite [16, 17, 52, 75].

In applications where fitting derivative information is needed, one needs a more general definition of positive definiteness. To avoid technicalities, we will assume that κ is C^∞. We will say that the kernel κ is positive definite on

$\mathbf{M}^m$ if and only if

$$Q(u) := (\bar{u} \otimes u, \kappa) \geq 0, \tag{1}$$

for every compactly supported distribution on $\mathbf{M}^m$. We remark that when $\mathbf{M}^m$ is a compact manifold, all distributions are compactly supported; the qualification only applies in the case of $\mathbb{R}^m$.

Let δ_p be the Dirac delta function located at p. If in (1) we let $u = \sum_{j=1}^n c_j \delta_{p_j}$, where the c_j's are arbitrary complex numbers, then $Q(u) \geq 0$ is equivalent to the matrix with entries $\kappa(p_j, p_k)$ being positive semi-definite. Thus being positive definite in the sense mentioned above implies that κ is pointwise positive definite. The converse is also true, as the following "extension theorem" shows.

Theorem 1. *Let κ be a C^∞ kernel that is pointwise positive definite. If u is any compactly supported distribution on $\mathbf{M}^m$, then $Q(u) = (\bar{u} \otimes u, \kappa) \geq 0$.*

Thus the two types of positive definiteness are equivalent. For Euclidean kernels, Madych and Nelson [47, §6] proved a version of this theorem. For compact manifolds and C^∞ kernels, Narcowich [52] obtained Theorem 1. A version this theorem that applies to the case when the kernels are in Sobolev spaces, and not necessarily C^∞, may be found in Dyn, Narcowich, and Ward [17, Theorem 2.1].

The theorem has a corollary that is useful for constructing positive definite kernels on manifolds embedded in $\mathbb{R}^m$.

Corollary 2. [16, Corollary 2.1]. *If $\kappa(p, q)$ is a smooth positive definite kernel on $\mathbb{R}^n$, and if $\mathbf{M}^m$ is embedded in $\mathbb{R}^n$, then the restriction of κ to $\mathbf{M}^m$ is positive definite.*

Proof: The matrix $[\kappa(p_j, p_k)]$ is positive semi-definite for any distinct set of points in $\mathbb{R}^n$, and it will therefore be positive definite when the points are restricted to $\mathbf{M}^m$. Since κ will also be smooth when restricted to $\mathbf{M}^m$, Theorem 1 applies and the result follows from it. □

For the sphere, the utility in this result is that one can construct positive definite kernels by restricting an RBF to points on the sphere.

2.2. Strictly Positive Definite Functions and Kernels

Again assume that κ is a C^∞ positive definite kernel on $\mathbf{M}^m$. We will say that κ is strictly positive definite if the equation $Q(u) = 0$ implies that the distribution $u = 0$. This has two immediate consequences.

Proposition 3. *If κ is a smooth, strictly positive definite kernel, then*

$$[\![u, v]\!] := (\bar{v} \otimes u, \kappa), \tag{2}$$

defines an inner product on the space of distributions.

Corollary 4. *If $\{u_1 \ldots u_n\}$ is a linearly independent set of distributions, then the $n \times n$ matrix A, with entries*

$$A_{j,k} = (\bar{u}_k \otimes u_j, \kappa), \tag{3}$$

is positive definite and hence invertible.

Strictly pd vs. pd. There is no extension theorem analogous to Theorem 1. If in Theorem 1 we replace "positive definite" by "strictly positive definite," the result is false. A counter example was given by Ron and Sun [67]. To see why this is so, let us look at the case of the circle $\mathbf{T}^1$. If $\kappa(p,q) = P(p-q)$ is a positive definite function on $\mathbf{T}^1$, then it has the expansion

$$P(\theta) = \sum_{n \in \mathbb{Z}} a_n e^{in\theta}, \ a_n \geq 0.$$

The associated quadratic form $Q(u)$ is

$$Q(u) = \sum_{n \in \mathbb{Z}} a_n |u_n|^2, \ u_n := (u, e^{-in\theta}).$$

It is easy to see that P is *strictly* pd in the distributional sense if and only if $a_n > 0$ for all n. Simply use $u = e^{in\theta}$ to get $a_n = Q(e^{in\theta}) > 0$. Conversely, if all $a_n > 0$ and $Q(u) = 0$ for a distribution u, then each $u_n = 0$, forcing $u = 0$. The Ron/Sun example shows that P may define a strictly positive definite kernel in the *pointwise* sense, even if some of the a_n's actually vanish. This follows because for $u = \sum_{j=1}^N c_j \delta_{p_j}$, the Fourier coefficients are $u_n = \sum_{j=1}^N c_j e^{-inp_j}$; these have support spread out over infinitely many n. Thus, even if a few of the a_n vanish, $Q(u)$ will still be positive.

Examples. We will now list a number of examples of strictly (conditionally) positive definite functions and kernels.

- Radial basis functions (RBFs)

  — Gaussians, Hardy multiquadric, and inverse Hardy multiquadric, and completely monotonic functions of r^2. Micchelli [51].

  — Compactly supported RBFs. Narcowich and Ward [58], Wendland [79], Schaback and Wu [73], and Wu [82].

- Periodic basis functions (PBFs)

  — "Wavians," $\exp(\sum_{j=1}^m \frac{\cos(\theta_j)-1}{\sigma_j^2})$.

  — Periodic functions having positive Fourier coefficients.

  — Bernoulli splines $(a_n = |n|^{-k}, n \neq 0, a_0 = 1)$.

- Spherical basis functions (SBFs)

  — $\exp(\cos(\theta))$. Here and below $\theta = \theta(p,q)$ is the geodesic distance on the m-sphere between points p and q.

  — $\sum_{\ell=0}^{\infty} \sum_{m=1}^{d_\ell} a_{\ell,j} Y_{\ell,j}(p)\overline{Y_{\ell,j}(q)}$, where the $Y_{\ell,j}$'s are spherical harmonics on $\mathbf{S}^m$, $a_{\ell,m} > 0$, and d_ℓ is the dimension of the space of spherical harmonics of order ℓ.

  — $\sum_{\ell=0}^{\infty} a_\ell P_\ell(m+1, \cos(\theta))$, $a_\ell > 0$. Xu and Cheney [83].

  — $[1 - 2\sin(\theta/2)]_+^5 (1 + 10\sin(\theta/2))$ Dyn, *et al.* [16].

- SPD kernels on compact manifolds

  - $\sum_{k=0}^{\infty} a_k F_k(p)\overline{F_k(q)}$, $a_k > 0$. The F_k's are the eigenfunctions of the Laplace-Beltrami operator on $\mathbf{M}^m$.
  - $\sum_{k=0}^{\infty} e^{-\lambda_k t} F_k(p)\overline{F_k(q)}$, where $0 = \lambda_0 < \lambda_1 \leq \cdots \leq \lambda_k < \cdots$ are the eigenvalues of the Laplace Beltrami operator on $\mathbf{M}^m$, counted according to multiplicity.

- SPD functions on $SO(3)$

  - $e^{\gamma \cos^2 \theta F(\varphi, \psi, \theta)}$, $\gamma > 0$. Gutzmer [31]. φ, θ, ψ are Euler angles, and $F(\varphi, \psi, \theta) := \cos \varphi \cos \psi - (1 - \sin \varphi \sin \psi) \cos \theta - \sin \theta (\sin \varphi + \sin \psi)$.

§3. Fitting Surfaces to Scattered Data

3.1. Generalized Hermite Interpolation

Hermite interpolation problems involve fitting a curve or surface to data that contains information involving derivatives. For splines, the standard reference on Hermite interpolation is the book by Lorentz, Jetter, and Riemenschneider [45]. By Generalized Hermite interpolation we will mean interpolating data arising from integrating a (Schwartz) distribution against a smooth, unknown function f. For example, measuring the value of f at a point p can be thought of as obtaining (δ_p, f). More generally, take $\mathcal{U} := \text{Span}\{u_j\}_{j=1}^{N}$ be a linearly independent (i.e., non-redundant) set of distributions on $\mathbf{M}^m$, and take the data to be

$$(\bar{u}_j, f) = \int_{\mathbf{M}^m} \bar{u}_j(p) f(p) d\mu(p) = d_j \in \mathbf{C}, \quad \text{for } j = 1, \ldots, N. \quad (4)$$

The form of the interpolant that we want here requires discussion. For the RBF case, with the distributions being delta functions and $\kappa(p, q) = F(\|p - q\|)$, the interpolant $\tilde{f}$ has the form

$$\tilde{f}(p) = \sum_j c_j F(\|p - p_j\|), \; p_j \in \mathcal{C}.$$

Rewriting this in terms of κ and $u = \sum_j c_j \delta_{p_j}$, we see that

$$\tilde{f}(p) = \underbrace{\int_{\mathbf{M}^m} \kappa(p, q) u(q) d\mu(q)}_{(\delta_p \otimes u, \kappa)}, \quad \text{where } u \in \mathcal{U}.$$

It is this expression that plays the rôle of interpolant in the general case. Let κ be a positive definite C^∞ kernel on $\mathbf{M}^m$, and let u be a distribution. Denote the function $p \mapsto (\delta_p \otimes u, \kappa)$

$$\kappa \star u(p) := (\delta_p \otimes u, \kappa) = \int_{\mathbf{M}^m} \kappa(p, q) u(q) d\mu(q). \quad (5)$$

We seek $u \in \mathcal{U}$ such that $\tilde{f} = \kappa \star u$ satisfies $(\bar{u}_j, \tilde{f}) = d_j$. When this problem can be solved uniquely for a specific set of linearly independent distributions $\mathcal{U}$, we will say that it is poised. If a problem is poised for every finite, linearly independent set of distributions $\mathcal{U}$, then we say the problem is well poised.

In [37], Jetter, Riemenschneider, and Shen considered the problem of cardinal Hermite interpolation on the lattice $\mathbb{Z}^d$. They employed a matrix-valued symbol to establish that certain classes of interpolation problems were well poised, and in certain cases they obtained approximation orders. Wu [81] was the first to show that Hermite interpolation of scattered data using RBFs was well poised; he used Kriging methods to do so. Independently and using methods different from Wu, although later, Sun [76] also discussed conditions under which Hermite scattered-data problems are well poised.

Narcowich and Ward [59] developed a distributional framework to investigate a class of generalized Hermite interpolation problems in which the interpolants are allowed to be divergence free, and the data set is generated by a collection of arbitrary, linearly independent distributions. We will discuss this in more detail below. We wish to point out that the introduction to [59] incorrectly states that in [81] the number of centers used must equal the number of linear functionals used. In fact, no such restriction is made by Wu in [81].

The well-poisedness of a generalized Hermite interpolation on a C^∞, compact, Riemannian manifold was treated by Narcowich in [52]. Both the m-sphere and m-torus are such manifolds.

3.2. Variational Formulation

We can rewrite the problem described above in terms of the inner product defined in Equation (2). Again, let $\mathcal{U} = \text{Span}\left(\{u_j\}_{j=1}^N\right)$. Assume that there is a distribution v such that the unknown function f satisfies $f = \kappa \star v$. Of course, v is not known either. The point is that we can regard v as the object we wish to interpolate by a distribution in the finite dimension space $\mathcal{U}$. The set of data is then $d_j = (\bar{u}_j, f) = (\bar{u}_j \otimes v, \kappa) = [\![v, u_j]\!]$. The interpolant $\tilde{f} = \kappa \star u$, $u \in \mathcal{U}$, satisfies $d_j = [\![u, u_j]\!]$. If we put $u = \sum_{k=1}^N c_k u_k$, then this becomes

$$d_j = \sum_{k=1}^N [\![u_k, u_j]\!] c_k$$
$$= \sum_{k=1}^N (\bar{u}_j \otimes u_k, \kappa) c_k = \sum_{k=1}^N A_{k,j} c_j,$$

where, by Corollary 4, $A = [A_{k,j}]$ is positive definite and hence invertible. Thus, there exists a unique $u \in \mathcal{U}$ that solves the interpolation problem; it also satisfies the following minimization principle.

Proposition 5. [17, Theorem 3.1]. *The distribution $v - u$ is orthogonal to $\mathcal{U}$ with respect to the inner product (2). In addition, if we let $[\![\cdot]\!]$ be the associated norm, then*

$$[\![v]\!]^2 = [\![v - u]\!]^2 + [\![u]\!]^2.$$

Finally, if $v \neq u$,

$$\|u\| < \|v\|.$$

Proof: Standard inner product space methods. □

While we do not need (or want!) to complete the set of distributions comprising the inner product space above—so we are not dealing with a Hilbert space—, the variational principles and results above are essentially those from the theory of reproducing kernel Hilbert spaces [30]. Such spaces have been used extensively in connection with RBFs, SBFs, and similar basis functions, albeit directly in terms of the functions rather than the distributions; they are called native spaces [71] or associated spaces [16] for κ. For RBFs, Madych and Nelson [46] first used them. In the case of the m-sphere, they have been used extensively by many researchers [17, 26, 29, [42]] to get estimates of rates of approximation. One useful tool in getting such rates is the hypercircle inequality [30], which is formulated below in our notation.

Corollary 6. [17, Proposition 3.2]. *If w is a distribution, then*

$$|(\bar{w}, f - \tilde{f})| = |\|v - u, w\|| \leq \mathrm{dist}(w, \mathcal{U}) \mathrm{dist}(v, \mathcal{U}), \qquad (6)$$

where the distances on the right are computed using the norm $\| \cdot \|$.

Proof: This follows from Schwarz's inequality together with the observation that $v - u$ being orthogonal to $\mathcal{U}$ implies that we may replace w by $w - u'$, where $u' \in \mathcal{U}$ is arbitrary. Minimizing over u' then yields the result. □

Let us put this into a more familiar setting. Set $w = \delta_x$, and note that $\mathrm{dist}(v, \mathcal{U}) \leq \|v\|$. Using these in (6) above yields $|f(x) - \tilde{f}(x)| \leq \|v\| \mathrm{dist}(\delta_x, \mathcal{U})$. In the case where $\mathbf{M}^m$ is Euclidean and the u_j's are point evaluations, the function $P(x) = \mathrm{dist}(\delta_x, \mathcal{U})$ is precisely what Schaback calls the *power function* [70], which plays a role in his "uncertainty principle." (See Section 3.3 below.)

By analogy, in the general case we define the function

$$\wp(w) := \mathrm{dist}(w, \mathcal{U}), \qquad (7)$$

which satisfies the following inequality.

Corollary 7. *If A_w is the interpolation matrix for κ with data generated by the distributions $\{u_1, \ldots, u_N, w\}$, then*

$$\wp(w)^2 \|A_w^{-1}\| \geq 1. \qquad (8)$$

Proof: (The proof we give follows along lines of that in [70]). Because $\wp(w)$ is the distance from w to the finite dimension space $\mathcal{U}$, we can find a linear combination of the u_j's for which $\wp(w) = \|w - \sum_{j=1}^{N} c_j u_j\|$. Letting $u_{N+1} := w$, and $b_j = -c_j$, $j = 1, \ldots, N$, and setting $b_{N+1} = 1$, we see that

$$\wp(w)^2 = \| \sum_{j=1}^{N+1} b_j u_j \|^2 = \sum_{j,k=1}^{N+1} b_j \bar{b}_k \|u_j, u_k\| = b^* A_w b.$$

Since the lowest eigenvalue of A_w is the reciprocal $\|A_w^{-1}\|$, we have that $b^* A_w b \geq \|A_w^{-1}\|^{-1} \|b\|^2$. In addition, because $b_{N+1} = -1$, $\|b\| \geq 1$. Combining these two inequalities and doing a little algebra yield (8). □

3.3. Uncertainty Principles

A stability vs. goodness-of-fit trade-off occurs in many approximation schemes. For instance, those employing finitely many translates of a single function $g(x)$, whether g is an RBF, a spline, or a sigmoidal function exhibit it. Intuitively, this is because if $g(x)$ itself can be approximated well by a linear combination of $g(x - \xi_1), \ldots, g(x - \xi_n)$, then interpolation matrices associated with g will be nearly singular.

Corollary 7 quantifies this for generalized Hermite interpolation. $\|A_w^{-1}\|$ measures the stability of the interpolation method, whereas $\wp(w)$ is a universal control on the error.

From the hypercircle inequality (6), we see that the distance function $\wp(w)$ is one of two factors bounding the error on the left in (6). The other factor, $\operatorname{dist}(v, \mathcal{U})$, depends on the unknown function f, and is bounded above by $\|v\|$. Although $\operatorname{dist}(v, \mathcal{U})$ does play an important role in controlling the error [16, 17] for a given function, that role is not a universal one. The distance function $\wp(w)$ does *not* depend on f. If it can be made small, the overall error can be made *uniformly* small. Thus (8) expresses quantitatively what we said above.

In the case of ordinary interpolation with RBFs, Schaback [70, 71] quantified this phenomenon, proving the inequality (8) for the Euclidean case and point evaluations. For obvious reasons, he called it the "uncertainty principle." We prefer the term stability-fit trade-off.

There is a second type of "uncertainty principle," one which is far closer to Heisenberg's original position-momentum uncertainty relation. In signal processing this is expressed as a trade-off between a signal's duration and its frequency content, and is used as a measure of how well various wavelets localize in time and frequency.

In connection with their investigation of wavelets on the circle and m-sphere, Narcowich and Ward [60, 61] constructed an uncertainty principle for the circle, which measures a trade-off between angular localization and frequency, and for $\mathbf{S}^2$, which measures the trade-off between localization on $\mathbf{S}^2$ and a "total" angular frequency, which is related to expansions in spherical harmonics.

3.4. Stability

We now turn to a discussion of how stable the interpolation matrix and its inverse are, as measured by norms of inverses and condition numbers.

Numerical results have long shown that the interpolation matrices associated with many RBFs were poorly conditioned. In [15], Dyn, Levin, and Rippa provided methods of preconditioning interpolation matrices for certain RBFs in special cases. However, RBFs and related basis functions are really

quite varied in the way they behave, and getting a general theory of preconditioners is not practical. To really deal with this issue requires quantitative
results on norms of inverses and condition numbers.

The first of these was a result by Ball [1], who studied the stability for
the order 1 RBF $D(r) := r$.

Theorem 8. (Ball [1]). *If s is an odd integer, then for any finite set of distinct
points $\{x_j\}_{j=1}^N \subset \mathbb{R}^s$, one has*

$$\|A^{-1}\| \leq \frac{3.55s}{2^s q} \binom{s-1}{\frac{1}{2}(s-1)}, \tag{9}$$

where $A_{j,k} = \|x_j - x_k\|$ and $q := \frac{1}{2} \min_{j \neq k} \|x_j - x_k\|$.

The estimate in (9) is for a special, simple case, and its proof relies on
features of that case. Nevertheless, the fact that (9) depends *only* on the
minimal separation distance for the data, and not on the number of data
points nor on any other aspect of the data, holds in much more general cases.
Using Fourier transform techniques, Narcowich and Ward [57, 58] developed a
general approach for making such estimates for a large class of order m RBFs,
with data in $\mathbb{R}^s$. To avoid technical detail, we will state only the $m = 0$
estimate. Define these:

$$\left.\begin{array}{l} \varphi_s(x) := x^{-s/2} e^{-1/x}, \ x > 0 \\[2mm] \gamma_s := 2^{s+1} \Gamma(\dfrac{s+2}{2}) \\[2mm] \delta_s := (3^{s-1} \pi \gamma_s^2)^{1/(s+1)} \end{array}\right\}. \tag{10}$$

The most general $m = 0$ order RBF has the form $F(r) := \int_0^\infty e^{-tr^2} d\eta(t)$,
where $d\eta$ is a nonnegative Borel measure on $[0, \infty)$ such that F is noncon-
stant and continuous on $[0, \infty)$. For the set of points $\{x_j\}_{j=1}^N \subset \mathbb{R}^s$, the
interpolation matrix A associated with F is an $N \times N$ matrix with entries
$A_{j,k} := F(\|x_j - x_k\|)$.

Theorem 9. [58, Corollary 2.5]. *With the notation of Theorem 8 and equation (10), we have*

$$\|A^{-1}\| \leq \left(\frac{q^s}{\gamma_s} \int_0^\infty \varphi_s(\frac{q^2 t}{\delta_s}) d\eta(t)\right)^{-1}. \tag{11}$$

The condition number of A is the product $c_A = \|A^{-1}\|\|A\|$. If $|A_{j,k}| \leq M$,
then it is easy to show that $\|A\| \leq NM$, where N is the number of centers [57,
§VII]. Combining this and (11) yields an N *dependent* bound on c_A. Are there
cases for which c_A is *independent* of N? For 0-order RBFs of the form used
in Theorem 9, Narcowich, Sivakumar, and Ward found that one has bounds
that are *independent* of the number of centers if and only if $\int_0^1 t^{-s/2} d\eta(t) < \infty$
[54].

The estimates in Theorem 9 are independent of any characteristics of the set of centers, except for q, the separation radius, and s, the dimension of the underlying space. How good are they?

Ball, Sivakumar, and Ward [2] showed that in the case where $q > 0$ is small, one can find a finite set of centers with minimal separation $2q$ such that for many RBFs the estimates in (11) are best possible or nearly so; these centers lie on the regular lattice $2q\mathbb{Z}^s$.

The situation is different if the separation distance is fixed. Baxter [3, Theorem 3.10 & Corollary 3.11] found bounds for $\|A^{-1}\|$ when the centers are on the lattice $\mathbb{Z}^s$ (q=1/2), and there are mild restrictions on the RBF:

$$\|A^{-1}\| \le \left(\sum_{k \in \mathbb{Z}^s} |\hat{\varphi}(\|\pi e + 2\pi k\|)| \right)^{-1}, \quad e = (1, 1, \cdots, 1) \in \mathbb{Z}^s, \qquad (12)$$

where $\hat{\varphi}$ is the s-dimensional Fourier transformation of $F(\|x\|)$. If the symbol function $\xi \in \mathbb{R}^s \mapsto \sum_{k \in \mathbb{Z}^s} \hat{\varphi}(\|\xi + 2\pi k\|_2)$ is continuous, then the bound (12) is best possible. For example, for the Gaussian $F(r) = e^{-tr^2}$ in the $s = 1$ case, the right side of (12) is $\sqrt{\frac{t}{\pi}} \vartheta_2(0, \exp(-\frac{\pi^2}{4t}))^{-1}$, where ϑ_2 is a type-2 theta function [80, p. 464]. For t small, this is approximately $\sqrt{\frac{t}{4\pi}} \exp(\frac{\pi^2}{4t})$. For the same parameters ($q = 1/2$, $s = 1$), the bound in (11) is $\sqrt{2t} \exp(\frac{8\pi}{t})$, which is much larger.

This itself raises the question of whether one can *improve* stability if the placement of centers is at one's disposal. Schaback [69] gave a set of bounds from *below* for $\|A^{-1}\|$. Although bounds from below for $\|A^{-1}\|$ were given in the papers [2, 3], they still relied on gridded data; Schaback's do not. They apply to any finite set of points in $\mathbb{R}^s$, and to a wide class of RBFs. What they give are asymptotic estimates when the number of centers N is large, and simultaneously, the set of centers has a fixed diameter. Knowing these, Schaback showed that choosing gridded centers nearly optimizes stability.

Concerning questions of stability for RBFs with centers in $\mathbb{R}^s$, we wish to mention that preconditioning using a difference operator has been studied by Quak, Sivakumar, and Ward [66] and by Jetter and Stöckler [38]. Finally, the stability of interpolation matrices associated with certain Hermite interpolation problems was investigated in [59, §VII].

The stability of scattered-data interpolation using SBFs has recently become of interest. Levesley, Luo, and Sun [43], Menagatto [48], and Narcowich, Sivakumar, and Ward [55] independently have obtained upper bounds for $\|A^{-1}\|$ similar to those discussed above for RBFs. The methods used, the SBFs to which they apply, and even the formulation of the bounds are considerably different. Here is a nice example worked out by Levesley, Luo, and Sun.

Example 10. [43, Example 3.3]. *Let* $\Phi(\cos(\theta)) := 1 - \sqrt{(1 - \cos(\theta))/2}$. *If* $q = \frac{1}{2} \min_{j \neq k} \theta(p_j, p_k)$ *is the (spherical) separation radius for the set of centers* $\{p_1, \ldots, p_N\} \subset \mathbf{S}^2$, *then*

$$\|A^{-1}\| \leq 2 \csc^2(q/2).$$

3.5. Rates of Approximation

In $\mathbb{R}^s$, rates of approximation using translates of a single RBF have been obtained in several ways. Duchon [19] and Madych/Nelson [46, 47] used reproducing-kernel Hilbert spaces to obtain rates of approximation for scattered data RBF interpolants to a function f that belonged to the particular RBF's native space. These native spaces become smaller (and smoother) as the smoothness of the RBF increases.

De Boor, DeVore, and Ron [8] obtained L^2 rates of convergence for least-squares approximation from spaces generated by the translates of an RBF on a scaled lattice in $\mathbb{R}^s$ to functions in various Sobolev spaces. Here, the RBF may very well be smooth, and the function one wishes to approximate may be non-smooth. When the RBF has a Fourier transform that is singular at the origin, Dyn and Ron [18] obtained rates of convergence in the scattered-center case; the scattered-center case for RBFs with non-singular Fourier transform— Gaussians, for example—is still not well understood from this point of view. Finally, on the scaled lattice, Jetter, Riemenschneider, Shen [37] have obtained rates when the data set includes information about derivatives (Hermite data).

The question of how fast translates of RBFs can approximate a smooth function in the unit cube in $\mathbb{R}^s$, given the freedom to place N centers anywhere in that cube, was investigated by Mhaskar [49] for Gaussian RBFs and by Mhaskar and Prestin [50] for a broader class of functions similar to RBFs.

In non-Euclidean settings, most of the work has been directed towards obtaining rates for interpolation of data on the circle or the unit sphere $\mathbf{S}^m$. Freeden, Schreiner, and Franke [26] discuss scattered-data point-interpolation on the 2-sphere, and obtain linear rates. In [16, 17], Dyn, Narcowich, and Ward provided the variational approach described earlier to set up a framework for obtaining rates of approximation in non-traditional settings and data–i.e., compact manifolds and generalized Hermite data—, and used this framework to derive approximation rates for various classes of functions on the circle and the 2-sphere. For the circle, the rates in [17] were for a limited class Hermite problems, but included the case of quasi-uniform, scattered data. On the 2-sphere, they obtained rates substantially better than linear, but for point-interpolation of gridded data. Using this framework and a Markov-type inequality on $\mathbf{S}^m$, Jetter, Stöckler, and Ward [39] handled the scattered-data case, obtaining explicit rates that depend on the smoothness of the SBF and of the function to be interpolated, and the mesh size of the data. Taking a different approach, still variational, but closer in spirit to that of Madych and Nelson, von Golitschek and Light [29] obtained rates for conditionally positive definite kernels, with a fixed number of spherical harmonics being added to the interpolant. The rates obtained by the two groups are complementary. For a detailed discussion, we refer the reader to the survey article by Fasshauer and Schumaker [21].

§4. Analysis of Scattered Data

Originally, RBFs were used for surface reconstruction, either directly or via neural networks, in other words, for *synthesizing* scattered data. Two recently developed methods employ RBFs and other positive definite kernels to do *analysis* of such data as well.

4.1. Wavelets

The first method involves constructing wavelets out of non-lattice "translates" of positive definite functions. Without a dyadic lattice, one cannot expect to be able to directly employ Mallat's definition of a multiresolution analysis (MRA) to create such wavelets. The idea that is needed is a *nonstationary* MRA, where the approximation spaces $\mathcal{V}_j$, which are spans of translates of RBFs, satisfy the usual inclusion relations, but the strict scaling requirements are dropped.

These had been used earlier with non-dyadic lattices by Buhmann and Micchelli [10] and de Boor, De Vore, and Ron [7]. Chui, Jetter, Stöckler, and Ward [12, 13] used them to construct RBF wavelets capable of analyzing scattered data in Euclidean space.

The m-sphere, for $m \geq 2$, has no arbitrarily large gridded lattices. For example, for the 2-sphere, the largest lattice of regular points comprises the twenty vertices of the dodecahedron. Thus, to construct wavelets that are intrinsically defined, one needs to use some kind of nonstationary wavelet. There are a number of approaches to this problem. The two making use of SBFs are by Freeden and Windheuser [27], who discretize the continuous wavelet transform, and by Narcowich and Ward [60, 61], who construct classes of nonstationary wavelets by techniques similar to ones used in [12, 13].

4.2. Multilevel Methods

We want to briefly describe a multilevel method that uses strictly positive definite kernels to do pointwise interpolation of large data sets on $\mathbf{M}^m$. When the data sets are very large, on the order of several thousand data points, the interpolation (or least-squares) matrices generated by many SPDKs are full; solving interpolation equations involving them is especially time consuming. Because of well-known conditioning problems [2, 57, 58, 70] that occur when the distance between points becomes small, it may be very difficult to reliably solve interpolation equations in such cases. Even if a solution is available, evaluation of the SPDKs themselves can add to computation time.

The evaluation problem is currently being addressed by Baxter and Roussos [4], Beatson and Newsam [6], and Powell [64, 65] who are working toward fast evaluation and interpolation of large data sets using thin plate splines and other RBFs.

Multilevel methods address what really is a problem of scale. The one described below for RBFs was introduced by Floater and Iske [22].

We have available the data set $\{(\xi_1, f(\xi_1)) \cdots (\xi_N, f(\xi_N)\}$, where N is large. The data sites $\mathcal{C} = \{\xi_1, \ldots, \xi_N\}$ serve to provide the sets of "centers" to be used at the various levels. In general, $\mathcal{C}$ comprises distinct, scattered points. (For the important case of least-squares surface fitting, one could simply choose $\mathcal{C}$ to be a subset of the data sites, or some other suitable point set.)

The central idea is to extract from the original set $\mathcal{C}$ a finite sequence of m nested sets, $\mathcal{C}_1 \subset \cdots \subset \mathcal{C}_m := \mathcal{C}$, each $\mathcal{C}_j$ corresponding to a level j. The $\mathcal{C}_j$'s are chosen to reflect the natural scales occurring in the data sites. Thus, at any level j one has points that are neither too far apart, nor too close together. (In [22], a thinning algorithm was used for this purpose.) For the jth level, one picks a basis of the form $\{\Phi_j(\cdot, \xi)\}_{\xi \in \mathcal{C}_j}$, where Φ_j is an SPDK appropriate for the level. The interpolant is then found as follows.

1) Find $f_1^* \in \text{Span}\{\Phi_1(\cdot, \xi)\}_{\xi \in \mathcal{C}_1}$ that interpolates f on $\mathcal{C}_1$.

2) Find $g_2^* \in \text{Span}\{\Phi_2(\cdot, \xi)\}_{\xi \in \mathcal{C}_2}$ that interpolates the residual $g_2 := f - f_1^*$ on $\mathcal{C}_2$. The level 2 interpolant is then $f_2^* = f_1^* + g_2^*$.

3) For $j = 3, \ldots, m$, find $g_j^* \in \text{Span}\{\Phi_j(\cdot, \xi)\}_{\xi \in \mathcal{C}_j}$ that interpolates the residual $g_j := g_{j-1} - g_{j-1}^*$ on $\mathcal{C}_j$. The level j interpolant is $f_j^* = f_1^* + g_2^* + \cdots + g_j^*$.

We assume that f belongs to a space $\mathcal{W}_0$ of smooth functions, a Sobolev space for example. The native space $\mathcal{W}_j$ associated with each Φ_j (see Section 3.2) changes with level; smoother spaces capture trends, and "spikier" ones, details. These spaces are nested, with $\mathcal{W}_0$ being a subspace of them all.

$$\mathcal{W}_0 \subset \underbrace{\mathcal{W}_1}_{smooth} \subset \mathcal{W}_2 \subset \cdots \subset \mathcal{W}_{m-1} \subset \underbrace{\mathcal{W}_m}_{spikey}.$$

The Floater-Iske technique [22] is to pick the lowest (smoothest) level basis function Φ, and scale it: $\Phi_j(r) := \Phi(\alpha^{j-1}r)$, where $\alpha > 0$.

In Figure 1, we have used the well-known Franke function to generate data to illustrate the first few functions in a multilevel analysis. The Φ_j's were taken to be scaled versions of one of the compactly supported C^2 functions described in [79].

The multilevel method does not dictate the choice for the Φ_j's. Narcowich, Schaback, and Ward [53], who provided a framework with which to analyze rates, and in addition did so in specific cases, start with the highest ("spikey") level $\Phi_m = \Phi$, and then convolve (in the $\star$-product sense): $\Phi_j = \Phi \star \Phi_{j+1}$, $j = 1, \ldots, m-1$. It is both an interesting and practical question as to which ones should be selected for a given application. Fasshauer and Jerome [20] modify the method described above by adding in a smoothing step at each iteration, and obtain improved rates of convergence.

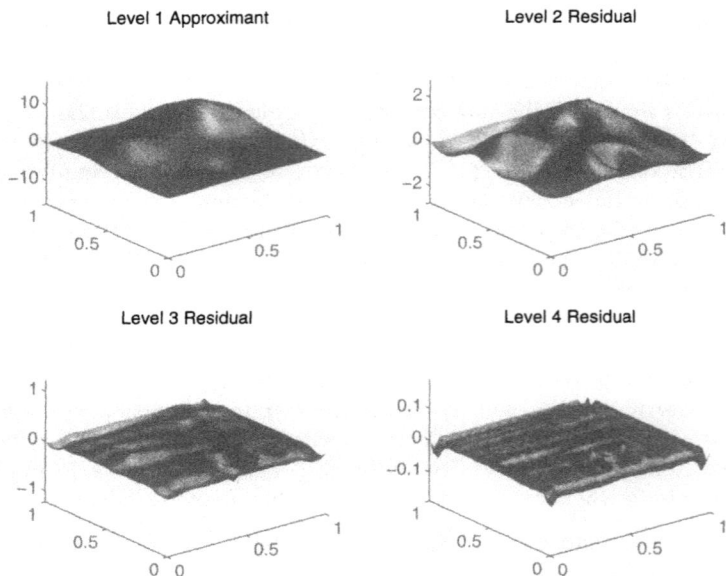

Fig. 1. Multilevel analysis for the Franke function.

§5. Applications

5.1. Phased Array Antennas

A phased array radar antenna works by detecting the phase differences between the voltages of adjacent antenna elements. For a perfectly tuned array and a single source, all of the phase differences are the same angle (modulo 2π), which is itself proportional to the sine of the angle of arrival (AOA) of an incoming electromagnetic plane wave. Unlike the familiar dish antenna, a phased array has no moving parts, and is thus less susceptible to breakdown. However, constructing perfectly matched, well-tuned antenna elements is costly. Minor damage to an element can change the relationship between its output and the direction of an incoming signal.

To overcome these drawbacks, Southall, Simmers, and O'Donnell [74] took a neural-network approach based on the premise that even when antennas are neither well-tuned nor perfectly matched, the information in the phase differences is still sufficient to determine the AOA of a single incoming source. The approach described in [74] uses a feed-forward artificial neural network with Gaussian RBFs in the processing nodes. In the language of approximation, this network creates a Gaussian RBF fit from scattered, naturally periodic input data. Because the input variables are periodic and Gaussians are not, the fit one gets can suffer from "edge effects." The remedy for this is to use the periodic basis functions mentioned earlier. See the paper by Southall and O'Donnell in this collection.

One unique feature of the network in [74] is the way the fit is arrived at. The network uses neither least squares nor simple interpolation. Instead, it employs a form of "local" Lagrange interpolation, which works well even though far too few centers are used for there to be a good fit under normal circumstances. This is not explicitly discussed in [74], so we now describe it, but in a general setting.

Suppose that we are given a region $\Omega \subset \mathbf{M}^m$ and a set of centers $\mathcal{C} = \{\xi_1, \ldots, \xi_N\} \subset \Omega$, and an SPD kernel $\kappa(p, q)$. The idea is that most of the points we are interested in are contained in Ω, and the centers $\mathcal{C}$, while scattered, are a quasi-uniform sample of Ω. If $A = [\kappa(\xi_k, \xi_j)]$ is the interpolation matrix for κ, then the ordinary Lagrange interpolants are

$$\Lambda_k(\cdot) = \sum_{j=1}^{N} \left(A^{-1}\right)_{j,k} \kappa(\cdot, \xi_j). \tag{13}$$

These satisfy $\Lambda_k(\xi_j) = \delta_{j,k}$. If at each ξ_k one had measured the value d_k, then the standard Lagrange interpolant would be

$$\tilde{f}_{global}(\xi) = \sum_{k=1} d_k \Lambda_k(\xi). \tag{14}$$

In *local* Lagrange interpolation, to find the value that one should assign to a point $\xi \in \Omega$, one first finds the two indices corresponding to the two largest values in the set $\{\Lambda_k(\xi)\}_{k=1}^{N}$. Let these be k_1 and k_2. The interpolant in this case is then

$$\tilde{f}_{local} = d_{k_1} \frac{\Lambda_{k_1}(\xi)}{\Lambda_{k_1}(\xi) + \Lambda_{k_2}(\xi)} + d_{k_2} \frac{\Lambda_{k_2}(\xi)}{\Lambda_{k_1}(\xi) + \Lambda_{k_2}(\xi)}. \tag{15}$$

The two functions multiplying d_{k_1} and d_{k_2} may not be defined for all $\xi \in \Omega$. They *are*, however, Lagrange interpolants near ξ_{k_1} and ξ_{k_2}, whence the name local.

5.2. Incompressible Fluids and Magnetic Fields

In regions where there are neither sources nor sinks, an incompressible fluid, water for example, has a velocity field $\vec{v}(\vec{x})$ that is divergence free,

$$\nabla \cdot \vec{v} = 0. \tag{16}$$

Magnetic fields also satisfy (16), because there are no magnetic charges. Standard RBFs can be, and have been [34], used to fit a vector field to data taken from a fluid. Such fits do not, however, preserve the divergence-free nature of the field.

In [59], we introduced "matrix-valued" RBFs, with the expressed purpose of doing divergence-free fits to scattered data from the velocity field of an incompressible fluid or a magnetic field. Of course, one can have s-dimensional divergence free vector fields defined on $\mathbb{R}^s$, $s \geq 2$. An $s \times s$ matrix-valued RBF related to a Gaussian with spread parameter $t > 0$ is

$$h_t^{s \times s}(x) = \left\{ (2(s-1)t - 4t^2 \|x\|^2)I + 4t^2 xx^T \right\} e^{-t\|x\|^2}, \qquad (17)$$

where I is the $s \times s$ identity, x is regarded as a column vector, and x^T is the transpose of x.

For example, if one wants to interpolate data $\{(\xi_j, \vec{d}_j)\}_{j=1}^N$, where ξ_j is a point (bound vector, in older terminology) in $\mathbb{R}^3$ and $\vec{d}_j$ is a (free) vector in $\mathbb{R}^3$, then we seek a vector field of the form

$$\vec{v}(x) = \sum_{j=1}^N h_t^{3 \times 3}(x - \xi_j)\vec{c}_j \qquad (18)$$

for which $\vec{v}(\xi_k) = \vec{d}_k$. (The multiplication in (18) involves matrices, and does not commute.) The interpolation matrix involved is always positive definite, and thus the equations can be inverted. The interpolant in (18) will also automatically satisfy (16).

5.3. Partial Differential Equations

RBFs have been used in solving PDEs in several ways: as a straight colloca-tion method [40, 41], in connection with a variant of the boundary element method [28], and via a different RBF-based collocation method, one that uses RBF-Hermite interpolation [23]. Indeed in [23], Franke and Schaback showed that the Hermite-based method converges to solutions for a variety of PDEs. Recently, several researchers [20, 72] have employed multilevel methods either in direct application to solving PDEs or in related matters.

There are two other situations where RBFs techniques might be helpful. First, although the finite element method is a well-established method for solving PDE's in two and three dimensions, there are situations in which alternative methods would be welcome. For example, in the field of neutron transport, nodal methods are used in their place [35]. It might be possible to couple multilevel methods with nodal methods to obtain smoother solutions to transport equations. Second, the multilevel method has potential as a collocation tool for PDEs in higher dimensions. One such equation arises in finance, where Black-Scholes models [36, Chapter 10] might be analyzed using the multilevel method.

5.4. Tomography

Recently, RBFs have been successfully used by the Medical Image Processing Group at the University of Pennsylvania as a tool for reconstructing images from data obtained via *positron emission tomography* (PET) [44].

There are several interesting features that arise here. The data sets are huge, on the order of millions, and noisy. They are non-traditional, because they amount to line integrals of an unknown distribution function. The distribution function that one is trying to reconstruct is nonnegative, so the reconstruction task involves shape-preservation, and it is continuous, but has edges and corners and is not differentiable.

§6. New Directions

We have discussed many, but certainly not all, aspects of research involving positive definite functions and kernels. Along the way, we have pointed out a number of problems and areas for future research. There are two other new directions for future research. The first is in finding and dealing with non-traditional applications, such as the one involving positron emission tomography. The second is attacking a range of *nonlinear* problems, for instance the shape preservation problem inherent in simply keeping fits to positive densities positive. Another example is the nonlinear optimization problem encountered if one is given N possible centers along with the freedom to place them anywhere. Preliminary results may be found in [49] and [56].

Acknowledgments. I wish to thank Professors Chui and Schumaker for the opportunity to speak at the conference, and for their support. I also wish to thank Professor Ward for many helpful suggestions. This work is partially supported by the Air Force Office of Scientific Research, USAF, under grant number F49620-98-1-0204. The U.S. Government is authorized to reproduce and distribute reprints for Governmental, notwithstanding any copyright notation thereon.

References

1. Ball, K., Eigenvalues of Euclidean distance matrices, J. Approx. Theory **68** (1992), 74–82.

2. Ball, K., N. Sivakumar, and J. D. Ward, On the sensitivity of radial basis interpolation to minimal data separation distance, Constr. Approx. **8** (1992), 401–426.

3. Baxter, B. J. C., Norm estimates for inverses of Toeplitz distance matrices, J. Approx. Theory **79** (1994), 222–242.

4. Baxter, B. J. C. and G. Roussos, On the fast evaluation of conditionally positive definite functions via the fast Gauss transform, *IMACS RBF Conference*, Monterey, California, May 1997, pp. 27–30.

5. Beale, R. and T. Jackson, *Neural Computing,* Adam Hilger Press, Bristol, 1990.

6. Beatson, R. K. and G. N. Newsam, A far field expansion for generalised multiquadrics in $\mathbb{R}^n$, *Computational Techniques and Applications: CTAC97*, Adelaide, Australia, September 1997, p. 30.

7. de Boor, C., R. DeVore, and A. Ron, On the construction of multivariate (pre) wavelets, Constr. Approx. **9** (1993), 123–166.

8. de Boor, C., R. DeVore, and A. Ron, Approximation from shift-invariant subspaces of $L_2(R^d)$, Trans. Amer. Math. Soc. **341** (1994), 787–806.

9. Buhmann, M. D., New developments in the theory of radial basis function interpolation, in *Multivariate Approximation: From CAGD to Wavelets*, K. Jetter and F. I. Uteras (eds.), World Scientific, Singapore, 1993, pp. 35–75.

10. Buhmann, M. D. and C. A. Micchelli, Spline prewavelets for nonuniform knots, Numer. Math. **61** (1992), 455–474.

11. Cheney, W., Approximation using positive definite functions, in *Approximation Theory VIII, Vol. 1: Approximation and Interpolation*, C. K. Chui and L. L. Schumaker (eds.), World Scientific Publishing Co., Inc., Singapore, 1995, pp. 145–168.

12. Chui, C. K., K. Jetter, J. Stöckler, and J. D. Ward, Wavelets for analyzing scattered data: An unbounded operator approach, Appl. Comput. Harmonic Anal. **3** (1996), 254–267.

13. Chui, C. K., J. Stöckler, and J. D. Ward, Analytic wavelets generated by radial functions, Adv. Comp. Math. **5** (1996), 95–123.

14. Dyn, N., Interpolation and approximation by radial and related functions, in *Approximation Theory VI*, C. K. Chui, L. L. Schumaker, and J. Ward (eds.), Academic Press, New York, 1989, pp. 211–234.

15. Dyn, L., D. Rippa, and S. Rippa, Numerical procedures for surface fitting of scattered data by radial functions, SIAM J. Sci. Statist. Comput. **7** (1986), 639–659.

16. Dyn, N., F. J. Narcowich, and J. D. Ward, A framework for interpolation and approximation on Riemannian manifolds, in *Approximation Theory and Optimization*, M. D. Buhmann and A. Iserles (eds.), Cambridge University Press, 1997, pp. 133–144.

17. Dyn, N., F. J. Narcowich, and J. D. Ward, Variational principles and Sobolev-type estimates for generalized interpolation on a Riemannian manifold, Constr. Approx., to appear.

18. Dyn, N. and A. Ron, Radial basis function approximation: from gridded centers to scattered centers, J. London Math. Soc. **71**(3) (1996), 76–108.

19. Duchon, J., Interpolation des fonctions de deux variables suivant le principe de la flexion des plaques minces, Rairo Analyse Numerique **10** (1976), 5–12.

20. Fasshauer, G. E. and J. Jerome, Multistep approximation algorithms: improved convergence rates through postconditioning with smoothing kernels, 1997, preprint.

21. Fasshauer, G. E. and L. L. Schumaker, Scattered Data Fitting on the Sphere, in *Mathematical Methods for Curves and Surfaces II*, M. Dæhlen, T. Lyche and L. L. Schumaker (eds.), Vanderbilt Univ. Press, Nashville & London, 1998, pp. 117–166.

22. Floater, M. S. and A. Iske, Multistep scattered data interpolation using compactly supported radial basis functions, J. Comput. Appl. Math. **73**(1-2) (1996), 65–78.

23. Franke, C. and R. Schaback, Solving partial differential equations by collocation using radial basis functions, 1997, preprint.

24. Franke, R., Scattered data interpolation: tests of some methods, Math. Comp. **38** (1982), 181–199.

25. Freeden, W., On spherical spline interpolation and approximation, Math. Meth. Appl. Sci. **3** (1981), 551–575.

26. Freeden, W., M. Schreiner, and R. Franke, A survey on spherical spline approximation, Surveys Math. Indust. **7** (1997), 29–85.

27. Freeden, W. and U. Windheuser, *Spherical Wavelet Transform and its Discretization*, Berichte der Arbeitsgruppe Technomathematik von Universität Kaiserlautern # 125, 1995.

28. Goldberg, M. A. and C. S. Chen, The theory of radial basis functions applied to the BEM for inhomogeneous partial differential equations, Boundary Elements Communications, **5** (1994), 57–61.

29. Golitschek, M. von and W. A. Light, Interpolation by polynomials and radial basis functions, 1997, preprint.

30. Golomb, M. and H. F. Weinberg, Optimal approximation and error bounds, in *on Numerical Approximation*, by R. E. Langer (ed.), Madison, 1959, pp. 117–190.

31. Gutzmer, T., Interpolation by positive definite functions on locally compact groups with application to $SO(3)$, Results in Mathematics **29** (1996), 69–77.

32. Hardy, R. L., Multiquadric equations of topography and other irregular surfaces, J. Geophys. Res. **76** (1971), 1905–1915.

33. Hardy, R. L., Theory and applications of the multiquadric-biharmonic method, Comput. Math. Appl. **19** (1990), 163–208.

34. Hassan, Y., T. Blanchat, and C. Seeley, Jr., Simultaneous velocity measurements of both components of a two phase flow using particle image velocimetry, Int. J. Multiphase Flow **18** (1992), 1–23.

35. Hennart, J.-P. and E. del Valle, On nodal transport methods, in *highlights of MAFELAP 1996*, Brunel University, June 1996.

36. Hull, J. C., *Options, Futures, and Other Derivative Securities*, 2nd edition, Prentice Hall, Englewood Cliffs, NJ, 1993.

37. Jetter, K, S. Riemenschneider, and Z. Shen, Hermite interpolation on the lattice $\mathbb{Z}^d$, SIAM J. Math. Anal. **25** (1994), 962–975.

38. Jetter, K. and J. Stöckler, A generalization of de Boor's stability result and symmetric preconditioning, Adv. Comput. Math. **3** (1995), 353–367.

39. Jetter, K., J. Stöckler, and J. D. Ward, Error estimates for scattered data interpolation on spheres, Math. Comp., to appear.

40. Kansa, E. J., Multiquadrics—a scattered data approximation scheme with applications to computational fluid dynamics—I. Surface approximations and partial derivative estimates, Computers Math. Applic. **19** (1990), 127–145.

41. Kansa, E. J., Multiquadrics—a scattered data approximation scheme with applications to computational fluid dynamics—II. Solutions to parabolic, hyperbolic and elliptic partial differential equations, Computers Math. Applic. **19** (1990), 147–161.

42. Levesley, J., W. Light, D. Rogozin, and X. Sun, Variational theory for interpolation on spheres, 1997, preprint.

43. Levesley, J., Z. Luo, and X. Sun, Norm estimates of interpolation matrices and their inverses associated with strictly positive definite functions, 1997, preprint.

44. Lewitt, R. M., S. Matej, and G. T. Herman, Discretization and iterative solution of inverse problems in 3D tomography using bell-shaped radial basis functions having compact support, IMACS RBF Conference, Monterey, California, May 1997, pp. 27–30.

45. Lorentz, G. G., K. Jetter, and S. Riemenschneider, Birkhoff interpolation, Encyclopedia of Mathematics and its Applications, vol 19, Addison-Wesley, Reading, 1983.

46. Madych, W. R. and S. A. Nelson, Multivariate interpolation and conditionally positive definite functions, Approx. Theory Appl. **4** (1988), 77–79.

47. Madych, W. R. and S. A. Nelson, Multivariate interpolation and conditionally positive definite functions II, Math. Comp. **54** (1990), 211–230.

48. Menegatto, V. A., Condition numbers associated with radial-function interpolation on spheres, 1997, preprint.

49. Mhaskar, H. N., Versitile Gaussian networks, in *Proc. of the IEEE Workshop on Nonlinear Image and Signal Processing*, I. Pitas (ed.), Halkidiki, Greece, June 1995, IEEE, pp. 70–73.

50. Mhaskar, H. N. and J. Prestin, On a choice of sampling nodes for optimal approximal of smooth functions by generalized translation networks, 1997, preprint.

51. Micchelli, C. A., Interpolation of scattered data: distances, matrices, and conditionally positive definite functions, Constr. Approx. **2** (1986), 11–22.

52. Narcowich, F. J., Generalized Hermite interpolation and positive definite kernels on a Riemannian manifold, J. Math. Anal. Applic. **190** (1995), 165–193.

53. Narcowich, F. J., R. Schaback, and J. D. Ward, *Multilevel Interpolation and Approximation*, Center for Approximation Theory Report # 379, Department of Mathematics, Texas A&M University, 1997.

54. Narcowich, F. J., N. Sivakumar, and J. D. Ward, On condition numbers associated with radial-function interpolation, J. Math. Anal. Appl. **186** (1994), 457–485.

55. Narcowich, F. J., N. Sivakumar, and J. D. Ward, Stability results for scattered-data interpolation on Euclidean spheres, Adv. Comp. Math., to appear.

56. Narcowich, F. J., P. W. Smith, and J. D. Ward, Density of translates of radial functions on compact sets, in *Approximation Theory VIII, Vol. 1: Approximation and Interpolation*, C. K. Chui and L. L. Schumaker (eds.), World Scientific Publishing Co., Inc., Singapore, 1995, pp. 435–442.

57. Narcowich, F. J. and J. D. Ward, Norms of inverses and condition numbers for matrices associated with scattered data, J. Approx. Theory **64** (1991), 69–94.

58. Narcowich, F. J. and J. D. Ward, Norm estimates for the inverses of a general class of scattered-data radial-function interpolation matrices, J. Approx. Theory **69** (1992), 84–109.

59. Narcowich, F. J. and J. D. Ward, Generalized Hermite interpolation via matrix-valued conditionally positive definite functions, Math. Comp. **63** (1994), 661–688.

60. Narcowich, F. J. and J. D. Ward, Wavelets associated with periodic basis functions, Appl. Comput. Harmonic Anal. **3** (1996), 40–56.

61. Narcowich, F. J. and J. D. Ward, Nonstationary wavelets on the *m*-sphere for scattered data, Appl. Comp. Harm. Anal. **3** (1996), 324–336.

62. Poggio, T. and F. Girosi, A theory of networks for approximating and learning, MIT Artificial Intelligence Laboratory and Center for Biological Information Processing (Whitaker College), A.I. Memo No. 1140, C.B.I.P. Paper No. 31, 1989.

63. Powell, M. J. D., The theory of radial basis approximation in 1990, in *Advances in Numerical Analysis II: Wavelets, Subdivision and Radial Functions*, W. Light (ed.), Oxford University Press, 1990, pp. 105–210.

64. Powell, M. J. D., Truncated Laurent expansions for the fast evaluation of thin plate splines, Numer. Algorithms **5** (1993), 99–120.

65. Powell, M. J. D., Thin plate spline interpolation in two dimensions, *Computational Techniques and Applications: CTAC97*, Adelaide, Australia, September 1997, p. 29.

66. Quak, E., N. Sivakumar, and J. D. Ward, Least squares approximation by radial functions, SIAM J. Math Anal. **24** (1993), 1043–1066.

67. Ron A. and X. Sun, Strictly positive definite functions on spheres, Math. Comp. **65** (1996), 1513–1530.

68. Schoenberg, I. J., Positive definite functions on spheres, Duke Math. J. **9** (1942), 96–108.

69. Schaback, R., Lower bounds for norms of inverses of interpolation matrices for radial basis functions, J. Approx. Theory **79** (1994), 287–306.

70. Schaback, R., Error estimates and condition numbers for radial basis function interpolation, Adv. Comp. Math. **3** (1995), 251–264.

71. Schaback, R., Multivariate Interpolation and Approximation by Translates of a Basis Function, in *Approximation Theory VIII, Vol. 1: Approximation and Interpolation*, C. K. Chui and L. L. Schumaker (eds.), World Scientific Publishing Co., Inc., Singapore, 1995, pp. 491–514.

72. Schaback, R., Solving partial differential equations using radial basis functions, IMACS RBF Conference, Monterey, California, May 1997, pp. 27–30.

73. Schaback, R., and Z. Wu, Operators on radial functions, J. Comput. Appl. Math. **73**(1-2) (1996), 1–17.

74. Southall, H. L., J. A. Simmers, and T. H. O'Donnell, Direction finding in phased arrays with a neural network beamformer, IEEE Transactions on Antennas and Propagation **43** (1995), 1369–1374.

75. Stewart, J., Positive definite functions and generalizations, an historical survey, Rocky Mountain J. Math. **6** (1976), 409–434.

76. Sun, X., Scattered Hermite interpolation using radial basis functions, Linear Algebra Applic. **207** (1994), 135-146.

77. Wahba, G., Spline interpolation and smoothing on the sphere, SIAM J. Sci. Statist. Comput. **2** (1981), 5–16.

78. Wahba, G., Surface fitting with scattered noisy data on Euclidean d-space and on the sphere, Rocky Mountain J. Math. **14** (1984), 281–299.

79. Wendland, H., Piecewise polynomial, positive definite and compactly supported radial functions of minimal degree, Adv. Comp. Math. **4** (1995), 389–396.

80. Whittaker, E. T. and G. N. Watson, *A Course in Modern Analysis*, Cambridge University Press, Cambridge, 1965.

81. Wu, Z., Hermite-Birkhoff interpolation of scattered data by radial basis functions, Approx. Theory & Its Appl. **8** (1992), 1–10.

82. Wu, Z., Multivariate compactly supported positive definite radial functions, Adv. Comp. Math. **4** (1995), 283–292

83. Xu, Y. and E. W. Cheney, Strictly positive definite functions on spheres, Proc. Amer. Math. Soc. **116** (1992), 977–981.

Francis J. Narcowich
Department of Mathematics
Texas A&M University
College Station, TX 77843
fnarc@math.tamu.edu
http://www.math.tamu.edu/~francis.narcowich/

Simulation-Based Modeling

Alexa Nawotki and Hans Hagen

Abstract. Simulation-based modeling means the generation of CAD models from simulation results. It is efficient to split this process into several steps, such as collection, triangulation, and segmentation of the data and surface reconstruction. The last step is solved by a physically based approach, which enables interaction intuitively, and we prove the existence and uniqueness of the solution for this case.

§1. The Modeling Pipeline

In many applications one wants to represent a real object in a CAD-system, for example, a first design sketch in clay has to be converted into a CAD-object. This proceeding is called reverse engineering, because in the standard case, a physical workpiece is created from a CAD-model. The challenge is the conversion of the starting data—in general nothing but a large cloud of unstructured points—into a surface representation of the real object, which is of high quality and can be manipulated easily.

It is useful to split this transformation into several steps:

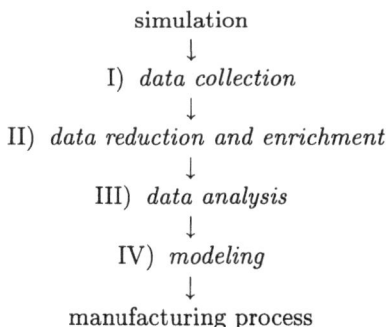

simulation

↓

I) *data collection*

↓

II) *data reduction and enrichment*

↓

III) *data analysis*

↓

IV) *modeling*

↓

manufacturing process

We outline the approaches for each problem in the following pages, where the emphasis is on the modeling part. In particular we prove the existence and uniqueness for a special 'physical' solution.

Let us first discuss how the data are assembled.

Approximation Theory IX, Volume 2: Computational Aspects   243
Charles K. Chui and Larry L. Schumaker (eds.), pp. 243–250.

§2. Data Collection

The data-source is a simulation, i.e., the computation of a technical process or the scanning of a design-prototype or a to-be-improved workpiece. Physical objects can be digitized using manual devices, CNC-controlled coordinate measuring machines or laser range scanning systems. The data generation and reception is highly dependent on the special situation, and has to be solved individually. The machine dependent errors have to be suppressed in a specific filtering process.

In the following, we assume that we have a large set of data points, referring to surface points of the desired object. We assume the most general case, i.e., the data points are arbitrarily distributed and there is no information at all besides the position-coordinates.

§3. Data Reduction and Enrichment

It is efficient in most cases to reduce this big set $P := \{p_1, \ldots, p_m\} \subset \mathbb{R}^n$. This can be attained by a cluster analysis, i.e., the points are grouped by minimizing a cost function K, and in each cluster C_h, one point is determined vicarious for all other cluster points. A suitable cost function for this problem is

$$K_h := \sum_{i,j \in I_h, i < j} \|p_i - p_j\|^2$$

with $I_h = \{i \in \mathbb{N} | p_i \in C_h\}$ and $h = 1, \ldots, q < m$.

Thus, we have to minimize

$$\sum_{h=1}^{q} K_h = \sum_{h=1}^{q} \left(\sum_{i,j \in I_h, i < j} \|p_i - p_j\|^2 \right) \to \min,$$

which is equivalent to

$$\sum_{h=1}^{q} \sum_{i,j \in I_h} \|p_i - S_h\|^2 \to \min,$$

where S_h is the center of gravity for the cluster C_h.

The minimum can also be specified by the following partition of the space:

Definition. *Let* $P = \{p_1, \ldots, p_m\} \subset \mathbb{R}^n$ *be a set of points and* $d(.,.)$ *a metric. Then* $V(i) = V(P_i) = \{p \in \mathbb{R}^n | d(p, p_i) \leq d(p, p_j), i \neq j\}$ *is called the* Voronoi-region *of* p_i *and the collection of all Voronoi-regions* $Vor(P) = \bigcup_{i=1}^{m} V(i)$ *is the* Voronoi-diagram.

The cost function is obviously minimal if the clusters coincide with the Voronoi-regions around the centers of gravity (with $\|p - q\| = d(p, q)$). Unfortunately this problem is *np*-complete and we get sub-optimal solutions only by alternating these two criteria. More details are included in [9].

For the following steps it is handy to have also neighboring information, which is given for example by a triangulation. This information can very easily be deduced from the Voronoi-diagram.

Definition. *Let* $P = \{p_1, \ldots, p_m\} \subset \mathbb{R}^n$ *be a set of points and* $V(i)$ *the Voronoi-regions. The connecting lines of all points* P_i, P_j *with* $V(i) \cap V(j) \neq \oslash$ *are a* triangulation, *the so-called* Delaunay-triangulation.

If we need the triangulation for a parametrization, we have to project the points in a suitable plane first and if more than $n+1$ Voronoi-regions meet in a single point, this construction is not unique and some extra work is necessary (see [8]). With this, we have reached the last data-preparation step.

§4. Data Analysis

A segmentation is necessary, because it is neither efficient nor feasible for realistic objects to construct merely one surface from all points. Thus the points are subdivided by means of curvature discontinuities and each group represents a separate surface in the final model to make it as smooth as possible. Hamann's approach is based on an approximation of the curvature at a point with an 'osculating' paraboloid, which interpolates the point of interest and approximates the neighbors with a paraboloid (see [5]).

This can be extended by using a general polynomial. Another possibility is to group the points according to their approximated normal vectors and minimal curvatures. The pros and cons of these methods are discussed in [1].

§5. Modeling

For every segment we have to construct a smooth surface. Variational design is a very popular method, where a functional is minimized while it has to meet some boundary conditions (see [2]). This proceeding is justified by the natural law of minimal energy (Hamiltonian Principle of Mechanics, [6]). One option is the functional of the 'clamped plate'

$$
F(u) = \frac{h^3}{24} \int \frac{E}{1 - \nu^2}
$$
$$
\times \left(\left(\frac{\partial^2 u}{\partial x^2} + \frac{\partial^2 u}{\partial y^2} \right)^2 - 2(1 - \nu) \left(\frac{\partial^2 u}{\partial x^2} \frac{\partial^2 u}{\partial y^2} - \left(\frac{\partial^2 u}{\partial x \partial y} \right)^2 \right) \right) \, dx dy, \quad (1)
$$

where the energy of a thin, homogeneous surface is minimized. (The physical deduction is included in [4]). In this formula h denotes the thickness of the plate, u the deformation and E and ν are material constants. They determine how a specific substance deforms under forces. These parameters can be used as intuitive interaction tool, if it is possible to change their values *locally*. For that we have to prove that the variational problem with functional (1) and variable material parameters, in the following called 'polytropic' functional, is still uniquely solvable. It is worth to investigate E and ν first, because their bounds are reflected in theoretical limits, resp. These bounds allow a mathematical existence and uniqueness proof.

E is called the modulus of elasticity and is defined as the proportional factor between the strain $\frac{\Delta l_i}{l_i}$ and the stress P_i, i.e.,

$$P_i =: E \cdot \frac{\Delta l_i}{l_i}.$$

Thus the dimension of E is $[E] = [P_i] = [\frac{force}{area}]$.

A stressed body changes its size not only parallel (l), but also orthogonal (b) to the force direction. This is expressed by the second constant

$$\nu := -\frac{\frac{\Delta b}{b}}{\frac{\Delta l}{l}}, \tag{2}$$

where ν is called Poisson's ratio and has no dimension.

Now we deduce the physical bounds for these constants. It follows that $E \geq 0$, because pressure (negative sign) shortens a body, and tension (positive sign) makes it longer. Here, ν is nonnegative on account of the minus sign in definition (2), because an extension in one direction means a shortening in the other and vice versa.

By the following considerations an upper bound for ν can be derived. The change of the volume for a rectangular parallelepiped with width b and length l under tension is (for small deformation)

$$\Delta V = (b + \Delta b)^2 \cdot (l + \Delta l) - b^2 l$$
$$= b^2 \Delta l + 2b \Delta b\, l + 2b\, \Delta b\, \Delta l + (\Delta b)^2\, l + (\Delta b)^2 \Delta l$$
$$\approx b^2 \Delta l + 2bl \Delta b.$$

Therefore we get

$$\frac{\Delta V}{V} = \frac{b^2 \Delta l + 2bl \Delta b}{b^2 l}$$
$$= \frac{\Delta l}{l} + 2\frac{\Delta b}{b}$$
$$= \frac{\Delta l}{l}(1 - 2\nu).$$

Assuming, that under tension the volume of a body cannot get smaller, i.e., $0 \leq \frac{\Delta V}{V}$ it follows that

$$\nu \leq \frac{1}{2}.$$

With these bounds we can prove the existence and uniqueness of the solution of the variational problem $F \to \min$ for the polytropic functional F.

§6. Existence and Uniqueness of the Solution

Let $H_0^m(\Omega)$ denotes the mth order Sobolev space of functions with support in Ω, whereby Ω is a nonempty, open, and bounded subset of $\mathbb{R}^n$, and $\|v\|_m :=$ $\sqrt{\sum_{|\alpha| \le m} \int_\Omega (\partial^\alpha v)^2 \, dx}$ is the belonging Hilbert-space-norm. Besides this, we need the seminorm $|v|_m := \sqrt{\sum_{|\alpha| = m} \int_\Omega (\partial^\alpha v)^2 \, dx}$.

The limitation of Ω is essential. For example, it allows the following statement to be checked easily (just partial integration):

Lemma. *Let Ω be a bounded, nonempty, open subset of $\mathbb{R}^n$. For all $v \in$ $H_0^2(\Omega)$,*

$$|v|_2 = \|\triangle v\|_{L^2(\Omega)} = |\triangle v|_0.$$

The existence and uniqueness of the solution of the polytropic functional of the clamped plate can be shown using the theorem of Lax-Milgram:

Lax-Milgram Theorem. *Let H be a Hilbert-space and $a : H \times H \to \mathbb{R}$ an H-elliptic bilinear form (i.e., $a(.,.)$ is continuous and $a(v,v) \ge C\|v\|$ for a positive constant C). The variational problem*

$$J(v) = \frac{1}{2}a(v,v) - l(v) \to \min_{v \in H}$$

has exactly one solution for every $l \in H^*$.

Thus, we only have to prove that $a(.,.)$ is continuous and $H_0^2(\Omega)$-elliptic.

Corollary. *Let $0 \le \nu(x) < 1$, $\frac{E(x)h^3}{24(1-\nu^2(x))} > 0$ be continuous and real-valued functions on $\mathbb{R}^n$, and Ω a bounded, open, nonempty subset of $\mathbb{R}^n$.*

Then the bilinear form

$$\mathrm{a}(u,v) = \int_\Omega \frac{E(x)h^3}{24(1-\nu^2(x))} \left(\nu(x) \triangle u \triangle v + (1-\nu(x)) \sum_{i,j=1}^n u_{ij}v_{ij} \right) dx$$

is continuous and $H_0^2(\Omega)$-elliptic in $H_0^2(\Omega)$.

Proof: It is sufficient to establish the statements for $u, v \in C_0^\infty(\Omega)$, because this set is dense in $H_0^2(\Omega)$. In the following, let x_i denote a point in Ω and C_i a positive constant.

Continuity: First we use the triangle-inequality for the separation of the fraction. Then we use the Mean Value Theorem for integrals, which holds for positive, continuous functions, to move this term in front of the integral.

$$|a(u,v)| = \left| \int_\Omega \frac{E(x)h^3}{24(1-\nu^2(x))} \left(\nu(x) \triangle u \triangle v + (1-\nu(x)) \sum_{i,j=1}^n u_{ij}v_{ij} \right) dx \right|$$

$$\le \int_\Omega \left| \frac{E(x)h^3}{24(1-\nu^2(x))} \right| \cdot \left| \nu(x) \triangle u \triangle v + (1-\nu(x)) \sum_{i,j=1}^n u_{ij}v_{ij} \right| dx$$

$$= \underbrace{\left| \frac{E(x_1)h^3}{24(1-\nu^2(x_1))} \right|}_{:=C_1} \int_\Omega \left| \nu(x) \triangle u \triangle v + (1-\nu(x)) \sum_{i,j=1}^n u_{ij}v_{ij} \right| dx$$

Now we repeat our arguments. We separate the remaining terms which contains material parameters using the triangle inequality and use the Mean Value Theorem for integrals to move these terms in front of the integrals:

$$|a(u,v)| \le C_1 \int_\Omega \left(\nu(x)|\triangle u||\triangle v| + (1-\nu(x)) \sum_{i,j=1}^n |u_{ij}||v_{ij}| \right) dx$$

$$= C_1 \left(\nu(x_2) \int_\Omega |\triangle u||\triangle v| \, dx + (1-\nu(x_3)) \int_\Omega \sum_{i,j=1}^n |u_{ij}||v_{ij}| \, dx \right).$$

With the inequality of Cauchy-Schwarz we can rewrite this relation using the $L^2(\Omega)$-norm and there we neglect the absolute values, because inside the norm we have real-valued squared functions.

$$|a(u,v)| \le C_2 \|\triangle u\|_{L^2(\Omega)} \|\triangle v\|_{L^2(\Omega)} + C_3 \sum_{i,j=1}^n \|u_{ij}\|_{L^2(\Omega)} \|v_{ij}\|_{L^2(\Omega)}$$

For the first term we use the lemma from above and the second is estimated with the seminorm.

$$|a(u,v)| \le C_2 |u|_2 |v|_2 + C_3 n^2 |u|_2 |v|_2$$
$$\le C_4 \|u\|_2 \|v\|_2$$

Therefore, $a(\,.\,,.\,)$ is continuous.

Ellipticity: Since $\partial^i v \in H_0^1(\Omega)$, we get by the Poincaré–Friedrich inequality constants $F_i > 0$ such that

$$\sqrt{\sum_{j=1}^n \|\partial^j(\partial^i v)\|_{L^2(\Omega)}^2} = |\partial^i v|_1 \ge F_i \|\partial^i v\|_{L^2(\Omega)}.$$

Let $F := \min_{i=1,\dots,n} F_i^2$. Then the above estimate and Poincaré-Friedrich yield

$$|v|_2^2 = \sqrt{\sum_{i=1}^{n}\sum_{j=1}^{n}\|\partial^j(\partial^i v)\|_{L^2(\Omega)}^2}^{-2}$$

$$\geq \sum_{i=1}^{n} F_i^2\|\partial^i v\|_{L^2(\Omega)}^2$$

$$\geq F\sum_{i=1}^{n}\|\partial^i v\|_{L^2(\Omega)}^2 \tag{3}$$

$$= F|v|_1^2$$

$$\geq G\,\|v\|_{L_2(\Omega)}^2 \tag{4}$$

for some constant $G > 0$. Altogether, we get a positive constant H with

$$|v|_2^2 = \frac{1}{3}\left(|v|_2^2 + |v|_2^2 + |v|_2^2\right)$$

$$\geq \frac{1}{3}\sum_{i,j=1}^{n}\|\partial^{ij}v\|_{L^2(\Omega)}^2 + \underbrace{\frac{1}{3}F\sum_{i=1}^{n}\|\partial^i v\|_{L^2(\Omega)}^2 + \underbrace{\frac{1}{3}G\|v\|_{L^2(\Omega)}^2}_{(4)}}_{(3)}$$

$$\geq H\|v\|_2^2. \tag{5}$$

We use the Mean Value Theorem for integrals for the material terms again, yielding

$$a(v,v) = \frac{E(x_4)h^3}{24(1-\nu^2(x_4))}\left(\nu(x_5)\int_{\Omega}(\triangle v)^2\,dx + (1-\nu(x_6))\int_{\Omega}\sum_{i,j=1}^{n}v_{ij}^2\,dx\right)$$

$$= C_5\,\|\triangle v\|_{L^2(\Omega)}^2 + C_6|v|_2^2.$$

With the lemma and (5), it follows that

$$a(v,v) = C_5\,|v|_2^2 + C_6|v|_2^2 \geq C_7\|v\|_2^2.$$

This completes the proof. $\square$

Now we know that it is admissible to use the material parameters as forming tools for the surface, because of the above Corollary for $n = 2$. The practical realization verifies that the elongation due to internal stress can be raised making the material locally softer or weakened by making it stiffer. Pictures and more details are included in [7].

Acknowledgments. We would like to thank Prof. Becker for sharing the idea of using the material parameters as interaction tools.

References

1. Hagen, H., S. Hahmann, and T. Schreiber, Visualisation and computation of curvature behaviour of freeform curves and surfaces, Computer Aided Design **27** (1995), 545–552.

2. Hagen, H., S. Heinz, and A. Nawotki, Variational design with boundary conditions and parameter optimized surface fitting, in *Geometric Modelling: Theory and Practice*, Strasser, Klein, and Rau (eds.), Springer, 1997, pp. 3–13.

3. Hagen, H., S. Heinz, M. Thesing, and T. Schreiber, Simulation based modelling, preprint.

4. Hagen, H. and A. Nawotki, Variational design and parameter optimized surface fitting, Computing, to appear.

5. Hamann, B., *Visualization and Modelling Contours of Trivariate Functions*, Dissertation, Arizona State University, 1991.

6. Nawotki, A. and H. Hagen, Physically based modeling, to be published in *Proceedings of the workshop 'Creating Fair and Shape Preserving Curves and Surfaces'*, TU Berlin, 1997.

7. Nawotki, A. and H. Hagen, Surface generation in a simulation based modelling process, preprint.

8. Schreiber, T., *Analyse und Approximation von unstrukturierten Daten und Freiformflächen*, Dissertation, Univ. Kaiserslautern, 1994.

9. Schreiber, T., A Voronoi diagram based adaptive k-means-typed clustering algorithm for multidimensional weighted data, in *Lecture Notes in Computer Science*, H. Bieri and H. Noltemeier (eds.), Springer, 1991, pp. 265–275.

Hans Hagen
Department of Computer Science
University of Kaiserslautern
Postfach 3049
D-67653 Kaiserslautern, Germany
hagen@informatik.uni-kl.de

Alexa Nawotki
Department of Computer Science
University of Kaiserslautern
Postfach 3049
D-67653 Kaiserslautern, Germany
nawotki@informatik.uni-kl.de

A Multiscale Method for the Evaluation
of Wiener Integrals

Erich Novak, Klaus Ritter, and Achim Steinbauer

Abstract. Many applications require approximate values of Wiener integrals. A typical approach is to approximate the path integral by a high dimensional integral and apply a Monte Carlo (randomized) method. Here we develop (deterministic) quadrature formulas for the Wiener measure, the 'knots' are piecewise linear functions. Our construction is based on polynomial interpolation and uses ideas of Smolyak as well as the multiscale decomposition of the Wiener measure, due to Lévy and Ciesielski. Numerical examples indicate that our method seems to work well if the integrand is smooth.

§1. Introduction

The Wiener measure w is a Borel probability measure on $C = C[0,1]$, equipped with the topology of uniform convergence. It is uniquely determined by the following properties, see, e.g., [8]:

1) $\int_C x(t)\,dw(x) = 0$ for every $t \in [0,1]$,

2) $\int_C x(s) \cdot x(t)\,dw(x) = \min(s,t)$ for all $s, t \in [0,1]$,

3) for all $d \in \mathbb{N}$ and $t_1, \ldots, t_d \in [0,1]$, the joint distribution of $x(t_1), \ldots, x(t_d)$ with respect to w is Gaussian.

The Wiener measure is probably the most widely used measure on an infinite-dimensional (function) space. It plays a central role in the theory of stochastic processes and it is used in many applications, see [5]. Integrals

$$I_\infty(f) = \int_C f(x)\,dw(x) \tag{1}$$

of w-integrable functionals f on C occur, for instance, in the theory of partial differential equations, in mathematical finance, and in several branches of physics.

Approximation Theory IX, Volume 2: Computational Aspects
Charles K. Chui and Larry L. Schumaker (eds.), pp. 251–258.
Copyright ℗ 1998 by Vanderbilt University Press, Nashville, TN.
ISBN 0-8265-1326-3.

These integrals usually cannot be evaluated explicitly, and therefore efficient quadrature formulas are needed. As in the finite dimensional case, we use the notion of quadrature formula for finite linear combinations of values of the integrand. Thus a quadrature formula

$$Q_\infty^n(f) = \sum_{i=1}^{n} a_i \cdot f(x_i) \qquad (2)$$

for the Wiener measure is defined by 'knots' $x_1, \ldots, x_n \in C$ and coefficients $a_1, \ldots, a_n \in \mathbb{R}$.

Typically, Wiener integrals are first approximated by d-dimensional integrals for large d. These high-dimensional integrals are then evaluated by Monte Carlo methods. Alternatively, quasi Monte Carlo methods are used, see [1]. The worst case complexity of integration with respect to the Wiener measure, and, more generally, of path integration for Gaussian measures, is studied in [15].

Since the Wiener measure is Gaussian, the high-dimensional integrals are Gaussian, too, for all linear discretizations. Interpolatory quadrature formulas for d-dimensional Gaussian integrals are constructed in [6, 10]. These formulas work well, say, up to dimension $d = 20$, but much larger dimensions are needed for the approximate calculation of Wiener integrals.

We present a modification of the approach from [10], based on a multiscale decomposition of the Wiener measure, due to Lévy and Ciesielski. One obtains for instance a formula with $n = 79\,537$ 'knots' and approximating dimension $d = 1024$. We demonstrate the efficiency of the new quadrature formulas by means of two examples, one of which is dealing with the valuation of a mortgage backed security.

§2. Approximation by Finite Dimensional Integrals

A natural approach for the approximate calculation of $I_\infty(f)$ is to replace the infinite dimensional integral (1) by a d-dimensional integral and then to apply a d-dimensional quadrature formula. Specifically, we use integrals with respect to joint distributions of $x(1/d), x(2/d), \ldots, x(1)$ for suitable dimensions d. For $Y \in \mathbb{R}^d$ let $\ell_d(Y)$ denote the piecewise linear function with breakpoints i/d and

$$\ell_d(Y)(i/d) = (Y_1 + \cdots + Y_i)/\sqrt{d}$$

for $i = 1, \ldots, d$ as well as $\ell_d(Y)(0) = 0$. Define

$$I_d(g) = \int_{\mathbb{R}^d} g(Y) \cdot \rho(Y_1) \cdots \rho(Y_d)\, dY, \qquad (3)$$

where

$$\rho(Y_i) = 1/\sqrt{2\pi} \cdot \exp(-Y_i^2/2).$$

Instead of $I_\infty(f)$ we consider the integral $I_d(f \circ \ell_d)$. The use of piecewise linear interpolation is motivated by the fact that $\ell_d(Y)(t)$ is the conditional mean of $x(t)$ with respect to w given $Y = (x(1/d), \ldots, x(1))$. Note that every quadrature formula Q_d^n for I_d yields a quadrature formula Q_∞^n for I_∞ by $Q_\infty^n(f) = Q_d^n(f \circ \ell_d)$. The latter formula is based on piecewise linear 'knots' x_i.

Let F denote the class of Lipschitz continuous functionals on C with Lipschitz constant bounded by one. Then

$$\limsup_{d \to \infty} \sup_{f \in F} |I_\infty(f) - I_d(f \circ \ell_d)| \cdot \sqrt{d/\ln d} \leq 1/\sqrt{2},$$

which follows from [12]. For $f(x) = \max(x) = \max_{t \in [0,1]} x(t)$ we have

$$\lim_{d \to \infty} |I_\infty(f) - I_d(f \circ \ell_d)| \cdot \sqrt{d} = c$$

with an explicitly known constant $c > 0$, see [2]. Therefore even $d = 100$ might only yield a modest accuracy.

Integrals (3) also arise from the Karhunen-Loève expansion. In this case piecewise linear functions are replaced by Fourier partial sums. See [15] for the analysis of a corresponding algorithm that is based on high-order derivatives of the functionals f.

The construction of approximating finite-dimensional integrals for functional integrals is studied in the monograph [5], see also [3, 9]. In [4, 7] the Wiener measure and functionals of a specific form are investigated.

§3. Quadrature Formulas for d-Dimensional Gaussian Integrals

For the approximate calculation of (3) we use quadrature formulas for d-dimensional Gaussian integrals, see [6] and [10]. First we select the Gauss-Hermite formulas U^i with $2^{i-1} + 1$ knots for $i > 1$ and a single knot for $i = 1$ to approximate $\int_{\mathbf{R}} g(y)\rho(y)\, dy$. Then we use Smolyak's construction to obtain formulas for the d-dimensional integral (3). The Smolyak formulas $A(q, d)$ are linear combinations of product formulas with the following key properties. Only products with a relatively small number of knots are used and the linear combination is chosen in such a way that an interpolation property for $d = 1$ is preserved for $d > 1$. The formula $A(q, d)$ is defined by

$$A(q, d) = \sum_{q-d+1 \leq |\mathbf{i}| \leq q} (-1)^{q-|\mathbf{i}|} \cdot \binom{d-1}{q-|\mathbf{i}|} \cdot (U^{i_1} \otimes \cdots \otimes U^{i_d}), \qquad (4)$$

where $q \geq d$, $\mathbf{i} \in \mathbf{N}^d$, and $|\mathbf{i}| = i_1 + \cdots + i_d$.

It is proved in [10] that $A(d + k, d)$ has polynomial exactness $2k + 1$ and uses, for $d \to \infty$, about $2^k/k! \cdot d^k$ knots. This dependence is almost optimal and also numerical tests show that the formulas $A(q, d)$ are very competitive if the integrand is smooth and if d is of moderate size, say $d = 10$ or $d = 20$.

Assume, however, that $d = 100$ which is not unrealistic, see Section 2. Then we can use only a few of the formulas $A(q, d)$, at most $A(104, 100)$.

For such small values of q we obtain completely wrong results for many functionals f because all 'knots' x_i in the respective quadrature formula $Q_\infty^n(f) = A(q, d)(f \circ l_d)$ have a too small L_∞-norm. For instance, if $d = 100$ and $q = 104$ then $\|x_i\|_\infty \leq 0.49$ for all i, while $I_\infty(\max) = \sqrt{2/\pi}$. Therefore we cannot use this method though it works well for $d \leq 20$. In the next section we apply a modified ('weighted') version of (4) to a transformed version of $I_d(f \circ l_d)$ to obtain much better formulas.

§4. Using the Lévy-Ciesielski Representation

For $\mu \in \mathbb{N}$ put $m(\mu) = 2^{\mu-2}$ if $\mu \geq 2$ and $m(1) = 1$. We define the Schauder functions $S_{\mu,\nu}$ for $\nu = 1, \ldots, m(\mu)$ in the usual way. Put $S_{1,1}(t) = t$. For $\mu \geq 2$ let $S_{\mu,\nu}$ denote the piecewise linear function with breakpoints $\xi/(2m(\mu))$ such that

$$S_{\mu,\nu}(\xi/(2m(\mu))) = \delta_{2\nu-1,\xi}$$

for $\xi = 0, \ldots, 2m(\mu)$. Finally, define $\sigma(1) = 1$ and

$$\sigma(\mu) = 2^{-\mu/2}$$

for $\mu \geq 2$. Henceforth, let $d = 2^k$ for $k \in \mathbb{N}_0$ and put

$$Y = (Y_{1,1}, \ldots, Y_{\mu,1}, \ldots, Y_{\mu,m(\mu)}, \ldots, Y_{k+1,m(k+1)}) \in \mathbb{R}^d.$$

We define

$$h_d(Y) = \sum_{\mu=1}^{k+1} \sigma(\mu) \sum_{\nu=1}^{m(\mu)} Y_{\mu,\nu} \cdot S_{\mu,\nu}.$$

In the limit, with a sequence of independent standard normal random variables $Y_{\mu,\nu}$, this so-called Lévy-Ciesielski construction yields a multiscale decomposition of the Wiener measure, see [8, Section 2.3.i]. For finite d we have

$$I_d(f \circ l_d) = I_d(f \circ h_d).$$

This representation was used in [1] to compute integrals by quasi-Monte Carlo methods, see also Section 6.

Observe that different variables $Y_{\mu,\nu}$ influence the integrand $f \circ h_d$ in a different way, according to the standard deviations $\sigma(\mu)$. Hence one could study weighted norms, see [13]. In the following we suggest a modification of (4). For simplicity we use the same Gauss-Hermite formulas U^i as in Section 3. However, a multiindex $\boldsymbol{i} \in \mathbb{N}^d$ now corresponds to a product formula $U^{i_1-\beta_1} \otimes \cdots \otimes U^{i_d-\beta_d}$. The sequence $\beta_1, \beta_2, \ldots$ is essentially defined by the following conditions. For the second variable $Y_{2,1}$ we take $\beta_2 = 0$. Suppose that $j = 2^{\mu-2} + \nu$ with $1 \leq \nu \leq m(\mu)$ or that $j = \mu = 1$. Then $U^{i_j-\beta_j}$ should use about $\sigma(\mu)/\sigma(2)$ times as many knots as U^{i_j}. After suitable rounding we end up with the definition

$$\beta_1 = -1, \qquad \beta_2 = \beta_3 = \beta_4 = 0, \qquad \beta_{2^{\mu-2}+\nu} = \lfloor \mu/2 \rfloor - 1.$$

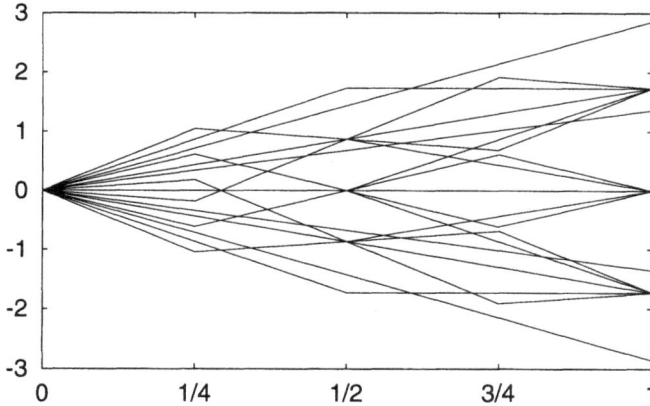

Fig. 1. The 25 'knots' of $A(1)$.

We obtain, for $d = 2^k$,

$$A(q, d) = \sum_{q-d+1 \leq |i| \leq q} (-1)^{q-|i|} \cdot \binom{d-1}{q-|i|} \cdot (U^{i_1 - \beta_1} \otimes \cdots \otimes U^{i_d - \beta_d}),$$

where $i \in \mathbb{N}^d$ and $U^j = U^1$ for $j < 1$.

The quadrature formulas $Q_\infty^n(f) = A(d+k, d)(f \circ h_d)$ do not depend on d if $d \geq 4^k$. This follows from the fact that

$$(U^{i_1} \otimes \cdots \otimes U^{i_d} \otimes U^1 \otimes \cdots \otimes U^1)(f \circ h_{2d}) = (U^{i_1} \otimes \cdots \otimes U^{i_d})(f \circ h_d).$$

We define

$$A(k)(f) = A(4^k + k, 4^k)(f \circ h_d)$$

for $k \in \mathbb{N}_0$ and obtain a sequence of quadrature formulas for the Wiener measure with $n(k)$ 'knots'. One gets $n(0) = 3$, $n(1) = 25$, $n(2) = 201$, $n(3) = 1561$, $n(4) = 11\,393$, and $n(5) = 79\,537$, and so on. In Figure 1 we show the 25 'knots' of $A(1)$.

§5. A Numerical Example

As an example, we study

$$f(x) = \exp\left(2 \int_0^1 x(t)\, dt\right). \tag{5}$$

The random variable $\int_0^1 x(t)\, dt$ is normally distributed with mean zero and variance $1/3$. Hence $I_\infty(f) = \exp(2/3)$ and

$$\sigma^2(f) = \exp(8/3) - \exp(4/3) \approx 10.6$$

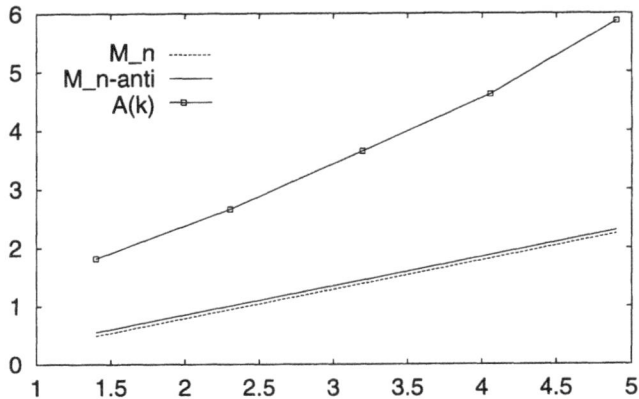

Fig. 2. Test results for the functional (5).

for the variance $\sigma^2(f)$ of f. An idealized form of the classical Monte Carlo method is

$$M_n(f) = \frac{1}{n} \sum_{i=1}^{n} f(x_i),$$

where the 'knots' x_i are chosen independently according to the Wiener measure. This method has a standard deviation ('error') $\sigma(f)/\sqrt{n}$.

In Figure 2 we show the accuracy of $A(k)$ and M_n. The accuracy ('number of correct digits') is defined by the negative logarithm (to the basis 10) of the relative error. This accuracy is plotted against the logarithm of the number n of 'knots'. We also show the accuracy of antithetic sampling, which is only slightly better than the 'crude' Monte Carlo method.

§6. An Example from Finance

High dimensional integration is needed to compute the value of financial derivatives, such as mortgage backed securities, see [1, 11]. The geometric Brownian motion is the canonical model for the interest rate, and the value of the security is given as an expectation. For mortgage backed securities the integrand is rather complicated, due to a prepayment option. Hence the integral cannot be computed analytically. A discretization is used, and a typical security of length 360 months yields a 360-dimensional integral.

Here we consider the example (4.5) from [1]. We use the method from Section 4, with minor modifications due to the fact that $d = 360$ is not a power of four. In Figure 3 we also present (as in Figure 2) results from [1] for four different methods. Namely, a crude Monte Carlo method (MC) and a Monte Carlo method with antithetic sampling (MC-anti), both being based on pseudo-random numbers. Furthermore, two quasi-Monte Carlo methods, (QR,LC) and (QR-anti,LC), that are based on the Sobol sequence and the

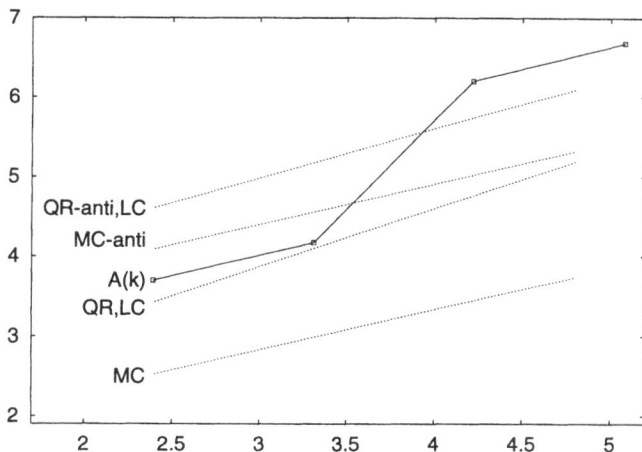

Fig. 3. An Example from Finance.

Lévy-Ciesielski representation. The method QR-anti,LC uses antithetic sampling, in addition.

We stress that even better results are obtained by taking into account the special structure of the integrand or by using second-order derivatives of the integrand. See [1, 14] for details. Here, however, we have only considered general purpose methods.

Acknowledgments. We thank William Morokoff for providing us the 'exact' value of the integrand in Figure 3.

References

1. Caflisch, R. E., W. Morokoff, and A. Owen, Valuation of mortgage backed securities using Brownian bridges to reduce effective dimension, J. Comput. Finance **1** (1997), 27–46.

2. Calvin, J. M., An asymptotically optimal non-adaptive algorithm for minimization of Brownian motion, in *The Mathematics of Numerical Analysis*, J. Renegar, M. Shub, and S. Smale (eds.), Amer. Math. Soc., Providence, 1996, pp. 157–163.

3. Cameron, R. H., A "Simpson's rule" for the numerical evaluation of Wiener integrals in function space, Duke Math. J. **18** (1951), 111–130.

4. Chorin, A. J., Accurate evaluation of Wiener integrals, Math. Comp. **27** (1973), 1–15.

5. Egorov, A. D., P. I. Sobolevsky, and L. A. Yanovich, *Functional Integrals: Approximate Evaluation and Applications*, Kluwer Academic Publishers, Dordrecht, 1993.

6. Genz, A. C. and B. D. Keister, Fully symmetric interpolatory rules for multiple integrals over infinite regions with Gaussian weight, J. Comput. Appl. Math. **71** (1996), 299–309.

7. Hald, O. H., Approximation of Wiener integrals, J. Comp. Phys. **69** (1987), 460–470.

8. Hida, T., *Brownian Motion*, Springer-Verlag, 1980.

9. Lobanov, Yu. Yu., Deterministic computation of functional integrals, Computer Physics Communications **99** (1996), 59–72.

10. Novak, E. and K. Ritter, Simple cubature formulas for d-dimensional integrals with high polynomial exactness and small error, preprint.

11. Paskov, S. H. and J. F. Traub, Faster valuation of financial derivatives, J. Portfolio Management **22** (1995), 113–120.

12. Ritter, K., Approximation and optimization on the Wiener space, J. Complexity **6** (1990), 337–364.

13. Sloan, I. H. and H. Woźniakowski, When are quasi-Monte Carlo algorithms efficient for high dimensional integrals?, J. Complexity **14** (1998), 1–33.

14. Steinbauer, A., Quadraturformeln für das Wienermaß, Erlangen, in preparation.

15. Wasilkowski, G. W. and H. Woźniakowski, On tractability of path integration, J. Math. Phys. **37** (1996), 2071–2088.

Erich Novak, Klaus Ritter, Achim Steinbauer
University of Erlangen and Nürnberg
Mathematical Institute
Bismarckstr. 1 1/2
D-91054 Erlangen
Germany
{novak,ritter,steinbau}@mi.uni-erlangen.de

Spline Interpolation on
Convex Quadrangulations

Günther Nürnberger and Frank Zeilfelder

Abstract. We describe an algorithm for constructing Lagrange and Hermite interpolation sets for bivariate spline spaces of smoothness one and two on quadrangulations with diagonals. The known methods are not applicable, since such triangulations do not belong to the class of crosscut partitions. Numerical examples are given.

§1. Introduction

Lagrange and Hermite interpolation sets for spaces $S_q^r(\Delta)$ of splines of degree q and smoothness r were constructed for crosscut partitions Δ, in particular for Δ^1 and Δ^2-partitions [1,2,6,10,11,12,16,17]. Results on the approximation order of these interpolation methods were given in [2,4,6,9,12,14,16,17]. Hermite interpolation schemes for $S_q^1(\Delta), q \geq 5$, where Δ is an arbitrary triangulation, were given in [3,7].

The aim of this paper is to construct interpolation sets for spaces $S_q^r(\Delta)$ for $r = 1, 2$, where Δ is a quadrangulation with diagonals. By an inductive method, in each step, we choose interpolation points on the union of finitely many subtriangles forming a cone. Since these cones can be nonconvex, the known methods for crosscut partitions are not applicable.

§2. Main Results

We consider bivariate spline spaces of the following type. First, the space of bivariate polynomials of total degree q is denoted by $\tilde{\Pi}_q$ and the space of univariate polynomials of degree q is denoted by Π_q. Let $v_{\mu,\nu} \in \mathbb{R}^2, \mu = 1, \ldots, m, \nu = 1, \ldots, n$, be given distinct points such that for $\mu = 1, \ldots, m - 1, \nu = 1, \ldots, n - 1$, the points $v_{\mu,\nu}, v_{\mu,\nu+1}, v_{\mu+1,\nu}, v_{\mu+1,\nu+1}$ are the vertices of a closed, convex quadrangle $Q_{\mu,\nu}$. We assume that the intersection of any two quadrangles is empty, a common vertex or a common edge. We add to each quadrangle both diagonals and denote their intersection point by $w_{\mu,\nu}$.

Copyright © 1998 by Vanderbilt University Press, Nashville, TN.
ISBN 0-8265-1326-3.

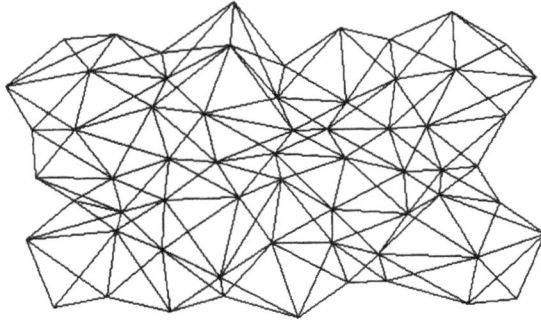

Fig. 1. A quadrangulation with diagonals.

Moreover, let $T^{[l]}_{\mu,\nu}, l = 1, \ldots, 4$, be the triangles in $Q_{\mu,\nu}$ in counterclockwise order, where $T^{[1]}_{\mu,\nu}$ is the left triangle in $Q_{\mu,\nu}$. The resulting triangulation of $\Omega = \bigcup_{\mu,\nu} Q_{\mu,\nu}$ is denoted by $\Delta = \{T^{[l]}_{\mu,\nu}\}$ (See Figure 1).

The spline spaces are defined as follows. Let r and q with $0 \le r < q$ be given integers. The space $S^r_q(\Delta)$ of all functions in $C^r(\Omega)$ such that the restriction to each triangle of Δ is in $\tilde{\Pi}_q$ is called the space of bivariate splines of degree q and smoothness r with respect to Δ.

We investigate interpolation by $S^r_q(\Delta)$. A set $\{z_1, \ldots, z_d\} \subset \Omega$, where $d = \dim S^r_q(\Delta)$ is called a Lagrange interpolation set for $S^r_q(\Delta)$ if for each function $f \in C(\Omega)$, a unique splines exists such that $s(z_i) = f(z_i), i = 1, \ldots, d$. Analogously, a set of points in Ω is called a Hermite interpolation set for $S^r_q(\Delta)$ if not only functional values but also partial derivatives of f are involved and the total number of Hermite interpolation conditions is d.

In the following, we construct Lagrange interpolation sets for $S^r_q(\Delta)$, where $q \ge 2$ if $r = 1$, and $q \ge 5$ if $r = 2$. Then we construct Hermite interpolation sets for $S^r_q(\Delta)$ by "taking limits." For $r = 2$, we always assume that the edges $[v_{\mu,\nu}, v_{\mu-1,\nu+1}]$, $[v_{\mu,\nu}, v_{\mu,\nu+1}]$, $[v_{\mu,\nu}, v_{\mu+1,\nu+1}]$, and $[v_{\mu,\nu}, v_{\mu+1,\nu}], \mu, \nu \ge 2$, have different slopes.

The Lagrange interpolation sets for $S^r_q(\Delta), r = 1, 2$, are constructed inductively. Therefore, we only have to describe some basic sets. For an arbitrary triangle $T \in \Delta$ with vertices v_1, v_2, v_3, exactly one of the following basic sets will be chosen.

Set Q : Choose $q+1$ disjoint line segments $p_1, \ldots, p_{q+1}$ in T. For $i = 1, \ldots, q+1$, choose $q + 2 - i$ distinct points on p_i.

Set A_r $(r = 1, 2)$: Choose $q - r$ disjoint line segments $a_1, \ldots, a_{q-r}$ in T. For $i = 1, \ldots, q - r$, choose $q + 1 - r - i$ distinct points on a_i.

Set B_r $(r = 1, 2)$: Choose $q - r - 1$ disjoint line segments $b_1, \ldots, b_{q-r-1}$ in T. For $i = 1, \ldots, q - r - 1$, choose $q + 1 - r - i$ distinct points on b_i.

Set C_2 : Choose $q - 4$ disjoint line segments $c_1, \ldots, c_{q-4}$ in T. For $i = 1, \ldots, q-4$, choose $q-1-i$ distinct points on c_i and choose a point on $[v_1, v_3]$ not lying on $c_i, i = 1, \ldots, q - 4$.

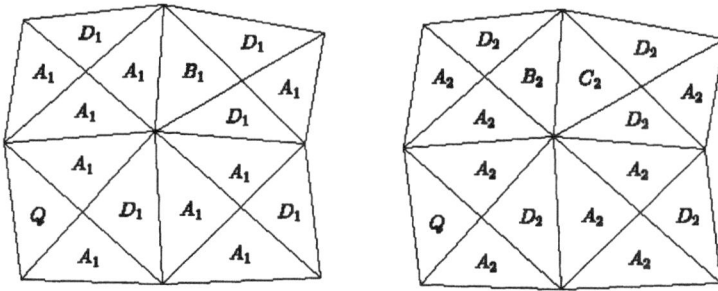

Fig. 2. Interpolation sets for $S_q^r(\Delta), r = 1, 2$.

Set D_r $(r = 1, 2)$: Choose $q - 2r - 1$ disjoint line segments $d_1, \ldots, d_{q-2r-1}$ in T. For $i = 1, \ldots, q - 2r - 1$, choose $q - 2r - i$ distinct points on d_i.

The line segments should be parallel to the edge $[v_2, v_3]$ and the chosen points should not lie on $[v_1, v_2]$. Given a quadrangulation Δ with diagonals, we choose the basic sets Q, A_1, B_1, D_1 for $S_q^1(\Delta), q \geq 2$, and Q, A_2, B_2, C_2, D_2 for $S_q^2(\Delta), q \geq 5$, as indicated in Figure 2. The basic sets are chosen row by row as described in the subsequent proofs. The union of these point sets is denoted by $\mathcal{L}_r, r = 1, 2$.

Theorem 1. *The set $\mathcal{L}_r$ is a Lagrange interpolation set for $S_q^r(\Delta)$, where $q \geq 2$ if $r = 1$, and $q \geq 5$ if $r = 2$.*

In the following, we construct Hermite interpolation sets. Let a sufficiently differentiable function $f \in C(\Omega)$ be given. For $z \in \Omega$, we set $D^\omega f(z) = (f_{x^i y^j}(z))_{i+j=\omega}$, and denote by $f_\tau(z)$ the derivative in the direction of the unit vector τ. For constructing Hermite interpolation sets for $S_q^r(\Delta), r = 1, 2$, we consider the following basic conditions. Let $T \in \Delta$ be an arbitrary triangle with vertices v_1, v_2, v_3 and denote by σ_k a unit vector parallel to the edge $[v_k, v_{k+1}], k = 1, 2, 3$, where $v_4 = v_1$. We impose exactly one of the following conditions on the polynomial $p = s|_T \in \tilde{\Pi}_q$.

Condition Q : $D^\omega p(v_3) = D^\omega f(v_3), \omega = 0, \ldots, q$.
Condition $A_r (r = 1, 2)$: $D^\omega p(v_3) = D^\omega f(v_3), \omega = 0, \ldots, q - r - 1$.
Condition $B_r (r = 1, 2)$: $D^\omega p(v_3) = D^\omega f(v_3), \omega = 0, \ldots, q - r - 1$, and
$$p_{\sigma_3^{q-r-1}}(v_3) = f_{\sigma_3^{q-r-1}}(v_3).$$
Condition C_2 : $D^\omega p(v_3) = D^\omega f(v_3), \omega = 0, \ldots, q - 3$, and
$$p_{\sigma_3^{q-3}}(v_3) = f_{\sigma_3^{q-3}}(v_3), p_{\sigma_3^{q-4} \sigma_2}(v_3) = f_{\sigma_3^{q-4} \sigma_2}(v_3).$$
Condition $D_r (r = 1, 2)$: $p_{\sigma_1^i \sigma_2^j}(v_1) = f_{\sigma_1^i \sigma_2^j}(v_1), j = r + 1, \ldots, q - r - 1 - i$,
$$i = 0, \ldots, q - 2r - 2.$$

Given a quadrangulation Δ with diagonals, the distribution of Hermite interpolation conditions to the subtriangles is the same as for Lagrange interpolation and is indicated in Figure 2. For $S_q^1(\Delta), q \geq 2$, we choose the basic conditions Q, A_1, B_1 and D_1. For $S_q^2(\Delta), q \geq 5$, we choose the basic conditions Q, A_2, B_2, C_2 and D_2. The resulting set of points (together with the interpolation conditions) is denoted by $\mathcal{H}_r, r = 1, 2$.

Theorem 2. *The set $\mathcal{H}_r$ is a Hermite interpolation set for $S_q^r(\Delta)$, where $q \geq 2$ if $r = 1$, and $q \geq 5$ if $r = 2$.*

Remark 3. Before giving the proof, we note that a simple computation shows that the number of chosen points (respectively conditions) is $mn(2q^2 - 6q + 5) + (m + n)(-2q^2 + 8q - 7) + 2q^2 - 10q + 12$, if $r = 1$, and $mn(2q^2 - 12q + 19) + (m + n)(-2q^2 + 15q - 25) + 2q^2 - 18q + 37$, if $r = 2$. It is easy to verify that these numbers are equal to the lower bounds for the dimension of $S_q^r(\Delta)$ in [15]. Therefore, it follows from the proofs in Section 3 that the number of chosen points (respectively conditions) is equal to the dimension of $S_q^r(\Delta), r = 1, 2$.

Building on these results, we construct, in a forthcoming paper [13], interpolation sets for $S_q^r(\Delta), r = 1, 2$, for the case when Δ is a triangulation of given points. (In addition, for the space $S_2^1(\Delta)$, we subdivide the triangles of Δ by a simple rule such that the only interpolation points for $S_2^1(\Delta)$ are the vertices of Δ.)

§3. Proof of the Main Results

Let $s \in S_q^r(\Delta), r = 1, 2$, satisfy the homogenous Lagrange (respectively Hermite) interpolation conditions. For $r = 1$, we define the following sets. For $\mu, \nu \geq 2$, let $\Delta_{\mu,\nu}^* \subseteq \Delta$ be the triangulation which consists of the triangles $T_{\mu_1,\nu_1}^{[l]}, \mu_1 \geq 1, \nu_1 \leq \nu - 1$, and $T_{\mu_1,\nu}^{[l]}, \mu_1 \leq \mu - 1, l = 1, \ldots, 4$, and set $\Omega_{\mu,\nu}^* = \{\bar{T} : \bar{T} \in \Delta_{\mu,\nu}^*\}$. Moreover, set $\tilde{\Delta}_{\mu,\nu} = \Delta_{\mu,\nu}^* \cup T_{\mu,\nu}^{[1]} \cup T_{\mu,\nu}^{[2]}$ and let $\tilde{\Omega}_{\mu,\nu}$ be the corresponding domain. In Section 2, we have chosen the set (respectively condition) B_1 in $T_{\mu,\nu}^{[1]}$ and D_1 in $T_{\mu,\nu}^{[2]}$.

Lemma 4. *Let $r = 1$ and $q \geq 2$. If $s = 0$ on $\Delta_{\mu,\nu}^*$, then $s = 0$ on $\tilde{\Delta}_{\mu,\nu}$.*

Proof: Let τ_1 (respectively τ_2) be a unit vector parallel to the edge $e^{[1]} = [v_{\mu,\nu}, v_{\mu,\nu+1}]$ (respectively $e^{[2]} = [v_{\mu,\nu}, v_{\mu+1,\nu}]$). It follows from the C^1 property of s that the polynomials $p^{[l]} = s|_{T_{\mu,\nu}^{[l]}} \in \tilde{\Pi}_q, l = 1, 2$, satisfy $p_{\tau^i \tau^j}^{[l]}|_{e^{[l]}} = 0, i \geq 0, j = 0, 1$, where τ is an arbitrary unit vector. Let τ_3 be a unit vector parallel to the edge $e^{[3]} = [v_{\mu,\nu}, w_{\mu,\nu}]$ and let $\gamma_1, \gamma_2 \in \mathbb{R} \setminus \{0\}$ be such that $\tau_3 = \gamma_1 \tau_1 + \gamma_2 \tau_2$. (Note that the slopes of $e^{[1]}$ and $e^{[2]}$ are different.) Since $p_{\tau_3^2}^{[1]}(v_{\mu,\nu}) = \gamma_1 p^{[1]}{}_{\tau_1 \tau_3}(v_{\mu,\nu}) + \gamma_2 p^{[2]}{}_{\tau_2 \tau_3}(v_{\mu,\nu}) = 0$, we have

$$D^\omega p^{[1]}(v_{\mu,\nu}) = 0, \quad \omega = 0, 1, 2. \tag{1}$$

First, we consider Lagrange interpolation. Let $b_{\alpha,\beta}, \beta = 1, \ldots, q - \alpha$, be the chosen points on the line segments $b_\alpha, \alpha = 1, \ldots, q - 2$, in $T_{\mu,\nu}^{[1]}$ parallel to $e^{[4]} = [w_{\mu,\nu}, v_{\mu,\nu+1}]$. We claim that $p^{[1]}|_{b_\alpha} = 0$. We prove this by induction on α. Since $p^{[1]}|_{b_1} \in \Pi_q$ and its first derivative parallel to b_1 vanish at the intersection point of b_1 and $e^{[1]}$, and $s(b_{1,\beta}) = 0, \beta = 1, \ldots, q - 1$, the claim holds for $\alpha = 1$. We assume that the claim holds for $\alpha \in \{1, \ldots, \alpha_1\}, \alpha_1 \leq q - 3$, and show that the claim holds for $\alpha_1 + 1$. It follows from the induction hypothesis that $q^{[1]} \in \tilde{\Pi}_{q-\alpha_1}$ exists such that $p^{[1]} = b_1 \ldots b_{\alpha_1} q^{[1]}$ holds on $T_{\mu,\nu}^{[1]}$. Since $q^{[1]}|_{b_{\alpha_1+1}} \in \Pi_{q-\alpha_1}$ and its first derivative parallel to b_{α_1+1} vanish at the intersection point of b_{α_1+1} and $e^{[1]}$, and $s(b_{\alpha_1+1,\beta}) = 0, \beta = 1, \ldots, q - \alpha_1 - 1$, it follows that $q^{[1]}|_{b_{\alpha_1+1}} = 0$. Thus, $p^{[1]}|_{b_{\alpha_1+1}} = 0$. This confirms the claim. Hence, from (1) we conclude $p^{[1]} = 0$. It follows that $D^\omega p^{[2]}(v_{\mu,\nu}) = 0, \omega = 0, \ldots, 3$. Let $d_\alpha, \alpha = 1, \ldots, q - 3$, be the chosen line segments in $T_{\mu,\nu}^{[2]}$ parallel to $e^{[5]} = [w_{\mu,\nu}, v_{\mu+1,\nu}]$. Since $p^{[2]}|_{d_\alpha}$ and its first derivative parallel to d_α vanish at the intersection points of d_α with $e^{[2]}$ and $e^{[3]}$, similar arguments as above show that $p^{[2]} = 0$. Hence, $s = 0$ on $\tilde{\Delta}_{\mu,\nu}$.

Now, we consider Hermite interpolation. We claim that $p_{\tau_3^\alpha}^{[1]}|_{e^{[4]}} = 0, \alpha = 0, \ldots, q - 3$. We prove this by induction on α. Let τ_4 be a unit vector parallel to the edge $e^{[4]}$. Since $p_{\tau_4^i}^{[1]}(v_{\mu,\nu+1}) = 0, i = 0, 1$, and $s_{\tau_4^\beta}(w_{\mu,\nu}) = 0, \beta = 0, \ldots, q - 2$, the claim holds for $\alpha = 0$. We assume that the claim holds for $\alpha \in \{0, \ldots, \alpha_1\}, \alpha_1 \leq q - 4$, and show that the claim holds for $\alpha_1 + 1$. Since $p_{\tau_1^{\alpha_1+1} \tau_4^i}^{[1]}(v_{\mu,\nu+1}) = 0, i = 0, 1$, we have

$$p_{\tau_3^{\alpha_1+1} \tau_4^i}^{[1]}(v_{\mu,\nu+1}) = \sum_{j=1}^{\alpha_1+1} \kappa_j p_{\tau_1^{\alpha_1+1-j} \tau_4^{i+j}}^{[1]}(v_{\mu,\nu+1}), \qquad i = 0, 1,$$

where $\kappa_j \in \mathbb{R}$. Moreover, for $j \in \{1, \ldots, \alpha_1 + 1\}$, we have

$$p_{\tau_1^{\alpha_1+1-j} \tau_4^{i+j}}^{[1]}(v_{\mu,\nu+1}) = \sum_{k=0}^{\alpha_1+1-j} \kappa_{j,k} p_{\tau_3^{\alpha_1+1-j-k} \tau_4^{i+j+k}}^{[1]}(v_{\mu,\nu+1}), \qquad i = 0, 1,$$

where $\kappa_{j,k} \in \mathbb{R}$. Therefore, it follows from the induction hypothesis that $p_{\tau_3^{\alpha_1+1} \tau_4^i}^{[1]}(v_{\mu,\nu+1}) = 0, i = 0, 1$. Since $s_{\tau_3^{\alpha_1+1} \tau_4^\beta}(w_{\mu,\nu}) = 0, \beta = 0, \ldots, q - 3 - \alpha_1$, we obtain $p_{\tau_3^{\alpha_1+1}}^{[1]}|_{e^{[4]}} = 0$. This confirms the claim. Hence, from (1), we conclude $p^{[1]} = 0$. Similar arguments as above show that $p_{\tau_3^\alpha}^{[2]}|_{e^{[5]}} = 0, \alpha = 0, \ldots, q - 4$, and since $D^\omega p^{[2]}(v_{\mu,\nu}) = 0, \omega = 0, \ldots, 3$, we conclude $p^{[2]} = 0$. Thus, $s = 0$ on $\tilde{\Delta}_{\mu,\nu}$. This proves Lemma 4. $\square$

We now consider the case $r = 2$ and define the following sets. For $\mu, \nu \geq 2$, let $\Delta_{\mu,\nu}^* \subseteq \Delta$ be the triangulation which consists of the triangles $T_{\mu_1,\nu_1}^{[l]}, \mu_1 \geq 1, \nu_1 \leq \nu - 1, T_{\mu_1,\nu}^{[l]}, \mu_1 \leq \mu - 2, l = 1, \ldots, 4, T_{\mu-1,\nu}^{[l]}, l = 1, 2$, and denote by $\Omega_{\mu,\nu}^* = \{\bar{T} : \bar{T} \in \Delta_{\mu,\nu}^*\}$. Moreover, set $\tilde{\Delta}_{\mu,\nu} = \Delta_{\mu,\nu}^* \cup T_{\mu-1,\nu}^{[3]} \cup T_{\mu,\nu}^{[1]} \cup T_{\mu,\nu}^{[2]}$ and let $\tilde{\Omega}_{\mu,\nu}$ be the corresponding domain. In Section 2, we have chosen the set (respectively condition) B_2 in $T_{\mu-1,\nu}^{[3]}$, C_2 in $T_{\mu,\nu}^{[1]}$ and D_2 in $T_{\mu,\nu}^{[2]}$.

Lemma 5. Let $r = 2$ and $q \geq 5$. If $s = 0$ on $\Delta^*_{\mu,\nu}$, then $s = 0$ on $\tilde{\Delta}_{\mu,\nu}$.

Proof: Let τ_l be a unit vector parallel to the edge $e^{[l]}, l = 1, \ldots, 4$, where $e^{[1]} = [v_{\mu,\nu}, w_{\mu-1,\nu}]$, $e^{[2]} = [v_{\mu,\nu}, v_{\mu,\nu+1}]$, $e^{[3]} = [v_{\mu,\nu}, w_{\mu,\nu}]$, and $e^{[4]} = [v_{\mu,\nu}, v_{\mu+1,\nu}]$. It follows from the C^2 property of s that the polynomials $p^{[1]} = s|_{T^{[3]}_{\mu-1,\nu}}, p^{[3]} = s|_{T^{[2]}_{\mu,\nu}} \in \tilde{\Pi}_q$, satisfy $p^{[1]}_{\tau_1^i \tau^j}|_{e^{[1]}} = p^{[3]}_{\tau_4^i \tau^j}|_{e^{[4]}} = 0, i \geq 0, j = 0, 1, 2$, where τ is an arbitrary unit vector. Let $p^{[2]} = s|_{T^{[1]}_{\mu,\nu}} \in \tilde{\Pi}_q$. We claim that

$$D^\omega p^{[1]}(v_{\mu,\nu}) = D^\omega p^{[2]}(v_{\mu,\nu}) = 0, \ \omega = 0, \ldots, 3. \tag{2}$$

It suffices to show that (2) holds for $p^{[2]}$ and $\omega = 3$. Let $\gamma_l \in \mathbb{R} \setminus \{0\}, l = 1, \ldots, 4$, be such that $\tau_2 = \gamma_1 \tau_1 + \gamma_3 \tau_3$, and $\tau_3 = \gamma_2 \tau_2 + \gamma_4 \tau_4$. (Note that by assumption $e^{[1]}$ and $e^{[3]}$ (respectively $e^{[2]}$ and $e^{[4]}$) have different slopes.) The following equations hold:

$$p^{[2]}_{\tau_2^3}(v_{\mu,\nu}) = \gamma_1 p^{[1]}_{\tau_1 \tau_2^2}(v_{\mu,\nu}) + \gamma_3 p^{[2]}_{\tau_3 \tau_2^2}(v_{\mu,\nu})$$

$$p^{[2]}_{\tau_2^3}(v_{\mu,\nu}) = -\gamma_1^2 p^{[1]}_{\tau_1^2 \tau_2}(v_{\mu,\nu}) + 2\gamma_1 p^{[1]}_{\tau_1 \tau_2^2}(v_{\mu,\nu}) + \gamma_3^2 p^{[2]}_{\tau_3^2 \tau_2}(v_{\mu,\nu})$$

$$p^{[2]}_{\tau_3^3}(v_{\mu,\nu}) = \gamma_2 p^{[2]}_{\tau_2 \tau_3^2}(v_{\mu,\nu}) + \gamma_4 p^{[3]}_{\tau_4 \tau_3^2}(v_{\mu,\nu})$$

$$p^{[2]}_{\tau_3^3}(v_{\mu,\nu}) = \gamma_2^2 p^{[2]}_{\tau_2^2 \tau_3}(v_{\mu,\nu}) + 2\gamma_4 p^{[3]}_{\tau_4 \tau_3^2}(v_{\mu,\nu}) - \gamma_4^2 p^{[3]}_{\tau_4^2 \tau_3}(v_{\mu,\nu}) .$$

Since the derivatives of $p^{[1]}$ and $p^{[3]}$ are zero, we obtain a homogenous system for the derivatives of $p^{[2]}$. The corresponding determinant $\gamma_2 \gamma_3 (\gamma_2 \gamma_3 - 1)$ is different from zero, since $e^{[1]}$ and $e^{[4]}$ have different slopes. This implies (2). Similarly as in the proof of Lemma 4 one can see that the C^2 property of s at $e^{[1]}$, the interpolation conditions in $T^{[3]}_{\mu-1,\nu}$ and (2) imply $p^{[1]} = 0$. Now, we claim that $p^{[2]} = 0$. For Lagrange interpolation, let $c_\alpha, \alpha = 1, \ldots, q-4$, be the chosen line segments in $T^{[1]}_{\mu,\nu}$ parallel to $e^{[5]} = [v_{\mu,\nu+1}, w_{\mu,\nu}]$. Similarly as in the proof of Lemma 4 one can see that the C^2 property of s at $e^{[2]}$ and the interpolation conditions in $T^{[1]}_{\mu,\nu}$ imply $p^{[2]}|_{c_\alpha} = 0, \alpha = 1, \ldots, q-4$, for Lagrange (respectively $p^{[2]}_{\tau_3^\alpha}|_{e^{[5]}} = 0, \alpha = 0, \ldots, q-5$, for Hermite) interpolation. For Lagrange interpolation, we have $s(c) = 0$ for an interpolation point $c \in e^{[3]}$ not lying on $c_\alpha, \alpha = 1, \ldots, q-4$. For Hermite interpolation, we have $s_{\tau_3^{q-4}}(w_{\mu,\nu}) = 0$. In both cases, it follows that $p^{[2]}|_{e^{[3]}} = 0$, and therefore, $p^{[2]}_{\tau_4^2}(v_{\mu,\nu}) = 0$. We set $\tau_3 = \gamma_2 \tau_2 + \gamma_4 \tau_4$. Since $p^{[2]}_{\tau_3^4}(v_{\mu,\nu}) = -\gamma_2^2 p^{[2]}_{\tau_2^2 \tau_3^2}(v_{\mu,\nu}) + 2\gamma_2 p^{[2]}_{\tau_2 \tau_3^3}(v_{\mu,\nu}) + \gamma_4^2 p^{[3]}_{\tau_4^2 \tau_3^2}(v_{\mu,\nu})$, it follows that $p^{[2]}_{\tau_2 \tau_3^3}(v_{\mu,\nu}) = 0$. Thus, $D^\omega p^{[2]}(v_{\mu,\nu}) = 0, \omega = 0, \ldots, 4$. Hence, $p^{[2]} = 0$. Finally, we consider $p^{[3]}$. For Lagrange interpolation, let $d_\alpha, \alpha = 1, \ldots, q-5$, be the chosen line segments in $T^{[2]}_{\mu,\nu}$ parallel to $e^{[6]} = [w_{\mu,\nu}, v_{\mu+1,\nu}]$. Similarly as in the proof of Lemma 4 one can see that the C^2 property of s at $e^{[3]}$ and $e^{[4]}$ and the interpolation conditions in $T^{[2]}_{\mu,\nu}$ imply $p^{[3]}|_{d_\alpha} = 0, \alpha = 1, \ldots, q-5$, for Lagrange (respectively $p^{[3]}_{\tau_3^\alpha}|_{e^{[6]}} = 0, \alpha = $

$0, \ldots, q - 6$, for Hermite) interpolation. Since $D^\omega p^{[3]}(v_{\mu,\nu}) = 0, \omega = 0, \ldots, 5$, we conclude $p^{[3]} = 0$. Thus, $s = 0$ on $\tilde{\Delta}_{\mu,\nu}$. This proves Lemma 5. $\square$

We now prove Theorems 1 and 2.

Proof of Theorems 1 and 2 : We have to show that $s = 0$ on Ω. First, we consider $s|_{Q_{1,1}}$. It follows from the interpolation conditions Q that $s|_{T_{1,1}^{[1]}} = 0$. It is easy to see that the C^r property of s and the interpolation condition A_r of s in $T_{1,1}^{[2]}$ imply $s|_{T_{1,1}^{[2]}} = 0$. Analogously, $s|_{T_{1,1}^{[4]}} = 0$. Similarly as in the proof of Lemma 4 (respectively Lemma 5), it follows that the C^r property of s and the interpolation condition D_r of s in $T_{1,1}^{[3]}$ imply $s|_{T_{1,1}^{[3]}} = 0$. By induction and similar arguments as above, it is easy to verify that $s|_{\Omega_{2,2}^*} = 0$, where $\Omega_{2,2}^*$ is chosen as in Lemma 4 for $r = 1$ (respectively Lemma 5 for $r = 2$). It follows from Lemma 4 (respectively Lemma 5) that $s|_{\tilde{\Omega}_{2,2}} = 0$. Here, $\tilde{\Omega}_{2,2}$ is chosen as in Lemma 4 for $r = 1$ (respectively Lemma 5 for $r = 2$). By induction, similar arguments as above and Lemma 4 (respectively Lemma 5) imply $s|_{\Omega_{2,3}^*} = 0$. By proceeding with this method, we obtain $s = 0$ on Ω. This proves Theorems 1 and 2. $\square$

§4. Numerical Examples

Finally, we give numerical examples. As a test, we interpolate Franke's test function $f(x,y) = \frac{3}{4}e^{-\frac{(9x-2)^2 + (9y-2)^2}{4}} + \frac{3}{4}e^{-\frac{(9x+1)^2}{49} - \frac{(9y+1)}{10}} + \frac{1}{2}e^{-\frac{(9x-7)^2 + (9y-3)^2}{4}} - \frac{1}{5}e^{-(9x-4)^2 - (9y-7)^2}$ by splines in $S_q^r(\Delta)$, where Δ results from a given Δ^2-partition on $[0,1] \times [0,1]$ deformed by a randomizer. Our results for the Hermite interpolating spline $s_f \in S_q^r(\Delta)$ are as follows:

$[S_2^1 \mid 1763 \mid 2.03 * 10^{-2}], [S_3^1 \mid 8323 \mid 9.06 * 10^{-5}], [S_4^1 \mid 21283 \mid 5.70 * 10^{-6}]$

$[S_5^1 \mid 40643 \mid 7.17 * 10^{-7}], [S_6^1 \mid 66403 \mid 5.98 * 10^{-8}], [S_7^1 \mid 98563 \mid 4.51 * 10^{-8}]$

$[S_7^2 \mid 55488 \mid 3.53 * 10^{-8}], [S_8^2 \mid 84258 \mid 7.54 * 10^{-9}],$

where we set $[S_q^r \mid$ number of interpolation conditions $\mid$ error $\|f - s_f\|_\infty]$. The interpolating splines are computed by passing from one triangle to the next and by solving several small systems instead of one large system. Therefore, the complexity of the algorithm is $O(N)$, where N is the number of triangles.

References

1. Adam, M. H., *Bivariate Spline-Interpolation auf Crosscut-Partitionen*, Dissertation, Mannheim, 1995.

2. Chui, C. K. and T. X. He, On location of sample points in C^1 quadratic bivariate spline interpolation, in *Numerical Methods of Approximation Theory*, L. Collatz, G. Meinardus, and G. Nürnberger (eds.), ISNM 81, Birkhäuser, Basel, 1987, pp. 30–43.

3. Davydov, O. V., Locally linearly independent basis for C^1 bivariate splines of degree $q \geq 5$, in *Mathematical Methods for Curves and Surfaces II*, M. Daehlen, T. Lyche, and L. L. Schumaker (eds.), Vanderbilt University Press, Nashville, London, 1998, pp. 1–7.

4. Davydov, O. V., G. Nürnberger, and F. Zeilfelder, Approximation order of bivariate spline interpolation for arbitrary smoothness, J. Comput. Appl. Math., to appear.

5. Davydov, O. V., G. Nürnberger, and F. Zeilfelder, Interpolation by cubic splines on triangulations, this volume.

6. Jeeawock-Zedek, F., Interpolation scheme by C^1 cubic splines on a non uniform type-2 triangulation of a rectangular domain, C.R. Acad. Sci. Ser. I Math. **314** (1992), 413–418.

7. Morgan, J. and R. Scott, A nodal basis for C^1 piecewise polynomials of degree $n \geq 5$, Math. Comp. **29** (1975), 736–740.

8. Nürnberger, G., *Approximation by Spline Functions*, SpringerVerlag, Berlin, Heidelberg, New York, 1989.

9. Nürnberger, G., Approximation order of bivariate spline interpolation, J. Approx. Theory **87** (1996), 117–136.

10. Nürnberger, G. and Th. Riessinger, Lagrange and Hermite interpolation by bivariate splines, Numer. Func. Anal. Optim. **13** (1992), 75–96.

11. Nürnberger, G. and Th. Riessinger, Bivariate spline interpolation at grid points, Numer. Math. **71** (1995), 91–119.

12. Nürnberger, G., O. V. Davydov, G. Walz, and F. Zeilfelder, Interpolation by bivariate splines on crosscut partitions, in *Multivariate Approximation and Splines*, G. Nürnberger, J. W. Schmidt, and G. Walz (eds.), Birkhäuser Verlag, Basel, 1997, pp. 189–204.

13. Nürnberger, G. and F. Zeilfelder, in preparation.

14. Nürnberger, G. and G. Walz, Error analysis in interpolation by bivariate C^1-splines, IMA J. Numer. Anal. **18** (1998), to appear.

15. Schumaker, L. L., Bounds on the dimension of spaces of multivariate piecewise polynomials, Rocky Mountain J. Math. **14** (1984), 251–264.

16. Sha, Z., On interpolation by $S_2^1(\Delta_{m,n}^2)$, Approx. Theory Appl. **1** (1985), 71–82.

17. Sha, Z., On interpolation by $S_3^1(\Delta_{m,n}^2)$, Approx. Theory Appl. **1** (1985), 1–18.

Günther Nürnberger and Frank Zeilfelder
Universität Mannheim
Fakultät für Mathematik und Informatik
68163 Mannheim
Germany
nuernberger@math.uni-mannheim.de
zeilfeld@fourier.math.uni-mannheim.de

Covariance Matrix Reduction Techniques for Neural Multi-Source Direction Finding

Teresa H. O'Donnell and Hugh L. Southall

Abstract. Neural networks are often used to approximate non-linear functions for performance and speed enhancements. However, when implemented on serial hardware, potential speed benefits are limited unless the network size is constrained. In this paper, we present a neural network direction finder which processes received antenna array signals and locates multiple targets. We define the problem of multiple source direction finding (DF) and simulate realistic received signals in array antennas. We describe several covariance matrix preprocessing techniques which reduce the network input dimensionality and we present multiple source DF results.

§1. Introduction

There has been much work applying neural networks (NNs) to electromagnetics problems [2, 5, 6], including the approximation of phased array antenna signal processing. We apply neural networks to a digital beam-forming (DBF) receiving antenna to estimate angles of arrival (AOA) of multiple sources (or targets) within the array's field of view, also called direction finding (DF). We call these neural networks neural beam-formers (NBFs). Single source DF NBFs have compared favorably with traditional methods [5] and multiple source NBFs have also been created [2, 6]. However, this problem is much more complex than single source DF, and networks require vast amounts of training data to achieve limited results.

§2. Neural Beam-Former Architecture for Direction Finding

The neural beam-former consists of a Radial Basis Function (RBF) NN surrounded by preprocessing and postprocessing. RBF networks used for NBFs include linear Gaussian node networks trained by matrix inversion [5] and linear periodic basis functions (PBFs) RBFs [3]. By linear, we mean that each

Approximation Theory IX, Volume 2: Computational Aspects   267
Charles K. Chui and Larry L. Schumaker (eds.), pp. 267–274.
Copyright © 1998 by Vanderbilt University Press, Nashville, TN.
ISBN 0-8265-1326-3.

training point creates a hidden layer node centered at the point, and the network weights are found by solving a system of linear equations [5].

For multiple source DF, the large number of training points makes it impractical to allocate RBF nodes at every training point. Thus, our multiple source DF neural beam-former uses the Moody-Darken (MD) RBF network with norm-cum-delta training as implemented in NeuralWare Professional II/Plus [4]. This network allocates a fixed number of hidden layer RBF Gaussian nodes to minimize the sum of the squares of the distances between each training vector, x_l, and its closest cluster center, c_k. After the hidden nodes have been allocated via iterative K-means clustering, each cluster is assigned a σ spread based on its distance to the P nearest cluster centers. Following this, the network output weights are solved by norm-cum-delta backprop training, which accumulates weight changes, normalizes them by the epoch size, and updates at the end of each epoch. More details can be found in [4].

To train and test the network, we collect or simulate antenna measurements (amplitude and phase) for target scenarios (single targets or target combinations) at different AOAs within the antenna field of view ($-80°$ to $+80°$). After preprocessing, a subset of the input vectors is used for training. The network inputs stimulate the hidden Gaussian processing nodes, k, such that the output of k depends on the radial distance between input vector x_l and node center c_k, and on σ, the spread parameter, i.e.,

$$\phi_{kl} = \exp \sum_{i=1}^{M} \frac{-|x_{li} - c_{ki}|^2}{2\sigma^2}, \tag{1}$$

where $k = 1, 2, \ldots, q$, for q hidden layer nodes, and M is the preprocessed input dimension. The Gaussian node centers, c_k, are determined during adaptive K-clustering. The output node response for input x_l is a weighted sum of the processing nodes, $y_{jl} = \sum_{k=1}^{q} w_{jk}\phi_{kl}$, where $j = 1, 2, \ldots, r$ for r output nodes. The weights, w_{jk}, are determined during norm-cum-delta network training. For multiple source DF, each output node indicates the AOA (in degrees) of a signal and no postprocessing is required.

§3. Multiple Source Neural Beam-Forming

For single source DF, only one array measurement is required to estimate the AOA [5]. However, multiple sources require many time samples (*looks* or *snapshots*) unless the signals are perfectly correlated and unmodulated.

3.1. Signal Representation

For an N-element uniform linear array receiving a single source at angle θ, the signal (column) vector is

$$\underline{x}(t)_{N \times 1} = \underline{V}(\theta)_{N \times 1} s(t)_{1 \times 1} + \underline{n}(t)_{N \times 1}, \tag{2}$$

where the components of $\underline{x}(t) = [x_1(t) \quad x_2(t) \ldots x_N(t)]^T$ are the time varying array element signals, $\underline{n}(t)$ the additive noise, $s(t)$ the signal modulation, and

$\underline{V}(\theta)$ the array response vector [1]. We represent the signal, $s(t)$, by a zero-mean complex Gaussian random variable with power equal to SNR times the Gaussian noise power, $n(t)$.

For K sources and L time samples, the columns of $\underline{V}(\theta_1, \theta_2, \ldots, \theta_K)$ are the array response vectors at AOAs θ_1, θ_2, ..., θ_K, and the signal matrix is

$$\underline{\underline{x}}_{N \times L} = \underline{\underline{V}}_{N \times K} \, \underline{\underline{s}}_{K \times L} + \underline{\underline{n}}_{N \times L}. \tag{3}$$

3.2. Training Complexity

Besides requiring multiple looks, the multiple source DF function space is much larger than the single source DF space. There are many possible signal combinations, hence many training samples are required for generalization. The number of sources may vary from 1 up to the limit of $N - 1$ (for an N-element array). The sources may have different relative power levels, various levels of correlation (fully correlated to uncorrelated), and be modulated. Finally, sources may be located at many different angular combinations. This results in a very large function space requiring many combinations of training points. To date, multiple source neural DF results have been either limited to two-source scenarios or sources with fixed angular separations. [2,6].

3.3. Covariance Matrix

Starting with the signal matrix, $\underline{\underline{x}}_{N \times L}$ in equation (3), the covariance matrix, $R = E\{\underline{x} \, \underline{x}^*\}$, where $*$ denotes conjugate transpose. For the expectation operation, $\overline{E}$, we actually use a time average over the L looks. With *uncorrelated signals*, $E\{s_i s_j^*\} = 0$ and $E\{s_i s_i^*\} = E\{|s_i|^2\} = P_i$ (the power). Under zero noise condition (which is best for training), the *zero-noise* covariance matrix, shown for a three element array ($N = 3$) receiving two uncorrelated signals at θ_1 and θ_2, would be (for $u_1 = \sin\theta_1$ and $u_2 = \sin\theta_2$):

$$R = \begin{bmatrix} P_1 + P_2 & P_1 e^{-jku_1} + P_2 e^{-jku_2} & P_1 e^{-2jku_1} + P_2 e^{-2jku_2} \\ P_1 e^{jku_1} + P_2 e^{jku_2} & P_1 + P_2 & P_1 e^{-jku_1} + P_2 e^{-jku_2} \\ P_1 e^{2jku_1} + P_2 e^{2jku_2} & P_1 e^{jku_1} + P_2 e^{jku_2} & P_1 + P_2 \end{bmatrix}. \tag{4}$$

3.4. Input Dimensionality Reduction

The covariance matrix, R, is of size N^2, for an N-element array. However, $r_{ij} \in \mathbb{C}$, and the RBF NN requires $x_l \in \mathbb{R}$. We, therefore, separate r_{ij} into real and imaginary parts, and use $2N^2$ values when all of R are used as inputs.

However, Equation (4) shows that the zero-noise R main diagonal contains no angular information. Thus, one common preprocessing method, RminusD, removes the main diagonal from the inputs [2, 6], reducing the dimensionality M, to $2N(N-1)$ real values, shown in Table 1(a). (Bold-face table $\boldsymbol{r_{ij}}$ entries are those preprocessed into real/imaginary NN inputs.) Equation (4)

$$\begin{array}{ccc}
\text{RminusD:} & \text{uphalf:} & \text{toprow:} \\
2N(N-1)\text{inputs} & N(N-1)\text{inputs} & 2(N-1)\text{inputs}
\end{array}$$

$$
\begin{bmatrix}
r_{11} & \mathbf{r_{12}} & \mathbf{r_{13}} & \mathbf{r_{14}} \\
\mathbf{r_{21}} & r_{22} & \mathbf{r_{23}} & \mathbf{r_{24}} \\
\mathbf{r_{31}} & \mathbf{r_{32}} & r_{33} & \mathbf{r_{34}} \\
\mathbf{r_{41}} & \mathbf{r_{42}} & \mathbf{r_{43}} & r_{44}
\end{bmatrix}
\qquad
\begin{bmatrix}
r_{11} & \mathbf{r_{12}} & \mathbf{r_{13}} & \mathbf{r_{14}} \\
r_{21} & r_{22} & \mathbf{r_{23}} & \mathbf{r_{24}} \\
r_{31} & r_{32} & r_{33} & \mathbf{r_{34}} \\
r_{41} & r_{42} & r_{43} & r_{44}
\end{bmatrix}
\qquad
\begin{bmatrix}
r_{11} & \mathbf{r_{12}} & \mathbf{r_{13}} & \mathbf{r_{14}} \\
r_{21} & r_{22} & r_{23} & r_{24} \\
r_{31} & r_{32} & r_{33} & r_{34} \\
r_{41} & r_{42} & r_{43} & r_{44}
\end{bmatrix}
$$

$$\begin{array}{ccc}
\text{(a)} & \text{(b)} & \text{(c)}
\end{array}$$

$$\begin{array}{ccc}
\text{updiag:} & \text{topndiag:} & \text{sumdiag:} \\
2(N-1)\text{inputs} & 2(2N-3)\text{inputs} & 2(N-1)\text{inputs}
\end{array}$$

$$
\begin{bmatrix}
r_{11} & \mathbf{r_{12}} & r_{13} & r_{14} \\
r_{21} & r_{22} & \mathbf{r_{23}} & r_{24} \\
r_{31} & r_{32} & r_{33} & \mathbf{r_{34}} \\
r_{41} & r_{42} & r_{43} & r_{44}
\end{bmatrix}
\qquad
\begin{bmatrix}
r_{11} & \mathbf{r_{12}} & \mathbf{r_{13}} & \mathbf{r_{14}} \\
r_{21} & r_{22} & \mathbf{r_{23}} & r_{24} \\
r_{31} & r_{32} & r_{33} & \mathbf{r_{34}} \\
r_{41} & r_{42} & r_{43} & r_{44}
\end{bmatrix}
\qquad
\begin{bmatrix}
r_{11} & \mathbf{r_{12}} & \mathbf{r_{13}} & \mathbf{r_{14}} \\
r_{21} & r_{22} & \mathbf{r_{23}} & \mathbf{r_{24}} \\
r_{31} & r_{32} & r_{33} & \mathbf{r_{34}} \\
r_{41} & r_{42} & r_{43} & r_{44}
\end{bmatrix}
$$

$$\begin{array}{ccc}
\text{(d)} & \text{(e)} & \text{(f)}
\end{array}$$

Tab. 1. Covariance Matrix Reductions.

also shows that R contains similar θ_i information above and below the diagonal. This suggests that either half of R contains sufficient information to determine the target angles. A second preprocessing method, uphalf, selects the upper half of R, shown in Table 1(b).

We may consider using a reference element, such as #1, and selecting covariance matrix entries which relate other elements to it. This toprow approach in Table 1(c) relates the array element values to element #1, i.e., $r_{1,j}$ where $1 < j \leq N$, yielding $2(N-1)$ real inputs. Another alternative is to consider the inter-element covariance matrix entries on the first diagonal above the main diagonal of R, i.e., $r_{i,i+1}$, where $1 \leq i < N$. This is similar to single source DF preprocessing [5] which used phase differences between consecutive elements. This updiag method yields $2(N-1)$ real inputs, shown in Table 1(d). A fifth input reduction approach, topndiag, combines toprow with updiag which yields $2(2N-3)$ real inputs, shown in Table 1(e).

In another approach suggested by the author of [3], we combine similar "inter-element" covariances by summing each of the $N-1$ diagonals above the main R diagonal. This sumdiag method, shown in Table 1(f), sums the three diagonals above the main diagonal independently. The complex summations are separated into real and imaginary parts, yielding $2(N-1)$ real inputs.

§4. Results for Different Preprocessing Techniques

We trained and optimized six RBF neural networks for *limited* two-source direction finding using the different preprocessing techniques. The networks were trained with a limited combination of sources, in a manner similar to [6], and were not expected to generalize to all possible target combinations. Our goal was to test the preprocessing methods to determine their generalization between training points and performance with additive noise.

For training, two sources 3° apart were moved together through the antenna field of view (−80° to +80°). Source #1 traveled from −80° to +77°

		Zero Noise			10dB SNR	3dB SNR
Preprocessing	3° (TS)	3° offset	1.5°	5°	3°	3°
RminusD	0.7041	0.7183	1.2091	1.1943	1.9067	10.9652
RminusD*	0.6904	0.7109	1.1809	1.1795	1.0806	4.9953
uphalf	0.3755	0.3707	0.8570	0.9081	1.7653	12.2977
toprow	0.3985	0.4248	0.6850	1.0945	1.7354	11.1502
toprow*	0.3252	0.3796	0.7882	1.0775	1.1012	5.5219
updiag	0.6935	0.7804	1.1036	1.1565	11.6411	26.9988
topndiag	0.6342	0.6542	1.1905	1.2831	1.5364	10.4258
sumdiag*	0.3367	0.3318	0.8949	1.1616	0.5655	2.6569

* : With amplitude modulated data.

Tab. 2. Network Results: Angular RMS error (in degrees).

while source #2 traveled 3° away from −77° to +80°. Data was collected at 1° increments, yielding 158 samples. The two sources had equal power and no noise. We trained with and without amplitude modulation. Simulated data was received by an ideal 8-element linear array, with $\lambda/2$ element spacing. The NeuralWare Professional II/Plus *Moody-Darken* RBF network [4] was trained with norm-cum-delta backprop. The input layer was variable size, depending on preprocessing. We used 50 hidden Gaussian nodes and 2 output linear summation nodes, which represented the target angles (in degrees). Each network was trained for 254 epochs, with 30 epochs used for K-clustering.

Following training, the networks were tested on various data sets. A summary of results for the different preprocessing techniques and testing sets are shown in Table 2. Each row represents a different network, trained and optimized for a specific preprocessing technique. The columns indicate the testing data sets. The first four columns represent sources with equal powers and no noise. Column five represents equal power sources at 10dB SNR each, while column six sets the two sources both to 3dB SNR. Table values indicate the average RMS angular error (in degrees) in estimating the two AOAs.

The first column, 3°(TS), tests the networks on the training set. The 3° offset data set consists of sources spaced 3° apart but offset 0.5° degrees from the training set. These points fall exactly in between the original training set. Columns three and four show network results for sources 1.5° and 5° apart. Columns five and six test the networks with sources at the training points (i.e., 3° apart) in the presence of additive noise. The * in the row descriptors indicate networks trained and tested with amplitude modulated signals. We discovered that the effect of modulation is similar to noisy data; the Gaussian σs are broadened and the networks performed better in the presence of noise.

As indicated in Table 2, the preprocessing techniques had variable results and, surprisingly, some reduction methods improved performance. All networks could direct find the trained region with superresolution and resolve sources located within an array beamwidth (12.7°). However, note the differences in handling noise or generalizing to different angular spacings. In

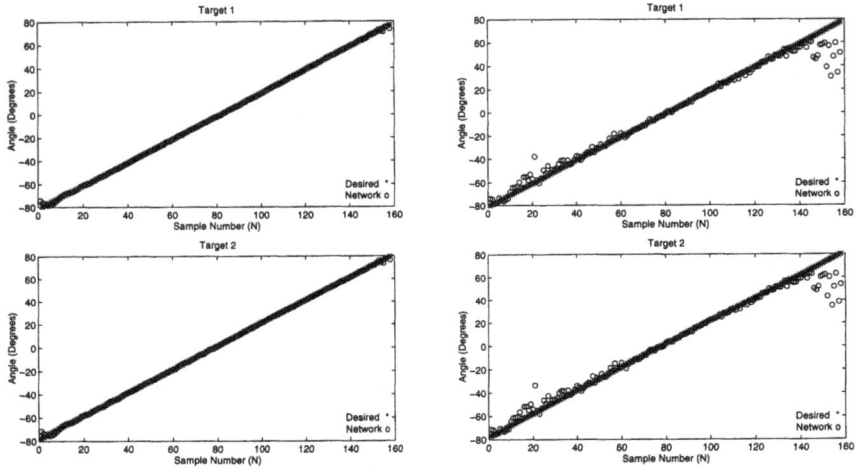

Fig. 1. AOA estimates for two equal sources, sumdiag preprocessing, trained at 3° separation. Tested with (a) 3° separation (offset from TS by 0.5°), no noise, and (b) 3° separation, noise, 3dB SNR.

particular, "sum of diagonals" proved especially robust to noise. Networks using the upper diagonal proved weak at handling noise, as did NBFs using RminusD.

Some examples of the angular error will illustrate areas where the NN *generalized* the function and how it reverted to a *classifier* or produced a *no match* output when utilized beyond the trained region. Figures 1 (a) and (b) compare desired target angles and network outputs for the sumdiag NN with amplitude modulated inputs. The upper plot indicates target #1; the lower plot target #2. Figure 1(a) shows excellent AOA estimation of the 3° offset testing data across the angular region, with minor deviations at the edges. In Figure 1(b), the original 3° spaced training set is contaminated with noise half the power of the source and the network still estimates AOAs accurately.

None of the networks generalized away from the trained data. The networks could *interpolate* around the training points, but could not *extrapolate*. Performance with AOA separations of 10° and beyond was poor, as shown in Figures 2 (a) and (b). In 2(a), equal power sources separated by 10° are moved across the field of view. Note that the angular error between the two sources is fairly evenly distributed. However, if the sources have unequal power, the higher power source dominates, shown in Figure 2(b). The higher power source (target #1 on top) has almost no error. The lower power source (target #2, bottom plot) is predicted to be about 3° away from the AOA of target #1 instead of the actual 10° separation. Unfortunately, the NN is classifying this AOA combination back into the training set rather than generalizing.

Figure 3 shows a dramatic illustration of this failure mode. Here two sources move across the field of view in opposite directions. Target #1 travels from −80° to +80°, while target #2 travels from +80° to −80°. This is very

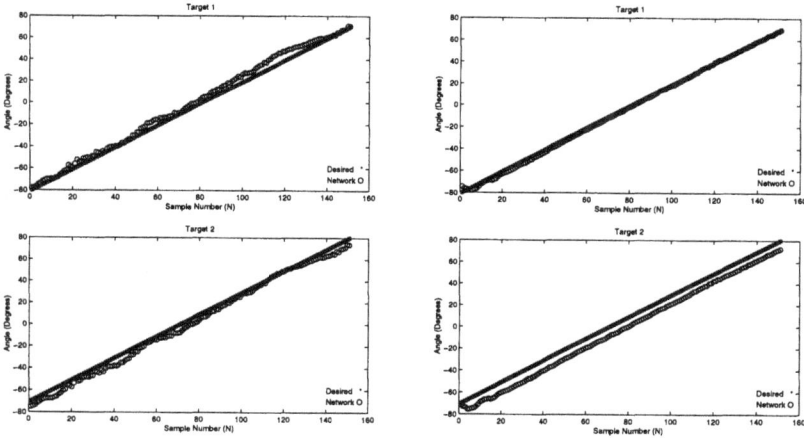

Fig. 2. AOA estimates, two sources, `sumdiag` preprocessing, trained at 3° separation. (a) 10° separation, equal power, no noise. (b) 10° separation, unequal power, 30dB, and 20dB.

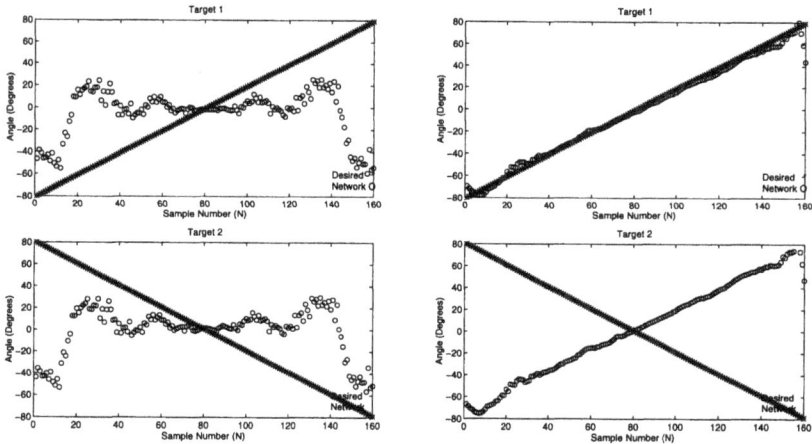

Fig. 3. AOA estimates, two sources, `sumdiag` preprocessing, trained at 3° separation. (a) Variable separations, crossing at 0°, equal power, no noise. (b) Same AOAs, unequal powers (30dB and 25dB) with noise.

different from any training samples and we expect very little network output, as the input distances to the node centers should be large and none of the Gaussian nodes should respond strongly. This "no match" response is shown in Figure 3(a) when the sources have equal power; the predicted target angles are close to 0°. However, when the sources have slightly unequal power, target #1 dominates and the network predicts target #2 once again 3° away from target #1, shown in Figure 3(b). Rather than the "no match" zero output, the network strongly classifies this case based on the higher power signal.

§5. Conclusions

Although the RBF NN shows promise for multiple source DF, it is clear that the networks must be trained extensively throughout target combinations spanning their operational region. When tested within the trained region (equal power, close separation), they *approximated* correctly. However, they incorrectly *classified* the inputs when stimulated with unequal power targets. Since the multiple source function space is large, this network approximation approach may be limited to special cases, where information about the number of sources, relative power, or angular directions/separations is known a priori. The alternative is to extensively train large networks across the entire function space up to some limiting number of signals.

Acknowledgments. The authors would like to thank the Air Force Office of Scientific Research (AFOSR) for their sponsorship of this research.

References

1. Litva J. and T. Lo, *Digital Beamforming in Wireless Communications*, Artech House, Boston, 1996, pp. 16–17.

2. Lo, T., H. Leung, and J. Litva, Radial basis function neural network for direction-of-arrivals estimation, IEEE Signal Processing Ltrs. **1** (1994), 45–47.

3. Narcowich F., Generalized Hermite interpolation and positive definite kernels on a Riemannian manifold, J. Math Anal. Comp. **190** (1995), 165–193.

4. NeuralWare Inc., *Neural Computing: A Technology Handbook for Professional II/Plus and NeuralWorks Explorer*, 1995, pp. 263–275.

5. Southall H., J. Simmers, and T. O'Donnell, Direction finding in phased arrays with a neural network beamformer, IEEE Trans. Antennas and Propagation **43** (1995), 1369–1374.

6. Zooghby A., C. Christodoulou, and M. Georgiopoulos, Performance of radial basis function networks for direction of arrival estimation with antenna arrays, IEEE Trans. Antennas and Propagation **45** (1997), 1611–1617.

Teresa Hohol O'Donnell, ARCON Corporation
260 Bear Hill Road, Waltham, MA 02154
terry@arcon.com, www.arcon.com

Hugh Southall, Electromagnetics Technology Division
US Air Force Research Laboratory
31 Grenier Street, Hanscom AFB, MA 01731-3010
southall@maxwell.rl.plh.af.mil

Multigrid Prolongations and Matrix Subdivision

Peter Oswald

Abstract. Attention is drawn to the connection between theoretical properties of prolongation operators in multigrid methods, on the one hand, and the regularity problem for solutions of matrix subdivision schemes, on the other. Both problems can be reduced to spectral properties of an associated matrix transfer operator. An application to multilevel preconditioners for the Stokes problem is discussed.

§1. Introduction

We consider multigrid methods for iteratively solving linear systems

$$Ax = f \qquad (1)$$

arising from a variational problem

$$\text{find} \quad u \in V : \quad a(u,v) = F(v) \qquad \forall v \in V, \qquad (2)$$

in a finite-dimensional real Hilbert space V. Roughly speaking, multigrid methods are based on embedding (2) into a sequence of analogous problems

$$\text{find} \quad u_j \in V_j : \quad a_j(u_j, v_j) = F_j(v_j) \qquad \forall v_j \in V_j, \qquad (3)$$

on real Hilbert spaces V_j, $j = 0, \ldots, J$ ($V \equiv V_J$, $a \equiv a_J$, etc.), of increasing dimension, and constructing an iterative method for solving (1) by combining approximate solvers (smoothers) S_j for the problems (3) and prolongation/restriction operations between different V_j. In the simplest version (multilevel preconditioning), the resulting iterative method takes the form

$$x^{(n+1)} = x^{(n)} - \omega C(Ax^{(n)} - f), \quad n = 0, 1, \ldots, \qquad (4)$$

where $x^{(0)}$ is an arbitrary starting vector, $\omega \neq 0$ a relaxation parameter, and the multilevel preconditioner $C = C_J$ is recursively defined as

$$C_0 = S_0, \qquad C_j = P_j C_{j-1} R_j + S_j, \quad j = 1, \ldots, J. \qquad (5)$$

Approximation Theory IX, Volume 2: Computational Aspects
Charles K. Chui and Larry L. Schumaker (eds.), pp. 275–282.
Copyright © 1998 by Vanderbilt University Press, Nashville, TN.
ISBN 0-8265-1326-3.

The matrices P_j resp. R_j represent the prolongation $P_j : V_{j-1} \to V_{j-1}$ and restriction operators $R_j : V_j \to V_{j-1}$. For simplicity, we restrict ourselves to the symmetric case where $A = A^T$, $S_j = S_j^T$, and $R_j = P_j^T$. In finite element applications, the matrix representations of all these operators are extremely sparse due to the local support property of the standard bases in finite element spaces.

The convergence properties of the method (4)–(5) solely depend on the choices of S_j and P_j. A particularly important design problem occurs if $V_{j-1} \not\subset V_j$ (this is called *nonnestedness* in multigrid theory). While under the assumption $V_{j-1} \subset V_j$ one naturally accepts the natural embedding as P_j, for non-nested spaces the choice of the prolongations P_j becomes nontrivial. An application where nonnestedness occurs are discretizations of elliptic boundary problems by nonconforming finite elements. Mixed finite element discretizations of the Stokes system lead to nonnested sequences of approximately divergence-free finite element spaces. Previous work [1] indicates that for nonnested spaces the energy norm behavior of the *iterated prolongations* $\tilde{P}_{j,J} \equiv P_J \ldots P_{j+1} : V_j \to V_J$ is crucial for the convergence rate of (4)–(5). More precisely, the constants $c_{j,J}$ in

$$a_J(\tilde{P}_{j,J}u_j, \tilde{P}_{j,J}u_j) \leq c_{j,J} b_j(u_j, u_j) \qquad \forall\, u_j \in V_j, \quad j < J, \qquad (6)$$

need to be controlled. In practice, the forms a_j represent discrete versions of Sobolev norms and the forms $b_j(u_j, u_j) \equiv \omega_j(S_j^{-1}x_j, x_j)$, $u_j \in V_j$, are related to scaled L_2-norms. Choosing appropriate scaling factors ω_j, we may assume that an inverse inequality

$$a_j(u_j, u_j) \leq b_j(u_j, u_j) \qquad \forall\, u_j \in V_j, \quad j \geq 0, \qquad (7)$$

is satisfied. Although estimates for the prolongations

$$a_j(P_j u_{j-1}, P_j u_{j-1}) \leq c_j a_{j-1}(u_{j-1}, u_{j-1}) \qquad \forall\, u_{j-1} \in V_{j-1}, \quad j \geq 1, \qquad (8)$$

are easy to obtain and imply $c_{j,J} \leq c_{j+1} \ldots c_J$, $j < J$, this trivial upper estimate is usually a crude overestimation. Proofs of sharper, optimal estimates in (6) have been obtained so far only for two particular cases of nonconforming finite element schemes [2, 3]. More details and references on multigrid theory with nonnested spaces and numerical examples for the behavior of the $c_{j,J}$ for several nonconforming finite element schemes can be found in [1, 4].

Recently, in [5], we have shown that under simplifying assumptions, for sequences of generic finite element spaces on dyadic sequences of uniform partitions of $\mathbb{R}^d$ and homogeneous norms, with the strong assumption of *invariance w.r.t. dyadic shifts and dilations* for all spaces and operators, asymptotically exact estimates for the constants in (7) can be derived from solving a finite-dimensional eigenvalue problem for an associated matrix transfer operator. Similar reductions have shown their usefulness for estimating the regularity of multiwavelets [6]. The analogy comes from the following observation: The shift-dilation-invariant setting assumed that the standard basis for the spaces V_0 associated with non-conforming schemes will necessarily consist of the integer shifts of $L > 1$ *scaling functions* $\phi^1, \ldots, \phi^L$. Representing the operator $D_{1/2}P_1 : V_0 \to V_0$ in this basis (the dilation D_t, $t > 0$ is defined by $D_t u(x) = u(tx)$), it becomes a matrix subdivision operator with compactly supported masks. For details we refer to [5]. In this note, we wish to demonstrate the same approach for investigating multigrid prolongations acting between spaces of approximately divergence-free finite element vector fields. The difference with [5] lies in the incorporation of constraints. In Section 2, the necessary notation is introduced, and the connection between norm estimates for iterated prolongations and the above mentioned eigenvalue problem is derived. The theory is applied to the low-order rotated Q1-Stokes element in Section 3. We propose prolongation operators that preserve the discrete divergence-free constraints and lead to nearly optimal asymptotic estimates in (6).

§2. Theory

Let H be a shift- and dilation-invariant Hilbert function space on $\mathbb{R}^d$ with shift-invariant norm $\|T_s u\|_H = \|u\|_H$, where $T_s u(x) = u(x-s)$, $s \in \mathbb{R}^d$. Moreover, $\|D_t u\|_H = t^{-s_0} \|u\|_H$ holds for some fixed $s_0 \in \mathbb{R}$. Given $\phi^1, \ldots, \phi^L \in H$, assume that the system

$$\Phi_j = \{\phi^l_{j,\alpha} \equiv 2^{j s_0} D_{2^j} T_\alpha \phi^l, \ \alpha \in \mathbb{Z}^d, \ l = 1, \ldots, L\} \qquad (9)$$

forms a Riesz basis in its closed linear hull $V_j \subset H$, $j \geq 0$. By definition, $2^{j s_0} D_{2^j} : V_0 \to V_j$ is an isometry, and the Riesz basis property of Φ_j yields an isomorphism between V_j and $\ell_2(\mathbb{Z}^d)^L$. Thus, operators acting on V_j induce equivalent operators acting on $\ell_2(\mathbb{Z}^d)^L$ (or, after going to Fourier series, on $L_2(\mathbb{T}^d)^L$). We need to introduce three basic operators: a *constraint operator* $\hat{G} : V_0 \to \ell_2(\mathbb{Z}^d)^M$, a *norm operator* $\hat{E} : V_0 \to \ell_2(\mathbb{Z}^d)^N$, and a *prolongation operator* $\hat{P}_1 : V_0 \to V_1$ which all are assumed to be invariant w.r.t. $\mathbb{Z}^d$-shifts. In particular, $\hat{P}_1 T_\alpha v_0 = T_\alpha \hat{P}_1 v_0$ for all $\alpha \in \mathbb{Z}^d$ and $v_0 \in V_0$. If we use the isomorphism between V_0 and $\ell_2(\mathbb{Z}^d)^L$, then the operators $\hat{G}$, $\hat{E}$, and $\hat{S} = D_{1/2}\hat{P}_1$ turn into bounded linear matrix operators between $\ell_2(\mathbb{Z}^d)^k$ spaces ($k = L, M, N$) denoted by G, E, and S, resp. By the above shift-invariance assumptions, S corresponds to a dyadic matrix subdivision scheme, i.e.,

$$(Sy)^k_\beta = \sum_{l=1}^{L} \sum_{\alpha} s^{kl}_{\beta - 2\alpha} y^l_\alpha, \ \beta \in \mathbb{Z}^d, \ k = 1, \ldots, L, \ y \in \ell_2(\mathbb{Z}^d)^L, \qquad (10)$$

while G, E are Toeplitz. In terms of Fourier series,

$$y \in \ell_2(\mathbb{Z}^d)^L \longleftrightarrow y(\theta) \equiv (y^l(\theta) = \sum_{\alpha} y^l_\alpha e^{-i\alpha\theta}, \ l = 1, \ldots, L)^T \in L_2(\mathbb{T}^d)^L,$$

these operators become multiplication operators characterized by matrix functions $G(\theta) = (g^{ml}(\theta))$, $E(\theta) = (e^{nl}(\theta))$, and $S(\theta) = (s^{kl}(\theta))$:

$$(Gy)(\theta) = G(\theta)y(\theta), \quad (Ey)(\theta) = E(\theta)y(\theta), \quad (Sy)(\theta) = S(\theta)y(2\theta). \qquad (11)$$

Throughout the paper, we assume that the entries of all these matrices are polynomials (this corresponds to situations where norms, constraints, and prolongations are computed by local schemes). For brevity, we call this *compact support assumption*.

We will study the constant $\hat{c}_k$ in

$$\|ES^k y\|^2_{\ell_2(\mathbb{Z}^d)^M} \leq \hat{c}_k \|y\|^2_{\ell_2(\mathbb{Z}^d)^L}, \qquad y \in Z \equiv \operatorname{Ker} G \subset \ell_2(\mathbb{Z}^d)^L, \ k > 0. \quad (12)$$

One easily detects the connection of (12) with (6) if the following assumptions are accepted. Suppose that two-sided inequalities

$$b_j(u_j, u_j) \approx 2^{2s_1 j} \|u_j\|^2_H \approx 2^{2s_1 j} \|x_j\|^2_{\ell_2(\mathbb{Z}^d)^L}, \tag*{}$$
$$a_j(u_j, u_j) \approx 2^{2s_2 j} a_0(D_{2^{-j}} u_j, D_{2^{-j}} u_j) \approx 2^{2s_2 j} \|\hat{E} D_{2^{-j}} u_j\|^2_{\ell_2(\mathbb{Z}^d)^M}, \quad (13)$$

hold for some s_1 and $s_2 = s_1 - s_0$, uniformly in $u_j \in V_j$ and $j \geq 0$. Define $\hat{P}_j :$ $V_{j-1} \to V_j$ by $\hat{P}_j u_{j-1} = D_{2^{j-1}} \hat{P}_1 D_{2^{-j+1}} u_{j-1}$, and introduce the subspaces $Z_j = \{u_j \in V_j : \hat{G} D_{2^{-j}} u_j = 0 \ (\iff G x_j = 0)\}$. We assume that $\hat{S}$ maps Z_0 into Z_0 which implies that $\hat{P}_j$ maps Z_{j-1} into Z_j for all $j > 0$. Then, since

$$\tilde{P}_{j,J} = \hat{P}_J \ldots \hat{P}_{j+1} = D_{2^J}(D_{1/2}\hat{P}_1 \ldots D_{1/2}\hat{P}_1)D_{2^{-j}} = D_{2^J} \hat{S}^{J-j} D_{2^{-j}},$$

from (12)–(13) we have for arbitrary $0 \leq j < J < \infty$ that

$$a_J(\tilde{P}_{j,J} u_j, \tilde{P}_{j,J} u_j) \leq C 2^{2s_2(J-j)} \hat{c}_{J-j} b_j(u_j, u_j) \quad \forall u_j \in Z_j. \quad (14)$$

Note that without the restriction to Z (i.e., as a problem on $\ell_2(\mathbb{Z}^d)^L$ which then would be directly related to (6)), (12) was studied in [5]. The key observation is that, when going to Fourier series, we have

$$\|ES^k z\|^2_{\ell_2(\mathbb{Z}^d)^M} = ((\mathcal{L}^k_{S^*,S}B)(\theta)y(\theta), y(\theta))_{L_2(\mathbb{T}^d)^L}, \qquad y(\theta) \in L_2(\mathbb{T}^d)^L, \quad (15)$$

where $B(\theta) = E^*(\theta)E(\theta)$, and

$$(\mathcal{L}_{S^*,S}X)(\theta) = 2^{-d} \sum_{e \in \{0,\pi\}^d} S^*\left(\frac{\theta}{2}+e\right) X\left(\frac{\theta}{2}+e\right) S\left(\frac{\theta}{2}+e\right) \quad (16)$$

is the matrix transfer operator that was mentioned in Section 1. The operator (16) is well-defined on the set $\mathcal{H}$ of $L \times L$ non-negative definite Hermitean matrix functions $X(\theta)$ with polynomial entries. Such transfer operators play a central role in characterizing the regularity of multiwavelets (see [6] for a recent survey). Due to the compact support assumption, the Krylov space $\mathcal{K} \subset \mathcal{H}$ generated from B by repeatedly applying $\mathcal{L}_{S^*,S}$ is finite-dimensional. Let $\{\lambda_r\}$, $\{B_r(\theta)\}$ be a complete set of eigenvalues and eigen'matrices' for $\mathcal{L}_{S^*,S}|_{\mathcal{K}}$ (i.e., we assume diagonalizability). Write $B(\theta) = \sum_r \beta_r B_r(\theta)$ and choose Λ as the maximum of all $|\lambda_r|$ such that $\beta_r \neq 0$ and that

$$\exists z \in Z \quad (B_r(\theta)z(\theta), z(\theta))_{L_2(\mathbb{T}^d)^L} \neq 0. \quad (17)$$

Condition (17) is the only place where the constraints need to be taken into account. This derivation implies the following result which, in combination with (14), yields the desired sharp estimates for the norm behavior of $\tilde{P}_{j,J}$.

Theorem. *Under the above conditions, we have*

$$\hat{c}_k \approx \Lambda^k, \qquad k \to \infty. \tag{18}$$

If diagonalizability is not assumed, an additional factor k^s may be necessary in (18), with $s > 0$ depending on the Jordan blocks associated with the leading λ_r, on B, and on Z in a more complicated way. Such a case has never appeared in any of our applications. The determination of Λ is now a problem of numerical linear algebra. In our examples, direct eigenvalue solvers can be used. For more details, see [5] and the next section.

§3. Example

Consider the Stokes problem

$$-\Delta \boldsymbol{u} + \nabla p = \mathbf{f}, \qquad \nabla \cdot \boldsymbol{u} = 0, \tag{19}$$

for the velocity field $\boldsymbol{u} \in \boldsymbol{V} \equiv (H_0^1(\Omega))^2$ and the pressure $p \in L_2(\Omega)\backslash\{const.\}$ in a two-dimensional domain Ω. If Ω can be represented as a union of rectangles (or, more general, quadrilaterals), a low-order discretization by nonconforming rotated Q1 finite elements for the velocity and P0 pressure elements has been introduced in [7]. We describe only the definition of the resulting spaces $\boldsymbol{Z}_j$ corresponding to the shift-dilation invariant setting of the previous section (i.e., the associated spaces of approximately divergence-free rotated Q1 vector fields on nested uniform partitions $\mathcal{R}_j$ of $\mathbb{R}^2$ into squares of side-length 2^{-j}, $j \geq 0$). A vector field $\boldsymbol{u} = (u_1, u_2) \in H \equiv L_2(\mathbb{R}^2)^2$ belongs to $\boldsymbol{Z}_j$ if it is componentwise in the space W_j of scalar rotated Q1 elements and satisfies the discrete divergence-free condition

$$\int_R \nabla u \, dx = 0 \qquad \forall R \in \mathcal{R}_j. \tag{20}$$

A function $u \in L_2(\mathbb{R}^2)$ belongs to W_j if its restriction to any $R \in \mathcal{R}_j$ belongs to span$\{1, x_1, x_2, x_1^2 - x_2^2\}$, and if the average value of u along any edge e associated with $\mathcal{R}_j$ is the same when taken from either side of that e. The local interpolation problem for W_j is defined by prescribing *edge averages*, compare [7] for alternatives. Accordingly, functions in $\boldsymbol{Z}_j \subset \boldsymbol{V}_j \equiv W_j \times W_j$ are uniquely determined from their *edge average vectors*. Obviously, $\boldsymbol{V}_0$ will then be generated by $L = 4$ functions $\boldsymbol{\phi}^l$. Their supports and defining nonzero edge average vectors (of unit length) are schematically depicted in Figure 1. Thus, an arbitrary $\boldsymbol{u} \in \boldsymbol{V}_0$ is uniquely representable as

$$\boldsymbol{u} = \sum_{l=1}^{4} \sum_{\alpha \in \mathbb{Z}^2} x_\alpha^l T_\alpha \boldsymbol{\phi}^l, \quad \|\boldsymbol{u}\|^2_{L_2(\mathbf{R}^2)^2} \approx \sum_{l=1}^{4} \sum_{\alpha \in \mathbb{Z}^2} (x_\alpha^l)^2.$$

From (20), it follows by Green's Theorem that $\boldsymbol{u} \in \boldsymbol{Z}_0$ is equivalent to

$$x_\alpha^4 + x_\alpha^2 - x_{\alpha+e^1}^4 - x_{\alpha+e^2}^2 = 0 \qquad \forall \, \alpha \in \mathbb{Z}^2 \quad (e^1 = (1,0), e^2 = (0,1)). \tag{21}$$

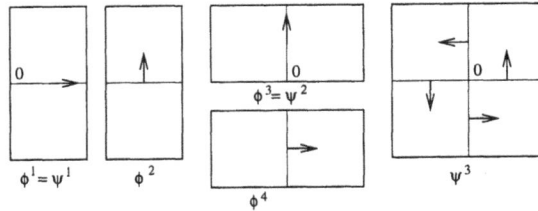

Fig. 1. Supports and edge average vectors for ϕ^l and ψ^l.

This defines the constraint operators $\hat{G}$ resp. G (here $M = 1$). We note that Z_0 possesses an explicit basis generated from the integer shifts of 3 locally supported functions $\psi^1 = \phi^1$, $\psi^2 = \phi^3$, and a so-called vertex function ψ^3 (see Figure 1). Since it is not $L_2(\mathbb{R}^2)^2$-stable, we will not use it any further.

The energy norm to control is a discrete $(H^1)^2$-seminorm (for discontinuous $\boldsymbol{u} \in \boldsymbol{V}_0$, $|\boldsymbol{u}|^2_{H^1(\mathbb{R}^2)^2}$ is defined by summing up the element-wise well-defined H^1-seminorms of its components). We have (see [3, 5] for all needed formulae)

$$|u_1|^2_{H^1(\mathbb{R}^2)} \approx \sum_{\alpha \in \mathbb{Z}^2} (x^1_\alpha - x^1_{\alpha+e^2})^2 + (x^4_\alpha - x^4_{\alpha+e^1})^2 + t(x^1_\alpha + x^1_{\alpha+e^2} - x^4_\alpha - x^4_{\alpha+e^1})^2,$$

analogously for the second component u_2 of $\boldsymbol{u}$. Here, $\approx$ can be replaced by $=$ if $t = 3/2$. In the computations below, we have used this value to determine the norm operator $\hat{E}$ resp. E (with $N = 3$) and the entries of the matrix function $B(\theta)$. Obviously, in this example we have $s_0 = 1$ while $s_1 = 1$ results from the anticipated inverse inequality (7). Consequently, $s_2 = 0$.

Finally, we need to describe the prolongations, i.e., $\hat{P}_1$ resp. $\hat{S}$. Prolongations with uniformly bounded H^1-norm growth for the scalar case, i.e., for $\{W_j\}$, are based on natural averaging procedures and described in [3, 5]. However, they do not preserve property (20)–(21) when applied to vector fields componentwise. We describe one approach of constructing prolongations suitable for the vector case. According to the abstract theory from [1] applied to the Stokes discretizations under consideration, a desirable property of $\hat{P}_1$ is that it preserves the average of the vector fields on the edges of $\mathcal{R}_0$. I.e., if the coefficients of $\hat{S}\boldsymbol{u}$ w.r.t. $\boldsymbol{\Phi}_0$ are denoted by $\{y^l_\beta\}$ we require that for all $\alpha \in \mathbb{Z}^2$

$$y^l_{2\alpha} + y^l_{2\alpha+e^1} = 2x^l_\alpha, \ l = 1,2, \quad y^l_{2\alpha} + y^l_{2\alpha+e^2} = 2x^l_\alpha, \ l = 3,4.$$

There are two natural ways to achieve this: *trivial extension*

$$y^l_{2\alpha} = y^l_{2\alpha+e^1} = x^l_\alpha, \ l = 1,2, \quad y^l_{2\alpha} = y^l_{2\alpha+e^2} = x^l_\alpha, \ l = 3,4, \tag{22}$$

or *averaging* as in the scalar case. The latter leads to (compare [3])

$$y^1_{2\alpha} = x^1_\alpha + \theta^1_\alpha/8, \quad y^1_{2\alpha+e^1} = x^1_\alpha - \theta^1_\alpha/8, \tag{23}$$

where $\theta_\alpha^1 = x_\alpha^4 + x_{\alpha-e^2}^4 - x_{\alpha+e^1}^4 - x_{\alpha+e^1-e^2}^4$, analogously for $l = 2, 3, 4$. The remaining values y_β^l corresponding to edges in $\mathcal{R}_1$ interior to squares of the coarser partition $\mathcal{R}_0$ have been obtained by satisfying (20) (under the assumption $\boldsymbol{u} \in \boldsymbol{Z}_0$) and minimizing the discrete expression for the $(H^1)^2$-seminorm of $\hat{S}\boldsymbol{u}$, with various values of the parameter t. Note that this reduces locally to an 8-dimensional constrained minimization problem, with a rank-deficient set of constraints (solvability is guaranteed only if (20) is satisfied). In general, the constraints are satisfied in a least-squares sense. Lack of space prevents us from presenting the detailed formula.

As explained in [5], it is easy to determine a relatively low-dimensional space $\hat{\mathcal{K}}$ of Hermitean matrix functions with polynomial entries with given spectra of Fourier coefficients such that it contains the Krylov space $\mathcal{K}$ of interest. Choosing the natural basis in $\hat{\mathcal{K}}$, the matrix representation $\hat{\mathcal{L}}$ for the operator $\mathcal{L}_{S^*, S}|_{\hat{\mathcal{K}}}$ can be computed. Solving the complete eigenvalue problem for $\hat{\mathcal{L}}$ and for $\hat{\mathcal{L}}^T$, one is able to find all λ_r and β_r of interest (see Section 2 for the notation). Our Matlab implementation (for details, see [5]) monitors both conditioning of the eigenvalues and rank of eigenspaces in the case of multiple eigenvalues (such safety measures become necessary, since in the computations below we have $\dim \hat{\mathcal{K}} > 400$). To verify (17), we find a $L \times M$ polynomial matrix function $H(\theta)$ and a scalar polynomial $\Delta(\theta)$ such that $G(\theta)H(\theta) = \Delta(\theta) \cdot Id_M$ and set $R(\theta) = \Delta(\theta) \cdot Id_L - H(\theta)G(\theta)$. Then $R^*(\theta)B_r(\theta)R(\theta) \neq 0$ is sufficient for (17) to hold. In our example, since $M = 1$ we can take $H(\theta) = G^*(\theta)$ and obtain $\Delta(\theta) = 4(\sin^2 \theta_1/2 + \sin^2 \theta_2/2)$. Again, this is simple linear algebra and easy to implement.

Table 1 shows, for various t, the computed leading eigenvalues of $\hat{\mathcal{L}}$ for the prolongation operators based on (23). Their algebraic dimension (which coincided with the geometric dimension in all cases) is shown in parentheses. Bold-faced λ indicates that not all associated β_r have vanished. The above sufficient condition for (17) to hold was satisfied for at least one associated $B_r(\theta)$ for all λ shown. The last row of Table 1 shows the computed value of Λ. The minimal $\Lambda = 1.0074$ was obtained for $t \approx 0.992$, and not for $t = 3/2$. Analogous testing of the $(L_2)^2$-norm behavior of $\hat{P}_{j,J}$ lead to the optimal value $\Lambda = 1$ for all t. The replacement of (23) by the simpler formula (22) produced similar but significantly worse results (here, $\Lambda \geq 1.8006$ was observed for the same range of t).

Note, that values $\Lambda < 1.05$ are quite acceptable from a more practical point of view. This holds, according to our tests, for all $0.95 < t < 1.7$, and hints at a certain robustness of the construction principle. Thus, a generalization to quadrilateral partitions seems to be promising and could lead to a new well-performing (even though not asymptotically optimal) algorithm for the linear system associated with the $\boldsymbol{Z}_j$-discretization of the Stokes problem by approximately divergence-free Q1-rotated vector fields. Clearly, this claim needs verification since the results presented here are asymptotic and only serve the case of uniform dyadic refinement. In practice, one deals with the range $J \leq 15$, spaces are defined on bounded domains and have to include essential boundary conditions. Also, the constants in the asymptotic results become of interest. In this respect, the natural value $t = 3/2$ seems to lead to slightly better performance than $t \approx 0.992$ which corresponds to the optimal Λ.

t	0.0	0.5	1.0	1.5	2.0
	4.0000 (10)	4.0000 (3)	4.0000 (3)	4.0000 (3)	4.0000 (3)
	2.0000 (18)	2.6667 (4)	2.0000 (6)	2.0000 (2)	2.0000 (2)
λ	**1.0000** (34)	2.0000 (2)	**1.0076** (1)	1.6000 (4)	1.3333 (4)
		1.7778 (3)	1.0000 (20)	**1.0362** (1)	**1.0644** (1)
		1.3333 (8)	0.8051 (1)	**1.0000** (9)	**1.0000** (9)
		1.0623 (1)	**0.6289** (1)	0.8000 (8)	0.7486 (1)
Λ	4.0000	1.7778	1.0076	1.0362	1.0644

Tab. 1. Leading eigenvalues $\hat{\mathcal{L}}$ for the prolongation operators.

References

1. Oswald, P., Intergrid transfer operators and multilevel preconditioners for nonconforming discretizations, Appl. Numer. Math. **23** (1997), 139–158.

2. Oswald, P., On a hierarchical basis multilevel method with nonconforming P1 elements, SIAM J. Numer. Anal. **62** (1992), 189–212.

3. Chen, Z. and P. Oswald, Multigrid and multilevel methods for nonconforming rotated Q1 elements, Math. Comp. **67** (1998), 667–693.

4. Bramble, J., *Multigrid Methods*, Longman, Harlow, 1993.

5. Oswald, P., On norm bounds for iterated intergrid transfer operators, Arbeitspapiere GMD 1079, Sankt Augustin, June 1997.

6. Jia, R.-Q., S. D. Riemenschneider, and D.-X. Zhou, Smoothness of multiple refinable functions and multiple wavelets, preprint.

7. Rannacher, R., and S. Turek, Simple nonconforming quadrilateral Stokes element, Numer. Meth. PDEs **8** (1992), 97–111.

Peter Oswald
Bell Laboratories, Lucent Technologies
600 Mountain Av., Rm. 2C403
Murray Hill, NJ 07974, USA
poswald@research.bell-labs.com
http://cm.bell-labs.com/who/poswald

Wavelets and Their Associated Operators

Amos Ron

Abstract. This article is devoted to the study of wavelets based on the *theory of shift-invariant spaces*. It consists of two, essentially disjoint, parts. In the first part, the fiberization of the analysis operator of a shift-invariant system is discussed. That fiberization applies to wavelet systems via the notion of *quasi-wavelet* systems, and leads to the theory of *wavelet frames*. Highlights in this theory are the *unitary and mixed extension principles*, and the MRA construction of *framelets*. The second part of the article is devoted to the study of the cascade/transfer operators and the subdivision operator associated with a refinable function. The analysis there is primarily based on the interpretation of the cascade operator as a special quasi-interpolation scheme. This leads to a surprisingly simple analysis of certain properties of refinable functions, including their *smoothness* and the *convergence of the cascade and subdivision algorithms*. In particular, it follows that these latter algorithms, if handled properly, always converge.

1. Preface: Wavelets and Their Associated Operators

This article advocates the analysis of wavelet systems via the study of their associated operators. The goal is neither to survey the current state-of-the-art in this area, nor to provide the reader with in-depth comprehensive analysis of any of the issues addressed. Rather, my attempt is to provide a glimpse into various contemporary aspects of wavelets, in a way that may whet the reader's appetite for further reading. Based on this philosophy, I have chosen setups that simplify the discussion even in cases when the simplification is purely notational.

The notion of 'the operators associated with a wavelet system' is so broad that it allows me to discuss two essentially disjoint topics. The first topic concerns the intrinsic operators of the wavelet system: analysis and synthesis, with the main aim being to review the recent developments in the area of wavelet frames (cf. [66, 67, 68, 69, 70, 71, 72, 34, 22, 30]). The second topic is the analysis of the corresponding refinable/scaling function(s), a topic that is also pertinent to the area of *uniform subdivision algorithms* (cf. [56, 33, 15, 32]). The relevant operators in this discussion are the subdivision and the

Approximation Theory IX, Volume 2: Computational Aspects   283
Charles K. Chui and Larry L. Schumaker (eds.), pp. 283–317.
Copyright ℗ 1998 by Vanderbilt University Press, Nashville, TN.
ISBN 0-8265-1326-3.

cascade. This second part is written as a short monograph, aiming to single out the few underlying principles, and to demonstrate the elegance and the simplicity of the resulting theory.

The *analysis and synthesis operators* are the two basic operators related to any setup where one represents functions in some function space with the aid of a basis or, more generally, a countable subset (referred to hereafter as a 'system') from that space. The study of the structure of the synthesis and analysis operators of a wavelet system is, probably, the most basic one. We show how the *analysis* operator of a wavelet system can be *fiberized*, *i.e.*, represented and thereby understood and analyzed with the aid of a collection of much simpler operators ('fibers'). It leads to a complete characterization of *wavelet frames* in terms of a certain collection of infinite-order non-negative definite matrices. The theory is based on an interplay between the wavelet system and a new (special) type of shift-invariant system: the quasi-wavelet system. As a result, simple *extension* techniques for constructing a wavelet system from a given MRA (multiresolution analysis, cf. [54, 55, 27, 48]) become possible: the highlight of this part is that the refinable function need not satisfy any particular property, *i.e.*, its shifts need neither form a Riesz basis nor form a frame. The wavelet frames constructed by these principles are termed *framelets* and the example of B-spline framelets is discussed.

Another pair of operators associated with a wavelet system are the *cascade/transfer operator and the subdivision operator*. In contrast with the general synthesis/analysis operators, these operators are not associated directly with the wavelet system but rather with the underlying refinable function (known also as the scaling function, which may be either a scalar function or a vector-valued function). Thus, one assumes that the wavelets are constructed via MRA and wishes to understand first the properties of the generator(s) of the MRA. Of relevance here are the *stability and linear independence* of the shifts of the scaling function, its *smoothness*, and the *convergence of the corresponding subdivision and cascade algorithms*. The goal in this part is to demonstrate the relative simplicity of the theory that is obtained by using this approach: in fact, a single identity that involves the cascade operator, when combined with the existing knowledge concerning *quasi-interpolation schemes* provides a unified approach for the study of all these aspects!

I forgo discussing and treating the approximation orders of the scaling functions and wavelets. There are general treatments of approximation orders of shift-invariant spaces (cf. [4, 7]), and applying these theories to the case of a single scaling function is quite straightforward, hence does not require a special exposition. The technique for dealing with the vector case is primarily based on superfunction theory (cf. [5, 7, 60]), a technique that does not invoke, at least not in an explicit way, any operator-based approach. The approximation orders of framelets are also derived from the general theory (cf. [30]), and while they do rely on the structure of the synthesis operator of a shift-invariant system, they seem to be a bit beyond the scope of this article. The relations between the approximation orders of the scaling function and its smoothness are discussed in [64].

2. The Analysis and Synthesis Operators

2.1. The Analysis and Synthesis Operators: General

Let X be a countable set in a Hilbert space $\mathcal{H}$. We can use X either in order to decompose or to reconstruct other elements in $\mathcal{H}$. Here, *reconstruction* means that we assemble functions from discrete data with the relevant operator, then the synthesis operator

$$T_X : \ell_2(X) \to \mathcal{H} : c \mapsto \sum_{x \in X} c(x) x.$$

When this operator is well-defined and bounded we say that the system X is a Bessel system. The complementary use of X is for *decomposition*, *i.e.*, using X as a collection of linear functionals. The corresponding operator is then the analysis operator T_X^* which is the adjoint of T_X:

$$T_X^* : \mathcal{H} \to \ell_2(X) : f \mapsto T_X^* f := (\langle f, x \rangle)_{x \in X}.$$

Being the adjoint of T_X, the analysis operator is well-defined and bounded if and only if X is a Bessel system. There are situations when we restrict T_X^* to a closed subspace $H \subset \mathcal{H}$, and there is no need to assume then that $X \subset H$.

For most examples of interest, the Bessel property of the system X is easily verified. For example, if $\Phi \subset L_2(\mathbb{R}^d)$ is a finite set of functions, and if we let $E(\Phi)$ be the collection of shifts of Φ:

$$E(\Phi) := (E^\alpha \phi : \phi \in \Phi, \ \alpha \in \mathbb{Z}^d), \quad E^\alpha : f \mapsto f(\cdot - \alpha), \quad (2.1.1)$$

then $E(\Phi)$ is always Bessel, provided, say, that the bracket product

$$[\widehat{\phi}, \widehat{\phi}] := \sum_{j \in 2\pi \mathbb{Z}^d} E^j (|\widehat{\phi}|^2)$$

is continuous for every $\phi \in \Phi$. That continuity is implied by a mild decay condition on ϕ at ∞: since the Fourier coefficients of $[\widehat{\phi}, \widehat{\phi}]$ are given by the inner products $(\langle E^\alpha \phi, \phi \rangle)_{\alpha \in \mathbb{Z}^d}$, it suffices, for example, to require that these coefficients lie in $\ell_1(\mathbb{Z}^d)$.

However, one almost always would like to boundedly invert the analysis and synthesis operators, and these additional requirements turn out to be highly non-trivial. For example, even if we require the sequence Φ in the previous example to belong to the space of compactly supported C^∞ test functions, we cannot conclude that any of the two operators of interest is boundedly invertible.

Definition 2.1.2. Let X *be a Bessel system in Hilbert space* $\mathcal{H}$*, and let* H *be a closed subspace of* $\mathcal{H}$*. We say that:*

1) X *is a stable system in* $\mathcal{H}$ *(or that* X *forms a Riesz basis in* $\mathcal{H}$*) whenever the synthesis operator* T_X *is boundedly invertible;*

2) X *is a frame for* H *if the restriction of* T_X^* *to* H *is boundedly invertible.*

The frame bounds are the numbers $\|T_X^*\|^2$ (upper frame bound, may be referred to as the 'Bessel bound' if X is merely a Bessel system) and $\|T_X^{*-1}\|^{-2}$ (lower frame bound).

It is not hard to see that, given a stable basis X in $\mathcal{H}$ and a closed subspace $H \subset \mathcal{H}$, X is a frame for H if (and only if) T_X^* is injective on H.

Some of the basics concerning stable bases and frames are collected in the following proposition:

Proposition 2.1.3. *Let* X *be a Bessel system in* $\mathcal{H}$. *Then*

1) X *is a stable basis in* $\mathcal{H}$ *if and only if there exists a map* $\mathrm{R} : X \to \mathcal{H}$ *such that* $\mathrm{R}X$ *is a dual basis: it is a Bessel system, and* $T_{\mathrm{R}X}^* T_X = \mathrm{id}$, *i.e.,*

$$\langle x', \mathrm{R}x \rangle = \delta_{x,x'}, \quad \forall x, x' \in X.$$

2) X *is a frame for* $H \subset \mathcal{H}$ *if and only if there exists a map* $\mathrm{R} : X \to \mathcal{H}$ *such that* $\mathrm{R}X$ *is a dual system: it is a Bessel system and* $T_{\mathrm{R}X} T_X^* = \mathrm{id}$ *on* H, *i.e.,*

$$\sum_{x \in X} \langle f, \mathrm{R}x \rangle \, x = f, \quad \forall f \in H.$$

As said, the Bessel property of a system X is usually easy to obtain and analyze. In contrast, the frame and Riesz basis properties are by far more demanding, and are also more challenging for mathematical analysis. There are two different possible approaches here: (a) an intrinsic analysis of the system X, and (b) an analysis of a pair $(X, \mathrm{R}X)$. The above proposition indicates that the second approach may be simpler: an intrinsic analysis of X requires one to check whether a certain operator is bounded below, while an analysis of the pair $(X, \mathrm{R}X)$ requires one to know whether a certain operator is the identity. On the other hand, the second approach requires one to augment first the given system X by a suitable system $\mathrm{R}X$, *i.e.*, a system which is a 'good candidate' for being dual to X, something that may not be simple at all. It is then important to emphasize the case a dual system is given for free:

Definition 2.1.4. *Let* X *be a system in a Hilbert space* H. *We say that* X *is a tight frame for* H *if the analysis operator* T_X^* *is unitary, or equivalently, if the condition*

$$T_X T_X^* = \mathrm{id}$$

holds (in H).

The advantage of tight frames over other frames is obvious: the same system X may be used for reconstruction and for decomposition. That may be attractive for two different reasons: first, it eliminates the need to *find* a dual system. Second, even in the case when a dual system is easy to find, its properties may not be as good as those of the original X. For example, if H is the subspace of $L_2(\mathbb{R})$ consisting of cardinal splines of order 2 (*i.e.*, continuous piecewise-linear functions with integer breakpoints), then, with B the hat function, $E(B)$ is a stable basis for H. However, the dual basis in H for $E(B)$ lacks the compact support.

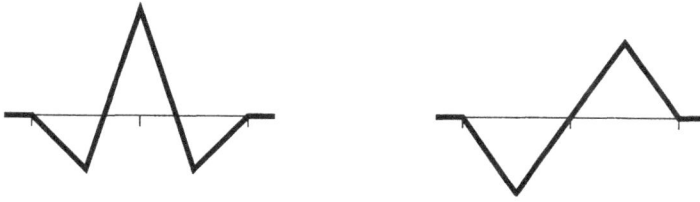

Fig. 1. The generators of the piecewise-linear tight frame.

In fact, our first example of a tight frame involves piecewise-linear functions as well.

Example 2.1.5. Let Φ be the set of the two piecewise-linear functions depicted in Figure 1 (the support of each is $[-1, 1]$, and the max-norms are 1 for the function on the left and $\frac{\sqrt{2}}{2}$ for the function on the right.) Let

$$X := \cup_{k \in \mathbb{Z}} \mathcal{D}^k E(2^{k/2}\Phi),$$

with $\mathcal{D}$ the dyadic dilation operator:

$$\mathcal{D} : f \mapsto f(2\cdot).$$

Then X is a tight (wavelet) frame for $L_2(\mathbb{R})$. $\square$

I do not believe that it is straightforward to verify that the system in the above example is indeed a tight frame. It is my intent in the rest of this section to briefly review the theory that leads to this construction as well as to many other more involved ones: the theory of wavelet frames. That theory, which is detailed in [68, 69], is based on the *fiberization of shift-invariant systems*, [5, 66].

Fiberization. The idea behind fiberization is to analyze a complicated operator S with the aid of a collection $(S_\omega)_{\omega \in \Omega}$ of operators of simpler structure (each of which is a 'fiber'). For the fiberization to be useful, the fiber operators need, at least, help in determining whether S is bounded and/or invertible, and need also be of help for computing or estimating the norms of S and S^{-1}.

Example 2.1.6. The fiberization of the analysis and synthesis operators of a shift-invariant system, [66]. Let Φ be a countable subset of $L_2(\mathbb{R}^d)$, and let X be the collection $E(\Phi)$ of all the shifts of Φ. For (almost) every $\omega \in \mathbb{R}^d$, let J_ω be the pre-Gramian of X: the matrix whose rows are indexed by $2\pi\mathbb{Z}^d$, whose columns are indexed by Φ, and whose $(\alpha, \phi) \in 2\pi\mathbb{Z}^d \times \Phi$ entry is

$$J_\omega(\alpha, \phi) := \widehat{\phi}(\omega + \alpha).$$

The pre-Gramian J_ω is considered as a map from $\ell_2(2\pi\mathbb{Z}^d)$ into $\ell_2(\Phi)$.

The collection $(J_\omega)_{\omega \in \mathbb{R}^d}$ fiberize the synthesis operator T_X of X. The reference [66] contains detailed information as to the exact meaning and the possible value of such fiberization. In particular, we have that, with $X :=$ $E(\Phi)$, and $c \in \ell_2(X)$,

$$\widehat{T_X c}\big|_{\omega + 2\pi \mathbb{Z}^d} = J_\omega \widehat{c}(\omega),$$

where

$$\widehat{c}(\omega) := (\widehat{c_\phi}(\omega))_{\phi \in \Phi}, \quad \widehat{c_\phi}(\omega) := \sum_{j \in \mathbb{Z}^d} c(E^j \phi) e^{-ij \cdot \omega}.$$

If follows [66] that

$$\|T_X\|^2 = \| \, \|J_\omega\| \, \|_{L_\infty(\mathbb{R}^d)}, \tag{2.1.7}$$

$$\|T_X^{-1}\|^2 = \| \, \|J_\omega^{-1}\| \, \|_{L_\infty(\mathbb{R}^d)}. \tag{2.1.8}$$

In a similar manner, the Gramian matrices $(J_\omega^* J_\omega)_\omega$ fiberize the self-adjoint operator $T_X^* T_X$. Note that $J_\omega^* J_\omega$ is an operator from $\ell_2(\Phi)$ into itself, and its $\phi \times \varphi \in \Phi \times \Phi$ entry is the bracket product $[\widehat{\varphi}, \widehat{\phi}](\omega)$:

$$[\widehat{\varphi}, \widehat{\phi}] := \sum_{j \in 2\pi \mathbb{Z}^d} \widehat{\varphi}(\cdot + j) \overline{\widehat{\phi}(\cdot + j)}.$$

Consequently, we can study the Bessel property and the Riesz basis property of a shift-invariant $X = E(\Phi)$ via the above Gramian fibers.

Fiberization of the analysis operator of $X = E(\Phi)$ is also possible, but is significantly more complicated (than that of the synthesis operator) unless we assume that X is fundamental in $L_2(\mathbb{R}^d)$ (*i.e.*, that the finite span of X is dense in that space). With the fundamentality assumption in hand, however, we get results as simple as in the synthesis case. Specifically, the matrices $(J_\omega^*)_\omega$ provide now fiberization for T_X^*, the dual Gramian matrices $(J_\omega J_\omega^*)_\omega$ fiberize $T_X T_X^*$, and results similar to (2.1.7) and (2.1.8) hold. Note that each dual Gramian fiber is a non-negative operator from $\ell_2(2\pi \mathbb{Z}^d)$ into itself, and its (α, β)-entry $((\alpha, \beta) \in 2\pi \mathbb{Z}^d \times 2\pi \mathbb{Z}^d)$ is

$$\sum_{\phi \in \Phi} \widehat{\phi}(\omega + \alpha) \overline{\widehat{\phi}(\omega + \beta)}.$$

Fiberizations of the analysis operator are useful in the study of the Bessel property and the frame property of the original X. $\square$

Here is an example that demonstrates the usefulness of the fiberization approach. Suppose that we would like to determine whether a system $X = E(\Phi)$ is a tight frame for $L_2(\mathbb{R}^d)$. X is a tight frame iff $T_X T_X^* = \mathrm{id}$ iff $J_\omega J_\omega^* = \mathrm{id}$ for almost every $\omega \in \mathbb{R}^d$. After suppressing some obvious repetitions (*e.g.*, different fibers that represent essentially the same operator), we obtain that X is a tight frame for $L_2(\mathbb{R}^d)$ if and only if, for every $j \in 2\pi \mathbb{Z}^d$,

$$\sum_{\phi \in \Phi} \widehat{\phi} \, \overline{E^j \widehat{\phi}} = \delta_j, \quad \text{a.e.} \tag{2.1.9}$$

Some readers may be able to find a simple direct proof for this result. Indeed, the tool of fiberization in the study of tight frames (as well as in the study of orthonormal bases and system-dual system setups) is not so essential: the fiber matrices in these cases are identity matrices, hence entrywise characterizations analogous to (2.1.9) are available. It is then plausible to claim that such characterizations can be obtained directly without assembling first the fiber matrices. Fiberization, however, is a powerful tool whenever the fiber matrices do not have an especially simple structure. □

Here is a striking example for the utility of the above fiberization. It is the simplest example of the so-called *duality principle of Weyl-Heisenberg systems*, [67].

Example 2.1.10. Self-adjoint Weyl-Heisenberg systems. Given $g \in L_2(\mathbb{R}^d)$, let

$$X := (e_{ij}E^k g : (j,k) \in 2\pi\mathbb{Z}^d \times \mathbb{Z}^d), \quad e_{ij} : \omega \mapsto e^{ij \cdot \omega}.$$

We have then that $X = E(\Phi)$, with $\Phi := (e_{ij}g)_{j \in 2\pi\mathbb{Z}^d}$. Indexing Φ by $2\pi\mathbb{Z}^d$, we obtain that the pre-Gramian fiber J_ω has the entries

$$J_\omega(k,j) = \widehat{g}(\omega + k + j).$$

This means that the pre-Gramians are self-adjoint (up to conjugation), hence that the condition that characterizes stability in terms of $(J_\omega)_\omega$ is identical to the condition that characterizes frames for L_2 in terms of $(J_\omega^*)_\omega$. This recovers the fact (cf. *e.g.*, [26, 1]) that the above X is a stable basis if and only if it is a fundamental frame.

Wavelets. We want to focus now on the main theme of the discussion: wavelet systems. In order to simplify notation, we will mostly assume that we employ dyadic dilations (the general theory allows arbitrary dilations, for as long as the entries of the dilation matrix are integers, and the spectrum of the dilation matrix lies outside the closed unit disc). To recall, given a finite $\Psi \subset L_2(\mathbb{R}^d)$ of mother wavelets, the wavelet system $X(\Psi)$ generated by Ψ is the collection of all dyadic dilations of the shift-invariant $E(\Psi)$:

$$X(\Psi) := \cup_{k \in \mathbb{Z}} \mathcal{D}^k E(2^{kd/2}\Psi), \quad \mathcal{D} : f \mapsto f(2\cdot). \tag{2.1.11}$$

One observes that a wavelet system is not shift-invariant: the k-scale $\mathcal{D}^k E(\Psi)$ of $X(\Psi)$ is invariant under $2^{-k}\mathbb{Z}^d$-shifts, and these shifts become indefinitely sparse as $k \to -\infty$.

The attempt to apply the shift-invariant fiberization techniques to the almost shift-invariant wavelet system led in [68] to the introduction of a link between wavelet systems and shift-invariant systems in the form of quasi-wavelet systems.

Definition 2.1.12. Quasi-wavelet systems. *Given a collection of mother wavelets* Ψ, *the* quasi-wavelet system generated by Ψ *is*

$$X^q(\Psi) := \cup_{k=0}^{\infty} \mathcal{D}^k E(2^{kd/2}\Psi) \bigcup \cup_{k=-\infty}^{-1} E(2^{kd}\mathcal{D}^k\Psi).$$

As one observes, the quasi-wavelet system is obtained from the wavelet system by oversampling the negative scales of the latter. For instance, in the (-1)-scale, the even shifts of the function $2^{-d/2}\psi(\frac{\cdot}{2})$ are replaced by the integer shifts of the re-normalized function $2^{-d}\psi(\frac{\cdot}{2})$.

Theorem 2.1.13. [68, 69]. *Let* $X := X(\Psi)$ *be a wavelet system,* X^q *its quasi-wavelet system counterpart.*

1) X *is a Bessel system iff* X^q *is a Bessel system. The two systems have the same Bessel bound.*
2) X *is a frame for* $L_2(\mathbb{R}^d)$ *if and only if* X^q *is a frame for that space. The two systems have the same frame bounds.*
3) *Suppose* X, X^q *are frames for* $L_2(\mathbb{R}^d)$, *and let* $\mathrm{R} : \Psi \to L_2(\mathbb{R}^d)$ *be some map. Then* $Y := X(\mathrm{R}\Psi)$ *is a frame dual to* X *if and only if* Y^q *is a frame dual to* X^q.

We remark that a smoothness assumption (a mild one: it is satisfied, *e.g.*, by the univariate and multivariate Haar functions) is imposed on Ψ in [68, 69]. In [22] it is shown that the first two statements in the above result hold even without that assumption.

Since the quasi-wavelet system X^q is shift-invariant, it admits a fiberization. Thanks to the above theorem, the so-obtained fibers can be used to characterize the Bessel property and the frame property of the *original* wavelet system X.

The fiberization of wavelet systems. In order to describe the dual Gramian fibers of $X := X(\Psi)$ (more precisely: the dual Gramian fibers of the shift-invariant X^q), we introduce first the affine product:

$$\Psi[\omega, \omega'] := \sum_{k=\kappa(\omega-\omega')}^{\infty} \sum_{\psi \in \Psi} \widehat{\psi}(2^k\omega)\overline{\widehat{\psi}(2^k\omega')}, \qquad (2.1.14)$$

where

$$\kappa(\omega) := \inf\{k \in \mathbb{Z} : 2^k\omega \in 2\pi\mathbb{Z}^d\}. \qquad (2.1.15)$$

(Thus, for example, $\kappa = \infty$ off the 2π-dyadic numbers, $\kappa \leq 0$ on $2\pi\mathbb{Z}^d$, and $\kappa(\omega) = -\infty$ iff $\omega = 0$.) Then, the $(\alpha, \beta) \in 2\pi\mathbb{Z}^d \times 2\pi\mathbb{Z}^d$-entry of the dual Gramian fiber $J_\omega J_\omega^*$ of X^q is [68],

$$\Psi[\omega + \alpha, \omega + \beta].$$

Thus we have the following fundamental result:

Theorem 2.1.16. [68]. *Let $X(\Psi)$ be a wavelet system. For each $\omega \in \mathbb{R}^d$, let S_ω be the operator from $\ell_2(2\pi\mathbb{Z}^d)$ to $\ell_2(2\pi\mathbb{Z}^d)$ defined by*

$$(S_\omega c)(\alpha) = \sum_{\beta \in 2\pi\mathbb{Z}^d} \Psi[\omega + \alpha, \omega + \beta]c(\beta).$$

Then

(a) *$X(\Psi)$ is a Bessel system if and only if the function $\omega \mapsto \|S_\omega\|$ is essentially bounded. Furthermore, $\|T_X^*\|^2 = \| \|S_\omega\| \|_{L_\infty(\mathbb{R}^d)}$.*

(b) *Assume that $X(\Psi)$ is Bessel. Then $X(\Psi)$ is a frame for $L_2(\mathbb{R}^d)$ if and only if the map $\omega \mapsto \|S_\omega^{-1}\|$ is essentially bounded. Also, the lower frame bound is then $1/\| \|S_\omega^{-1}\| \|_{L_\infty(\mathbb{R}^d)}$.*

(c) *$X(\Psi)$ is a tight frame for $L_2(\mathbb{R}^d)$ if and only if almost all the fibers S_ω are the identity operators.*

We conclude this section with a variety of examples.

Example 2.1.17. Tight frames. In view of the above result, $X(\Psi)$ is a tight frame if and only if $\Psi[\omega + \alpha, \omega + \beta] = \delta_{\alpha,\beta}$ for every $\alpha, \beta \in 2\pi\mathbb{Z}^d$, and almost every $\omega \in \mathbb{R}^d$. Replacing ω by $\omega + \alpha$, we may assume $\alpha = 0$. Since the affine product is also dilation-invariant (*i.e.*, $\Psi[2\omega, 2\omega'] = \Psi[\omega, \omega']$), we may assume that, unless $\beta = 0$, $\kappa(\beta) = 0$, *i.e.*, that $\beta \in 2\pi(\mathbb{Z}^d\backslash 2\mathbb{Z}^d)$. We then obtain that $X(\Psi)$ is a tight frame if and only if, for a.e. ω,

$$\Psi[\omega, \omega] = 1, \quad \Psi[\omega, \omega + \beta] = 0, \quad \forall \beta \in 2\pi(\mathbb{Z}^d\backslash 2\mathbb{Z}^d).$$

This result was established independently by others (cf. [38] where the univariate dyadic case of the above is proved, and [35] where the general case is established.) □

Example 2.1.18. Univariate band-limited diagonal wavelets.
Assume that supp $\widehat{\psi} \subset [-2\pi, 2\pi]\backslash(-\pi, \pi)$, for every $\psi \in \Psi$. Then one easily confirms that $\Psi[\omega + \alpha, \omega + \beta] = 0$ (for $\alpha, \beta \in 2\pi\mathbb{Z}^d$) unless $\alpha = \beta$. This means that the fiber matrices S_ω are all diagonal with diagonal entries

$$\Psi[\omega, \omega] = \sum_{\psi \in \Psi, k \in \mathbb{Z}} |\widehat{\psi}(2^k\omega)|^2. \tag{2.1.19}$$

It follows that $X(\Psi)$ is a frame if and only if the function $P : \omega \mapsto \Psi[\omega, \omega]$ is essentially bounded together with its reciprocal. This recovers a well-known result.

Moreover, given an arbitrary wavelet system $X(\Psi)$, since the values of the above P comprise the diagonal entries of the fibers matrices, and since each fiber matrix is non-negative definite, the boundedness of P and $1/P$ is always a *necessary* condition for X to be a frame (since the norm of any non-negative operator is bounded below by the largest element on its diagonal, while the norm of its inverse is bounded below by the reciprocal of the smallest element on that diagonal). Again, this recovers a known result (cf. [26, 27, 19]). In fact, almost all the results that provide estimates on the frame bounds of a wavelet frame can be in retrospect understood as attempts to estimate norms and inverse norms of non-negative definite matrices in terms of their entries. □

Example 2.1.20. **Univariate band-limited block diagonal systems with 2×2 blocks.**
We assume that, for each $\psi \in \Psi$, supp $\widehat{\psi} \subset [-8\pi/3, 8\pi/3] \backslash (-2\pi/3, 2\pi/3)$, and examine the affine product $\Psi[\omega + \alpha, \omega + \beta]$, $\beta \neq \alpha$. A direct computation shows that if $|\omega + \alpha| < 2\pi/3$ all the summands in the affine product vanish; otherwise, due to symmetry considerations and to the invariance properties of $\Psi[\cdot, \cdot]$, we may assume that $2\pi/3 \leq \omega < 4\pi/3$, that $\alpha = 0$, and that $\beta \in 2\pi\mathbb{Z}$. This implies that $\widehat{\psi}(2^k \omega) = 0$, unless $k = 0, 1$ and that, for $\beta \in 2\pi\mathbb{Z} \backslash 0$ and $k \in \{0, 1\}$, $\widehat{\psi}(2^k(\omega + \beta)) \neq 0$ only if $\beta = -2\pi$ (the 'magic' in this example is the fact that it is the same β for the two different values of k). As a result, it follows that S_ω is block diagonal, with all the blocks being (at most) 2×2, and with each block of the form

$$
\begin{pmatrix}
|a|^2 + |b|^2 & \langle a, c \rangle + \langle b, d \rangle \\
\langle c, a \rangle + \langle d, b \rangle & |c|^2 + |d|^2
\end{pmatrix},
\tag{2.1.21}
$$

with $a := (\widehat{\psi}(\omega))_{\psi \in \Psi}$, $b := (\widehat{\psi}(2\omega))_{\psi \in \Psi}$, $c := (\widehat{\psi}(\omega - 2\pi))_{\psi \in \Psi}$, $d := (\widehat{\psi}(2(\omega - 2\pi)))_{\psi \in \Psi}$, and with $|\cdot|$ being the ℓ_2-norm, and $\omega \in [2\pi/3, 4\pi/3)$. Thus, such a wavelet system is a frame for $L_2(\mathbb{R}^d)$ if and only if each matrix of the form (2.1.21) is invertible, and, in addition, the norms as well as inverse norms of these matrices are bounded independently of ω. Of particular interest is the case when Ψ is a singleton $\{\psi\}$. In this case, each matrix in (2.1.21) is of the form BB^*, with B the *square* matrix

$$
B = \begin{pmatrix}
\widehat{\psi}(\omega) & \widehat{\psi}(2\omega) \\
\widehat{\psi}(\omega - 2\pi) & \widehat{\psi}(2\omega - 4\pi)
\end{pmatrix}.
$$

Obviously, the result can now be stated directly in terms of the norms and inverse norms of the matrices of the form B. This case was studied in [75], where it was observed that the frame property is equivalent here to the (seemingly stronger) Riesz basis property. (From the point of view of the current discussion, that can be attributed to the fact that each B is square, hence the norms and inverse norms of BB^* are identical to those of B^*B). □

Example 2.1.22. Oversampling. In order to explore this case we need also to consider systems that are shift-invariant with respect to a superlattice of $\mathbb{Z}^d$. For simplicity, we consider only lattices that are scalar scales of an integer lattice (see [68] for the general case). Thus, we let

$$
E_n(\Phi) := (E^\alpha \phi : \phi \in \Phi, \ \alpha \in \mathbb{Z}^d/n),
$$

with n some integer, and define the wavelet system $X_n(\Psi)$ similarly to (2.1.11), with $E(\Psi)$ there replaced by $E_n(\Psi)$ here. Following [18], we consider the new system as an oversampling of $X(\Psi)$. The study of relations between properties of $X(\Psi)$ and properties of $X_n(\Psi)$ was originated in the work of Chui and Shi (cf. [18, 20, 21]).

In order to analyze the oversampling via the fiberization theory, we need to know the fibers of the oversampling system. These fibers are simple (and straightforward) variants of the fibers in the integer case: this time, each fiber $S_{n,\omega}$ is indexed by $2\pi n \mathbb{Z}^d \times 2\pi n \mathbb{Z}^d$, and its (α, β)-entry is

$$n^d \Psi_n[\omega + \alpha, \omega + \beta],$$

with the only difference between $\Psi_n[,]$ and $\Psi[,]$ in the definition of the valuation κ: here κ is replaced by

$$\kappa_n(\omega) := \inf\{k \in \mathbb{Z} : 2^k \omega \in 2\pi n \mathbb{Z}^d\}.$$

One then observes that $\kappa_n = \kappa$ on $2\pi n \mathbb{Z}^d$ whenever n is odd. This means that for an odd n, $n^{-d} S_{n,\omega}$ is a submatrix of S_ω. Since S_ω is non-negative definite, it follows that

$$\|S_{n,\omega}\| \le n^d \|S_\omega\|.$$

Hence, the fiberization theory yields that

$$\|T^*_{X_n(\Psi)}\| \le n^{d/2} \|T^*_{X(\Psi)}\|,$$

and a similar argument provides an analogous estimate on the other frame bound. It follows then that $X_n(\Psi)$ is tight if $X(\Psi)$ is.

The particular oversampling result above is due to Chui and Shi (and was proved originally by different means). We note that the situation is different if the oversampling rate is even (cf. [18, 68]). □

2.2. Wavelets: Extension Principles, Framelets, B-Spline Framelets

The fiberization of wavelet systems, when combined with the vehicle of *multiresolution analysis* (MRA), leads to new ways for constructing wavelet frames.

To recall, a function $\phi \in L_2(\mathbb{R}^d)$ is (dyadically) refinable if there exists a 2π-periodic mask function τ_0 such that

$$\widehat{\phi}(2\cdot) = \tau_0 \widehat{\phi}. \tag{2.2.1}$$

We assume that $\widehat{\phi}(0) = 1$, but do not impose any other standard assumption; in particular, the shifts of ϕ need not form a Riesz basis. For notational convenience, we set $\psi_0 := \phi$.

Let V_0 be the PSI space generated by ϕ, *i.e.*, the smallest closed subspace of $L_2(\mathbb{R}^d)$ that contains $E(\phi)$. The refinability assumption (2.2.1) is equivalent to the condition that $V_1 := DV_0$ is a superspace of V_0 (cf. [6]). We then attempt to construct a wavelet frame that is generated by finitely many mother wavelets from V_1:

$$\Psi := (\psi_1, \ldots, \psi_n) \subset V_1.$$

This means that each ψ_i, $i = 1, \ldots, n$ satisfies a relation of the form

$$\widehat{\psi_i}(2\cdot) = \tau_i \widehat{\psi_i},$$

for some 2π-periodic wavelet mask τ_i. Since the wavelet system $X(\Psi)$ is completely determined by choice of the wavelet masks, we attempt to construct the wavelet frame by an appropriate selection of the wavelet masks. As one encounters in the Riesz basis case (cf. [55, 48]), the construction is based on *matrix extensions*. However, in stark contrast with Riesz bases constructions, no *a priori* assumption need be imposed on the refinable ϕ.

We start with the discussion of tight wavelet frames. [68] contains a complete characterization of all tight wavelet frames that can be constructed by using the above MRA approach. That characterization easily leads to the following *unitary extension principle*.

Theorem 2.2.2. The unitary extension principle. [68]. *Let $X(\Psi)$ be a wavelet system constructed by the above MRA recipe, and assume further that all the masks involved are bounded. Then $X(\Psi)$ is a tight frame for $L_2(\mathbb{R}^d)$ if the following condition holds: for almost every $\omega \in \mathbb{R}^d$, and for every $\gamma \in \{0, \pi\}^d$,*

$$\sum_{i=0}^{n} \tau_i(\omega)\overline{\tau_i(\omega + \gamma)} = \begin{cases} 1, & \gamma = 0, \\ 0, & otherwise. \end{cases}$$

Discussion. One may interpret the condition in Theorem 2.2.2 as a matrix extension. Let v_i be the row vector $v_i = (\tau_i(\cdot + \gamma))_{\gamma \in \{0,\pi\}^d}$, and let V be the matrix whose rows are $v_0, \ldots, v_n$. The refinable ϕ determines the vector v_0, and selecting the wavelets $(\psi_1, \ldots, \psi_n)$ is tantamount to selecting the additional rows $v_1, \ldots, v_n$. The above extension principle asserts that $X(\Psi)$ is a tight frame once the columns of V are orthonormal for almost every ω. Since the number of columns is fixed (viz. 2^d), then by selecting in an appropriate manner a large number of wavelets, it is plausible that we can find a suitable unitary extension *without imposing any further condition on the refinement mask τ_0*.

B-spline framelets. We refer to a wavelet frame that is constructed from MRA by a matrix extension principle (either the above unitary one, or the mixed extension principle detailed in the sequel) as framelet. The simplest construction of framelets are those derived from the B-spline MRA. For notational convenience, we discuss here the construction of B-spline framelets of even order; the odd case is treated similarly.

The mask of a centered B-spline of order n, n even, is

$$\tau_0(\omega) = \cos^n(\omega/2).$$

Thus, τ_0^2 is the first term in the binomial expansion of

$$1 = (\cos^2(\omega/2) + \sin^2(\omega/2))^n. \tag{2.2.3}$$

Let τ_j, $j = 1, 2, \ldots, n$, be the squareroots of the other terms in this expansion, i.e.,

$$\tau_j(\omega) = \sqrt{\binom{n}{j}} \cos^j(\omega/2) \sin^{n-j}(\omega/2).$$

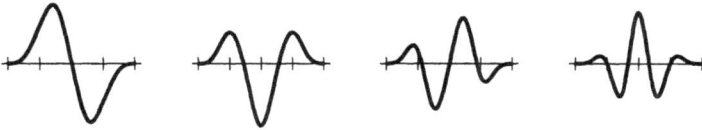

Fig. 2. The generators of the cubic B-spline tight framelet.

Setting $\tau := (\tau_j)_{j=0}^n$, (2.2.3) ensures that $\tau(\omega)$ is a unit vector for every ω. The only other required condition is that, for every ω, $\tau(\omega)$ be orthogonal to $\tau(\omega + \pi)$: since $\tau_j(\omega)\tau_j(\omega + \pi) = (\cos(\omega/2)\sin(\omega/2))^n \binom{n}{j}(-)^j$, that additional requirement follows from the fact that $(1 - 1)^n = 0$.

One can now verify that the mother wavelet set in Figure 1 corresponds to the case $n = 2$. In Figure 2 we show the case $n = 4$. $\square$

We refer the reader to [70, 34, 30] for further construction methods that are based on the unitary extension principle. It is worth mentioning that in more than one variable, dyadically refinable functions may not be the prime candidates for framelet constructions. The refinement mask on the one hand has a relatively large spectrum, and the relevant extension matrix has, on the other hand, 2^d columns, forcing one to use many mother wavelets in the construction. Thus, dilation matrices with small determinants may be preferred. For example, the Powell-Zwart element ϕ (which is a bivariate C^1 piecewise-quadratic spline supported in an octagon that lies in $[0, 3]^2$, cf. [9]) is refinable with respect to the dilation matrix

$$\begin{pmatrix} 1 & -1 \\ 1 & 1 \end{pmatrix}.$$

The mask has only four terms (and is identical to the dyadic mask of the support function of $[0, 1]^2$), and there are only two columns in the extension matrix. In [70], two different wavelet systems are derived from that function: one with three mother wavelets and the other with two mother wavelets. It is worth mentioning that the shifts of this ϕ do not form a Riesz basis, and this explains the essential lack of prior constructions based on this function (the only exception I know is the 4-direction frames that were constructed in [17] by oversampling. However, the dual systems of those frames do not have compact support). $\square$

An added flexibility to the construction of framelets is obtained when one uses MRA to construct bi-frames *i.e.*, a wavelet frame together with a dual wavelet frame. The setup and development is similar to the 'tight case.' One starts with *two* refinable functions ϕ and φ with refinement masks τ_0 and t_0, and attempts to extend each τ_0 and t_0 to an n-column vector (τ and t, respectively). Two wavelet systems are then constructed. The first is $X(\psi_1, \ldots, \psi_n)$ with

$$\widehat{\psi_i}(2\cdot) := \tau_i \widehat{\phi}, \ i = 1, \ldots, n,$$

and the second is $X(\mathrm{R}\psi_1, \dots, \mathrm{R}\psi_n)$, where

$$\widehat{\mathrm{R}\psi_i}(2\cdot) := \widehat{t_i}\,\widehat{\varphi}.$$

The mixed extension principle: Constructing bi-framelets. Suppose that we construct two wavelet systems $X(\Psi)$ and $X(\mathrm{R}\Psi)$ via the extensions of τ_0 and t_0 as above. Assuming that both $X(\Psi)$ and $X(\mathrm{R}\Psi)$ are Bessel systems (and imposing a mild smoothness condition on ϕ and φ), it is proved in [69] that $(X(\Psi), X(\mathrm{R}\Psi))$ forms a bi-framelet (*i.e.*, a framelet together with a dual framelet) if the following bi-orthogonality relation holds: for every $\gamma \in \{0, \pi\}^d$

$$\sum_{i=0}^{n} \tau_i \,\overline{t_i(\cdot + \gamma)} = \begin{cases} 1, & \gamma = 0, \\ 0, & \text{otherwise.} \end{cases}$$

See [69, 71, 30] for more details as well as specific examples of bi-framelets.

Quasi-wavelet systems may not be a mere theoretical tool. These systems were introduced in [68] for the sole purpose of the eventual fiberization of the wavelet operators. However, experiments that were done (independently) with 'translation-invariant' wavelet systems [14] revealed superior results compared to the standard wavelet systems. Since those latter systems are (essentially) dilations of quasi-wavelet systems, there might be intrinsic promise in quasi-wavelet systems. At present, we lack a theory that explains the results of [14]. □

Framelets and their extension principles possess great potential in several areas of applications: feature detection, noise removal, image compression, and possibly for solving PDEs. All these applications require the implementation of the system with the aid of a *fast transform*. Most of these applications require the system to have good *approximation order*. These aspects of framelets are dealt with in [30].

It is worthwhile to note that a different type of wavelet fiberization technique appears in [75, 76]. The results there appear helpful in analyzing oversampling systems, when the oversampling ratio is even.

3. The Cascade/Transfer and the Subdivision Operators

This part is devoted to the study of refinable functions via the exploitation of two relevant operators: the cascade and the subdivision (with the transfer operator being the 'Hermitian form' of the cascade operator, hence suitable for efficient treatment of L_2-problems; cf. Section 3.4 for the precise meanings of that). There is already a very rich literature devoted to this approach, and it is beyond the scope of this section to review that literature to any extent. Instead, my goal here is to highlight a specific approach: a treatment that is based on the existing tools from and the acquired knowledge in the theory of shift-invariant spaces: first and foremost, quasi-interpolation basics (cf. [77, 10, 2, 3, 11, 46, 9]). I do not claim any of the results in this part to be novel, although assume that some of them *are* new. The proofs given (whenever given) are not borrowed from elsewhere, but, again, similar arguments might already exist somewhere in the literature.

As stated before, we prefer to carry out the analysis under simplifying assumptions:

1) *The dilation is dyadic.* The extensions of the results to other dilations are almost entirely notational. If a general dilation matrix s is employed, one should use dilation by s on the original domain, and by s^* on the Fourier domain. The group $\Gamma = \{0, 2\pi\}^d = 2\pi(\mathbb{Z}^d/2\mathbb{Z}^d)$ should be replaced by $2\pi(\mathbb{Z}^d/s^*\mathbb{Z}^d)$, and the dyadic lattices

$$\mathcal{Z}_k := \mathbb{Z}^d/2^k \qquad (3.0.1)$$

should be replaced by the s-adic lattices $s^{-k}\mathbb{Z}^d$. The only place I am aware of where the results do not carry over to general dilations is in the context of *smoothness*: there, one needs to assume that the dilation is isotropic, *i.e.*, that all the eigenvalues of s have the same modulus. Without this assumption, one can only get upper and lower bounds on the smoothness (cf. [16]).

2) *The refinable ϕ is scalar-valued* (rather than vector-valued). This assumption simplifies substantially the *notations*; however, it should be stressed that many of the arguments here can be easily carried over to that setup (specifically, the treatment of convergence and the treatment of smoothness).

3) *The refinable ϕ is compactly supported.* Some of the results are valid without this assumption; however, one then loses a major component of the analysis, viz. that the underlying spaces of interest are *finite dimensional.* So, this assumption should be regarded as essential.

Unless explicitly stated, we do *not* assume that the mask τ (cf. (3.1.1)) is a trigonometric polynomial. With rare exceptions, such an assumption leads neither to improved results nor to simpler arguments. It is worth noting that, since we assume ϕ to have compact support, the mask τ is necessarily a rational trigonometric polynomial (cf. [63]). Throughout the entire analysis we do assume (without further mentioning) that τ is *bounded*, and that $\widehat{\phi}(0) = \tau(0) = 1$.

3.1. The Transfer Operator: Stability and Related Properties

Let ϕ be a compactly supported refinable distribution, *i.e.*,

$$\widehat{\phi}(2\cdot) = \tau\widehat{\phi}, \qquad (3.1.1)$$

for some 2π-periodic τ. Set

$$m := |\tau|^2.$$

The (Fourier transform version of the) transfer operator $\mathcal{T} := \mathcal{T}_m$ is defined as

$$\mathcal{T} : f \mapsto \sum_{\gamma \in \Gamma} E^\gamma \mathcal{D}^{-1}(mf),$$

with

$$\Gamma := \{0, 2\pi\}^d, \quad E^t : f \mapsto f(\cdot - t), \quad \mathcal{D}^{-1}f = f(\cdot/2).$$

Thus, in one variable, for instance,

$$T f = (mf)(\frac{\cdot}{2}) + (mf)(\frac{\cdot}{2} + \pi).$$

Before we begin the discussion, we remind the reader of the following basic consequence of Poisson's summation formula:

Lemma 3.1.2. *Let f be a function in the Wiener algebra $A(\mathbb{R}^d)$, i.e., $\widehat{f} \in L_1(\mathbb{R}^d)$. Then the 2π-periodization $\sum_{j \in 2\pi \mathbb{Z}^d} E^j \widehat{f}$ of $\widehat{f}$ lies in the span of $\{e_{i\alpha} : f(\alpha) \neq 0\}$. Here,*

$$e_\theta : \omega \mapsto e^{\theta \cdot \omega}. \tag{3.1.3}$$

Proof: The Fourier coefficients of the 2π-periodization of $\widehat{f}$ are, up to a multiplicative constant, the values of f at the integers. $\square$

The most useful property of the transfer operator in wavelet analysis is the following lemma. It is somewhat awkward to state its most general case, hence I put instead two separate statements (that together suffice for the subsequent applications):

Lemma 3.1.4. *Let T be the transfer operator of a refinable compactly supported distribution ϕ.*

1) *Let ν be a compactly supported distribution, and assume that $\widehat{\nu}|\widehat{\phi}|^2 \in L_1(\mathbb{R}^d)$. Let $\Phi_\nu \in L_1(\mathbb{T}^d)$ be the 2π-periodization of $\widehat{\nu}|\widehat{\phi}|^2$*

$$\Phi_\nu := \sum_{j \in 2\pi \mathbb{Z}^d} E^j (\widehat{\nu}|\widehat{\phi}|^2). \tag{3.1.5}$$

Then, for every k, $T^k(\Phi_\nu)$ is the 2π-periodization of $\widehat{\nu}(\frac{\cdot}{2^k})|\widehat{\phi}|^2$.

2) *Assume $\phi \in L_2(\mathbb{R}^d)$. Let $t \in L_\infty(\mathbb{R}^d)$. Then, with F_t the 2π-periodization of $t|\widehat{\phi}|^2$, the function $T^k(F_t)$ is the 2π-periodization of $t(\frac{\cdot}{2^k})|\widehat{\phi}|^2$.*

3) *Assume $\phi \in L_2(\mathbb{R}^d)$. Let Φ be the 2π-periodization of $|\widehat{\phi}|^2$. Then $(1, \Phi)$ is an eigenpair of T.*

We omit the simple proof of the first part; the second part is merely another variant of the first (with the same proof). Of course, the third part is the special case of the first one corresponding to the choice $\nu := \delta$.

Here are some illustrations of the power of the above lemma:

Corollary 3.1.6. [73]. *Let T be the transfer operator of the compactly supported refinable distribution ϕ. Let $Z_\phi := \mathbb{Z}^d \cap \text{supp}\,\varphi$, with φ the autocorrelation of ϕ, i.e., the compactly supported distribution whose Fourier transform is $|\widehat{\phi}|^2$. Set (cf. (3.1.3))*

$$H := \text{span}\{e_{i\alpha} : \alpha \in Z_\phi\}. \tag{3.1.7}$$

Then, given a function of the form Φ_ν *(as defined in* (3.1.5)*) there exists* k_0 *(that depends only on* diam supp ν*) such that* $T^k(\Phi_\nu) \in H$ *for every* $k \geq k_0$.

Proof: By Lemma 3.1.4, $T^k(\Phi_\nu)$ is the 2π-periodization of $\widehat{\nu}(\frac{\cdot}{2^k})|\widehat{\phi}|^2$. Note that the inverse transform of that latter function is (up to a constant) the convolution $\mathcal{D}^k \nu * \varphi$. By choosing a sufficiently large k, we can ensure that the support of $\mathcal{D}^k \nu$ lies in a sufficiently small neighborhood of the origin. The result then follows from Lemma 3.1.2. $\square$

We will use in the next result, as well as in some subsequent results, the following definition:

Definition 3.1.8. The E-condition and the weak E-condition. *We say that a linear endomorphism* S *on a finite dimensional space satisfies the* weak E-condition *if the spectral radius of* S *is 1, and all its eigenvalues on the unit circle are non-defective. We further say that* S *satisfies the* E-condition *if, in addition, 1 is the unique eigenvalue on the unit circle and is simple.*

Under the polynomiality assumption on the mask, the next result can be found in [52].

Theorem 3.1.9. *Let* T *be the transfer operator of a compactly supported refinable distribution* ϕ. *Let*

$$H_\phi \tag{3.1.10}$$

be the largest T-*invariant subspace of* H, *and let*

$$T_\phi$$

be the restriction of T *to* H_ϕ. *If* T_ϕ *satisfies the weak E-condition then* $\phi \in L_2(\mathbb{R}^d)$.

Remarks. 1) Note that the result does not assume the mask to be a (trigonometric) polynomial. 2) The non-defectivity assumption on the dominant eigenvalues is necessary: The first derivative ϕ' of Daubechies' first scaling function ϕ [25, 27] is a suitable example ($\phi \notin W_2^1(\mathbb{R})$), while the spectral radius of $T_{\phi'}$ is 1. The 'culprit' is the eigenvalue 1 which is defective). 3) We also note that the converse of this statement is false. [73] shows that the spectral radius of the transfer operator T_ϕ of a compactly supported refinable $\phi \in L_2(\mathbb{R})$ can be as large as one wishes (the corresponding masks in the examples there are polynomial; in the case where the mask is polynomial and the dilation is dyadic, we have that $H = H_\phi$).

Proof: Note that the weak E-condition guarantees that, given any $f \in H_\phi$, the sequence $(T^k f)_k$ is bounded (in any norm, since H_ϕ is finite dimensional).

Let ν be a compactly supported function such that (i) $\widehat{\nu}|\widehat{\phi}|^2 \in L_1(\mathbb{R}^d)$, (ii) $\widehat{\nu}(0) = 1$, (iii) $\widehat{\nu} \geq 0$ everywhere, (iv) $\widehat{\nu} \in L_\infty$. Then, as $k \to \infty$,

$$\|\widehat{\nu}(\frac{\cdot}{2^k})|\widehat{\phi}|^2\|_{L_1(\mathbb{R}^d)} \to \|\widehat{\phi}\|^2_{L_2(\mathbb{R}^d)}.$$

However, Lemma 3.1.4 together with the non-negativity of $\widehat{\nu}$ implies that

$$\|\widehat{\nu}(\frac{\cdot}{2^k})|\widehat{\phi}|^2\|_{L_1(\mathbb{R}^d)} = \|T^k(\Phi_\nu)\|_{L_1(\mathbb{T}^d)},$$

while Corollary 3.1.6 guarantees $T^k(\Phi_\nu)$ to lie, for all sufficiently large k, in the domain H_ϕ of T_ϕ. Hence, by the weak E-condition, $(T^k \Phi_\nu)_k$ is bounded in H_ϕ (say, in the $L_1(\mathbb{T}^d)$-norm). Thus $\|\widehat{\phi}\|_{L_2(\mathbb{R}^d)} < \infty$. $\square$

The next result was proved first in [51], under the assumption that the mask is polynomial; see also [16]. We remind the reader that for a compactly supported $\phi \in L_2(\mathbb{R}^d)$, the stability of $E(\phi)$ is characterized by the positivity *everywhere* of the function Φ from Lemma 3.1.4 (cf. [77, 24, 47, 48, 5]).

Theorem 3.1.11. [65]. *Let ϕ be a compactly supported refinable distribution. Then the following conditions are equivalent:*

(i) *$\phi \in L_2(\mathbb{R}^d)$ and the shifts $E(\phi)$ of ϕ are stable.*

(ii) *T_ϕ satisfies the E-condition, and there is an eigenvector of the eigenvalue 1 which is positive everywhere.*

(iii) *T_ϕ satisfies the weak E-condition, 1 is a simple eigenvalue of T_ϕ, and an eigenvector of it is positive everywhere.*

Proof: We prove that (i)$\Longrightarrow$(ii)$\Longrightarrow$(iii)$\Longrightarrow$(i).

The implication (ii)$\Longrightarrow$(iii) is trivial. Also, assuming (iii), we conclude from Theorem 3.1.9 that $\phi \in L_2(\mathbb{R}^d)$; hence (Lemma 3.1.4) that $(1, \Phi)$ is an eigenvector of T. In view of the assumption in (iii), this implies that $c\Phi > 0$ for some constant c, which must be positive since $\Phi \geq 0$.

Now, assume (i) and let f be any trigonometric polynomial such that $f(0) = 0$. Since $\Phi > 0$ everywhere, we can write $f = t\Phi$, with $t \in C(\mathbb{T}^d)$ and $t(0) = 0$. By Lemma 3.1.4, $T^k f$ is the 2π-periodization of $t(\frac{\cdot}{2^k})|\widehat{\phi}|^2$, the latter converges to 0 pointwise (since t is continuous at the origin and vanishes there). The dominated convergence theorem then implies that $\|t(\frac{\cdot}{2^k})|\widehat{\phi}|^2\|_{L_1(\mathbb{R}^d)} \to 0$, hence that $\|T^k f\|_{L_1(\mathbb{T}^d)} \to 0$. Since the subspace of all trigonometric polynomials that vanish at 0 has co-dimension 1 in the space of all trigonometric polynomials, this implies that at most one eigenvalue of T_ϕ lies outside the open unit disc, and that this eigenvalue, if it exists, must be algebraically simple. The E-condition then follows from the fact that, Lemma 3.1.4, $(1, \Phi)$ is an eigenpair of T_ϕ. The positivity of the eigenvector follows directly from the stability assumption. $\square$

3.2. The Cascade and Subdivision Algorithms Always Converge

Let

$$\mathcal{Q}_k$$

be the space of all complex-valued sequences defined on $\mathcal{Z}_k := \mathbb{Z}^d/2^k$ (*i.e.*, $\mathcal{Q}_k = \mathbb{C}^{\mathcal{Z}_k}$). Also set

$$\mathcal{Q} := \mathcal{Q}_0.$$

Given a sequence λ defined (at least) on the lattice $\mathcal{Z}_k$, and a compactly supported distribution f, we define the k-scale semi-discrete convolution as

$$f *'_k \lambda := \sum_{j \in \mathbb{Z}^d} \lambda(2^{-k}j)f(2^k \cdot -j).$$

Note that $f *'_k \lambda$ is actually a linear combination of the $\mathcal{Z}_k$-shifts of $\mathcal{D}^k f$, with coefficients λ. Also,

$$f *' \lambda := f *'_0 \lambda.$$

We have preferred in the previous section to carry out the analysis on the Fourier domain. Since the results of this section target L_p-norms for $p \neq 2$, we must switch to the time/space domain. Thus, the refinability assumption now reads as

$$\phi = \phi *'_1 a, \tag{3.2.1}$$

with a a sequence defined on the half-integers $\mathcal{Z}_1$, still referred to the mask (and which essentially comprises the Fourier coefficients of the previous mask τ).

The transfer operator is a 'folded version' of another operator (the connection is made explicit in Section 3.4). Since the iterations of the latter operator form the *cascade algorithm* (see below), we have chosen to name the operator the *cascade operator*.

The cascade algorithm. The cascade algorithm aims at computing the refinable ϕ much in the same way the power method computes an eigenvector of a matrix. Starting with some initial compactly supported function f_0, it generates a sequence of functions $(f_k)_k$ by applying the fixed point iteration

$$f_k := C^k f_0 = Cf_{k-1} := f_{k-1} *'_1 a,$$

with a the mask of ϕ. The cascade operator is then the map

$$C : f \mapsto f *'_1 a.$$

Given $1 \leq p \leq \infty$ and $\alpha > 0$, we say that the cascade algorithm converges in the p-norm on the function set F at a rate α if, for every $f \in F$,

$$\|C^k f - \phi\|_{L_p(\mathbb{R}^d)} \leq \text{const}_f \, 2^{-\alpha k}.$$

In order to analyze the cascade algorithm, I find it convenient to assume that the initial function f above is of the form $f := g *' u$, with u some finitely supported element of $\mathcal{Q}$ and g some fixed function. For example, g can be taken to be the support function of $[0,1]^d$, or a tensor product spline, a box spline, etc. The cascade algorithm is intimately tied to the issue of *quasi-interpolation from shift-invariant spaces* (cf. [11, 9]). It is not hard to see that

$$C^k(g *' u) = g *'_k C^k u, \tag{3.2.2}$$

with

$$\mathcal{C} : u \mapsto a * \mathcal{D}u,$$

i.e., $\mathcal{C}u(j/2) = \sum_{n \in \mathbb{Z}^d} a(n/2)u(j - n)$ and $\mathcal{C}(\mathcal{Q}_k) \subset \mathcal{Q}_{k+1}$. Approximation theory basics then tell us that, in order for (3.2.2) to approximate ϕ at a rate α, three conditions should be satisfied:

(i) The shifts $E(g)$ of g should provide high approximation order. For *convergence* only, approximation order 1 (*i.e.*, partition of unity) suffices. If we are interested in an α-rate of convergence, we should choose g to provide approximation order $\geq \alpha$.

(ii) We can get approximation rate α in the p-norm only if $\phi \in W_p^\alpha(\mathbb{R}^d)$. We discuss the smoothness of refinable functions in the next subsection. (Warning: for general anisotropic dilations, the rate of convergence may only be a fraction of the smoothness parameter).

(iii) The coefficient sequence u_k in the approximation $\phi \approx g *'_k u_k$ should be selected according to a *quasi-interpolation rule*, [11, 9]. One selects a compactly supported ν such that $1 - \widehat{\nu}\widehat{g}$ has a zero at the origin of order $\geq \alpha$ and that ν is sufficiently smooth (so that $\mathcal{D}^k\nu * \phi \in A(\mathbb{R}^d)$), and defines u_k as the restriction to $\mathcal{Z}_k$ of $2^{kd}\mathcal{D}^k\nu * \phi$.

The first condition is entirely benign: after all, the selection of g is within our control. The second condition belongs to another topic, viz., the smoothness of refinable functions; it establishes the actual upper bound on any attempt to get fast convergence with the cascade algorithm. The interesting condition is the last one, and one may initially look at this requirement with utmost despair. After all, the sequence u_k is determined by the cascade algorithm as $\mathcal{C}^k u$, and we can only control the initial u. Fortunately, the counterpart of Lemma 3.1.4 changes despair into joy: the cascade iterations respect the rules of quasi-interpolation.

Lemma 3.2.3. *Let ϕ be a refinable distribution with mask a. Let ν be a compactly supported distribution such that $\nu * \phi \in A(\mathbb{R}^d)$, and set $u := (\nu * \phi)_{|_{\mathbb{Z}^d}}$. Then $\mathcal{C}^k u$ is the restriction to $\mathcal{Z}_k$ of $2^{kd}\mathcal{D}^k\nu * \phi$.*

Thus we have the following result:

Theorem 3.2.4. The cascade algorithm always converges fast [59].
Let ϕ be a refinable function that lies in $W_p^\alpha(\mathbb{R}^d)$ for some $1 \leq p \leq \infty$ and some $\alpha > 0$. Let ν be some compactly supported distribution so that $\widehat{\nu}\widehat{\phi} \in L_1(\mathbb{R}^d)$, and set $u := (\nu * \phi)_{|_{\mathbb{Z}^d}}$. Let g be a compactly supported bounded function whose shifts provide approximation order $n \geq \alpha$, and let β be the order of the zero $1 - \widehat{\nu}\widehat{g}$ has at the origin. Then, choosing the initial seed to be $g *' u$, the cascade algorithm converges in the p-norm to ϕ at a rate $\min\{\alpha, \beta\}$. Moreover, given any $m < \alpha$ and assuming $g \in W_\infty^m(\mathbb{R}^d)$, the cascade algorithm converges on $g *' u$ to ϕ in the $W_p^m(\mathbb{R}^d)$-norm at a rate $\min\{\alpha - m, \beta\}$.

The usefulness of the above result depends on the ability to compute a good initial sequence u. If, for example, ϕ is continuous, we can choose ν to be supported on $\mathbb{Z}^d$, in a way that $1 - \widehat{g}\widehat{\nu}$ has a zero of order n at the origin (and with n, say, any number $\geq \alpha$). Then, in order to implement the above theorem we need to find $u_0 := \phi_{|_{\mathbb{Z}^d}}$. Note the $(1, u_0)$ is an eigenpair of the operator

$$\mathcal{C}_0 : u \mapsto (\mathcal{C}u)_{|_{\mathbb{Z}^d}};$$

(this eigenpair is analogous to the eigenpair $(1, \Phi)$ in Lemma 3.1.4).

Discussion 3.2.5. The idea that the convergence of the cascade iterations can be accelerated by a 'smart' choice of the initial function is not entirely new. First, in [58], it is shown how, for a box spline ϕ, the convolution of $C^k \delta$ with $(\mathcal{D}^k \phi)_{|_{\mathbb{Z}^d}}$ removes undesired artifacts from the surface obtained (the actual discussion there is in terms of the subdivision operator). Second, in [28, 29], it was shown that if an initial seed f for the cascade algorithm is chosen in a way that $f - \phi$ vanishes to order n at the integers, then the cascade algorithm converges to ϕ at a rate n (provided that f and ϕ are smooth, and that $E(f)$ provide approximation order n). That technique is more restrictive. As Theorem 3.2.4 asserts, given any g whose shifts provide high enough approximation order, one can accelerate the convergence of the cascade iterations by replacing g by a suitable element in the span of $E(g)$. In contrast, only for exceptional examples of (necessarily smooth) g, one can find in the span of $E(g)$ a function f that interpolates ϕ at the integers to a high order. $\square$

We close the discussion here with the following results in which we use

$$K_\phi := \ker(\phi *') := \{\lambda \in \mathcal{Q} : \phi *' \lambda = 0\}. \tag{3.2.6}$$

Proposition 3.2.7. *Let ϕ be a compactly supported distribution (not necessarily refinable) and let $u \in \mathcal{Q}$ be finitely supported. Then the following conditions are equivalent:*
(i) $u * K_\phi = 0$.
(ii) *There exists a smooth compactly supported function ν that $(\nu * \phi)_{|_{\mathbb{Z}^d}} = u$.*

Theorem 3.2.8. *Let $1 \leq p \leq \infty$, and let $\phi \in W_p^\alpha(\mathbb{R}^d)$. Let g be a bounded compactly supported function for which $\widehat{\phi} - \widehat{g} = O(|\cdot|^n)$ near the origin. Let $u \in \mathcal{Q}$ be a finitely supported sequence such that $u * K_\phi = 0$ and $1 - \widehat{u} = O(|\cdot|^\alpha)$. If the shifts of g provide approximation order $\geq \alpha$, then the cascade algorithm converges at rate $\min\{\alpha, n\}$ on $g *' u$. Moreover, if $g \in W_\infty^m(\mathbb{R}^d)$ for some positive m, then $(C^k(g *' u))_k$ converges in $W_p^m(\mathbb{R}^d)$ to ϕ at a rate $\min\{\alpha - m, n\}$.*

Note that, if $K_\phi = 0$ (e.g., if the shifts of ϕ are orthonormal, or form a Riesz basis with a compactly supported dual basis), then we may choose $u := \delta$ in the above theorem.

The subdivision algorithm. Given the mask a of a refinable function ϕ and a sequence $\lambda \in \mathcal{Q}$, the main aim of the subdivision algorithm is to produce fast a good approximation for $\phi *' \lambda$ without computing ϕ first.

The subdivision algorithm involves the iterations of the *subdivision operator*. The subdivision literature usually assumes that the subdivision operator is an endomorphism on $\mathcal{Q}$, with the only advantage in this description that the same operator S is employed in all the iterations. I find it more convenient to align the subdivision iterations with the cascade iterations, and to assume that after k iterations the sequence obtained lives on $\mathcal{Z}_k = \mathbb{Z}^d/2^k$. This entails that the kth order subdivision operator is not exactly the kth power of the 1st order one.

The kth order subdivision operator S_k. That operator maps $\mathcal{Q}$ into $\mathcal{Q}_k$ and is defined inductively as follows: S_0 is the identity, and

$$S_k \lambda := \mathcal{D}^{k-1} a * S_{k-1} \lambda, \tag{3.2.9}$$

i.e., $S_k \lambda(j) := \sum_{n \in \mathcal{Z}_{k-1}} a(2^{k-1}(j-n)) S_{k-1} \lambda(n)$.

The standard definition for 'convergence of subdivision' [32] is also somewhat inconvenient for analysis. I prefer the following (equivalent, at least for $p = \infty$) definition, in which we use

$$G_\alpha$$

for the collection of all compactly supported functions g whose shifts provide approximation order $\geq \alpha$, and whose Fourier transform is α-flat at the origin: $1 - \hat{g}$ has a zero at the origin of order $\geq \alpha$. Examples of functions in G_α include functions whose shifts are orthonormal, and cardinal interpolants, but there are many others. In fact, given any compactly supported f whose shifts provide approximation order $\geq \alpha$, there exists $g \in G_\alpha$ which is finitely spanned by $E(f)$. Given $g \in G_\alpha$, it is well-known that, if $\phi \in W_p^\beta(\mathbb{R}^d) \cap C(\mathbb{R}^d)$ and if $\beta \leq \alpha$, then $\|\phi - g *_k' \phi\|_{L_p(\mathbb{R}^d)} \leq \text{const } 2^{-\beta k}$.

Definition 3.2.10. Convergence of subdivision. *Let ϕ be refinable with mask a, and let $1 \leq p \leq \infty$ and $\alpha > 0$. Given a subset Q of $\mathcal{Q}$, we say that the subdivision converges on Q in the p-norm at α rate if for any $g \in G_\alpha$, for every $q \in Q$, and for every compact set K,*

$$\|(\phi *' q) - (g *_k' S_k q)\|_{L_p(K)} \leq \text{const}_{q,K} \, 2^{-\alpha k}.$$

Note that S_k commutes with (integer) shifts, and hence $S_k q = S_k \delta *' q$. Thus, the subdivision converges on the entire $\mathcal{Q}$ (for some fixed p and α) iff it converges on δ. As to the convergence of the subdivision on δ, we have the following simple observation:

Proposition 3.2.11. *Let ϕ be refinable with mask a. Then, for every k, $C^k \delta = S_k \delta$, and hence the subdivision converges in the p-norm at α-rate on the entire $\mathcal{Q}$ if and only if the cascade converges at that norm and at that rate on δ (i.e., on G_α).*

The above proposition has one immediate consequence: since we can force the cascade algorithm to converge, and even at fast rates, we can do the same with the subdivision:

Theorem 3.2.12. [59]. **The subdivision always converges, and converges fast.** *Let ϕ be a refinable function that lies in $W_p^\alpha(\mathbb{R}^d)$ for some $1 \le p \le \infty$ and some $\alpha > 0$. Let ν be some compactly supported distribution so that $\widehat{\nu}\widehat{\phi} \in L_1(\mathbb{R}^d)$. In addition, let $u := (\nu * \phi)|_{\mathbb{Z}^d}$ and β be the order of the zero that $1 - \widehat{\nu}$ has at the origin. Then, $(\mathcal{D}^k u * S_k \delta)_k$ converges in the p-norm at a rate $\min\{\alpha, \beta\}$.*

The theorem implies, in particular, that for $p = \infty$,

$$\|(\phi *' q)|_{\mathbb{Z}_k} - \mathcal{D}^k u * S_k q\|_{\ell_\infty(\mathbb{Z}_k)} \le \text{const } 2^{-\alpha k},$$

provided that $\phi \in W_\infty^\alpha(\mathbb{R}^d)$, and that u is selected as in the theorem (and that $\beta \ge \alpha$). A highlight here is that the sequence u depends only on ϕ and not on the initial sequence q of the subdivision.

Note that if the subdivision algorithm converges on δ (for some p and α), it converges on each $\lambda \in \mathcal{Q}$ (in that norm and rate); in particular, it converges on each $\lambda \in K_\phi$ to $\phi *' \lambda = 0$. One of the main results of [59] provides a converse for this result:

Theorem 3.2.13. *Let ϕ be a compactly supported refinable function with finitely supported mask a. Let $1 \le p \le \infty$, and let $0 < \alpha \le 1$. Assume that $\phi \in W_p^\alpha(\mathbb{R}^d)$. Then the following conditions are equivalent:*

(a) *The cascade algorithm converges in the p-norm at an α-rate on any compactly supported initial seed f whose shifts provides approximation order 1.*

(b) *The subdivision algorithm converges in the p-norm at an α-rate on all sequences in $\mathcal{Q}$.*

(c) *The subdivision converges in the p-norm at an α-rate to zero on K_ϕ.*

The theorem leads to several important consequences. For instance, if $\phi \in W_\infty^\alpha(\mathbb{R})$ for some $\alpha > 0$, and if the subdivision converges (on all sequences) in some norm at some rate, it must converge in all other p-norms (albeit at possibly different rates). We refer to [59, 13] for further discussions.

3.3. Smoothness of Refinable Functions

Lemma 3.1.4 and Lemma 3.2.3 are key tools in almost any analysis of refinable functions that is based on the transfer operator. Here is a very brief discussion of one of the major implications of the latter lemma: smoothness of refinable functions.

Definition 3.3.1. *Let ϕ be a compactly supported distribution. Given $1 \le p \le \infty$, we define the p-smoothness parameter of ϕ, $\alpha_p(\phi)$, as follows:*

$$\alpha_p(\phi) := \sup\{\alpha \in \mathbb{R} : \phi \in W_p^\alpha(\mathbb{R}^d)\}.$$

Note that $\alpha_p(\phi)$ may be negative.

Many articles in the literature exploit the cascade/transfer operator for the analysis of smoothness. The early, univariate, results were based on the idea of *factorization* (cf. *e.g.*, [39, 78]). Riemenschneider and Shen were probably the first to provide lower bounds on the $L_2(\mathbb{R}^d)$-smoothness without factorization, and Jia [43] used that approach to characterize the L_2-smoothness of a single scaling function ϕ in several variables, under the assumptions that $E(\phi)$ are stable, that the dilation is isotropic, and that the mask is a trigonometric polynomial. The treatment of L_2-smoothness for single $\phi \in L_2(\mathbb{R}^d)$ and for non-isotropic dilations (still under the stability and polynomiality assumptions) is contained in [16]: as said, one cannot compute exactly the smoothness parameter in that general setup. That Cohen et al. article stimulated [73] where the L_2-smoothness parameter was completely characterized (several variables, isotropic dilations, positive/negative smoothness, a vector ϕ, no stability and/or polynomiality assumptions). An algorithm (and software) that implements the results of [73] is contained in [74]. There are fewer treatments of L_p-smoothness, $p \neq 2$, and, in particular, I could find only one reference that characterizes the $L_p(\mathbb{R}^d)$-smoothness, $d > 1$, [53]. The reader is referred to [73, 49, 45] for further references and discussions.

In this section, I merely wish to explain in the simplest possible terms the natural connection between the cascade iterations and the smoothness of the refinable ϕ. The treatment here is purely theoretical, and one must keep in mind, [78, 74, 80], that implementing results of this type in practical situations is non-trivial.

The connection between the cascade iterations and smoothness is clear once we compare the basic Lemma 3.2.3 with the definition of smoothness in terms of difference operators (cf. *e.g.*, [31]). Specifically, the following is one of the possible equivalent definitions of smoothness:

Proposition 3.3.2. *Let ϕ be a compactly supported distribution. In addition, let $1 \leq p \leq \infty$ and β be some positive number. Then the following conditions are equivalent:*

(a) *The smoothness exponent $\alpha_p(\phi)$ is $\geq \beta$.*

(b) *Let n be any integer $\geq \beta$. Then, for any sufficiently smooth compactly supported ν, if $\widehat{\nu}$ has a zero at the origin of order n, then, for every $\alpha < \beta$,*
$$2^{kd}\|\mathcal{D}^k \nu * \phi\|_{L_p(\mathbb{R}^d)} = O(2^{-k\alpha}).$$

The smoothness problem can now be efficiently attacked by combining the above proposition with Lemma 3.2.3 and Proposition 3.2.7. As said, the result is also valid in the FSI (vector) case, as well as for general isotropic dilations (the technique handles also negative smoothness parameters). It is convenient here to normalize the ℓ_p-norm of sequences defined on $\mathcal{Z}_k$, so that the total mass of the points in the unit cube always equals 1:

$$\|c\|^p_{\ell_p(\mathcal{Z}_k)} := 2^{-kd} \sum_{j \in \mathcal{Z}_k} |c(j)|^p.$$

We recall [11, 12, 4, 41, 50] that if ϕ is compactly supported and its shifts provide approximation order n, then it satisfies the Strang-Fix conditions of order n, i.e., $\hat{\phi}$ has a zero of order n at each $j \in 2\pi\mathbb{Z}^d\backslash 0$. It follows then from Poisson's summation formula that, if ν is a smooth compactly supported function, and if $\hat{\nu}$ has a zero of order n' at the origin, then, with $u := (\nu * \phi)_{|_{\mathbb{Z}^d}}$, $\hat{u}$ has a zero at the origin of order $\geq \min\{n, n'\}$. Moreover, if n' is the exact order of the zero of $\hat{\nu}$ at the origin, and if $n' \leq n$, then n' is also the exact order of the zero $\hat{u}$ has at the origin (provided $\hat{\phi}(0) \neq 0$).

Theorem 3.3.3. *Let ϕ be a compactly supported refinable distribution with mask a and cascade operator C. Assume $\hat{\phi}(0) = 1$. Let $1 \leq p \leq \infty$ be given, and let N be the approximation order provided by $E(\phi)$. Let K be a compact set that contains $\operatorname{supp}\phi$, and let $U \subset \mathcal{Q}$ be the space of all sequences u that satisfy the following conditions:*
  (i) *$\operatorname{supp} u \subset K$.*
  (ii) *$\hat{u}$ has an N-fold zero at the origin.*
  (iii) *$u * K_\phi = 0$.*
Let $\tilde{\alpha}$ be the supremum of all α that satisfy the following condition: '$\|C^k\| = O(2^{-k\alpha})$ as a map from U to $\ell_p(\mathcal{Z}_k)$'. Then $\tilde{\alpha} = \alpha_p(\phi)$.

Proof: We prove first the lower bound inequality $(\alpha_p(\phi) \geq \tilde{\alpha})$. Let n be any number larger than $\max\{N, \alpha_p(\phi)\}$. Also, let ν be a smooth compactly supported function of small support, whose Fourier transform has a zero of order n at the origin. By Proposition 3.2.7 and the discussion preceding the current theorem, the sequence $u_t := (E^t\nu * \phi)_{|_{\mathbb{Z}^d}}$ lies in U (provided that t lies in small neighborhood of the origin). Now, suppose that $\|C^k\| = O(2^{-k\alpha})$ as a map from U to $\ell_p(\mathcal{Z}_k)$ (note that U is finite dimensional hence the choice of its norm is immaterial here). Since $\nu * \phi$ is continuous, the sequences $(u_t)_t$ lie in some bounded subset of U, hence $\|C^k u_t\|_{\ell_p(\mathcal{Z}_k)} = O(2^{-k\alpha})$, uniformly in $t \in [0,1]^d$. But Lemma 3.2.3 asserts that $C^k u_t = 2^{kd}((\mathcal{D}^k E^t\nu) * \phi)_{|_{\mathcal{Z}_k}}$, and thus $C^k u_t$ is the restriction to $t/2^k + \mathcal{Z}_k$ of $2^{kd}\mathcal{D}^k\nu * \phi$. So we can integrate our estimate over $[0,1]^d/2^k$ to conclude that $2^{kd}\|\mathcal{D}^k\nu * \phi\|_{L_p(\mathbb{R}^d)} = O(2^{-k\alpha})$. Thus, by Proposition 3.2, $\alpha_p(\phi) \geq \alpha$.

For the converse, assume that $\phi \in W_p^\alpha(\mathbb{R}^d)$ for some $\alpha > 0$. Since 'smoothness implies approximation orders' (cf. [15, 64, 44]), we know that $\alpha < N$. We need to show that $\alpha \leq \tilde{\alpha}$, too. Since U is finite dimensional, it suffices to prove that, for each $u \in U$,

$$\|C^k u\|_{\ell_p(\mathcal{Z}_k)} = O(2^{-k\alpha}).$$

Fix such u, and let ν be a smooth compactly supported function for which $u = (\nu * \phi)_{|_{\mathbb{Z}^d}}$ (Proposition 3.2.7). By the remarks preceding this theorem, since $u \in U$, $\hat{\nu}$ must have an N-fold zero at the origin. Let B be a compactly supported function whose shifts provide approximation order N, and let μ be a compactly supported smooth function such that $1 - \hat{B}\hat{\mu}$ has an N-fold zero at the origin. Then, quasi-interpolation basics tell us that the approximation scheme $f \approx B *'_k (2^{kd} \mathcal{D}^k \mu * f)$ provides approximation order N (cf. [9]), hence, by standard interpolation arguments (and since $N \geq \alpha$), we have that

$$\|\phi - B *'_k (2^{kd} \mathcal{D}^k \mu * \phi)\|_{L_p(\mathbb{R}^d)} = O(2^{-k\alpha}).$$

The above also holds when μ is replaced by $\mu + \nu$, hence we obtain that

$$\|B *'_k (2^{kd} \mathcal{D}^k \nu * \phi)\|_{L_p(\mathbb{R}^d)} = O(2^{-k\alpha}).$$

However, by Lemma 3.2.3, $B *'_k (2^{kd} \mathcal{D}^k \nu * \phi) = B *'_k C^k u$. Assuming, without loss, that the shifts of B are stable in the p-norm (cf. [47]), we conclude that $\|C^k u\|_{\ell_p(\mathcal{Z}_k)} = O(2^{-k\alpha})$. $\square$

As a nice application, we obtain the following result. The case $n = 1$ in this result is essentially very well known (cf. [33, 42, 57, 36]).

Corollary 3.3.4. *Let ϕ be compactly supported and refinable and assume that $\widehat{\phi}(0) = 1$. Let n be a positive integer such that (i) $E(\phi)$ provide approximation order $\geq n$, and (ii) $1 - \widehat{\phi}$ has a zero of order $\geq n$ at the origin. Let K be a compact set that contains $\mathrm{supp}\,\phi$, and let $U_0 \subset \mathcal{Q}$ be the space of all sequences u that satisfy the following conditions:*
(i) $\mathrm{supp}\,u \subset K$.
(ii) $\widehat{u}$ has a zero of order $\geq n$ at the origin.
Let $0 < \alpha \leq n$ and $1 \leq p \leq \infty$ be given. Then the subdivision converges in the p-norm at any rate $< \alpha$ on the entire sequence space $\mathcal{Q}$ whenever the following condition holds: "With C the cascade operator associated with ϕ, we have that $\|C^k\| = O(2^{-k\alpha})$, when considering C^k as a map from U_0 to $\ell_p(\mathcal{Z}_k)$."

Proof: Let $\alpha' < \alpha$. We need to show that for each $g \in G_\alpha$, $(g *'_k C^k \delta)_k$ converges in the p-norm at a rate α' to ϕ.

First, since the space U_0 here is a superspace of the space U of Theorem 3.3.3, we can invoke the latter to conclude that, under the current assumptions, $\phi \in W_p^{\alpha'}(\mathbb{R}^d)$. Now, let u be a sequence supported on K such that $1 - \widehat{u}$ has a zero of order n at the origin, and such that $u * K_\phi = 0$. (We are tacitly assuming that K contains such a sequence; a suitable K is *e.g.*, $\mathrm{supp}\,\phi + [0, n-1]^d$.) By Theorem 3.2.8, the cascade iterations converge on $g *' u$ to ϕ at a rate α' in the p-norm. On the other hand, $\delta - u \in U_0$, hence, by our assumption here $\|C^k(\delta - u)\|_{\ell_p(\mathcal{Z}_k)} = O(2^{-k\alpha})$, hence the cascade iterations converge to 0 on $g *' (\delta - u)$ at a rate α. Thus, those iterations converge to ϕ on $g = g *' \delta$ at a rate α'. $\square$

The cascade operator can be represented as the composition of 2^d basic operators: with $j \in \mathcal{Z}_1 \cap [0, 1)^d$, the jth component C_j of C is

$$C_j c := (E^{-j} C c)_{|_{\mathbb{Z}^d}}.$$

Using this approach, one may interpret the condition

$$\|C^k\| = O(2^{-k\alpha})$$

(with C^k viewed as an operator from U to $\ell_p(\mathcal{Z}_k)$) that appears in Theorem 3.3.3 as an equivalent statement on the joint spectral radius of these 2^d operators (acting on U: that joint spectral radius should be in the p-norm $< 2^{-\alpha}$). I forego providing further details in this regard, since, at the time this article is written, I am not convinced that the formulation of the previous theorem in that equivalent language provides a more efficient venue compared to the straightforward attempt of estimating $\|C^k u\|_{\ell_p(\mathcal{Z}_k)}$ for $k = 1, 2, 3, \ldots$, and with u varies over a basis for U.

Remark. In the above analysis, we took into account the fact that each point in $\mathcal{Z}_k$ represents a shift of the dilated cube $[0,1]^k/2^k$, and that the L_p-norm of that dilated cube is $2^{-k/p}$. We thus attached mass $2^{-k/p}$ to each point in $\mathcal{Z}_k$ when defining the $\ell_p(\mathcal{Z}_k)$-norm. For $p < \infty$, this removes the unnecessary artifact in the original definition of the p-joint spectral radius ([79, 42]). For example, in these terms, the subdivision algorithm converges on the entire $\mathcal{Q}$ space in the L_p-norm if and only if the p-joint spectral radius on U_0 is < 1, [42, 36].

Remark. The formulation of Theorem 3.3.3 in terms of the kernel K_ϕ is convenient, especially since we do not assume ϕ to be continuous, hence cannot restrict it or a translate of it to the integers. However, it might be hard, in general, to compute K_ϕ; the alternative is to invoke Proposition 3.2.7 and to compute instead all sequences of the form $(\nu * \phi)|_{\mathbf{Z}^d}$. In case ϕ is continuous, that may not be hard (cf. [40] and the discussion in the next section). Since, in general, we do not know in advance whether ϕ is continuous, we may instead try the following idea, which is an adaptation of the approach used in [53, 45], and which was suggested to me by D. X. Zhou [80]. If we choose ν to be a well-understood smooth refinable function (e.g., a box spline), and if we guarantee that $\nu * \phi$ *is* continuous, we should be able to compute $u_\nu := (\nu * \phi)|_{\mathbf{Z}^d}$ (cf. the discussion after Theorem 3.2.4). By varying the above ν (and applying suitable difference operators to each so-obtained u_ν), one may hope to get a spanning set for the space U in Theorem 3.3.3. A rigorous treatment using this approach has yet to be found.

3.4. Miscellaneous Results

The connection between the transfer operator and the cascade operator. We started the discussion of the second part of this article with L_2-analysis via the transfer operator. We then presented the L_p-approach via the cascade operator. As said, the two operators are intimately related. Indeed, it is straightforward to prove the following (where we define $\widehat{a}(\omega) := \sum_{j \in \mathcal{Z}_1} a(j) e^{ij \cdot 2\omega}$):

Proposition 3.4.1. *Let* ϕ *be refinable with mask* a *and set* $m := 2^{-d}|\widehat{a}|^2$. *Let* C *be the cascade operator associated with* a *and let* T *be the transfer operator associated with* m. *Given any* $c \in \ell_2(\mathbb{Z}^d)$, *we have*

$$\|C^k c\|^2_{\ell_2(\mathcal{Z}_k)} = \|T^k(|\widehat{c}|^2)\|_{L_1(\mathbb{T}^d)}.$$

In view of the fact that T is an endomorphism on the space H_ϕ (cf. Section 3.1), while C is not an endomorphism of any non-trivial space, it is preferable to use the transfer operator for studies in the L_2-norm. For example, the contractivity assumption on the space U in Corollary 3.3.4 is equivalent, when $p = 2$, to the condition

$$\|(T_{|H_0})^k\| = O(2^{-2k\alpha}),$$

with $H_0 := \{f \in H : f(0) = 0\}$ (with H as in (3.1.7)). This eventually leads to the characterization of the convergence of the cascade/subdivision algorithms in terms of the E-condition on T (cf. [52]). Similar remarks can be made with respect to the smoothness problems. Usually it is easier to get the L_2-results directly from the transfer operator, compared to the alternative of converting the cascade operator results via Proposition 3.4.1.

Linear independence. The recent papers [37, 40] suggest an interesting way for analyzing the *local linear independence* of the shifts of a refinable function ϕ. To recall, given a compactly supported (not necessarily refinable) ϕ, and an open set $\Omega \subset \mathbb{R}^d$, we say that the shifts of ϕ are independent on Ω if, whenever

$$\phi *' q = 0, \quad \text{on } \Omega$$

for some $q \in \mathcal{Q}$, we have that $q(j)\phi(\cdot - j) = 0$ on Ω, for every $j \in \mathbb{Z}^d$. We assume, for simplicity, that $\phi \in C(\mathbb{R}^d)$ (this allows us to choose $\Omega := [0, 1]^d$, though that set is not open).

Let us look closer at this problem. Let

$$Z_\Omega := \{j \in \mathbb{Z}^d : \Omega \cap (j + \operatorname{supp} \phi) \neq \emptyset\}.$$

Note that $q(j)\phi(\cdot - j) = 0$ on Ω unless $j \in Z_\Omega$. Now, if $\phi *' q = 0$ on Ω, then for every $x \in \Omega$,

$$\sum_{j \in Z_\Omega} q(j)\phi(x - j) = 0.$$

So, with

$$\phi_x : Z_\Omega \to \mathbb{C} : j \mapsto \phi(x - j),$$

the question is whether or not $(\phi_x)_{x \in \Omega}$ span

$$U_\Omega := \mathbb{C}^{Z_\Omega}.$$

In [37, 40] the following idea was devised for finding the local dependence relations of ϕ on Ω. If we find first the vector $u := \phi_{|\mathbb{Z}^d}$ (cf. the discussion after Theorem 3.2.4), then, by Lemma 3.2.3, $C^k u = \phi_{|z_k}$. The iterations thus provide us eventually with the sequences ϕ_x, x dyadic, which suffice here since ϕ is continuous. The only remaining practical problems are: (i) to compute the initial u, something that usually is not hard when ϕ is continuous, and (ii) determining a stopping criterion for the iterations: the tree structure of the cascade operator (which we have largely ignored) entails that, for $\Omega := [0, 1]^d$, we stop exactly when

$$\operatorname{span}\{\phi_x : x \in \mathcal{Z}_k \cap \Omega\} = \operatorname{span}\{\phi_x : x \in \mathcal{Z}_{k+1} \cap \Omega\}.$$

One can also devise a stopping criterion when, *e.g.*, Ω is a box whose corners lie in some $\mathcal{Z}_k$. I do no know of a strategy for choosing a stopping criterion for a general Ω.

The local linear independence is nicely connected with the problem of *global* linear independence: this is the case when $K_\phi = 0$ (cf. (3.2.6)). We recall [23, 62] that the shifts of a compactly supported distribution ϕ are globally linearly dependent if there exists an exponential $\xi \in K_\phi$. Here,

$$\xi : j \mapsto \xi^j, \quad j \in \mathbb{Z}^d,$$

and $\xi \in (\mathbb{C}\backslash 0)^d$. The following connection between global independence and local independence is a consequence of that characterization:

Lemma 3.4.2. *Let ϕ be a compactly supported continuous function. Let $\Omega := [0,1]^d$, and let Λ be all the local dependence relations of $E(\phi)$ on Ω:*

$$\Lambda := \{q \in \mathcal{Q} : \operatorname{supp} q \subset \Omega + \operatorname{supp}\phi, \ (\phi *' q)_{|\Omega} = 0\}.$$

Then the shifts of ϕ are globally linearly independent if and only if Λ contains a sequence that coincides on $\mathbb{Z}^d \cap (\Omega + \operatorname{supp}\phi)$ with an exponential ξ.

Proof: If q and ξ coincide on $\mathbb{Z}^d \cap (\Omega + \operatorname{supp}\phi)$, and if $q \in \Lambda$, then, on Ω, $\phi *' \xi = \phi *' q = 0$. But since ξ is an exponential, it is obvious that $\phi *' \xi = 0$ on any integer translate of Ω. Those integer translates cover $\mathbb{R}^d$, hence $\phi *' \xi = 0$ everywhere. The converse is trivial. $\square$

This leads to the following result [40]:

Corollary 3.4.3. *Let ϕ be a compactly supported continuous function, and let V be any spanning set for $\operatorname{span}(\phi_x)_{x \in [0,1]^d}$. Then $E(\phi)$ are linearly dependent if and only if there exists an exponential ξ such that $v \perp \xi$, for every $v \in V$, i.e., such that $\sum_{j \in \mathbb{Z}^d} v(j)\xi^j = 0$, for every $v \in V$.*

4. A Conjecture

I want to close this article with the following conjecture concerning the convergence of the cascade algorithm. At the time this article is written, I strongly believe it to be true, but cannot say whether it is easy or hard to resolve it. The conjecture is proved in [59] under various additional assumptions (for example, under the assumption that $K_\phi = 0$, and under the weaker assumption that $f = g *' u$, where $u * K_\phi = 0$. These results readily imply that the conjecture is true in one dimension.)

Conjecture. *Let ϕ be a compactly supported refinable function in $W_p^\alpha(\mathbb{R}^d)$, $d \geq 2$, and assume $\widehat{\phi}(0) \neq 0$ (the mask need not be finite). Let g be a compactly supported bounded function that satisfies the following three conditions:*
(a) *The shifts of g provide approximation order $\geq \alpha$ (and $\widehat{g}(0) = 1$).*
(b) *$\widehat{\phi} - \widehat{g} = O(|\cdot|^\alpha)$ near the origin.*
(c) *$K_\phi \subset K_g$.*
Then the cascade algorithm converges on g to ϕ in the p-norm at rate α.

312   *A. Ron*

Acknowledgments. I am indebted to Mike Neamtu, Ingrid Daubechies, Zuowei Shen, Joachim Stöckler, and Ding-Xuan Zhou who, each, made many helpful suggestions on an earlier draft of this article. I also thank Carl de Boor for his critical reading of the final version of this article, Ron DeVore for a communication concerning the notion of smoothness, and (my former advisor) Nira Dyn for discussions concerning subdivision schemes. This work is partially sponsored by the National Science Foundation under Grant DMS-9626319, and by the United States Army Research Office under Contract DAAH04-95-1-0089.

References

1. Benedetto, J. J. and D. F. Walnut, Gabor frames for L^2 and related spaces, in *Wavelets: Mathematics and Applications*, J. Benedetto and M. Frazier (eds.), CRC Press, Boca Raton, FL, 1994, pp. 97–162.

2. de Boor, C., The polynomials in the linear span of integer translates of a compactly supported function, Constructive Approximation **3** (1987), 199–208.

3. de Boor, C., Quasiinterpolants and approximation power of multivariate splines, in *Computation of Curves and Surfaces*, M. Gasca and C. A. Micchelli (eds.), Kluwer Academic, Dordrecht, Netherlands, 1990, pp. 313–345.

4. de Boor, C., R. A. DeVore, and A. Ron, Approximation from shift-invariant subspaces of $L_2(\mathbb{R}^d)$, Trans. Amer. Math. Soc. **341** (1994), 787–806. Ftp: ftp://ftp.cs.wisc.edu/Approx file: l2shift.ps

5. de Boor, C., R. A. DeVore, and A. Ron, The structure of finitely generated shift-invariant subspaces of $L_2(\mathbb{R}^d)$, J. Functional Anal. **119** (1994), 37–78. Ftp: ftp://ftp.cs.wisc.edu/Approx file: several.ps

6. de Boor, C., R. A. DeVore, and A. Ron, On the construction of (pre)wavelets, Constructive Approximation, Special Issue on Wavelets **9** (1993), 123–166. Ftp: ftp://ftp.cs.wisc.edu/Approx file: wavelet.ps

7. de Boor, C., R. A. DeVore, and A. Ron, Approximation orders of FSI spaces in $L_2(\mathbb{R}^d)$, Constr. Approx., to appear. Ftp: ftp://ftp.cs.wisc.edu/Approx file: BDR4.ps

8. de Boor, C. and K. Höllig, B-splines from parallelepipeds, J. d'Anal. Math. **42** (1982/3), 99–115.

9. de Boor, C., K. Höllig, and S. D. Riemenschneider, *Box splines*, Springer Verlag, New York, 1993.

10. de Boor, C. and R. Q. Jia, Controlled approximation and a characterization of the local approximation order, Proc. Amer. Math. Soc. **95** (1985), 547–553.

11. de Boor, C. and A. Ron, The exponentials in the span of the integer translates of a compactly supported function: approximation orders and quasi-interpolation, J. London Math. Soc. **45** (1992), 519–535. Ftp: ftp://ftp.cs.wisc.edu/Approx file: quasi.ps

12. de Boor, C. and A. Ron, Fourier analysis of approximation orders from principal shift-invariant spaces, Constr. Approx. **8** (1992), 427–462. Ftp: ftp://ftp.cs.wisc.edu/Approx file: aoinfty.ps

13. de Boor, C. and A. Ron, Box splines revisited: convergence and acceleration methods for the subdivision and the cascade algorithms, 1998, manuscript.

14. Coifman, R. R. and D. L. Donoho, Translation-invariant de-noising, in *Wavelets in Statistics*, Springer Verlag, New York, 1995, pp. 125–150.

15. Cavaretta, A. S., W. Dahmen, and C. A. Micchelli, *Stationary subdivision*, Memoir Amer. Math. Soc. **#453**, Providence 1991.

16. Cohen, A., K. Gröchenig, and L. Villemoes, Regularity of multivariate refinable functions, 1996, preprint.

17. Chui, C. K., K. Jetter, and J. Stöckler, Wavelets and frames on the four-directional mesh, in *Wavelets: Theory, Algorithms and Applications*, C. K. Chui, L. Montefusco, and L. Puccio (eds.), Academic Press, New York, 1994, pp. 213–230.

18. Chui, C. K. and X. Shi, Bessel sequences and affine frames, Appl. Comput. Harmonic Anal. **1** (1993), 29–49.

19. Chui, C. K. and X. Shi, Inequality of Littlewood-Paley type of frames and wavelets, SIAM J. Math. Anal. **24** (1993), 263–277.

20. Chui, C. K. and X. Shi, $N \times$ oversampling preserves any tight affine-frame for odd N, Proc. Amer. Math. Soc. **121** (1994), 511–517.

21. Chui, C. K. and X. Shi, Inequalities on matrix-dilated Littlewood-Paley functions and oversampled affine operators, SIAM J. Math. Anal. **28** (1997), 213–232.

22. Chui, C. K., X. L. Shi, and J. Stöckler, Affine frames, quasi-affine frames, and their duals, Adv. Comp. Math. **8** (1998), 1–17.

23. Dahmen, W. and C. A. Micchelli, Translates of multivariate splines, Linear Algebra and Appl. **52/3** (1983), 217–234.

24. Dahmen, W. and C. A. Micchelli, Recent progress in multivariate splines, in *Approximation Theory IV*, C. K. Chui, L. L. Schumaker, and J. Ward (eds.), Academic Press, 1983, pp. 27-121.

25. Daubechies, I., Orthonormal bases of compactly supported wavelets, Comm. Pure and Appl. Math. **41** (1988), 909–996.

26. Daubechies, I., The wavelet transform, time-frequency localization and signal analysis, IEEE Trans. Inform. Theory **36** (1990), 961–1005.

27. Daubechies, I., *Ten Lectures on Wavelets*, CBMS Conference Series in Applied Mathematics, Vol. **61**, SIAM, Philadelphia, 1992.

28. Daubechies, I. and J. C. Lagarias, Two-scale difference equations I. Existence and global regularity of solutions, SIAM J. Math. Anal. **22** (1991), 1388–1410.

29. Daubechies I. and J. C. Lagarias, Two-scale difference equations II. Local regularity, infinite products of matrices and fractals, SIAM J. Math. Anal. **23** (1992), 1031–1079.

30. Daubechies, I., A. Ron, and Z. Shen, Construction and analysis of framelets in one and several dimensions, 1998, manuscript.

31. DeVore, R. A. and G. G. Lorentz, *Constructive approximation*, Springer-Verlag, Berlin, 1993.

32. Dyn, N., Subdivision schemes in CAGD, in *Advances in Numerical Analysis Vol. II: Wavelets, Subdivision Algorithms and Radial Basis Functions*, W. A. Light (ed.), Oxford University Press, 1992, pp. 56–109.

33. Dyn, N., J. A. Gregory, and D. Levin, Analysis of linear binary subdivision schemes for curve design, Constr. Approx. **7** (1991), 127–147.

34. Gröchenig, K. and A. Ron, Tight compactly supported wavelet frames of arbitrarily high smoothness, Proc. Amer. Math. Soc. **126** (1998), 1101–1107. Ftp: `ftp://ftp.cs.wisc.edu/Approx` file: `cg.ps`.

35. Han, B., On Dual wavelet tight frames, Appl. Comp. Harmonic Anal., to appear.

36. Han, B. and R. Q. Jia, Multivariate refinement equations and subdivision schemes, SIAM J. Math. Anal. **29** (1998), 1177–1199.

37. Hardin, D. P. and T. A. Hogan, Refinable subspaces of a refinable space, 1998, preprint.

38. Hernandez, E. and G. Weiss, *A First Course on Wavelets*, CRC Press, Boca Raton, FL, 1996.

39. Herve, L., Construction et régularité des fonctions d'échelle, SIAM J. Math. Anal. **26** (1995), 1361–1385.

40. Hogan, T. A. and R. Q. Jia, Dependency relations among the shifts of a multivariate refinable distribution, 1998, preprint.

41. Jia, R. Q., The Toeplitz theorem and its applications to approximation theory and linear PDEs, Trans. Amer. Math. Soc. **347** (1995), 2585–2594.

42. Jia, R. Q., Subdivision schemes in L_p spaces, Advances in Comp. Math. **3** (1995), 309–341.

43. Jia, R. Q., Characterization of smoothness of multivariate refinable functions in Sobolev spaces, Trans. Amer. Math. Soc., to appear.

44. Jia, R. Q., Q. Jiang, and S. L. Lee, Cascade algorithms and integrals of wavelets, 1998, preprint.

45. Jia, R. Q., K. S. Lau, and D. X. Zhou, L_p-solutions of vector refinement equations, 1998, preprint.

46. Jia, R. Q. and J. J. Lei, Approximation by multiinteger translates of functions having global support, J. Approx. Theory **72** (1993), 2–23.

47. Jia, R. Q. and C. A. Micchelli, On linear independence for integer translates of a finite number of functions, Proc. Edinburgh Math. Soc. **36** (1992), 69–85.

48. Jia, R. Q. and C. A. Micchelli, Using the refinement equation for the construction of pre-wavelets II: powers of two, in *Curves and Surfaces* P. J. Laurent, A. Le Méhauté, and L. L. Schumaker (eds.), Academic Press, New York, 1991, pp. 209–246.

49. Jia, R. Q., S. D. Riemenschneider, and D. X. Zhou, Smoothness of multiple refinable functions and multiple wavelets, SIAM J. Mat. Anal. Appl., to appear.

50. Johnson, M. J., On the approximation order of principal shift-invariant subspaces of $L_p(\mathbb{R}^d)$, J. Approx. Theory **91** (1997), 279–319.

51. Lawton, W., S. L. Lee, and Z. Shen, Stability and orthonormality of multivariate refinable functions, SIAM J. Math. Anal. **28** (1997), 999–1014.

52. Lawton, W., S. L. Lee, and Z. Shen, Convergence of multidimensional cascade algorithm, Numer. Math. **78** (1998), 427–438.

53. Ma, B. and Q. Sun, Compactly supported refinable distributions in Triebel-Lizorkin spaces and Besov spaces, J. Fourier Anal. Appl., to appear.

54. Mallat, S. G., Multiresolution approximations and wavelet orthonormal bases of $L^2(\mathbb{R})$, Trans. Amer. Math. Soc. **315** (1989), 69–87.

55. Meyer, Y., *Ondelettes et Opérateurs I: Ondelettes*, Hermann Éditeurs, 1990.

56. Micchelli, C. A. and H. Prautzsch, Refinement and subdivision for spaces of integer translates of compactly supported functions, in *Numerical Analysis*, D. F. Griffiths & G. A. Watson (eds.), Longman, 1987, pp. 192–222.

57. Micchelli, C. A. and T. Sauer, Regularity of multiwavelets, 1996, preprint.

58. Neamtu, M., Discrete simplex splines and subdivision, J. Approx. Theory **70** (1992), 358–374.

59. Neamtu, M., A. Ron, and Z. Shen, The cascade and subdivision algorithms, 1998, manuscript.

60. Plonka, G. and A. Ron, A new factorization technique of the matrix mask of univariate refinable functions, 1998, preprint.
    Ftp: `ftp://ftp.cs.wisc.edu/Approx` file: `ger.ps`

61. Riemenschneider, S. D. and Z. Shen, Multidimensional interpolatory subdivision schemes, SIAM J. Numer. Anal., to appear.

62. Ron, A., A necessary and sufficient condition for the linear independence of the integer translates of a compactly supported distribution, Constr. Approx. **5** (1989), 297–308. Ftp: `ftp://ftp.cs.wisc.edu/Approx` file: `csd1.ps`

63. Ron, A., Characterizations of linear independence and stability of the shifts of a univariate refinable function in terms of its refinement mask, CMS TSR #93-3, U. Wisconsin-Madison, 1992
Ftp: ftp://ftp.cs.wisc.edu/Approx file: stablemask.ps

64. Ron, A., Smooth refinable functions provide good approximation, SIAM J. Math. Anal. **28** (1997), 731–748. Ftp: ftp://ftp.cs.wisc.edu/Approx file: smoothwav.ps

65. Ron, A., Seven lectures on shift-invariant spaces and wavelets, Lecture Notes, Math887, University of Wisconsin-Madison, Spring 1998. Written by Olga Holtz. Ftp: http://www.cs.wisc.edu/~amos/887

66. Ron, A. and Z. Shen, Frames and stable bases for shift-invariant subspaces of $L_2(\mathbb{R}^d)$, Canad. J. Math. **47** (1995), 1051–1094.
Ftp: ftp://ftp.cs.wisc.edu/Approx file: frame1.ps

67. Ron, A. and Z. Shen, Weyl-Heisenberg frames and Riesz bases in $L_2(\mathbb{R}^d)$, Duke Math. J. **89** (1997), 237–282.
Ftp: ftp://ftp.cs.wisc.edu/Approx file: wh.ps

68. Ron, A. and Z. Shen, Affine systems in $L_2(\mathbb{R}^d)$: the analysis of the analysis operator, J. Functional Anal. **148** (1997), 408–447.
Ftp: ftp://ftp.cs.wisc.edu/Approx file: affine.ps

69. Ron, A. and Z. Shen, Affine systems in $L_2(\mathbb{R}^d)$ II: dual systems, J. Fourier Anal. Appl. **3** (1997), 618–637.
Ftp: ftp://ftp.cs.wisc.edu/Approx file: dframe.ps

70. Ron, A. and Z. Shen, Compactly supported tight affine spline frames in $L_2(\mathbb{R}^d)$, Math. Comp. **67** (1998), 191–207.
Ftp: ftp://ftp.cs.wisc.edu/Approx file: tight.ps

71. Ron, A. and Z. Shen, Construction of Compactly Supported Affine Frames in $L_2(\mathbb{R}^d)$, in *Advances in Wavelets*, K. S. Lau (ed.), Springer Verlag, 1998. Ftp: ftp://ftp.cs.wisc.edu/Approx file: hk.ps

72. Ron, A. and Z. Shen, Gramian analysis of affine bases and affine frames, in *Approximation Theory VIII, Vol. 2: Wavelets and Multilevel Approximation*, C. K. Chui and L. L. Schumaker (eds.), World Scientific Publishing, New Jersey, 1995, pp. 375–382. Ftp: ftp://ftp.cs.wisc.edu/Approx file: frame2.ps

73. Ron, A. and Z. Shen, The sobolev regularity of refinable functions, March 1997, preprint. Ftp: ftp://ftp.cs.wisc.edu/Approx file: reg.ps

74. Ron, A., Z. Shen, and K-C. Toh, Computing the Sobolev regularity of refinable functions by the Arnoldi Method, 1998, preprint.

75. Stöckler, J., A Laurent operator technique for multivariate frames and wavelet bases, in *Advanced Topics in Multivariate Approximation*, F. Fontanella, K. Jetter, and P.-J. Laurent (eds.), World Scientific, Singapore, 1996, pp. 339–354.

76. Stöckler, J., Preconditioning of the frame algorithm, in *Large Scale Scientific Computations of Engineering and Environmental Problems*, M. Griebel et al. (eds.), Vieweg Verlag, Braunschweig, 1998, pp. 338–346.

77. Strang G. and G. Fix, A Fourier analysis of the finite element variational method. C.I.M.E. II Ciclo 1971, in *Constructive Aspects of Functional Analysis*, G. Geymonat (ed.), 1973, pp. 793–840.

78. Villemoes, L. F., Wavelet analysis of refinable equations, SIAM J. Math. Anal. **25** (1994), 1433–1460.

79. Wang, Y., Two-Scale Dilation equations and the mean spectral radius, Random & Computational Dynamics **4** (1996), 49–72.

80. Zhou, D. X., private communication.

Amos Ron
Computer Sciences Department
University of Wisconsin - Madison
1210 West Dayton
Madison, WI 57311, USA
amos@cs.wisc.edu

A Tool for Approximation in Bivariate Periodic Sobolev Spaces

Frauke Sprengel

Abstract. We give a characterization of Sobolev spaces of bivariate periodic functions with dominating mixed smoothness properties in terms of Sobolev spaces of univariate functions. The mixed Sobolev norm is proved to be a uniform crossnorm. This property can be used as a powerful tool in approximation theory.

§1. Introduction

Beside the approximation of functions from the usual isotropic periodic Sobolev spaces, the approximation of bivariate periodic functions with dominating mixed smoothness properties has attracted more and more attention. Spaces of functions with dominating mixed smoothness properties are especially well suited for the approximation or numerical integration using sparse grids or a hyperbolic approximation approach (e.g., [1,8]).

In this paper we prove that the Sobolev spaces of bivariate periodic functions with dominating mixed smoothness properties are tensor products of Sobolev spaces of univariate periodic functions. Additionally, the corresponding mixed Sobolev norms turn out to be uniform crossnorms. In other words, that means that one can derive error estimates for functions from these spaces from error estimates for the univariate functions in the same simple way as in the Hilbert space case (see e.g., [5]).

§2. Periodic Sobolev Spaces

We recall the definitions of the periodic Sobolev spaces first. Let $\mathbb{T}^d$ denote the d-dimensional torus represented by the cube

$$\mathbb{T}^d := \left\{ x = (x_1, \ldots, x_d) \in \mathbb{R}^d \; ; \; |x_k| \leq \pi, \; k = 1, \ldots, d \right\}.$$

Approximation Theory IX, Volume 2: Computational Aspects   319
Charles K. Chui and Larry L. Schumaker (eds.), pp. 319–326.
Copyright ℚ 1998 by Vanderbilt University Press, Nashville, TN.
ISBN 0-8265-1326-3.

By $D'(\mathbb{T}^d)$ we denote the linear space of tempered periodic distributions (or generalized periodic functions) (see [3, 6]). The Fourier coefficients of a distribution $g \in D'(\mathbb{T}^d)$ are $c_k(g) := g(e^{-ik\cdot})$ for $k \in \mathbb{Z}^d$.

Let $1 < p < \infty$ and $\alpha, \beta \in \mathbb{R}$. The isotropic Sobolev (or Bessel potential) space $H_p^\alpha(\mathbb{T}^d)$ of order α is defined as

$$H_p^\alpha(\mathbb{T}^d) := \Big\{ f \in D'(\mathbb{T}^d) \, ; \, \| f \mid H_p^\alpha(\mathbb{T}^d) \| := $$
$$\Big\| \sum_{k \in \mathbb{Z}^d} (1 + |k|_2^2)^{\alpha/2} c_k(f) e^{ik\cdot} \mid L_p(\mathbb{T}^d) \Big\| < \infty \Big\}.$$

The Sobolev space $S_{p,p}^{\alpha,\beta} H(\mathbb{T}^2)$ of order (α, β) of bivariate generalized periodic functions with dominating mixed smoothness properties is given by

$$S_{p,p}^{\alpha,\beta} H(\mathbb{T}^2) := \Big\{ f \in D'(\mathbb{T}^2) \, ; \, \| f \mid S_{p,p}^{\alpha,\beta} H(\mathbb{T}^2) \| := $$
$$\Big\| \sum_{k \in \mathbb{Z}^2} (1 + k_1^2)^{\alpha/2} (1 + k_2^2)^{\beta/2} c_k(f) e^{ik\cdot} \mid L_p(\mathbb{T}^2) \Big\| < \infty \Big\}.$$

§3. Tensor Products of Reflexive Banach Spaces

Because the Sobolev spaces considered here are reflexive, we restrict our following introduction to tensor products to this case. In [4], one can find a general introduction.

Given two reflexive Banach spaces X, Y. We form the linear space of formal linear combinations

$$F(X,Y) := \Big\{ \sum_{k=1}^r \alpha_k \, (x_k, y_k) \, ; \, \alpha_k \in \mathbb{C}, \, x_k \in X, \, y_k \in Y, \, k = 1, \ldots, r, \, r \in \mathbb{N} \Big\}$$

of pairs $(x_k, y_k) \in X \times Y$. Furthermore, we denote by G the subspace of $F(X, Y)$ spanned by elements of the form

$$\sum_{k=1}^r \sum_{\ell=1}^s \alpha_k \, \beta_\ell \, (x_k, y_\ell) - \Big(\sum_{k=1}^r \alpha_k \, x_k \, , \, \sum_{\ell=1}^s \beta_\ell \, y_\ell \Big).$$

Then, the algebraic tensor product of X and Y is the quotient space

$$X \otimes Y := F(X, Y)/G.$$

The equivalence class generated by (x, y) is denoted by $x \otimes y$. In general, the algebraic tensor product $X \otimes Y$ equipped with a suitable norm is not complete. The completion of $X \otimes Y$ with respect to the norm λ yields the tensor product $X \otimes_\lambda Y$ of the spaces X and Y. Two Banach spaces will be identified if they are isomorphic to each other. A norm λ on $X \otimes Y$ with

$$\lambda(x \otimes y) = \| x \mid X \| \cdot \| y \mid Y \|$$

for all $x \otimes y \in X \otimes Y$, is termed a crossnorm.

A pair of crossnorms λ_1 on $X_1 \otimes Y_1$ and λ_2 on $X_2 \otimes Y_2$ is uniform if for all operators $P \in \mathcal{L}(X_1, X_2)$ and $Q \in \mathcal{L}(Y_1, Y_2)$ and all $x_k \in X_1$, $y_k \in Y_1$,

$$\lambda_2 \Big(\sum_{k=1}^{r} P x_k \otimes Q y_k \Big) \leq \|P \mid \mathcal{L}(X_1, X_2)\| \cdot \|Q \mid \mathcal{L}(Y_1, Y_2)\| \, \lambda_1 \Big(\sum_{k=1}^{r} x_k \otimes y_k \Big).$$

In the case of uniform crossnorms the tensor product $P \otimes Q$ has a unique continuous extension $P \hat{\otimes} Q$ with

$$\|P \hat{\otimes} Q \mid \mathcal{L}(X_1 \otimes_{\lambda_1} Y_1, X_2 \otimes_{\lambda_2} Y_2)\| = \|P \mid \mathcal{L}(X_1, X_2)\| \cdot \|Q \mid \mathcal{L}(Y_1, Y_2)\|.$$

Using this property of uniform crossnorms, one is able to obtain results for the tensor product spaces directly from the results for the underlying single spaces. An example for spaces with uniform crossnorms is a pair of tensor product spaces both equipped with the p-nuclear norm (see [7]).

For $1 \leq p \leq \infty$, $1/p + 1/q = 1$, the p-nuclear norm of $z \in X \otimes Y$ is given by

$$\tilde{\lambda}_p(z) := \inf\Big\{ \Big(\sum_{j=1}^{n} \|x_j \mid X\|^p \Big)^{1/p} \theta_q(y_1, \ldots, y_n) \, ; \, z = \sum_{j=1}^{n} x_j \otimes y_j \Big\},$$

where

$$\theta_q(y_1, \ldots, y_n) := \sup\Big\{ \Big\| \sum_{j=1}^{n} \eta_j \, y_j \mid Y \Big\| \, ; \, \|\{\eta_j\}_{j=1,\ldots,n} \mid \ell_p(\{1, \ldots, n\})\| = 1 \Big\}.$$

The infimum is taken with respect to all representations of z.

§4. Tensor Products of Sobolev Spaces

After all these preparations we are able to characterize the Sobolev spaces of bivariate functions with dominating mixed smoothness properties.

Theorem 1. *Let* $1 < p < \infty$, $\alpha, \beta \in \mathbb{R}$. *Then*

$$S_{p,p}^{\alpha,\beta} H(\mathbb{T}^2) = H_p^\alpha(\mathbb{T}) \otimes_{\tilde{\lambda}_p} H_p^\beta(\mathbb{T})$$

with the p-nuclear norm $\tilde{\lambda}_p$ and

$$\tilde{\lambda}_p(f) = \|f \mid S_{p,p}^{\alpha,\beta} H(\mathbb{T}^2)\|$$

for all $f \in S_{p,p}^{\alpha,\beta} H(\mathbb{T}^2)$.

Since the p-nuclear norms are uniform, error estimates for functions with higher order dominating mixed smoothness in a mixed norm of lower orders can be derived in a simple manner from the error estimates for the univariate functions. This can be done for all approaches using tensor products of linear operators, for instance interpolation on full or on sparse grids.

The proof of the theorem requires the following two lemmata. Here and in the sequel μ denotes the Lebesgue measure.

Lemma 2. If $f_1, \ldots, f_n \in H_p^\alpha(\mathbb{T})$, $1 < p < \infty$, $\alpha \in \mathbb{R}$,

$$\mathrm{supp}\left(\sum_{k \in \mathbb{Z}} (1+k^2)^{\alpha/2} c_k(f_j) e^{ik\cdot}\right) = E_j \subset \mathbb{T}$$

with $\mu(E_j \cap E_k) = 0$ for $j \neq k$, then

$$\theta_q(f_1, \ldots, f_n) = \max_j \|f_j \mid H_p^\alpha(\mathbb{T})\|.$$

Proof: In the following, we take the supremum over all $\|\eta \mid \ell_p\{1, \ldots, n\}\| = 1$ and compute

$$\theta_q(f_1, \ldots, f_n) = \sup_\eta \left\{ \left\| \sum_{j=1}^n \eta_j f_j \mid H_p^\alpha(\mathbb{T}) \right\| \right\}$$

$$= \sup_\eta \left\{ \int_{\mathbb{T}} \left| \sum_{j=1}^n \eta_j \sum_{k \in \mathbb{Z}} (1+k^2)^{\alpha/2} c_k(f_j) e^{ikx} \right|^p dx \right\}^{1/p}$$

$$= \sup_\eta \left\{ \sum_{\ell=1}^n \int_{E_\ell} \left| \sum_{j=1}^n \eta_j \sum_{k \in \mathbb{Z}} (1+k^2)^{\alpha/2} c_k(f_j) e^{ikx} \right|^p dx \right\}^{1/p}$$

$$= \sup_\eta \left\{ \sum_{j=1}^n \int_{E_j} \left| \eta_j \sum_{k \in \mathbb{Z}} (1+k^2)^{\alpha/2} c_k(f_j) e^{ikx} \right|^p dx \right\}^{1/p}$$

$$= \sup_\eta \left\{ \sum_{j=1}^n |\eta_j|^p \|f_j \mid H_p^\alpha(\mathbb{T})\|^p \right\}^{1/p}$$

$$= \sup_{\sum |\kappa_j| = 1} \left\{ \sum_{j=1}^n |\kappa_j| \|f_j \mid H_p^\alpha(\mathbb{T})\|^p \right\}^{1/p}$$

$$= \max_j \|f_j \mid H_p^\alpha(\mathbb{T})\|. \quad \square$$

Lemma 3. Let $1 < p < \infty$, $\alpha \in \mathbb{R}$. The functions of the form $\sum_{j=1}^N f_j$ with $N \in \mathbb{N}$, $f_j \in H_p^\alpha(\mathbb{T})$,

$$\mathrm{supp}\left(\sum_{k \in \mathbb{Z}} (1+k^2)^{\alpha/2} c_k(f_j) e^{ik\cdot}\right) = E_j \subset \mathbb{T},$$

$\mu(E_j \cap E_k) = 0$ for $j \neq k$, form a dense set in $H_p^\alpha(\mathbb{T})$.

Proof: If $f \in H_p^\alpha(\mathbb{T})$, $\varepsilon > 0$, then

$$\tilde{f} := \sum_{k \in \mathbb{Z}} (1+k^2)^{\alpha/2} c_k(f) e^{ik\cdot} \in L_p(\mathbb{T}).$$

The simple functions form a dense set in $L_p(\mathbb{T})$ (cf. [2], p. 125). Therefore, a function

$$\tilde{f}_N = \sum_{j=1}^{N} a_j \chi_{E_j}$$

with $a_j \in \mathbb{C}$, $E_j \subset \mathbb{T}$, $\mu(E_j \cap E_k) = 0$ for $j \neq k$, has to exist, such that

$$\|\tilde{f} - \tilde{f}_N \mid L_p(\mathbb{T})\| < \varepsilon.$$

Now, we define

$$g_j := a_j \sum_{k \in \mathbb{Z}} (1 + k^2)^{-\alpha/2} c_k(\chi_{E_j}) e^{ik\cdot} \in H_p^{\alpha}(\mathbb{T})$$

and conclude

$$\left\| f - \sum_{j=1}^{N} g_j \mid H_p^{\alpha}(\mathbb{T}) \right\| = \|\tilde{f} - \tilde{f}_N \mid L_p(\mathbb{T})\| < \varepsilon.$$

This proves the lemma. □

After having proved these two lemmata we are now able to proceed to the proof of the theorem.

Proof of Theorem 1. Denote $\lambda_{\alpha,\beta} := \| \cdot \mid S_{p,p}^{\alpha,\beta} H(\mathbb{T}^2)\|$. Let

$$h = \sum_{j=1}^{n} f_j \otimes g_j \tag{1}$$

be a representation of $h \in H_p^{\alpha}(\mathbb{T}) \otimes H_p^{\beta}(\mathbb{T})$. Then

$$\tilde{f}_j := \sum_{k \in \mathbb{Z}} (1 + k^2)^{\alpha/2} c_k(f_j) e^{ik\cdot} \in L_p(\mathbb{T}),$$

$$\tilde{g}_j := \sum_{\ell \in \mathbb{Z}} (1 + \ell^2)^{\beta/2} c_\ell(g_j) e^{i\ell\cdot} \in L_p(\mathbb{T}).$$

With the definition

$$\tilde{h} = \sum_{j=1}^{n} \tilde{f}_j \otimes \tilde{g}_j,$$

we obtain

$$\lambda_{\alpha,\beta}(h) = \left\| \sum_{j=1}^{n} \sum_{k,\ell \in \mathbb{Z}} (1 + k^2)^{\alpha/2}(1 + \ell^2)^{\beta/2} c_k(f_j) c_\ell(g_j) e^{ikx} e^{i\ell y} \mid L_p(\mathbb{T}^2) \right\|$$

$$= \left\| \sum_{j=1}^{n} \tilde{f}_j \otimes \tilde{g}_j \mid L_p(\mathbb{T}^2) \right\| = \|\tilde{h} \mid L_p(\mathbb{T}^2)\|.$$

Now, we use the tensor product property of

$$L_p(\mathbb{T}^2) = L_p(\mathbb{T}) \otimes_{\tilde{\lambda}_p} L_p(\mathbb{T})$$

and the equality
$$\tilde{\lambda}_p = \| \cdot \mid L_p(\mathbb{T}^2)\|$$

of the p-nuclear norm and the usual L_p-Norm (cf. [4], Chap. 1). That leads
to

$$\lambda_{\alpha,\beta}(h) = \inf\left\{ \left(\sum_{j=1}^n \|\hat{f}_j \mid L_p(\mathbb{T})\|^p\right)^{1/p} \theta_q(\hat{g}_1, \ldots, \hat{g}_n) \; ; \; \tilde{h} = \sum_{j=1}^n \hat{f}_j \otimes \hat{g}_j \right\},$$

where the infimum is taken over all representations of $\tilde{h}$.

The functions $\hat{f}_j, \hat{g}_j$ need not necessarily have a form like $\tilde{f}_j, \tilde{g}_j$ which
stems from the representation (1) of h. Therefore, we obtain

$$\lambda_{\alpha,\beta}(h) \le \inf\left\{ \left(\sum_{j=1}^n \|\tilde{f}_j \mid L_p(\mathbb{T})\|^p\right)^{1/p} \theta_q(\tilde{g}_1, \ldots, \tilde{g}_n) \; ; \; \tilde{h} = \sum_{j=1}^n \tilde{f}_j \otimes \tilde{g}_j \right\},$$

where we take the infimum over all representations of $\tilde{h}$ with particular func-
tions $\tilde{f}_j, \tilde{g}_j$ from a representation of h. Using the definitions of $\tilde{f}_j, \tilde{g}_j$ yields

$$\lambda_{\alpha,\beta}(h) \le \inf\left\{ \left(\sum_{j=1}^n \|f_j \mid H_p^\alpha(\mathbb{T})\|^p\right)^{1/p} \theta_q(g_1, \ldots, g_n) \; ; h = \sum_{j=1}^n f_j \otimes g_j \right\}$$
$$= \tilde{\lambda}_p(h).$$

It remains to prove that $\tilde{\lambda}_p(h) \le \lambda_{\alpha,\beta}(h)$.

Let h have the representation (1) where the functions g_j satisfy

$$\mathrm{supp}\left(\sum_{k\in\mathbb{Z}}(1+k^2)^{\alpha/2}c_k(g_j)e^{ik\cdot}\right) = E_j \subset \mathbb{T},$$

$\mu(E_j \cap E_k) = 0$ for $j \ne k$, and $\|g_j \mid H_p^\beta(\mathbb{T})\| = 1$. By Lemma 2,

$$\theta_q(g_1, \ldots, g_n) = 1$$

and hence,

$$\tilde{\lambda}_p(h) \le \left(\sum_{j=1}^n \|f_j \mid H_p^\alpha(\mathbb{T})\|^p\right)^{1/p}.$$

On the other hand,

$$
\begin{aligned}
\lambda_{\alpha,\beta}(h) &= \Big\| \sum_{j=1}^{n} \tilde{f}_j \otimes \tilde{g}_j \ \Big| \ L_p(\mathbb{T}^2) \Big\| \\
&= \Big(\sum_{k=1}^{n} \int_{\mathbb{T}} \int_{E_k} \Big| \sum_{j=1}^{n} \tilde{f}_j(x)\tilde{g}_j(y) \Big|^p \, dx dy \Big)^{1/p} \\
&= \Big(\sum_{j=1}^{n} \int_{\mathbb{T}} \int_{E_j} |\tilde{f}_j(x)\tilde{g}_j(y)|^p \, dx dy \Big)^{1/p} \\
&= \Big(\sum_{j=1}^{n} \| \tilde{f}_j \ | \ L_p(\mathbb{T}) \|^p \cdot \| \tilde{g}_j \ | \ L_p(\mathbb{T}) \|^p \Big)^{1/p} \\
&= \Big(\sum_{j=1}^{n} \| f_j \ | \ H_p^\alpha(\mathbb{T}) \|^p \Big)^{1/p}.
\end{aligned}
$$

That means, for functions h of this particular form we proved $\tilde{\lambda}_p(h) \le \lambda_{\alpha,\beta}(h)$. Because of Lemma 3 and the first part of this proof it holds for all functions $h \in H_p^\alpha(\mathbb{T}) \otimes H_p^\beta(\mathbb{T})$ that

$$
\tilde{\lambda}_p(h) = \lambda_{\alpha,\beta}(h).
$$

Since $S_{p,p}^{\alpha,\beta} H(\mathbb{T}^2)$ is complete, we finally have

$$
S_{p,p}^{\alpha,\beta} H(\mathbb{T}^2) = H_p^\alpha(\mathbb{T}) \otimes_{\lambda_{\alpha,\beta}} H_p^\beta(\mathbb{T}) = H_p^\alpha(\mathbb{T}) \otimes_{\tilde{\lambda}_p} H_p^\beta(\mathbb{T}),
$$

where the first equality was proved in [7] and isomorphic spaces are identified. □

Acknowledgments. The author would like to thank Dr. W. Sickel for helpful discussions and useful remarks which stimulated her work in this direction.

References

1. Delvos, F.-J. and W. Schempp, *Boolean Methods in Interpolation and Approximation*, Pitman Research Notes in Mathematics Series, Longman Scientific & Technical, Harlow, 1989.

2. Dunford, N. and J. T. Schwartz, *Linear Operators I*, Wiley, New York, 1988.

3. Edwards, R. E., *Fourier Series. A Modern Introduction*, vol. 1/2, Springer, New York, 1982.

4. Light, W. A. and E. W. Cheney, *Approximation Theory in Tensor Product Spaces*, Lecture Notes in Mathematics 1169, Springer, Berlin, 1985.

5. Pöplau, G. and F. Sprengel, Some error estimates for periodic interpolation on full and sparse grids, in *Curves and Surfaces with Applications in CAGD*, A. Le Méhauté, C. Rabut, and L. L. Schumaker (eds.), Vanderbilt Univ. Press, Nashville, 1997, pp. 355–362.

6. Schmeißer, H.-J. and H. Triebel, *Topics in Fourier Analysis and Function Spaces*, Akadem. Verlagsgesellschaft Geest & Portig, Leipzig, 1987.

7. Sprengel, F., *Interpolation and Wavelet Decomposition of Multivariate Periodic Functions*, Ph.D. Thesis, University of Rostock, 1997.

8. Temlyakov, V. N., *Approximation of Periodic Functions*, Nova Science, New York, 1993.

Frauke Sprengel
CWI
Kruislaan 413
P.O. Box 94079
NL-1090 GB Amsterdam
The Netherlands
frauke.sprengel@cwi.nl

Convergence of Biorthogonal
Coifman Wavelet Systems

Jun Tian and Raymond O. Wells, Jr.

Abstract. Biorthogonal Coifman wavelet (BCW) systems are biorthogonal wavelet systems with the vanishing of moments equally distributed between the scaling functions and wavelet functions. It has been shown that these wavelet systems provide an optimal wavelet sampling approximation with an exponential convergence rate, where the optimality is measured over all possible vanishing moments distributions. The scaling filters and wavelet filters are all dyadic rational, which means we can implement a very fast multiplication-free discrete wavelet transform. Recent work showed that the scaling filters of BCW systems converge to the sinc scaling filter in $l^2(\mathbb{Z})$, while the $L^2(\mathbb{R})$ convergence of the scaling functions remained open. In this work we will prove the scaling functions of BCW systems converges to the sinc function in $L^2(\mathbb{R})$. Following the pioneering work of Chui and Wang, we establish a quantitative relationship between the orthonormal Daubechies wavelet systems and the BCW systems. Based on the convergence property of the Daubechies systems, we are able to show the $L^2(\mathbb{R})$ convergence of BCW systems.

§1. Introduction

Wavelet analysis (see [3, 4, 10, 16, 19, 23, 31, 32], etc.) has been proven to be a very powerful tool in harmonic analysis, neural networks, numerical analysis, and signal processing, especially in the area of image compression [20, 27, 29, 33] and data denoising [8, 13, 14]. The theoretical work of orthogonal wavelets was done in the late eighties [9, 15, 16] and the framework of biorthogonal wavelets was established in the early nineties [5, 7, 30]. Wavelet theory is closely connected with subband coding, and it provides a functional space structure for subband coding, often leading to better understanding and tremendous improvement.

In this paper we will focus on biorthogonal Coifman wavelet (BCW) systems, a family of compactly supported biorthogonal wavelet systems with the vanishing of moments equally distributed between the scaling functions and

Approximation Theory IX, Volume 2: Computational Aspects
Charles K. Chui and Larry L. Schumaker (eds.), pp. 327–335.
Copyright © 1998 by Vanderbilt University Press, Nashville, TN.
ISBN 0-8265-1326-3.

wavelet functions. These wavelet systems provide a wavelet sampling approximation with an exponential convergence rate. Recent work ([19, 26]) showed that the scaling filters of BCW systems converge to the sinc scaling filter in $l^2(\mathbb{Z})$, where the $L^2(\mathbb{R})$ convergence of the scaling functions remained open. We will show that the scaling functions of BCW systems converge to the sinc function in $L^2(\mathbb{R})$.

§2. Biorthogonal Coifman Wavelet Systems

BCW systems are the generalization of orthogonal Coifman wavelet systems (or Coiflets, see [2, 11]) to the biorthogonal setting. Here we give a brief review of BCW systems, for more details, we refer to [1, 12, 17, 18, 21–26, 28].

Definition 1. *A biorthogonal wavelet system with compact support is called a biorthogonal Coifman wavelet (BCW) system of degree N if the following two conditions are satisfied,*
   1) *the vanishing moments of the synthesis scaling function $\tilde{\phi}(x)$ and wavelet function $\tilde{\psi}(x)$ are both of degree N,*

$$\mathrm{Mom}_p(\tilde{\phi}) := \int_{\mathbb{R}} x^p \tilde{\phi}(x)\, dx \;=\; 0, \qquad p = 1, \ldots, N,$$

$$\mathrm{Mom}_p(\tilde{\psi}) := \int_{\mathbb{R}} x^p \tilde{\psi}(x)\, dx \;=\; 0, \qquad p = 0, \ldots, N.$$

   2) *the vanishing moment of the analysis wavelet function $\psi(x)$ is of degree N,*

$$\mathrm{Mom}_p(\psi) \;:=\; \int_{\mathbb{R}} x^p \psi(x)\, dx \;=\; 0, \qquad p = 0, \ldots, N.$$

Lemma 2. *For a biorthogonal Coifman wavelet system of degree N, the vanishing moment of the analysis scaling function $\phi(x)$ is also of degree N,*

$$\mathrm{Mom}_p(\phi) \;:=\; \int_{\mathbb{R}} x^p \phi(x)\, dx \;=\; 0, \qquad p = 1, \ldots, N.$$

Theorem 3. *(Wavelet Sampling Approximation). For $f(x) \in C^N(\mathbb{R})$ with compact support, and the Nth derivative $f^{(N)}(x)$ being Lipschitz, let $j \in \mathbb{N}$, define*

$$S^j(f)(x) \;:=\; \sum_{k \in \mathbb{Z}} f\left(\frac{k}{2^j}\right) \phi(2^j x - k),$$

where $\phi(x)$ is the analysis (or synthesis) scaling function of a biorthogonal Coifman wavelet system of degree N. Then

$$\left\| f(x) - S^j(f)(x) \right\|_{L^2} \;\leq\; C 2^{-j(N+1)},$$

where C is a constant, depending only on f and ϕ. If, in addition, $\phi(x) \in C^n(\mathbb{R})$, then

$$\left\| f(x) - S^j(f)(x) \right\|_{H^n} \;\leq\; C 2^{-j(N+1-n)}.$$

It can be shown (see, for example, [26]) that for an optimal wavelet sampling approximation, vanishing moments should be equally distributed between the scaling function and the wavelet function. The closed form solution of the minimum length of a BCW system of degree N is given by the scaling filters $\{a_k\}$ and $\{\tilde{a}_k\}$:

- if N is even, $N = 2n$,

$$a_{2k+1} = \tilde{a}_{2k+1} = \frac{(-1)^k(2n+1)}{2^{4n-1}(2k+1)}\binom{2n-1}{n-1}\binom{2n}{n+k}, \tag{1}$$

- if N is odd, $N = 2n - 1$,

$$a_{2k+1} = \tilde{a}_{2k+1} = \frac{(-1)^k(2n-1)}{2^{4n-3}(2k+1)}\binom{2n-2}{n-1}\binom{2n-1}{n+k}, \tag{2}$$

and

$$\tilde{a}_{2k} = \delta_{0,k}, \tag{3}$$

$$a_{2k} = 2\delta_{0,k} - \sum_{l\in\mathbb{Z}} \tilde{a}_{2l+1}\tilde{a}_{2l+1-2k}. \tag{4}$$

In the remainder of this paper, a BCW system will always be referred to the one with minimum length, which is given by the above formula, unless it is otherwise stated. An attractive feature of the formula is that all a_k's, $\tilde{a}_k$'s are either zero or a dyadic rational number (see [26] for proof), that is, an integer divided by a power of 2. Thus, one can implement a very fast multiplication-free discrete wavelet transform, which consists of only addition and shift operations, on digital computers. Also from the above formula, when the degree N is odd, the BCW system is symmetric, i.e., $a_k = a_{-k}$, and $\tilde{a}_k = \tilde{a}_{-k}$. An interesting observation is that BCW systems and orthonormal Daubechies wavelet systems ([9, 10]) are closely related by their scaling filters. More precisely,

$$\tilde{m}_{2n+1}(\xi) = \left|m_n^D(\xi)\right|^2, \tag{5}$$

where m is defined by the scaling filter $\{a_k\}$,

$$m(\xi) := \frac{1}{2}\sum_{k\in\mathbb{Z}} a_k e^{-ik\xi},$$

and $\tilde{m}_{2n+1}$ corresponds to the synthesis part of the BCW system of degree $2n+1$, m_n^D corresponds to the orthonormal Daubechies wavelet system of degree n, whose wavelet function ψ_n^D has vanishing moments up to degree n,

$$\int_{\mathbb{R}} x^p \psi_n^D(x)\,dx = 0, \qquad p = 0,\ldots,n. \tag{6}$$

One can prove (see [19, 26]) that the scaling filters of BCW systems converge to the sinc scaling filter, which is defined by

$$a_{2k}^{\text{sinc}} := \delta_{0,k}, \qquad a_{2k+1}^{\text{sinc}} := \frac{(-1)^k 2}{(2k+1)\pi}.$$

The sinc scaling filter is the scaling filter of the sinc wavelet system, whose scaling function is exactly $\text{sinc}(\pi x) = \frac{\sin \pi x}{\pi x}$.

Lemma 4. *Suppose* $\{\tilde{a}_k^N, k \in \mathbb{Z}\}$ *to be the synthesis scaling filter of the BCW system of degree* N, *then*

$$\lim_{N \to \infty} \left\| \tilde{a}^N - a^{\text{sinc}} \right\|_{l^2} = \lim_{N \to \infty} \left(\sum_{k \in \mathbb{Z}} \left(\tilde{a}_k^N - a_k^{\text{sinc}} \right)^2 \right)^{1/2} = 0.$$

§3. L^2 Convergence Property

It had been conjectured (see [28]) for some time that the scaling functions of BCW systems converge to $\text{sinc}(\pi x)$ in $L^2(\mathbb{R})$. Such convergence could also explain the good sampling approximation property of BCW systems (see Theorem 3).

3.1. Chui and Wang's Convergence Theorem

In [6], Chui and Wang proved the following theorem for *orthonormal* wavelet systems.

Theorem 5. *(Chui-Wang). Let* ϕ_n *be orthonormal scaling functions defined by*

$$\phi_n(x) = \sum_k a_{n,k} \phi_n(2x - k),$$

$$m_n(\xi) = \frac{1}{2} \sum_k a_{n,k} e^{-ik\xi},$$

$$m_n(\xi) = \left(\frac{1 + e^{-ik\xi}}{2} \right)^n \mathcal{L}_n(\xi),$$

$$\mathcal{L}_n(\pi) \neq 0 \qquad \text{and} \qquad \mathcal{L}_n(0) = 1,$$

$$|\mathcal{L}_n(\xi)| \leq C 2^n \left| \sin^n \left(\frac{\xi}{2} \right) \right|, \qquad \frac{\pi}{2} \leq |\xi| \leq \pi,$$

for some absolute constant C *(independent of* n). *Then*

$$\lim_{n \to \infty} \left\| |\hat{\phi}_n| - \chi_{(-\pi,\pi)}(\xi) \right\|_{L^2} = 0.$$

As an immediate consequence, let ϕ_n^D be an (arbitrarily chosen) orthonormal Daubechies scaling function of degree n (see Equation (6)), then

$$\lim_{n \to \infty} \left\| |\hat{\phi}_n^D| - \chi_{(-\pi,\pi)}(\xi) \right\|_{L^2} = 0. \tag{7}$$

3.2. L^2 Convergence to $\operatorname{sinc}(\pi x)$

Since $\hat{\phi}(\xi) = \prod_{j=1}^{\infty} m(2^{-j}\xi)$, Equation (5) implies a relationship between the synthesis scaling functions of BCW systems and orthonormal Daubechies scaling functions,

$$\widehat{\tilde{\phi}_{2n+1}}(\xi) = \left|\widehat{\phi_n^D}(\xi)\right|^2.$$

Thus,

$$\left|\widehat{\tilde{\phi}_{2n+1}}(\xi) - \chi_{(-\pi,\pi)}(\xi)\right| = \left|\left|\widehat{\phi_n^D}(\xi)\right|^2 - \chi_{(-\pi,\pi)}(\xi)\right|$$

$$= \left|\left(|\widehat{\phi_n^D}| + \chi_{(-\pi,\pi)}(\xi)\right) \cdot \left(|\widehat{\phi_n^D}| - \chi_{(-\pi,\pi)}(\xi)\right)\right|$$

$$\leq 2\left||\widehat{\phi_n^D}| - \chi_{(-\pi,\pi)}(\xi)\right|,$$

since $|\widehat{\phi_n^D}| \leq 1$. By Equation (7) and Parseval's identity, it follows

$$\lim_{n\to\infty} \left\|\tilde{\phi}_{2n+1}(x) - \frac{\sin \pi x}{\pi x}\right\|_{L^2} = (2\pi)^{-1/2} \lim_{n\to\infty} \left\|\widehat{\tilde{\phi}_{2n+1}}(\xi) - \chi_{(-\pi,\pi)}(\xi)\right\|_{L^2}$$

$$= 0.$$

For the even degree part, from Equations (1) and (2), we have

$$\tilde{a}_{2n,2k+1} = \tilde{a}_{2n+1,2k+1} + \frac{(-1)^k}{2^{4n+1}}\binom{2n}{n}\binom{2n+1}{n+k+1},$$

where $\{\tilde{a}_{2n,k}|k \in \mathbb{Z}\}$ and $\{\tilde{a}_{2n+1,k}|k \in \mathbb{Z}\}$ are the synthesis scaling filters of BCW systems of degree $2n$ and of degree $2n + 1$, respectively. By summing up, one gets

$$\tilde{m}_{2n}(\xi) = \tilde{m}_{2n+1}(\xi) - \frac{(2n-1)!!}{(2n)!!}i\sin^{2n+1}\xi,$$

where

$$(2n-1)!! := \prod_{k=1}^{n}(2k-1), \qquad (2n)!! := \prod_{k=1}^{n}(2k) = 2^n n!.$$

Notice that

$$1 - C\sin^{2n}(\xi) \leq \left|m_n^D(\xi)\right|^2 \leq 1, \qquad \text{for } |\xi| \leq \frac{\pi}{2},$$

for some absolute constant C, which is independent of n. Choose a positive integer k such that $|\xi/2^k| < \frac{\pi}{2}$, then

$$1 - C\sin^{2n}(\xi/2^k) \leq \tilde{m}_{2n+1}(\xi/2^k) \leq 1.$$

So

$$
\begin{aligned}
\left| \tilde{m}_{2n}(\xi/2^k) \right| &= \left| \tilde{m}_{2n+1}(\xi/2^k) - \frac{(2n-1)!!}{(2n)!!} i \sin^{2n+1}(\xi/2^k) \right| \\
&= \left| \tilde{m}_{2n+1}(\xi/2^k) \right| \cdot \left| 1 - \frac{(2n-1)!!}{(2n)!!} i \frac{\sin^{2n+1}(\xi/2^k)}{\tilde{m}_{2n+1}(\xi/2^k)} \right|.
\end{aligned}
$$

Since

$$
\begin{aligned}
\left| \frac{(2n-1)!!}{2n!!} i \frac{\sin^{2n+1}(\xi/2^k)}{\tilde{m}_{2n+1}(\xi/2^k)} \right| &\leq \left| \frac{\sin^{2n+1}(\xi/2^k)}{\tilde{m}_{2n+1}(\xi/2^k)} \right| \leq \left| \frac{\sin^{2n+1}(\xi/2^k)}{1 - C \sin^{2n}(\xi/2^k)} \right| \\
&\leq \left| C \sin^{2n+1}(\xi/2^k) \right| \leq \left| C(\xi/2^k)^{2n+1} \right|,
\end{aligned}
$$

for n large enough, where C is an absolute constant independent of n. Thus,

$$
\lim_{n \to \infty} \left(\prod_{j=k+1}^{\infty} \left| 1 - \frac{(2n-1)!!}{(2n)!!} i \frac{\sin^{2n+1}(\xi/2^j)}{\tilde{m}_{2n+1}(\xi/2^j)} \right| \right) = 1.
$$

This implies

$$
\begin{aligned}
\lim_{n \to \infty} \left| \widehat{\tilde{\phi}_{2n}}(\xi) \right| &= \lim_{n \to \infty} \left(\prod_{j=1}^{\infty} \left| \tilde{m}_{2n}(\xi/2^j) \right| \right) \\
&= \lim_{n \to \infty} \left(\prod_{j=1}^{k} \left| \tilde{m}_{2n}(\xi/2^j) \right| \cdot \prod_{j=k+1}^{\infty} \left| \tilde{m}_{2n}(\xi/2^j) \right| \right) \\
&= \lim_{n \to \infty} \left(\prod_{j=1}^{k} \left| \tilde{m}_{2n+1}(\xi/2^j) \right| \cdot \prod_{j=k+1}^{\infty} \left| \tilde{m}_{2n+1}(\xi/2^j) \right| \right) \\
&= \lim_{n \to \infty} \left| \widehat{\tilde{\phi}_{2n+1}}(\xi) \right| = \chi_{(-\pi,\pi)}(\xi).
\end{aligned}
$$

Now by taking $\{2\widehat{\tilde{\phi}_{2n+1}}, n \in \mathbb{Z}\}$ as a dominating sequence, the sequence $\{\widehat{\tilde{\phi}_{2n}}, n \in \mathbb{Z}\}$ converge to $\chi_{(-\pi,\pi)}$ in L^2. By using the concept of *asymptotically linear phase* (see [6]), the sequence $\{\tilde{\phi}_{2n}, n \in \mathbb{Z}\}$ converge to $\frac{\sin \pi x}{\pi x}$ in $L^2(\mathbb{R})$.

For the analysis scaling function $\phi_n(x)$, again, based on Equations (1), (2), and (4), we have

$$
\begin{aligned}
m_n(\xi) &= \frac{1}{2} \sum_{k \in \mathbb{Z}} \left(\left(2\delta_{0,k} - \sum_{l \in \mathbb{Z}} \tilde{a}_{n,2l+1} \tilde{a}_{n,2l+1-2k} \right) e^{-2ik\xi} + \tilde{a}_{n,2k+1} e^{-i(2k+1)\xi} \right) \\
&= 2\tilde{m}_n(\xi) + \tilde{m}_n(-\xi) - 2\tilde{m}_n(\xi)\tilde{m}_n(-\xi).
\end{aligned}
$$

One can apply the previous approach to prove the convergence. Thus we derive

Theorem 6. *Let $\tilde{\phi}_n(x)$ be the synthesis (or analysis) scaling function of BCW system of degree n, then*

$$\lim_{n \to \infty} \left\| \tilde{\phi}_n(x) - \frac{\sin \pi x}{\pi x} \right\|_{L^2} = 0.$$

§4. Conclusions

We proved the scaling functions of BCW systems converge to $\text{sinc}(\pi x)$ in $L^2(\mathbb{R})$. This provides a family of smooth wavelet systems to approximate the sinc wavelet system. It can also be shown that the wavelet functions of BCW systems converge to the difference of two sinc functions. An open question concerning Coiflets is the existence of large degree Coiflets, and there is no explicit formula for Coiflets. If Coiflets do exist for large degrees, one would expect that they possess the same convergence property as of BCW systems.

Acknowledgments. Supported in part by DARPA/AFOSR F49620-97-1-0513.

References

1. Ansari, R., C. Guillemot, and J. F. Kaiser, Wavelet construction using Lagrange halfband filters, IEEE Trans. Circ. Syst. **38** (1991), 1116–1118.

2. Beylkin, G., R. Coifman, and V. Rokhlin, Fast wavelet transforms and numerical algorithms I, Commun. Pure Appl. Math. **44** (1991), 141–183.

3. Burrus, C. S., R. A. Gopinath, and H. Guo, *Introduction to Wavelets and Wavelet Transforms*, Prentice Hall, Englewood Cliffs, NJ, 1997.

4. Chui, C. K., *An Introduction to Wavelets*, Academic Press, Boston, MA, 1992.

5. Chui, C. K. and J. Z. Wang, A general framework of compactly supported splines and wavelets, J. Approx. Theory **71** (1992), 263–304.

6. Chui, C. K. and J. Z. Wang, High-order orthonormal scaling functions and wavelets give poor time-frequency localization, J. Fourier Anal. Appl. **2** (1996), 415–426.

7. Cohen, A., I. Daubechies, and J.-C. Feauveau, Biorthogonal bases of compactly supported wavelets,Commun. Pure Appl. Math. **XLV** (1992), 485–560.

8. Coifman, R. R. and D. L. Donoho, Translation-invariant de-noising, in *Wavelets and Statistics*, A. Antoniadis (ed.), Springer-Verlag, 1995.

9. Daubechies, I., Orthonormal bases of compactly supported wavelets, Commun. Pure Appl. Math. **XLI** (1988), 906–996.

10. Daubechies, I., *Ten Lectures on Wavelets*, SIAM, Philadelphia, PA, 1992.

11. Daubechies, I., Orthonormal bases of compactly supported wavelets II: variations on a theme, SIAM J. Math. Anal. **24** (1993), 499–519.

12. Deslauriers, G. and S. Dubuc, Symmetric iterative interpolation processes, Constr. Approx. **5** (1989), 49–68.

13. Donoho, D. L., De-noising by soft-thresholding, IEEE Trans. Info. Theory **41** (1995), 613–627.

14. Lang, M., H. Guo, J. E. Odegard, C. S. Burrus, and R. O. Wells, Jr., Noise reduction using an undecimated discrete wavelet transform, IEEE Signal Proc. Letters **3** (1996), 10–12.

15. Mallat, S. G., Multiresolution approximation and wavelet orthonormal bases of $L^2(\mathbb{R})$, Trans. AMS **315** (1989), 69–87.

16. Meyer, Y., *Wavelets and Operators*, Cambridge University Press, Cambridge, 1992.

17. Phoong, S., C. W. Kim, and P. P. Vaidyanathan, A new class of two-channel biorthogonal filter banks and wavelet bases, IEEE Trans. Signal Proc. **43** (1995), 649–665.

18. Reissell, L.-M., Wavelet multiresolution representation of curves and surfaces, Grap. Models Image Proc. **58** (1996), 198–217.

19. Resnikoff, H. L. and R. O. Wells, Jr., *Wavelet Analysis and the Scalable Structure of Information*, Springer-Verlag, New York, 1998.

20. Said, A. and W. A. Pearlman, A new fast and efficient image codec based on set partitioning in hierarchical trees, IEEE Trans. Circ. Syst. Video Tech. **6** (1996), 243–250.

21. Saito, N. and G. Beylkin, Multiresolution representation using the autocorrelation functions of compactly supported wavelets, IEEE Trans. Signal Proc. **41** (1993), 3584–3590.

22. Shensa, M. J., The discrete wavelet transform: wedding the à trous and Mallat algorithms, IEEE Trans. Signal Proc. **40** (1992), 2464–2482.

23. Strang, G. and T. Nguyen, *Wavelets and Filter Banks*, Wellesley Cambridge Press, Wellesley, MA, 1995.

24. Sweldens, W., The lifting scheme: A custom-design construction of biorthogonal wavelets, Appl. Comput. Harmonic Anal. **3** (1996), 186–200.

25. Tian, J. and R. O. Wells, Jr., *Vanishing Moments and Wavelet Approximation*, Technical Report, CML TR 95-01, Rice University, January 1995. (ftp://cml.rice.edu/pub/reports/CML9501.ps.Z)

26. Tian, J., *The Mathematical Theory and Applications of Biorthogonal Coifman Wavelet Systems*, Ph.D. Thesis, Rice University, February 1996.

27. Tian, J. and R. O. Wells, Jr., A lossy image codec based on index coding, in *Proc. Data Compression Conference*, J. A. Storer and M. Cohn (eds.), 1996. (ftp://cml.rice.edu/pub/reports/CML9515.ps.Z)

28. Tian, J., R. O. Wells, Jr., J. E. Odegard, and C. S. Burrus, Coifman wavelet systems: Approximation, smoothness, and computational algorithms, in *Computational Science for the 21st Century*, John Wiley & Sons Ltd, 1997, pp. 831–840.

29. Topiwala, P. (ed.), *Wavelet Image and Video Compression*, Klumer, 1998.

30. Vetterli, M. and C. Herley, Wavelets and filter banks: theory and design, IEEE Trans. ASSP **40** (1992), 2207–2232.

31. Vetterli, M. and J. Kovačević, *Wavelets and Subband Coding*, Prentice Hall, Englewood Cliffs, NJ, 1995.

32. Wickerhauser, M. V., *Adapted Wavelet Analysis from Theory to Software*, Wellesley, MA, 1993.

33. Xiong, Z., K. Ramchandran, and M. T. Orchard, Space-frequency quantization for wavelet image coding, IEEE Trans. Image Proc. **6** (1997), 677–693.

Jun Tian and Raymond O. Wells, Jr.
Computation Mathematics Laboratory
Rice University
Houston, TX 77005-1892
juntian@rice.edu, wells@rice.edu
http://cml.rice.edu/

Optimal Bivariate Crosscut Segmentations

Guido Walz

Abstract. We investigate the problem of finding an optimal segmentation of a given rectangle into subrectangles, thus continuing earlier work
on this subject published in [5] and [7]. In the present paper, we give
a direct approach to optimal so-called crosscut segmentations, which are
defined by vertical and horizontal crosscut lines through the rectangle. Finally, we prove comparison results for piecewise linear approximation on
a rectangle.

§1. Introduction and Preliminary Results

Let there be given a rectangle $\Omega = \{(x,y) \in \mathbb{R}^2 \; ; \; a \leq x \leq b \, ; \, c \leq y \leq d\}$ and
integers k and r. A crosscut segmentation of Ω is a set of rk subrectangles
$\{\Omega_{\mu,\nu}\}$, where, for $\mu = 1, \ldots, k$ and $\nu = 1, \ldots, r$,

$$\Omega_{\mu,\nu} \; = \; \{(x,y) \; ; \; \sigma_{\mu-1} \leq x \leq \sigma_\mu \; ; \; \tau_{\nu-1} \leq y \leq \tau_\nu\}$$

with knots $a = \sigma_0 \leq \sigma_1 \leq \cdots \leq \sigma_k = b$ and $c = \tau_0 \leq \tau_1 \leq \cdots \leq \tau_r = d$.
Throughout the paper we tacitly assume that the ranges of μ and ν are $\mu = 1, \ldots, k$ resp. $\nu = 1, \ldots, r$.

We consider the problem of finding an optimal segmentation of this type.
In order to formulate our optimality criterion, we need the following definition.

Definition 1. *Let $\mathcal{R}$ denote the set of all subrectangles of Ω. A functional
$d : \mathcal{R} \to \mathbb{R}$ is called a* regular functional, *if it satisfies*

$$
\begin{aligned}
d(R) &\leq d(\widetilde{R}) && \textit{if } R \subset \widetilde{R}, \; R, \widetilde{R} \in \mathcal{R}, \\
d(R) &= 0 && \textit{if } R \textit{ consists of one point}, \\
d(R_n) &\to d(R) && \textit{if } R_n \to R, \; R, R_n \in \mathcal{R} \textit{ for all } n.
\end{aligned}
$$

We now can define what an optimal crosscut segmentation should be.

Approximation Theory IX, Volume 2: Computational Aspects
Charles K. Chui and Larry L. Schumaker (eds.), pp. 337–344.
Copyright © 1998 by Vanderbilt University Press, Nashville, TN.
ISBN 0-8265-1326-3.

Definition 2. *Let d be a regular functional. A segmentation $\{\Omega_{\mu,\nu}^*\}$ is called optimal (w.r.t. d), if*

$$\max_{\mu,\nu} d(\Omega_{\mu,\nu}^*) = \inf_{\{\Omega_{\mu,\nu}\}} \max_{\mu,\nu} d(\Omega_{\mu,\nu}),$$

where the infimum is taken over all crosscut segmentations $\{\Omega_{\mu,\nu}\}$.

The problem of finding optimal crosscut segmentations is a reasonable generalization of the well-known univariate segment approximation problem, see, e.g., [1, 6, 8] and the references therein. The first time where this problem was successfully attacked appears to be the paper [5] from 1996. Later, in [7], Nürnberger could generalize these results to a more general class of functionals. In both approaches, one first has to solve two auxiliary problems, where the σ_μ's resp. τ_ν's are defined separately in each strip; see [5, 7] for details. The aim of the present paper is to give a more direct approach.

§2. Direct Approach to Optimal Crosscut Segmentations

Definition 3. *Let d be a regular functional. A segmentation $\{\Omega_{\mu,\nu}\}$ is called leveled (w.r.t. d), if the values $d(\Omega_{\mu,\nu})$, are the same for all μ and ν.*

The connection between the concepts of leveled and optimal segmentations is given in the next result.

Theorem 4. *Every leveled crosscut segmentation is optimal.*

Proof: For two arbitrary crosscut segmentations $\{\Omega_{\mu,\nu}\}$ and $\{\widetilde{\Omega}_{\mu,\nu}\}$ with $\{\Omega_{\mu,\nu}\} \neq \{\widetilde{\Omega}_{\mu,\nu}\}$, it is not difficult to see (cf. [5]) that there is an index pair (μ_1, ν_1) such that $\Omega_{\mu_1,\nu_1} \subset \widetilde{\Omega}_{\mu_1,\nu_1}$. If $\{\Omega_{\mu,\nu}\}$ is a leveled segmentation, there exists a real number m with $d(\Omega_{\mu,\nu}) = m$ for all μ and ν.

Assume now that there is a better segmentation $\{\widetilde{\Omega}_{\mu,\nu}\}$. Then

$$\max\{d(\widetilde{\Omega}_{\mu,\nu})\} < m.$$

But then, due to the properties of the functional d,

$$m = d(\Omega_{\mu_1,\nu_1}) \leq d(\widetilde{\Omega}_{\mu_1,\nu_1}) < m,$$

which is a contradiction. □

An analoguos result holds in univariate, and also in the auxiliary type bivariate problem considered in [5]. However, in these cases a leveled segmentation always exists. The situation for crosscut segmentations is much more complicated, as indicated by the following two examples:

Example. Let $\Omega = [0,1]^2$ and a function $\Psi \in C(\Omega)$ be given. For an arbitary subrectangle R of Ω of the form

$$R = \{(x,y) \; ; \; x_1 \leq x \leq x_2 \; ; \; y_1 \leq y \leq y_2\}, \tag{1}$$

we consider the regular functional

$$d(R) \; := \; \|\Psi\|_R \cdot \max\{(x_2 - x_1)\, , \, (y_2 - y_1)\} \, .$$

1) Let $\Psi(x,y) := |x + y - 1|$ and $k = r = 2$. In this case, the equidistant segmentation defined by $\sigma_1 = \tau_1 = 1/2$ is easily recognized to be the *unique optimal* one, but is obviously not leveled (cf. also [5]). This easy example already shows that a leveled crosscut segmentation does not always exists.
2) Let $\Psi(x,y) \equiv 1$, $k = r \geq 2$, and h some real number with $1/k < h < 1/(k-1)$. We consider the segmentation defined by the inner knots

$$\sigma_\mu \; := \; \tau_\mu \; := \; \mu h \, , \; \mu = 1, \ldots, k-1.$$

In this case, $d(\Omega_{\mu,\nu}) = h$ *for all* $(k^2 - 1)$ subrectangles except one, namely $\Omega_{k,k}$, where

$$d(\Omega_{k,k}) \; = \; 1 - (k-1)h \; < \; h.$$

On the other hand, this segmentation is clearly not optimal. This shows that even something like "nearly leveled" does not imply optimality of the segmentation.

We therefore investigate cases in which a leveled crosscut segmentation exists, such that Theorem 4 can be applied. To do this, we first need some results on special types of bivariate functions.

2.1. Best Approximation of Translational Functions

We consider best approximation of translational functions, i.e., bivariate functions of the form

$$f(x,y) \; = \; \varphi(x) + \psi(y) \tag{2}$$

(with suitable univariate functions φ and ψ) on a rectangle R of the form introduced in (1) by polynomials from the space $P_1 = \mathrm{span}\{1, x, y\}$.

Due to the structure of the functions defined in (2), there exists a unique polynomial p_I from the three-dimensional space P_1, which interpolates f in the four vertices of R. It can be written explicitly as

$$p_I(x,y) \; = \; a_I + b_I x + c_I y$$

with

$$a_I \; = \; \frac{x_2\varphi(x_1) - x_1\varphi(x_2)}{x_2 - x_1} + \frac{y_2\psi(y_1) - y_1\psi(y_2)}{y_2 - y_1} \, ,$$

$$b_I \; = \; \frac{\varphi(x_2) - \varphi(x_1)}{x_2 - x_1} \, , \quad \text{and} \quad c_I \; = \; \frac{\psi(y_2) - \psi(y_1)}{y_2 - y_1} \, . \tag{3}$$

Having recognized this, the best approximation of f is easy to get. We denote the error function of the interpolant p_I by δ_I, i.e.,

$$\delta_I(x,y) \; = \; a_I + b_I x + c_I y - \varphi(x) - \psi(y) \, .$$

Then the following result holds:

Proposition 5. Let the function $f \in C^2(R)$ be of the form defined in (2), and suppose that $\varphi' \cdot \psi' \neq 0$ and $\varphi'' \cdot \psi'' > 0$ in the interior of R. Then the best approximation of f from the space P_1 on R is the polynomial

$$p_f(x,y) \;=\; a_f + b_I x + c_I y$$

with b_I and c_I given in (3), and $a_f = a_I - \delta_I(\xi,\eta)/2$. Furthermore, the minimal deviation is

$$\rho(f, P_1) \;=\; |\delta_I(\xi,\eta)|/2 \;.$$

Here, (ξ,η) is the uniquely determined point in R where

$$\varphi'(\xi) \;=\; b_I \quad \text{and} \quad \psi'(\eta) \;=\; c_I . \tag{4}$$

Proof: Due to the mean value theorem there exist numbers ξ and η with the property (4), and since $\varphi'' \cdot \psi'' > 0$ in R, the point (ξ,η) is unique.

We now consider the error function $\delta_f(x,y) := p_f(x,y) - f(x,y)$. By construction, we have for the four vertices of R, i.e., for the points (x_i, y_j) with $i, j \in \{1,2\}$:

$$\delta_f(x_i, y_j) \;=\; a_I - \frac{\delta_I(\xi,\eta)}{2} + b_I x_i + c_I y_j - \varphi(x_i) - \psi(y_j) \;=\; -\frac{\delta_I(\xi,\eta)}{2} \;.$$

The gradient of δ_f, $(b_I - \varphi'(x), c_I - \psi'(y))$ vanishes only in (ξ,η), and since the Hessian

$$\begin{pmatrix} -\varphi''(x) & 0 \\ 0 & -\psi''(y) \end{pmatrix}$$

is either positive or negative definite, δ_f assumes its extremal value in (ξ,η), the value being

$$+ \delta_I(\xi,\eta)/2 \;.$$

Obviously, no nontrivial function $p \in P_1$ can have the same sign pattern as δ_f in these five points, so Kolmogorov's criterion (cf. e.g., [2, p. 15]) completes the proof. $\square$

We briefly discuss two examples for further use below.

Example. 1) Consider the square root function $f_1(x,y) = \sqrt{x} + \sqrt{y}$. Application of Proposition 5 yields that

$$\rho(f_1, P_1) \;=\; \frac{(\sqrt{x_1} - \sqrt{x_2})^2}{8\,(\sqrt{x_1} + \sqrt{x_2})} + \frac{(\sqrt{y_1} - \sqrt{y_2})^2}{8\,(\sqrt{y_1} + \sqrt{y_2})} \;, \tag{5}$$

which is well in line with the univariate result, cf. [4].

2) A second interesting special case is the function $f_2(x,y) = x^2 + y^2$. Here, Proposition 5 implies that

$$\rho(f_2, P_1) \;=\; \left((x_2 - x_1)^2 + (y_2 - y_1)^2\right)/8 \;. \tag{6}$$

Note that the minimal deviation in this example does not depend on the location of the rectangle R, but only on its sidelength.

We remark that each function f of the form $f(x,y) = x^\alpha + y^\beta$, where α and β are either both in $(0,1)$ or both greater than 1, also satisfies the assumptions of Proposition 5.

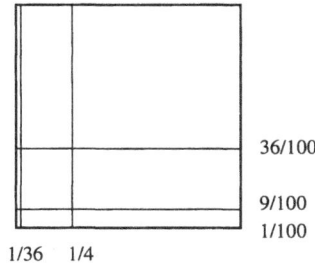

Fig. 1. Optimal segmentation of the unit square, $k = 3$, $r = 4$.

2.2. Applications

We now consider types of functionals which lead to optimal crosscut segmentations and illustrate this by using the results just proved. Again, R denotes a rectangle of the form introduced in (1).

Theorem 6. *Assume that the functional d is of the form*

$$d(R) := \Phi(\tilde{d}_1([x_1, x_2]), \tilde{d}_2([y_1, y_2])) , \tag{7}$$

where $\tilde{d}_1$ and $\tilde{d}_2$ are two regular functionals in the univariate sense, i.e., when applied to intervals, and where Φ is a continuous bivariate function having the properties

$$\begin{aligned} \Phi(0,0) &= 0 \quad \text{and} \\ \Phi(t_1, t_2) &\leq \Phi(u_1, u_2) \quad \text{for } t_1 \leq u_1 \text{ and } t_2 \leq u_2 . \end{aligned} \tag{8}$$

Then the functional d is regular. Moreover, there exists a leveled, hence optimal crosscut segmentation w.r.t. this functional.

Proof: The regularity of the bivariate functional d is a direct implication of the assumed regularity of $\tilde{d}_1$ and $\tilde{d}_2$ and the properties of the function Φ defined in (8).

Since $\tilde{d}_1$ and $\tilde{d}_2$ both are regular, it follows from [6] that there exists a leveled segmentation $\{\sigma_\mu^*\}$ of the interval $[a, b]$, and also a leveled segmentation $\{\tau_\nu^*\}$ of the interval $[c, d]$. But then $\{\Omega_{\mu,\nu}^*\}$, defined through

$$\Omega_{\mu,\nu}^* = \{(x, y) ; \sigma_{\mu-1}^* \leq x \leq \sigma_\mu^* ; \tau_{\nu-1}^* \leq y \leq \tau_\nu^*\}$$

for all μ and ν, establishes a leveled crosscut segmentation of Ω w.r.t d. $\square$

Example. Consider best approximation of the function $f_1(x, y) = \sqrt{x} + \sqrt{y}$ by linear polynomials. It was shown above that the minimal deviation, given in (5), is of the form (7). It follows from Theorem 6 in connection with the univariate result (see [4, 9]) that a leveled crosscut segmentation of the unit square is given by the choice

$$\sigma_\mu = \left(\frac{\mu(\mu+1)}{k(k+1)}\right)^2 \quad \text{and} \quad \tau_\nu = \left(\frac{\nu(\nu+1)}{r(r+1)}\right)^2 .$$

As expected, these optimal knot lines show a tendency to the origin, a singular point of the function f_1. For $k = 3$, $r = 4$, this is illustrated in Figure 1.

Corollary 7. *Consider the functional d in (7) for the choices $\tilde{d}_1([x_1, x_2]) = (x_2 - x_1)$ and $\tilde{d}_2([y_1, y_2]) = (y_2 - y_1)$. Then the equidistant segmentation is an optimal crosscut segmentation w.r.t. this functional.*

Proof: Since d only depends on the sidelength of R (and not on its location), the equidistant segmentation is leveled, hence optimal, due to Theorem 4. □

Example. 1) As shown above, the best approximation of the function $f(x, y) = x^2 + y^2$ on R w.r.t. P_1 is given by $((x_2 - x_1)^2 + (y_2 - y_1)^2)/8$, which is of the form desired.

2) The minimal deviation in approximating the function $f(x, y) = x^{n_1} y^{n_2}$ on R by polynomials of total degree $n_1 + n_2 - 1$ is given by

$$d(R) = \frac{(x_2 - x_1)^{n_1} \cdot (y_2 - y_1)^{n_2}}{2^{2(n_1 + n_2 - 1)}}$$

(cf. [5]); this is also a functional of right kind.

§3. Comparison Theorems

Usually it is not possible to obtain the minimal deviation in bivariate approximation explicitly, as done in the above examples. In those situations, comparison theorems as derived in this section can be helpful.

Theorem 8. *Again we consider a function $f \in C^2(R)$ of the form $f(x, y) = \varphi(x) + \psi(y)$ with $\varphi'' \geq 0$ and $\psi'' \geq 0$ in the interior of R. Let*

$$m_\varphi := \min_{x_1 \leq x \leq x_2} \varphi''(x) \ , \quad m_\psi := \min_{y_1 \leq y \leq y_2} \psi''(y) \ ,$$
$$M_\varphi := \max_{x_1 \leq x \leq x_2} \varphi''(x) \ , \quad M_\psi := \max_{y_1 \leq y \leq y_2} \psi''(y) \ .$$

Then the minimal deviation $\rho(f, P_1)$ is bounded by

$$\left((m_\varphi(x_2 - x_1)^2 + m_\psi(y_2 - y_1)^2 \right)/16 \ \leq \ \rho(f, P_1) \ \leq$$
$$\leq \ \left((M_\varphi(x_2 - x_1)^2 + M_\psi(y_2 - y_1)^2 \right)/16 \ .$$

Proof: For any C^2-function f, we denote the Hessian of f by H_f, i.e.,

$$H_f = \begin{pmatrix} f_{xx} & f_{xy} \\ f_{yx} & f_{yy} \end{pmatrix} .$$

Moreover, we define the auxiliary functions

$$g^{\min}(x, y) := (m_\varphi x^2 + m_\psi y^2)/2 \quad \text{and} \quad g^{\max}(x, y) := (M_\varphi x^2 + M_\psi y^2)/2 \ .$$

The Hessians of these functions are

$$H_{g^{\min}} = \begin{pmatrix} m_\varphi & 0 \\ 0 & m_\psi \end{pmatrix} \quad \text{and} \quad H_{g^{\max}} = \begin{pmatrix} M_\varphi & 0 \\ 0 & M_\psi \end{pmatrix} .$$

Due to our assumptions, the four matrices $H_{g^{\min}}$, $H_{f-g^{\min}}$, H_f, and $H_{g^{\max}-f}$ are all positive semidefinite in R, and so it follows from [3] that

$$\rho(g^{\min}, P_1) \leq \rho(f, P_1) \leq \rho(g^{\max}, P_1) .$$

As shown above (see (6)),

$$\rho(g^{\min}, P_1) = \left(m_\varphi(x_2 - x_1)^2 + m_\psi(y_2 - y_1)^2 \right) / 16$$

and

$$\rho(g^{\max}, P_1) = \left(M_\varphi(x_2 - x_1)^2 + M_\psi(y_2 - y_1)^2 \right) / 16 .$$

This completes the proof of Theorem 8. $\square$

For general functions $f \in C^2(R)$, inclusions of this type cannot be obtained so easily, but is possible to get close upper bounds.

Theorem 9. *Consider* $f \in C^2(R)$ *such that* H_f *is positive semidefinite. Let*

$$M_\varphi := \max_{x_1 \leq x \leq x_2} f_{xx}(x, y) , \quad M_\psi := \max_{y_1 \leq y \leq y_2} f_{yy}(x, y) .$$

Then the minimal deviation $\rho(f, P_1)$ *is bounded by*

$$\rho(f, P_1) \leq \left((M_\varphi(x_2 - x_1)^2 + M_\psi(y_2 - y_1)^2 \right) / 8 .$$

Proof: The proof is very similar to that of Theorem 8, where one now has to use the auxiliary function

$$g^{\max}(x, y) := M_\varphi x^2 + M_\psi y^2$$

with Hessian

$$H_{g^{\max}} = \begin{pmatrix} 2M_\varphi & 0 \\ 0 & 2M_\psi \end{pmatrix} . \quad \square$$

Example. For the function $f(x, y) = -\sqrt{x + y}$, we have

$$M_\varphi = M_\psi = (x_1 + y_1)^{-3/2}/4$$

for approximation on R $(x_1, y_1 > 0)$. Then Theorem 9 implies

$$\rho(f, P_1) \leq \frac{(x_2 - x_1)^2 + (y_2 - y_1)^2}{32 (x_1 + y_1)^{3/2}} . \tag{9}$$

A little computation shows that

$$\rho(f, P_1) = \frac{\left(\sqrt{x_1 + y_1} - \sqrt{x_2 + y_2} \right)^2}{8 \left(\sqrt{x_1 + y_1} + \sqrt{x_2 + y_2} \right)} . \tag{10}$$

For small rectangles, the bound in (9) is indeed asymptotically exact; this can be seen as follows. Let $x_2 = x_1 + \varepsilon$ and $y_2 = y_1 + \varepsilon$ with a small positive number ε. Then the right-hand side of (9) becomes

$$\frac{\varepsilon^2}{16 \cdot (x_1 + y_1)^{3/2}},$$

while, from (10)

$$
\begin{aligned}
\rho(f, P_1) &= \frac{\left(\sqrt{x_1 + y_1} - \sqrt{(x_1 + y_1) + 2\varepsilon}\right)^2}{8\left(\sqrt{x_1 + y_1} + \sqrt{(x_1 + y_1) + 2\varepsilon}\right)} \\[2mm]
&= \frac{\sqrt{x_1 + y_1}}{8} \cdot \frac{\left(-\frac{\varepsilon}{x_1 + y_1} + O(\varepsilon^2)\right)^2}{2 + O(\varepsilon)} \\[2mm]
&= \frac{\varepsilon^2}{16 \cdot (x_1 + y_1)^{3/2}} \cdot \left(1 + O(\varepsilon)\right).
\end{aligned}
$$

References

1. Lawson, C. L., Characteristic properties of the segmented rational minimax approximation problem, Numer. Math. **6** (1964), 293–301.

2. Meinardus, G., *Approximation of Functions: Theory and Numerical Methods*, Springer-Verlag, New York, 1967.

3. Meinardus, G., Zur abschätzung der minimalabweichung bei linearer approximation, Z. Angew. Math. Mech. **50** (1970), 509–514.

4. Meinardus, G., Some results in segmential approximation, Comput. Math. Appl. **33** (1997), 165–180.

5. Meinardus, G., G. Nürnberger, and G. Walz, Bivariate segment approximation and splines, Adv. Comp. Math. **6** (1996), 25–45.

6. Meinardus, G., G. Nürnberger, M. Sommer, and H. Strauß, Algorithms for piecewise polynomials and splines with free knots, Math. Comp. **53** (1989), 235–247.

7. Nürnberger, G., Optimal partitions in bivariate segment approximation, in *Curves and Surfaces with Applications in CAGD*, A. Le Méhauté, C. Rabut, and L. L. Schumaker (eds.), Vanderbilt Univ. Press, Nashville, 1997, pp. 271–278.

8. Nürnberger, G., M. Sommer, and H. Strauß, An algorithm for segment approximation, Numer. Math. **48** (1986), 463–477.

9. Powell, M. J. D., *Approximation Theory and Methods*, Cambridge University Press, Cambridge, 1981.

Guido Walz
Department of Mathematics and Computer Science
University of Mannheim
D-68131 Mannheim, Germany
walz@math.uni-mannheim.de

Variational Subdivision for
Natural Cubic Splines

Joe Warren and Henrik Weimer

Abstract. This paper explores the intrinsic link between natural cubic splines and subdivision. Natural cubic splines are defined via the variational problem of minimizing a simple approximation of bending energy. A subdivision scheme is derived which converges to the minimizer of this particular variational problem.

§1. Introduction

Geometric design is the study of the representation of shapes with mathematical models. Today, curved shapes are most commonly described using a parametric representation. These are based on the weighting of some number of control points with appropriate parametric basis functions. The most successful family of such basis functions is without any doubt the B-spline basis [7, 5].

More recently subdivision has evolved as a novel approach for representing shape [2, 3]. In this framework smooth shapes are represented as the limit of a repeated weighted averaging process of control points.

This paper exposes the intrinsic link between the two concepts: Starting with the variational definition of natural cubic spline curves, a representation by means of a subdivision process is derived. The principles underlying this derivation are applicable to other variational problems in perfect analogy. Thus, the method presented in this paper can be used to derive subdivision schemes which produce the minimizers of variational problems.

1.1. Natural Cubic Splines

Historically, a spline was a thin, flexible piece of wood used in drafting. The draftsman attached the spline to a sequence of anchor points on a drafting table. The spline was then allowed to slide through the anchor points and assume a smooth, minimum energy shape.

Approximation Theory IX, Volume 2: Computational Aspects
Charles K. Chui and Larry L. Schumaker (eds.), pp. 345–352.

In the 1940s and 50s, mathematicians realized that the behavior of a spline could be modeled mathematically. Let the shape of the spline be modeled by the graph of a function $F(t)$ over a domain interval $[a, b]$. The bending energy at parameter value $t \in [a, b]$ is roughly proportional to the value of the second derivative of F with respect to t. The total energy associated with the function F on the interval $[a, b]$ is thus approximately the integral of the square of the second derivative of F

$$\mathcal{E}[F] = \int_a^b F_{tt}(t)^2 dt. \tag{1}$$

The effects of the anchor points on the spline are modeled by constraining F to satisfy the additional interpolation conditions $F(t_i) = v_i$. If the vector T_0 denotes the parameter values for the interpolation conditions $\{t_0, t_1, \cdots, t_n\}$ and c_0 denotes the vector of the interpolation values $\{v_0, v_1, \cdots, v_n\}$, then these conditions can be stated more concisely in vector notation as

$$F(T_0) = c_0. \tag{2}$$

Functions that minimize (1) and satisfy (2) are called natural cubic splines.

1.2. Associated Nested Spline Spaces

As the interpolation values c_0 vary, the minimizing solutions $F(t)$ span a linear spline space $V(T_0)$ for a fixed knots sequence $T_0 \in [a, b]$. Natural cubic splines satisfy the simple differential equation

$$F_{tttt}(t) = 0 \quad \text{for all} \quad t \in [a, b] \tag{3}$$

which can be derived directly from (1) using the Euler-Lagrange equations. Therefore, the possible solutions are exactly the cubic polynomials in t.

The Euler-Lagrange condition (3) does not take into account the interpolation conditions (2) on F. If $F(t)$ is a natural cubic spline which satisfies the interpolation conditions (2) then F is actually a piecewise cubic polynomial function. The breaks between the polynomial pieces occur at the parameter values for the interpolation conditions, t_i. The polynomial pieces satisfy the differential equation (3) between two adjacent knots t_i and t_{i+1} but the differential equation may not hold at the actual knots.

Theorem 1. Let $V(T_0)$ and $V(T_1)$ be the spline spaces of minimizers of (1) with respect to the interpolation condition (2) over $[a, b]$ corresponding to knot sequences T_0 and T_1, respectively. Then, $T_0 \subset T_1 \Rightarrow V(T_0) \subset V(T_1)$.

Proof: Given any $F(t) \in V(T_0)$, consider $\hat{F}(t) \in V(T_1)$ with $\hat{F}(T_1) = F(T_1)$. Assuming $\mathcal{E}[F] > \mathcal{E}[\hat{F}]$ yields a contradiction to $\mathcal{E}[F]$ being minimal because $\hat{F}(T_0) = c_0$ and $\hat{F}$ has smaller energy, i.e., $F \notin V(T_0)$. Conversely, assuming $\mathcal{E}[F] < \mathcal{E}[\hat{F}]$ contradicts $\mathcal{E}[\hat{F}]$ being minimal because F itself interpolates $F(T_1)$, i.e., $\hat{F} \notin V(T_1)$. Therefore, $\mathcal{E}[F] = \mathcal{E}[\hat{F}]$. $\square$

Theorem 2. *Let $N_0(t)$ and $N_1(t)$ be vectors of basis functions for the spline spaces $V(T_0)$ and $V(T_1)$, respectively. Then $V(0) \subset V(T_1)$ implies $N_0(t) = N_1(t)S_0$ for some matrix S_0, called subdivision matrix.*

Proof: N_0 is a basis for $V(T_0)$. As $V(T_0) \subset V(T_1)$ we also have $N_0 \in V(T_1)$. As N_1 is a basis for $V(T_1)$ we can represent N_0 in terms of this basis. A column of S_0 contains the coefficients for basis functions in N_1 to represent one particular function in N_0. $\square$

Any function $F(t) \in V(T_0)$ can be represented in terms of the basis $N_0(t)$, $F(t) = N_0(t)p_0$ for some vector of coefficients p_0. Due to the nesting of the spline spaces $V(T_0)$ and $V(T_1)$, $F(t)$ can also be expressed in terms of the basis $N_1(t)$ for $V(T_1)$, $F(t) = N_1(t)p_1$. Because $N_0(t) = N_1(t)S_0$, it is possible to express the coefficients p_1 of $F(t)$ in terms of the coefficients p_0 as

$$p_1 = S_0 p_0. \tag{4}$$

Theorem 3. [6]. *If $T_0, T_1, T_2, \cdots$ defines an infinite sequence of finer and finer knot sequences which grows dense in $\mathbb{R}$ and $p_{k+1} = S_k p_k$, then the piecewise linear functions with values p_k over the knots T_k uniformly converge to $F(t)$ as $k \to \infty$.*

The B-Spline basis [1, 4, 5] is the only basis for natural cubic splines with local support. It has been studied to a great extent and was successful in many real world applications.

§2. Finite Element Solution

This section presents a derivation of natural cubic splines in terms of a finite element process starting from the specification of the variational problem (1). In particular we will derive B-splines over the domain $[0, 4]$ with initial knots at $T_0 = \{0, 1, 2, 3, 4\}$. The derivations presented here are general and can be used in the irregular case in perfect analogy. However, we chose to address the uniform, bounded case in this expository discussion for the sake of clearness and simplicity.

Solving a variational problem with finite elements involves three major steps:

- Choose a set of finite element basis functions $B_k(t) = \{b_1^k, b_2^k, \cdots\}$ and define a continuous version of the level k solution to $F_k(t) = B_k(t)p_k$ in vector notation where p_k denotes the set of coefficients of level k.
- Measure the energy of $F_k(t)$ via $\mathcal{E}[F_k] = p_k^T E_k p_k$.
- Minimize $p_k^T E_k p_k$ by solving $E_k p_k = 0$.

2.1. Finite Element Basis Functions

Due to the structure of the variational problem (1) the finite element basis functions have to lie in the Sobolev space $H_2([0, 4])$, i.e., they have to be piecewise quadratic functions with non-zero, square integrable derivatives up to order two.

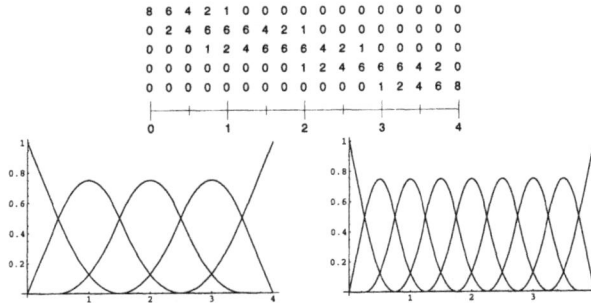

Fig. 1. Bézier control coefficients for the finite element basis functions of level 0 and finite element basis of levels 0 and 1 over the domain $[0, 4]$. The basis functions are piecewise quadratic and centered over the knots.

For reasons of computational stability and simplicity the basis functions should be centered over the knots, e.g., for level 0 the ith basis function should have maximum magnitude over the knot t_i.

A possible choice for these basis functions are quadratic piecewise Bézier curves. The Bézier control coefficients for the finite element basis $B_0(t)$ for level 0 are shown in Figure 1 together with the resulting basis functions for levels 0 and 1. Basis functions for finer grids are derived similarly using a refined knot vector.

These finite element basis functions can be used to define a continuous solution $F_k(t)$ for level k by weighting the coefficients p_k of the level k solution with the basis functions of level k, i.e.,

$$F_k(t) = B_k(t)p_k. \tag{5}$$

2.2. Energy Matrix and Inner Product

Using the continuous representation (5) of the level k solution of the finite element process, the energy of p_k can be assessed as $\mathcal{E}[F_k]$. This can be expressed as a quadratic form

$$\mathcal{E}[F_k] = p_k^T E_k p_k, \tag{6}$$

where E_k is a symmetric, positive definite matrix called the energy matrix. Defining an inner product

$$\langle f, g \rangle = \int_a^b f_{tt}(t) g_{tt}(t) dt \tag{7}$$

we get $E_k = (e_{ij}^k)_{n \times n}$ where $e_{ij}^k = \langle b_i^k(t), b_j^k(t) \rangle$. For the level 0, grid T_0 with knots at the integers over the domain $[0, 4]$,

$$E_0 = \begin{pmatrix} 1 & -2 & 1 & 0 & 0 \\ -2 & 5 & -4 & 1 & 0 \\ 1 & -4 & 6 & -4 & 1 \\ 0 & 1 & -4 & 5 & -2 \\ 0 & 0 & 1 & -2 & 1 \end{pmatrix}.$$

Energy matrices E_k for $k > 0$ have a similar structure. In particular, interior rows are scaled shifts of the center row of E_0.

2.3. Minimization with Interpolated Values

For $\mathcal{E}[F_k]$ to be minimal the control points p_k have to minimize $p_k^T E_k p_k$. This in turn implies that the derivative of this expressions, $2E_k p_k$ is zero.

However, p_k is not completely unknown. Some of the entries in p_k correspond to interpolation conditions c_0. Thus, p_k can be partitioned into a known part $p_k^n = c_0$ and an unknown part p_k^u as

$$p_k = \begin{pmatrix} p_k^n \\ p_k^u \end{pmatrix}. \tag{8}$$

Using this decomposition of p_k, Equation (6) becomes

$$\mathcal{E}[F_k] = p_k^T E_k p_k$$
$$= \begin{pmatrix} p_k^n & p_k^u \end{pmatrix} \begin{pmatrix} E_k^{nn} & E_k^{nu} \\ E_k^{un} & E_k^{uu} \end{pmatrix} \begin{pmatrix} p_k^n \\ p_k^u \end{pmatrix}$$
$$= p_k^n E_k^{nn} p_k^n + 2 p_k^u E_k^{un} p_k^n + p_k^u E_k^{uu} p_k^u.$$

E_k^{nn} contains the entries of E_k which have row and column indices corresponding to known grid values, E_k^{un} contains entries of E_k which have row indices of unknown and column indices of known coefficients and so on.

Now, $p_k^n E_k^{nn} p_k^n$ is constant, $2 p_k^u E_k^{un} p_k^n$ is linear in the unknowns p_k^u and $p_k^u E_k^{uu} p_k^u$ is a quadratic form in the unknowns. Therefore, the derivative of (6) with respect to the unknowns p_k^u can be expressed as $2 E_k^{uu} p_k^u + 2 E_k^{un} p_k^n$ and the minimizer of (6) is the solution p_k^u to

$$E_k^{uu} p_k^u + E_k^{uk} p_k^n = 0. \tag{9}$$

Using block matrix notation (9) can be rewritten as

$$\begin{pmatrix} E_k^{uu} & E_k^{un} \end{pmatrix} \begin{pmatrix} p_k^u \\ p_k^n \end{pmatrix} = 0.$$

Together with the interpolation constraints we get

$$\begin{pmatrix} E_k^{uu} & E_k^{un} \\ 0 & I \end{pmatrix} \begin{pmatrix} p_k^u \\ p_k^n \end{pmatrix} = \begin{pmatrix} 0 \\ c_0 \end{pmatrix} \tag{10}$$

where I is the identity matrix.

The solution $\begin{pmatrix} p_k^n & p_k^u \end{pmatrix}$ converges to the solution of the original variational problem as $k \to \infty$, see [6]. Furthermore, the original interpolation conditions c_0 are satisfied because p_k contains these values explicitly as p_k^n.

§3. Minimization with Interpolated Differences

The second row of the system (10) enforces p_k^n to interpolate the c_0. Therefore, all solutions to (10) interpolate the c_0. Unfortunately this leads to undesirable properties of the solution p_k to this system. In particular, changing just one of the entries in c_0 results in a global change in the solution. This section presents a new scheme which leads to solutions with much nicer properties.

To this effect, the interpolation constraints are replaced by constraints on the energy of the solution. Equation (10) can be rephrased to

$$\begin{pmatrix} E_k^{uu} & E_k^{un} \\ E_k^{nu} & E_k^{nn} \end{pmatrix} \begin{pmatrix} p_k^u \\ p_k^n \end{pmatrix} = \begin{pmatrix} 0 \\ E_0 p_0 \end{pmatrix}, \tag{11}$$

where p_0 is some initial set of control points centered over the initial knots T_0. Note that (11) forces the p_k to be chosen such that the energy values over the knots T_0 are the same as the original energy values at these knots, i.e., $E_0 p_0$. All remaining entries in $E_k p_k$ must be zero.

Condition (11) can be expressed more concisely using the notion of an upsampling matrix U_{k-1} which replicates coefficients associated with knots in T_{k-1} to the next finer grid T_k and forces zero coefficients at the knots $T_k - T_{k-1}$. Now (11) can be restated as

$$E_k p_k = U_{k-1} U_{k-2} \cdots U_0 E_0 p_0. \tag{12}$$

Again, Equation (12) states that the p_k are chosen such that $E_k p_k$ reproduces the energy $E_0 p_0$ over the original knots T_0 and forces zero energy at all knots in $T_k - T_0$.

Two such conditions (12) for levels k and $k + 1$ can be assembled as

$$E_k p_k = U_{k-1} \cdots U_0 E_0 p_0$$
$$E_{k+1} p_{k+1} = U_k \cdots U_0 E_0 p_0.$$

Combining them yields $E_{k+1} p_{k+1} = U_k E_k p_k$. Finally, applying the definition of the subdivision matrix, $p_{k+1} = S_k p_k$, yields

$$E_{k+1} S_k = U_k E_k. \tag{13}$$

Note that Equation (13) allows us to express the solution to the variational problem (1) in a subdivision scheme defined by the subdivision matrices S_k. Because the columns of E_{k+1} are not linearly independent this equation does not uniquely determine S_k. However, enforcing sparsity of S_k yields a unique solution, the subdivision matrices for uniform natural cubic splines.

The action of the subdivision matrices S_k satisfying (13) can be understood as follows: S_k produces control coefficients p_{k+1} for the knots T_{k+1} such that the differences $E_k p_k$ of the level k grid are maintained at the old knots in T_k and zero differences are forced at all new knots in $T_{k+1} - T_k$. Thus, telescoping this equation and taking the limit yields a surface which replicates

Fig. 2. Subdivision for natural cubic splines.

the differences $E_0 p_0$ of the original level 0 grid at the knots in T_0 and has zero difference everywhere else.

Expanding Equation (13) for $k = 0$ and solving for a sparse S_0 yields

$$S_0 = \frac{1}{8} \begin{pmatrix} 8 & 0 & 0 & 0 & 0 \\ 4 & 4 & 0 & 0 & 0 \\ 1 & 6 & 1 & 0 & 0 \\ 0 & 4 & 4 & 0 & 0 \\ 0 & 1 & 6 & 1 & 0 \\ 0 & 0 & 4 & 4 & 0 \\ 0 & 0 & 1 & 6 & 1 \\ 0 & 0 & 0 & 4 & 4 \\ 0 & 0 & 0 & 0 & 8 \end{pmatrix}.$$

Subdivision matrices for finer grids, $k > 0$ have a similar structure: interior columns of S_k are shifts of the center column of S_0. Note that S_k is sparse, i.e., the basis functions induced by the scheme are local. Furthermore, rows in S_k are all positive and sum to one. Hence, the scheme is affinely invariant and the limit curve lies in the convex hull of the control points. Finally, the first and last rows of S_k are unit vectors, i.e., the limit curve interpolates the first and last control points.

The subdivision scheme defined by the S_k is not interpolating. Initial control points p_0 must be chosen such that the limit curve satisfies (2). This can be accomplished using the interpolation matrix I_0 which contains samples of the basis functions induced by the subdivision scheme at the knots T_0. If c_0 is the set of interpolation conditions over T_0 then we get p_0 by $c_0 = I_0 p_0$. For the subdivision scheme presented here we get

$$I_0 = \frac{1}{6} \begin{pmatrix} 6 & 0 & 0 & 0 & 0 \\ 1 & 4 & 1 & 0 & 0 \\ 0 & 1 & 4 & 1 & 0 \\ 0 & 0 & 1 & 4 & 1 \\ 0 & 0 & 0 & 0 & 6 \end{pmatrix}.$$

An application of the scheme is shown in Figure 2.

§4. Conclusion and Future Work

This paper presented a binary subdivision scheme for uniform bounded B-splines. The scheme is stationary, i.e., the same subdivision masks are applied at different levels of the subdivision process.

The method presented here extends to higher order variational problems even in higher dimensions. The steps to construct a subdivision scheme which converges to the minimizer of a variational problem can be outlined as follows: First, find appropriate finite element basis functions. Next, compute the energy matrix based on the inner product (7). Finally, find subdivision matrices S_k by solving $E_{k+1}S_k = U_k E_k$. This system is usually rank-deficient and a subdivision scheme with particularly nice properties, e.g., sparseness, can be chosen among the possible solutions.

In the future we plan to derive a subdivision scheme for thin plate splines on irregular grids and link the resulting scheme to the representation of these surfaces in terms of radial basis functions. Also, we plan to investigate the link between variational subdivision methods and multiresolution analysis.

Acknowledgments. This work has been supported in part under National Science Foundation grant CCR-9500572, Texas Advanced Technology Program grant 003604-010 and by Western Geophysical.

References

1. Böhm, W., Inserting new knots into B-spline curves, Computer Aided Design **12** (1980), 199–201.

2. Catmull. E. and J. Clark, Recursively generated B-spline surfaces on arbitrary topological meshes, Computer Aided Design **10** (1978), 350–355.

3. Cavaretta A., W. Dahmen, and C. Micchelli, *Stationary Subdivision*, Memoirs of the AMS **453**, 1991.

4. Farin, G., *Curves and Surfaces for CAGD*, 3rd Edition, Academic Press, 1993.

5. Hoschek, J. and D. Lasser, *Fundamentals of Computer Aided Geometric Design*, A K Peters, 1993.

6. Oden, J. T. and J. N. Reddy, *An Introduction to the Mathematical Theory of Finite Elements*, John Wiley & Sons, 1976.

7. Schoenberg, I. J., Contributions to the problem of approximation of equidistant data by analytic functions, Quart. Appl. Math. **4** (1946) 45–99.

Joe Warren and Henrik Weimer
Department of Computer Science
Rice University
P.O. Box 1892
Houston, TX 77005-1892
{jwarren,henrik}@rice.edu

Fitting Conic Sections to Measured Data

G. Alistair Watson

Abstract. Fitting conic sections to data is an important problem with
many applications. Often the data are produced using a co-ordinate measuring machine with a touch probe, although conventional fitting methods
(orthogonal distance regression, algebraic fitting) take no account of this.
We consider here an approach which makes use of the measurement design,
and also allows standard statistical theory to apply. The connection with
conventional methods is indicated, algorithms are developed, and some
numerical results given.

§1. Introduction

There is great interest in the problem of fitting conic sections to given data
because there are many applications. These arise for example in computer
graphics, petroleum engineering, statistics, metrology, astronomy, reflectom-
etry etc: see, for example, [2, 3, 4, 8, 9, 10, 17]. Metrology is measurement
science, and of particular interest is the analysis of co-ordinate measurements
of a manufactured part, involving an assessment of the quality of the part by
comparing it with a theoretical model. The comparison is usually made by
generating data by identifying points on the surface of the part with respect to
a particular frame of reference; it is then necessary to fit the theoretical model
to these data in some way. Because of measurement errors or manufacturing
limitations, an exact fit is not usually achieved, and the question then arises
as to how best to achieve the fit.

The most commonly used measure of fit is the least squares norm, although if robustness is required in the presence of gross errors or wild points,
then other measures, for example the l_1 norm, may be more appropriate. On
the other hand, it should be pointed out that strict accept/reject tests may
require the use of the Chebyshev norm. In practice, the data on the manufactured part are normally obtained by using a co-ordinate measuring machine.
This is a device in which a probe moving in a particular direction identifies
a point on the part surface; the part is moved (for example rotated) with
respect to the machine and this operation is repeated. Hulting [7] has pointed

Approximation Theory IX, Volume 2: Computational Aspects   353
Charles K. Chui and Larry L. Schumaker (eds.), pp. 353–360.

out that a feature of the methods currently in use (for example minimizing orthogonal distances from the part to the model) is that they take no account of the way in which the data are obtained, although this is a fundamental part of the whole process.

The purpose of this paper is to consider the problem in a manner which takes specific account of the probe directions (relative to a particular frame of reference). The fitting criterion is due to Hulting [7], and he claims that the appeal of the approach lies not just in the use of the measurement design, but also in its compliance with traditional fixed-regressor assumptions (enabling standard inference theory to apply). This stands in contrast to, say, orthogonal least squares fitting.

We begin by introducing the mathematical problem, and we will show how the problem which arises relates to traditional approaches to fitting curves to data when the least squares norm is used. The main contribution of the paper lies in the consideration of methods of solving the problems in the special case when the curves to be fitted can be given in terms of a parameter. Methods for the more conventional approach to particular examples of these fitting problems have been studied recently in a series of papers by Späth [11, 12, 13, 14], and also in the papers [5, 16]. We will begin by considering the problem for a general curve or surface.

§2. The General Fitting Problem

Suppose that a manufactured part in n dimensions (in practice n equals 2 or 3, although from a mathematical point of view this is not a limitation) is required to take the shape of the model curve

$$F(x, \alpha) = 0,$$

where $x \in R^n$ is the vector of variables, and $\alpha \in R^p$ is a vector of free parameters characterizing the model. The problem is then the estimation of these parameters. Suppose that a point on the surface of the manufactured (or actual) part is measured as follows. A probe (a ball of radius ρ attached to the end of a rod) is positioned so that the rod passes through a fixed point, and is aligned with a given direction v which we assume to be a unit vector with $v^T v = 1$. The rod is then moved in the direction v (the "approach vector") until the ball makes contact with the part. The situation is as illustrated for an elliptical part in 2 dimensions in Figure 1; included is the model shown for a particular set of parameters α, and the probe is shown also making contact with the model part.

The process is repeated a number of times for different directions v_i (with respect to a particular frame of reference), so that for each i, the part is moved relative to the measuring tool. Let the co-ordinates of the centre of the probe head where it touches the part be a_i, and let c_i denote the corresponding centre of the probe where it touches the model. We will assume that the probe head touches the actual part at the point $a_i + \rho v_i$ and touches the model at the point $c_i + \rho v_i$ (although these are approximations in general). Then the error

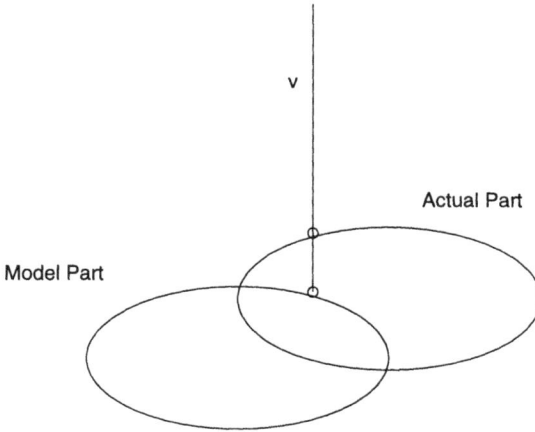

Fig. 1. The measuring process.

δ_i as measured by the difference between the point where the probe touches the actual part and the point where it touches the model is given by

$$\delta_i = (c_i - a_i)^T v_i, \quad i = 1, \ldots, m.$$

By assumption, the points $(c_i + \rho v_i)$ will lie on the surface of the model, and so must satisfy

$$F(c_i + \rho v_i, \alpha) = 0, \quad i = 1, \ldots, m,$$

or equivalently,

$$F(a_i + \mu_i v_i, \alpha) = 0, \quad i = 1, \ldots, m,$$

for some numbers μ_i. Further

$$\delta_i^2 = (\mu_i - \rho)^2, \quad i = 1, \ldots, m.$$

Thus the appropriate least squares minimization problem is

$$\text{minimize} \sum_{i=1}^{m} \delta_i^2 \tag{1}$$

$$\text{subject to } F(a_i + \rho v_i + \delta_i v_i, \alpha) = 0, \quad i = 1, \ldots, m,$$

where the variables are δ_i, $i = 1, \ldots, m$ and $\alpha \in R^p$.

Now consider the application of the technique of implicit orthogonal distance regression [1, 6, 15] to this problem. Data corresponding to values of a_i are obtained, assumed to be in error by ϵ_i, and it is required to solve the problem:

$$\text{minimize} \sum_{i=1}^{m} (\epsilon_i^T \epsilon_i)$$

$$\text{subject to } F(a_i + \epsilon_i, \; \alpha) \;=\; 0, \; i = 1, \ldots, m.$$

The above two problems are equivalent if

$$\rho = 0$$

and

$$\epsilon_i = \delta_i v_i, \; i = 1, \ldots, m.$$

Thus the present problem essentially *specifies* the direction of each ϵ_i (aligned with v_i) and minimizes the sum of squares of the distances of the observed points to the curve along these directions.

Now consider the special case when one of the variables can be thought of as the dependent variable. We can write for each i

$$a_i^T = (y_i, \; z_i^T),$$

where $y_i \in R$ is a value of the dependent variable, and we have the problem

$$\text{minimize } \sum_{i=1}^{m} \delta_i^2$$

$$\text{subject to } F(y_i + \delta_i, \; z_i, \; \alpha) \;=\; 0, \; i = 1, \ldots, m.$$

By assumption, this system of equations can be rewritten in the form

$$y_i + \delta_i \;=\; g(z_i, \; \alpha), \;\; i = 1, \ldots, m,$$

for some g. The constraints can then be removed in the usual way, and we have the traditional (explicit) least squares problem. This is just the present problem if

$$v_i = e_1, \;\; i = 1, \ldots, m,$$

and $\rho = 0$, that is e_1 represents the direction associated with the dependent variable (the "vertical" direction).

To summarize the different approaches:

(a) for the present problem formulation, all the v_i are prescribed as part of the problem formulation, as is ρ,

(b) for implicit least squares, or implicit orthogonal distance regression, all the directions v_i are left as parameters in the problem, to be calculated as part of the solution, and ρ is set to zero,

(c) for explicit least squares, all the directions v_i are prescribed to be e_1, the vertical direction (associated with the dependent variable), and ρ is set to zero.

§3. An Algorithm

Let $x_i = a_i + \rho v_i$, $i = 1, \ldots, m$. Then (1) is a special case of the more general problem

$$\text{minimize } \|\delta\| \qquad (2)$$

$$\text{subject to } F(x_i + \delta_i v_i, \ \alpha) \ = \ 0, \ i = 1, \ldots, m,$$

where the norm is a given norm on $\mathbb{R}^m$, and where the variables are δ_i, $i = 1, \ldots, m$ (the components of $\delta \in \mathbb{R}^m$), and $\alpha \in \mathbb{R}^p$. The components x_i, v_i, $i = 1, \ldots, m$ are given, with $v_i^T v_i = 1$, $i = 1, \ldots, m$. This is a constrained optimization (or best approximation) problem and as such it can be solved in many ways. We will focus attention on special cases of this problem, where structure can be exploited to facilitate computation.

Suppose that $n = 2$ and that a point lying on the curve $F(x, \ \alpha) = 0$ can be expressed parametrically in the form

$$x \ = \ A(t)\alpha \ + \ h(t), \qquad (3)$$

where $A(t)$ is a $2 \times p$ matrix which depends on the *scalar* parameter t which parameterizes the position of x. The general approach which we take based on the use of (3) is as follows. The constraints which must be satisfied are

$$x_i + \delta_i v_i \ = \ A(t_i)\alpha \ + \ h(t_i), \ \ i = 1, \ldots, m. \qquad (4)$$

Given α, for each i we can solve for t_i, resolving ambiguity by choosing the particular t_i which minimizes δ_i^2. In this way, we can interpret t_i as a function of α. Notice that premultiplying each equation of (4) by v_i, it follows that

$$\delta_i \ = \ v_i^T(A(t_i)\alpha \ + \ h(t_i) \ - x_i), \ \ i = 1, \ldots, m.$$

Thus we can regard $\|\delta\|$ as a function of α and define a descent direction to improve the current value. Systematic repetition of this process can lead to the required minimum. Let

$$f(\alpha) \ = \ \|\delta\|.$$

Then the problem is the minimization of f, a nonlinear best approximation problem. The Gauss-Newton method identifies directions d at the current value of α by solving the linearized problem

$$\text{find } d \in \mathbb{R}^p \text{ to minimize } \|\delta + Gd\|, \qquad (5)$$

where $G \in R^{m \times p}$ has ith row $\nabla_\alpha \delta_i$, for $i = 1, \ldots, m$, providing this is defined. It is then necessary to conduct a line search in the direction of the solution to get the next approximate value of α. A step length should be chosen to ensure a reduction in f: there are various possibilities, the simplest just being to start with a step length of 1, and successively multiply by one half until a reduction is obtained.

The matrix G will be defined provided that we can obtain derivatives of δ with respect to α, and this will follow if there is continuously differentiable dependence of t_i on α as previously defined using (4). Now if we can differentiate (4) with respect to α, we will have

$$v_i \nabla_\alpha \delta_i = A(t_i) + (A'(t_i)\alpha + h'(t_i))\nabla_\alpha t_i, \qquad i = 1, \ldots, m. \qquad (6)$$

For each i, this is a system of $2p$ linear equations for the $2p$ components of the row vectors of derivatives $\nabla_\alpha \delta_i$ and $\nabla_\alpha t_i$. In addition because the matrices which multiply the unknown derivatives are rank one, it separates into p 2×2 systems of equations for corresponding pairs of components of the unknown derivative vectors. Thus it can easily be solved, and some manipulation shows that we have for each i,

$$\nabla_\alpha \delta_i = C_i(A'(t_i)\alpha + h'(t_i))^T M A(t_i),$$

where

$$C_i = \frac{1}{\det[v_i : A'(t_i)\alpha + h'(t_i)]},$$

$$M = \begin{bmatrix} 0 & -1 \\ 1 & 0 \end{bmatrix},$$

and the dash denotes differentiation with respect to t. Provided that the determinants of the 2×2 matrices on the denominators of C_i are nonzero, these derivatives exist, and they can then be substituted into (5). Under these circumstances, it is then straightforward to implement a (damped) Gauss-Newton method for the minimization of f.

§3. An Example

We will illustrate the method by fitting a circle to one of the data sets previously used by other authors, using the least squares norm. Table 1 gives such a set, used by Späth [11], together with a set of direction vectors (components v_1, v_2 in un-normalized form) for each point.

x_1	8	3	2	7	6	6	4
x_2	1	6	3	7	1	10	0
v_1	2	-1	-1	1	1	1	-1
v_2	-3	1	-1	1	-4	4	-4

Tab. 1. Späth data.

With an initial approximation the circle with centre (2,2) and radius 4, the damped Gauss-Newton method converges to a solution in 7 iterations, using the stopping criterion

$$\|\nabla_\alpha f\|_2 < 0.0001.$$

The computed circle has centre (5.6333, 4.4107) and radius 3.9676. A picture of the initial and final approximations, together with the data points, is shown in Figure 2.

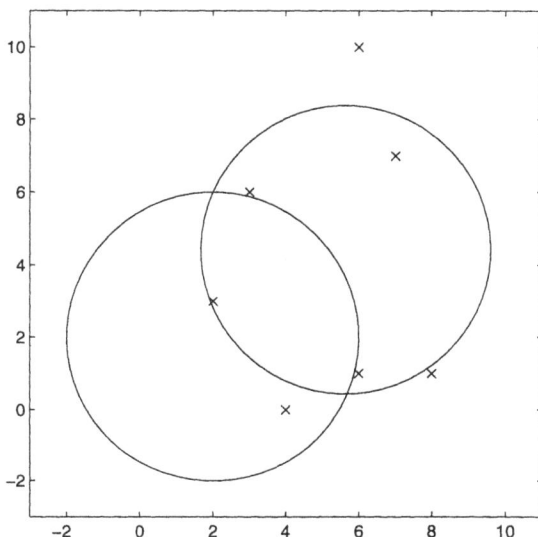

Fig. 2. Späth data, circle fit.

References

1. Boggs, P. T., R. H. Byrd, and R. B. Schnabel, A stable and efficient algorithm for nonlinear orthogonal distance regression, SIAM J. Sci. Stat. Comp. **8** (1987), 1052–1078.

2. Bookstein, F. L., Fitting conic sections to scattered data, Computer Graphics and Image Processing **9** (1979), 56–71.

3. Butler, B. P., A. B. Forbes, and P. M. Harris, Algorithms for geometric tolerance assessment, NPL Report DITC 228/94, 1994.

4. Forbes, A. B., Least-squares Best Fit Geometric Elements, in *Algorithms for Approximation II*, J. C. Mason and M. G. Cox (eds.), Chapman and Hall, London, 1990.

5. Gander, W., G. H. Golub, and R. Strebel, Fitting of circles and ellipses: least square solution, BIT **34** (1994), 556 –577.

6. Helfrich H.-P. and D. Zwick, A trust region method for implicit orthogonal distance regression, Numer. Alg. **5** (1993), 535–545.

7. Hulting, F. L., Discussion contribution to the paper by M. M. Dowling, P. M. Griffin, K-L. Tsui, and C. Zhou, Statistical issues in geometric feature inspection using coordinate measuring machines, Technometrics **39** (1997), 18–20.

8. Kasa, I., A circle fitting procedure and its error analysis, IEEE Trans. Instr. Measurement **25** (1976), 8–14.

9. Pratt, V., Direct least-squares fitting of algebraic surfaces, Computer Graphics **21** (1987), 145–152.

10. Rosin, P. L., A note on the least squares fitting of ellipses, Pattern Recognition Letters **14** (1993), 799–808.

11. Späth, H., Orthogonal least squares fitting by conic sections, in *Recent Advances in Total Least Squares Techniques and Errors-in-Variables Modeling*, S. Van Huffel (ed.), SIAM, Philadelphia, 1997, pp. 259–264.

12. Späth, H., Least squares fitting of ellipses and hyperbolas, Computational Statistics **12** (1997), 329–341.

13. Späth, H., Least squares fitting by circles, Computing, to appear.

14. Späth, H., Orthogonal squared distance fitting with parabolas, in *IMAGS-GAMM International Symposium on Numerical Methods and Error Bounds*, G. Alefeld and J. Herzberger (eds.), Akademie-Verlag, Berlin, 1996, pp. 261–269.

15. Strebel, R., D. Sourlier, and W. Gander, A comparison of orthogonal least squares fitting in coordinate metrology, in *Recent Advances in Total Least Squares Techniques and Errors-in-Variables Modeling*, S. Van Huffel (ed.), SIAM, Philadelphia, 1997, pp. 249–258.

16. Varah, J. M., Least squares data fitting with implicit functions, BIT **36** (1996), 842–854.

17. Zwick, D. S., Applications of orthogonal distance regression in metrology, in *Recent Advances in Total Least Squares Techniques and Errors-in-Variables Modeling*, S. Van Huffel (ed.), SIAM, Philadelphia, 1997, pp. 265–272.

G. Alistair Watson
Department of Mathematics
University of Dundee
Dundee DD1 4HN
Scotland
gawatson@mcs.dundee.ac.uk
http://www.mcs.dundee.ac.uk:8080/~gawatson/

Numerical Solution of Variational Problems by Radial Basis Functions

Holger Wendland

Abstract. In this paper we investigate the application of radial basis functions to solve variational problems as they appear in the weak formulation of elliptic partial differential equations. Two algorithms based on a multilevel scheme are introduced. Their numerical advantages and disadvantages are discussed and demonstrated on an example.

§1. Introduction

The classical Galerkin approach for the solution of elliptic partial differential equations leads to a variational problem, which can be solved by discretization. This is the way finite elements treat this kind of problems in a very successful way. The aim of this paper is to give first ideas how radial basis functions can be used in this setting. There are several good reasons to investigate this meshless method (cf. [1] for an overview on general meshless methods and the literature in [3] for collocation-based methods), e.g., the independence of space dimension (in contrast to classical finite element methods) and the independence of the underlying grid. The latter might be interesting in time dependent problems with moving boundaries, where classical finite element methods spend a lot of time for generating and adapting the mesh.

The paper [9] deals with radial basis functions in the context of Galerkin approximation. But in contrast to its theoretical approach, we now want to investigate the numerical side of the problem.

Therefore, we concentrate on the Helmholtz equation with natural boundary conditions:

$$-\Delta u + u = f \quad \text{in } \Omega,$$

$$\frac{\partial}{\partial \nu} u = 0 \quad \text{on } \partial\Omega.$$

Here $\Omega \subseteq \mathbb{R}^d$ is a bounded domain with a sufficiently smooth boundary $\partial\Omega$ and f is a given function. The outer unit normal vector is denoted by ν. See also [5] for the connection of radial functions and the Helmholtz equation.

Approximation Theory IX, Volume 2: Computational Aspects   361
Charles K. Chui and Larry L. Schumaker (eds.), pp. 361–368.
Copyright © 1998 by Vanderbilt University Press, Nashville, TN.
ISBN 0-8265-1326-3.

The Galerkin approach chooses the Sobolev space $V = H^1(\Omega)$ consisting of all functions $g \in L_2(\Omega)$ possessing weak derivatives of order one in $L_2(\Omega)$. It solves the problem

$$\text{find } u \in V \text{ with } a(u,v) = (f,v)_{L_2(\Omega)} \text{ for all } v \in V, \qquad (1)$$

where the bilinear form $a : V \times V \to \mathbb{R}$ is given by

$$a(u,v) = \int_\Omega (\nabla u \cdot \nabla v + uv) \; dx.$$

§2. Radial Basis Function Discretization

To solve the problem (1) numerically, one usually fixes a finite dimensional subspace $V_N \subseteq V$ and solves the discrete problem

$$\text{find } u_N \in V_N \text{ with } a(u_N,v) = (f,v)_{L_2(\Omega)} \text{ for all } v \in V_N. \qquad (2)$$

Classical finite elements, for example, choose a triangulation of the domain Ω and take V_N as the space of piecewise polynomials induced by the triangulation. Here, we want to use radial basis functions. Thus we select a (radial) function $\Phi : \mathbb{R}^d \to \mathbb{R}$, preferably compactly supported, and a set of pairwise distinct centers $X = \{x_1, \ldots, x_N\} \subseteq \Omega$ and form

$$V_N = \text{span}\{\Phi(\cdot - x_j) \; : \; 1 \le j \le N\}.$$

If the function Φ is positive definite and possesses a Fourier transform $\widehat{\Phi}$ which decays like $(1 + \| \cdot \|_2)^{-2\beta}$, the following result was proven in [9]:

Theorem 1. *Let the bounded domain Ω possess a C^1-boundary and let the solution u of (1) satisfy $u \in H^k(\Omega)$. Let h denote the fill distance $h = \sup_{x \in \Omega} \min_{1 \le j \le N} \|x - x_j\|_2$. If $\beta \ge k > \frac{d}{2} + 1$, the error between u and the discrete solution u_N of (2) can be bounded for sufficiently small h by*

$$\|u - u_N\|_{H^1(\Omega)} \le Ch^{k-1}\|u\|_{H^k(\Omega)}.$$

This reflects the expected order of convergence, comparable to that of classical finite element methods. But to achieve this order we have to work in the so-called stationary case. We have to fix a basis function, especially its support and let the data density approach zero. This means that the advantage of a compact support is annihilated, and the corresponding stiffness matrices lose their sparsity and become ill conditioned. On the other hand, if we work in the stationary case, which means that we scale the support proportional to the data density, the matrices keep their sparsity, have a uniformly bounded condition, but the process does not converge.

This problem is well known in the theory of interpolation by radial basis functions. There, a promising solution was initially given by Floater and Iske in [4]. The solution is based on a multilevel scheme where on each level the residual of the previous level is interpolated.

It is our aim to formulate a multilevel scheme based on these ideas for the Galerkin approximation.

§3. Multilevel Algorithms

The multilevel algorithm we want to introduce can be formulated in a more general way. Assume that V is a Hilbert space with inner product $(\cdot, \cdot)$ and assume further that we have on V a strictly V-elliptic bilinear form $a : V \times V \to \mathbb{R}$. Then the solution u of (1) is uniquely determined.

Next, we take a sequence of finite dimensional subspaces $V_1, V_2, \ldots \subseteq V$, thinking about V_j to represent more details, if j turns to infinity. But instead of solving the discrete problem (2) for each subspace V_j we solve the problem with different right-hand sides.

Algorithm 2.
1) *Set* $v_0 = 0$,
2) *For* $k = 1, 2, \ldots, m$
    a) *Find* $u_k \in V_k$ *with* $a(u_k, v) = (f, v) - a(v_{k-1}, v)$ *for all* $v \in V_k$.
    b) *Set* $v_k = v_{k-1} + u_k$.

The final solution v_m still satisfies

$$a(v_m, v) = (f, v) \text{ for all } v \in V_m, \tag{3}$$

but it is in contrast to the solution of (2) not an element of V_m but of $V_1 + \cdots + V_m$. Even if no convergence results are known so far, property (3) implies the stability result

$$\|u - v_m\|_a \le \|u - v_{m-1}\|_a \le \cdots \le \|u\|_a,$$

where $\| \cdot \|_a$ denotes the energy norm coming from the bilinear form $a(\cdot, \cdot)$. Numerical tests show that after the correction on the finest level is done, the error seems to be dominated by a global behavior (see Figure 1). This motivates us to go back to the coarsest level and to repeat the process. In this case a theoretical justification can be given. Let us denote with $P_j : V \to V_j$ the orthogonal projector with respect to $a(\cdot, \cdot)$. Then (3) and (1) imply $u_m = P_m(u - v_{m-1})$. This leads to

$$u - v_m = (E - P_m)(E - P_{m-1}) \cdots (E - P_1)u$$
$$=: R_m u,$$

where E denotes the identity on V. It is well known (cf. [2] and the literature therein) that an iterative application of R_m leads to linear convergence in the following sense:

Theorem 3. *Let* u^* *denote the best approximation to* u *from* $V_1 + \ldots + V_m$ *with respect to the energy norm* $\| \cdot \|_a$. *Let* $\tilde{u}_k$ *be given by* $R_m^k u = u - \tilde{u}_k$. *Then there exists a* $\theta \in (0, 1)$ *such that*

$$\|u^* - \tilde{u}_k\|_a \le \theta^k \|u\|_a.$$

The constant θ is determined by the angle between the subspaces V_j. The functions $\tilde{u}_k$ coincide with the functions v_k in Algorithm 2, if the index k runs farther than m and the spaces and projections are taken modulo m.

Algorithm 4.
  1) *Fix $m \in \mathbb{N}$, set $w_0 = 0$,*
  2) *For $k = 0, 1, \ldots$,*
    a) *Set $v_0 = w_k$.*
    b) *Apply Algorithm 2. Denote the solution with $v_m(w_k)$.*
    c) *Set $w_{k+1} = v_m(w_k)$.*

For a practical implementation, we only have to deal with m functions u_k and to update them.

Theoretically, the quality of both algorithms depends on the constant θ, i.e., on the angles between the subspaces, and on the approximation properties of $V_1 + \cdots + V_m$ for sufficiently large m. Numerically, the quality depends further on the possibility of a fast (numerical) evaluation of the projections. Actually, step 2a) of Algorithm 2 has often to be replaced by solving the problem

$$\text{find } u_k \in V_k \text{ with } a(u_k, v) = (f, v)_k - \sum_{j=1}^{k-1} a_j(u_j, v) \text{ for all } v \in V_k,$$

where $(\cdot, \cdot)_j$ and $a_j(\cdot, \cdot)$ denote numerical versions of $(\cdot, \cdot)$ and $a(\cdot, \cdot)$ on the jth-level. We will discuss this problem in the next section.

Finally, let us return to radial basis functions. To construct the subspaces V_j, we take a chain of sets of centers

$$X_1 \subseteq X_2 \subseteq \cdots \subseteq X_m \subseteq \Omega,$$

and a decreasing chain of support radii

$$\delta_1 \geq \delta_2 \geq \cdots \geq \delta_m.$$

Then the subspaces are given by

$$V_j := \{\Phi_{\delta_j}(\cdot - x) \; : \; x \in X_j\}$$

with the scaled basis function $\Phi_\delta(\cdot) = \Phi(\frac{\cdot}{\delta})$. For simplicity, we assume Φ always to have initial support in $B_1(0) = \{x \in \mathbb{R}^d \; : \; \|x\| \leq 1\}$. Note, that we actually do not need an increasing chain of set of centers, in contrast to the multilevel scheme in case of interpolation.

§4. Numerical Aspects

If we want to solve the partial differential equation of Section 1 using the defined subspaces and algorithms, we have to compute $a(\Phi_\mu(\cdot - u), \Phi_\varepsilon(\cdot - v))$ efficiently. In most cases, one of the supports of the shifted and scaled basis function is completely contained in the domain Ω. If Φ is in C^2 we then have

$$
\begin{aligned}
a(\Phi_\mu(\cdot - u), \Phi_\varepsilon(\cdot - v)) = &\int_{\mathbf{R}^d} (\nabla \Phi_\mu)(x - u) \cdot (\nabla \Phi_\varepsilon)(x - v) dx \\
&+ \int_{\mathbf{R}^d} \Phi_\mu(x - u)\Phi_\varepsilon(x - v) dx \\
= &-\mu^{d-2}(\Delta \Phi) \ast \Phi_{\frac{\varepsilon}{\mu}}\left(\frac{v - u}{\mu}\right) + \mu^d \Phi \ast \Phi_{\frac{\varepsilon}{\mu}}\left(\frac{v - u}{\mu}\right).
\end{aligned}
\tag{4}
$$

Here, $f \ast g$ denotes the usual convolution $f \ast g(x) = \int f(y)g(x - y)dy$. If, in addition, Φ is a radial function, so are $\Delta \Phi$ and the convolutions. Thus in the earlier mentioned case, $a(\Phi_\mu(\cdot - u), \Phi_\varepsilon(\cdot - v)) =: F_{\mu,\varepsilon}(\|u - v\|)$ is a radial function.

If the space dimension $d = 2n + 1$ is odd and the basis function $\Phi = \phi(\|\cdot\|)$ is radial, the function $F_{\mu,\varepsilon}(r)$ can be computed using only one dimensional operations (cf. [10]). If we denote the radial Laplacian of ϕ with ψ, i.e., $\psi(r) = \phi''(r) + \phi'(r)(d - 1)/r$, with I the operator $I\phi(r) := \int_r^\infty \phi(t)t dt$ and with D its inverse $D\phi(t) = -\frac{1}{t}\phi'(t)$, we have

$$
\begin{aligned}
(2\pi)^{\frac{1-d}{2}} F_{\mu,\varepsilon}(r) = &-\mu^{d-2} D^n \left\{ (I^n \psi) \ast_1 (I^n \phi_{\frac{\varepsilon}{\mu}}) \right\}(r/\mu) \\
&+ \mu^d D^n \left\{ (I^n \phi) \ast_1 (I^n \phi_{\frac{\varepsilon}{\mu}}) \right\}(r/\mu).
\end{aligned}
\tag{5}
$$

The subscript on the convolution operator indicates that this is a one-dimensional convolution. If ϕ is supposed to be a polynomial within its support, all operations can be done in an elementary way.

In case of even space dimension, the function $F_{\mu,\varepsilon}$ is still radial, and a formula similar to (5) is valid, but needs more complicated operators D and I (cf. [7]). But a different approach is also possible. The choice of the support on each level determines the ratio ε/μ. If we use $\delta_j = 2^{-j}\delta_0$ and take the symmetry of the problem into account, we can restrict ourselves to ratios of the form 2^{-j} with $0 \leq j \leq m$ (remember, m denotes the number of levels). Thus, if we tabulate $(\Delta \Phi) \ast \Phi_\delta$ and $\Phi \ast \Phi_\delta$ for these δ's on a sufficiently fine partition of $[0, 1 + \delta]$ (both are radial), we can compute an approximative value of $F_{\mu,\varepsilon}$ by linear interpolation. This is the way we work in space dimension two.

To tabulate the just mentioned functions and to evaluate the bilinear form in case of both supports overlapping the boundary, and to evaluate the right-hand side of (2), integrals of the form

$$
\int_\Omega \Phi(\omega)g(\omega)d\omega
$$

have to be computed numerically. To do this in $\mathbf{R}^2$, we use the obvious strategy of introducing polar coordinates:

$$
\int_\Omega \Phi(\omega)g(\omega)d\omega = \int_0^\infty r\phi(r) \int_0^{2\pi} g(r\cos\phi, r\sin\phi)\chi_\Omega(r\cos\phi, r\sin\phi)d\phi dr.
\tag{6}
$$

Here, χ_Ω denotes the characteristic function of the domain Ω. Now, we can use standard quadrature formulas like Gauss quadrature to calculate the iterated integrals. Formula (6) is the only way to handle arbitrary domains. For special domains improvements might be achieved by taking advantage of the special form of the domain.

Note, that Algorithm 4 does not need more numerical integrations than Algorithm 2 if a reasonable storage strategy is used. Furthermore, both algorithms can easily be parallelized.

Up to now, we have only reported on the problem of numerical integration and not on the problem of solving the discrete linear systems. But actually, this is no real problem. In our numerical tests (see next section) we used a standard conjugate gradient method without preconditioning and only few iterations on each level were needed. This is a consequence of the form of the stiffness matrix. It is known from the theory of interpolation with radial basis functions (cf. [6]) that if all matrix entries are given by (4) with $\varepsilon = \mu = \delta$, the condition number of the matrices depends only on the ratio h/δ, and is independent of the current level if we choose the supports proportional to the data density. It seems that for sufficiently nice domains the boundaries have no damaging influence on this fact.

§5. An Example

We have tested both algorithms on the partial differential equation of Section 1, using $\Omega = [-1,1]^2$ and $f(x,y) = \cos(\pi x)\cos(\pi y)$. The radial basis function was given by

$$\Phi(x) = (1 - \|x\|)_+^4 (4\|x\| + 1),$$

which is in C^2 and positive definite on $\mathbb{R}^2$ (cf. [8]). For the sets of centers X_j we took a finite uniform grid of N points and the initial support was chosen to achieve the bandwidth, given in the tables. We used a classical conjugate gradient method without preconditioning to solve the linear systems. The stopping criterion was $\|Res\| \leq 1e-16$.

The number of steps in which Algorithm 2 achieves an improvement in the error depends obviously on the bandwidth. But a real improvement is only given in the first steps. This might be, especially in case of a large bandwidth, a consequence of the inexactness of the numerical quadrature.

Table 1 contains the results for Algorithm 2. Here, N denotes the number of centers and CG gives the number of iterations of the conjugate gradient method. The error is measured on a fine 300×300 grid on $[-1,1]^2$. We have looked at the l_∞-error and the discrete l_2-error. The symbol $\tilde{l}_2$ denotes the l_2-error measured only on $[-0.5, 0.5]^2$. Figure 1 shows the error after 6 steps in case of bandwidth 21.

Algorithm 4 does not improve Algorithm 2 in the first steps. But the situation is completely different in the following steps. There, it achieves convergence which seems to to be linear at least.

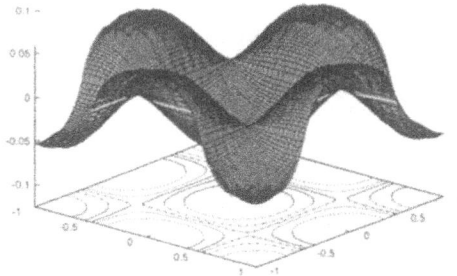

Fig. 1. Error of level 6, bandwidth 21.

N	25	81	289	1089	4225	16641
Bandwidth 21						
l_∞	3.2224e−1	1.4912e−1	8.0373e−2	6.6226e−2	6.4191e−2	6.3801e−2
l_2	2.0728e−1	8.9860e−2	5.2229e−2	4.2437e−2	4.0112e−2	3.9551e−2
$\tilde{l}_2$	6.9094e−4	2.9953e−4	1.7410e−4	1.4158e−4	1.3371e−4	1.3184e−4
CG	16	22	24	28	28	27
Bandwidth 45						
l_∞	1.0579e−1	1.2988e−2	5.6004e−3	4.3831e−3	3.4946e−3	3.0506e−3
l_2	3.5258e−2	4.5341e−3	1.1170e−3	9.5154e−4	8.6110e−4	8.1147e−4
$\tilde{l}_2$	1.1753e−4	1.5114e−5	3.7234e−6	3.1718e−6	2.8703e−6	2.7049e−6
CG	18	26	28	38	50	54

Tab. 1. Results for Algorithm 2.

N	25	81	289	1089	4225	16641
Bandwidth 21						
l_∞	3.2224e−1	1.4372e−1	5.8780e−2	2.4098e−2	9.7980e−3	3.8820e−3
l_2	2.0728e−1	8.2008e−2	2.9454e−2	1.0229e−2	3.5295e−3	1.2716e−3
$\tilde{l}_2$	6.9094e−4	2.7336e−4	9.8181e−5	3.4096e−5	1.1765e−5	4.2387e−6
Bandwidth 45						
l_∞	1.0579e−1	1.5621e−2	6.0450e−3	3.4579e−3	1.7878e−3	8.9663e−4
l_2	3.5258e−2	4.4971e−3	9.7278e−4	5.3662e−4	2.8713e−4	1.9330e−4
$\tilde{l}_2$	1.1753e−4	1.4990e−5	3.2426e−6	1.7888e−6	9.5711e−7	6.4434e−7

Tab. 2. Results for Algorithm 4.

Acknowledgments. Several parts of the program system we used were provided by R. Schaback.

References

1. Belytschko, T., Y. Krongauz, D. Organ, M. Fleming, and P. Krysl, Meshless methods: an overview and recent developments, Comp. Methods in Appl. Mech. Engin. **139** (1996), 3–47.

2. Deutsch, F., The method of alternating orthogonal projections, in *Approximation Theory, Spline Functions and Applications*, S. P. Singh (ed.), Kluwer Academic Publishers, Dordrecht, 1990, pp. 105–121.

3. Fasshauer, G., Solving partial differential equations by collocation with radial basis functions, in *Surface Fitting and Multiresolution Methods*, A. Le Méhauté, C. Rabut, and L. L. Schumaker (eds.), Vanderbilt Univ. Press, Nashville, 1997, pp. 131–138.

4. Floater, M. and A. Iske, Multistep scattered data interpolation using compactly supported radial basis functions, J. Comput. Appl. Math. **73** (1996), 65–78.

5. González–Casanova, P. and K. B. Wolf, Interpolation for solutions of the Helmholtz equation, Numerical Methods for Partial Differential Equations **11** (1995), 77–91.

6. Schaback, R., Error estimates and condition numbers for radial basis function interpolation, Adv. Comp. Math. **3** (1995), 251–264.

7. Schaback, R. and Z. Wu, Operators on radial functions, J. Comput. Appl. Math. **73** (1996), 257–270.

8. Wendland, H., Piecewise polynomial, positive definite and compactly supported radial functions of minimal degree, Adv. Comp. Math. **4** (1995), 389–396.

9. Wendland, H., Meshless Galerkin approximation using radial basis functions, Math. Comp., to appear.

10. Wu, Z., Multivariate compactly supported positive definite radial functions, Adv. Comp. Math. **4** (1995), 283–292.

Holger Wendland
Institut für Numerische und Angewandte Mathematik
Universität Göttingen
Lotzestr. 16-18
D-37083 Göttingen
Germany
wendland@math.uni-goettingen.de

Wavelet Denoising Using
Generalized Cross Validation

Norman Weyrich

Abstract. Smoothing of noisy data can be performed by inverting modified wavelet transform coefficients of the data. The different modification methods depend on additional parameters which control the amount of noise removed from the data. We describe two such modification methods, namely thresholding and regularization, and compare the results when in both cases the method of generalized cross validation is used to estimate the smoothing parameters. We apply the methods to noisy one- and two-dimensional signals.

§1. Introduction

Given are noisy data $\underline{y} = \underline{f} + \underline{\epsilon}$, $\underline{y}, \underline{f}, \underline{\epsilon} \in \mathbf{R}^N$ with $\epsilon_i \sim^{\text{iid}} N(0, \sigma^2)$ (white Gaussian noise). The goal is to find approximations for the clean data $\underline{f}$ despite the error variance σ^2 being unknown. In this paper we consider the application of wavelets to the noise reduction problem. We will restrict ourselves to the one-dimensional case, but want to remark that the methods also work well in higher dimensions, e.g., on images (see [10]). The number of given data will always be $N = 2^m$.

Let W be the matrix performing the one-dimensional discrete wavelet transformation (decomposition) $\hat{\underline{y}} := W\underline{y}$ and W^{-1} be its inverse performing the reconstruction $\underline{y} = W^{-1}\hat{\underline{y}}$. When using orthogonal wavelets we have $W^{-1} = W^T$. For this (orthogonal) case we can transform the noise reduction problem to the wavelet domain via $W\underline{y} = W\underline{f} + W\underline{\epsilon}$ or $\hat{\underline{y}} = \hat{\underline{f}} + \hat{\underline{\epsilon}}$ with $\hat{\underline{y}}, \hat{\underline{f}}, \hat{\underline{\epsilon}} \in \mathbf{R}^N$. The noise $\hat{\epsilon}_i \sim^{\text{iid}} N(0, \sigma^2)$ is still white Gaussian noise with the same unknown variance σ^2. We now need to find an approximation for $\hat{\underline{f}}$, i.e., we have to modify the wavelet coefficients $\hat{\underline{y}}$. Summarizing, the main steps of wavelet denoising are:

- Wavelet transformation of the given data (decomposition).
- Modification of the wavelet coefficients.

Approximation Theory IX, Volume 2: Computational Aspects
Charles K. Chui and Larry L. Schumaker (eds.), pp. 369–376.
Copyright © 1998 by Vanderbilt University Press, Nashville, TN.
ISBN 0-8265-1326-3.

- Inverse wavelet transformation (reconstruction).

We call this method modified wavelet reconstruction. The various noise reduction techniques using wavelets differ in the way in which the modification of the wavelet coefficients is done. We focus throughout the paper on the two most commonly used methods, that is, thresholding and modification of the wavelet coefficients by minimizing the classical regularization functional.

Both modification methods depend on additional parameters which control the amount of noise reduction. These parameters depend heavily on the unknown error variance σ^2. Good estimations for the parameters without prior explicit estimation of σ^2 can be obtained with the generalized cross validation (GCV) method. The main goal of this paper is to give a comparison between wavelet thresholding and the wavelet regularization method when using GCV in both cases. Therefore we describe the thresholding method in Section 2 and the regularization method in Section 3. In Section 3 we also show the common roots of both methods. In Section 4 we present numerical results comparing the performance of the two methods on Donoho's well established test functions ([3]) and on a test image.

§2. Wavelet Denoising with Soft Thresholding

Wavelet thresholding has been studied extensively by Donoho and Johnstone [2, 3]. We present here a slightly more general approach.

Let the thresholding function $tr_{\lambda,\,\delta}$ be defined by

$$tr_{\lambda,\,\delta}(x) := \begin{cases} 0 & \text{for } |x| \le \delta, \\ sign(x)(|x| - \lambda\delta) & \text{else.} \end{cases}$$

The parameter δ is called the thresholding parameter while λ defines a whole set of different thresholding methods. The most common choices are $\lambda = 0$, called hard thresholding, and $\lambda = 1$, called soft thresholding. It is important to note that for $\lambda \ne 1$ the thresholding function $tr_{\lambda,\,\delta}$ is discontinuous with jumps at $x = \pm\delta$. In general, hard thresholding gives better denoising results than soft thresholding. However, for hard thresholding no objective criterion for the estimation of the thresholding parameter δ has been found so far. This is the reason why soft thresholding is used in all applications.

Soft thresholding applied to wavelet denoising can be expressed as $\hat{\underline{y}}_\delta := D_\delta \hat{\underline{y}}$ with

$$D_\delta = D_\delta(\hat{\underline{y}}) := \operatorname{diag}\left(1, \ldots, 1, d_{\delta,\,m_\ell}, \ldots, d_{\delta,\,N-1}\right), \quad m_\ell := 2^{m-\ell}$$

and

$$d_{\delta,\,i} := \begin{cases} 0 & \text{for } |\hat{y}_i| < \delta, \\ 1 - \frac{\delta}{|\hat{y}_i|} & \text{else,} \end{cases} \quad (i = m_\ell, \ldots, N-1).$$

The index ℓ denotes the number of decomposition levels in the wavelet transformation. The first $2^{m-\ell}$ coefficients in the wavelet transformed data vector $\hat{\underline{y}}$

are the pure low pass filtered data and usually remain unchanged. This is necessary to preserve the mean of the clean signal (the additive noise has zero mean).

In order to perform the wavelet denoising it remains to estimate the thresholding parameter δ. For an overview about different methods see [6]. The methods proposed in [2, 3] need an apriori estimation of the noise variance σ^2. They also do not work too well if the error variance is very small. The generalized cross validation method, developed for wavelet thresholding in [8, 9], overcomes both drawbacks. We give here a short general description of the GCV method followed by its application to wavelet thresholding.

In general the GCV method to determine an unknown smoothing parameter δ works as follows: We are given a denoising problem $\underline{z}_{noisy} = \underline{z}_{clean} + \underline{\epsilon}$ ($\underline{z}_{noisy}, \underline{z}_{clean}, \underline{\epsilon} \in \mathbf{R}^N$), where the ϵ_i are independent identically distributed noise with zero mean and common unknown variance. Let $\underline{z}_{smooth}$ be the approximation of $\underline{z}_{clean}$. Then we need to find two matrices.

1) Determine the so-called influence matrix $A = A_\delta$ defined as

$$\underline{z}_{smooth} = A\underline{z}_{noisy}.$$

2) Determine the matrix $\tilde{A}$ with

$$\tilde{A} = (\tilde{a}_{ij})_{i,j=0}^{N-1}, \quad \tilde{a}_{ij} := \frac{\partial z_{smooth,\, i}}{\partial z_{noisy,\, j}}, \quad (i,j = 0, \ldots, N-1).$$

Note that if A is independent of the data $\underline{z}_{noisy}$, both matrices are identical ($\tilde{A} = A$). The GCV function is defined as

$$GCV(\delta) := \frac{\frac{1}{N} \|(I_N - A)\underline{\hat{y}}\|_2^2}{\left[\frac{1}{N} \mathrm{Tr}(I_N - \tilde{A})\right]^2},$$

with I_N being the identity matrix of dimension N. The optimality property of the GCV function is

$$\lim_{N \to \infty} \frac{R(\delta_G)}{R(\delta_R)} = 1 \tag{1}$$

with

$$R(\delta) := \frac{1}{N} \sum_{i=0}^{N-1} \left(z_{clean,\, i} - z_{noisy,\, i}\right)^2$$

and

$$\delta_G := \mathrm{argmin}\ GCV(\delta), \quad \delta_R := \mathrm{argmin}\ R(\delta).$$

The asymptotic optimality of the GCV function (1) has not been proven in full generality, but only for each special smoothing method separately. For the spline smoothing case see [7].

For wavelet thresholding we have $A = D_\delta$ and

$$\tilde{A} = \tilde{D}_\delta = \text{diag}(1, \ldots, 1, \tilde{d}_{\delta, m_\ell}, \ldots, \tilde{d}_{\delta, N-1})$$

with

$$\tilde{d}_{\delta, i} := \begin{cases} 0 & \text{if } |\hat{y}_i| \leq \delta, \\ 1 & \text{else,} \end{cases} \quad (i = m_\ell, \ldots, N-1).$$

This yields for the GCV function

$$GCV(\delta) = \frac{\frac{1}{N} \|\underline{\hat{y}} - \underline{\hat{y}}_\delta\|_2^2}{\left[\frac{1}{N} \text{Tr}(I_N - \tilde{A})\right]^2} = \frac{\frac{1}{N} \left(\sum_{i \in I_1} \hat{y}_i^2 + q\delta^2 \right)}{\left[1 - \frac{m_\ell + q}{N}\right]^2}$$

with

$$q := \#\{\hat{y}_i \ : \ |\hat{y}_i| > \delta \ (i = m_\ell, \ldots, N-1)\}$$

and

$$I_1 := \{i : \ m_\ell \leq i < N, \ |\hat{y}_i| \leq \delta\}.$$

Remarks

1) In contrast to the spline smoothing case, the GCV function is very easy to compute here.

2) The optimality of the GCV function (see (1)) has been shown in [4].

3) The minimizer of the GCV function is one of the wavelet coefficients $\hat{y}_i$ ($i = m_\ell, \ldots, 2^N$). This additional information can be used in the numerical minimization procedure. Hence a combination of global and local search strategy can be employed.

4) The GCV method can also be applied level dependently (see [10]), that is, on each decomposition level a different thresholding parameter is computed. This potentially results in better denoising. The drawback is that since the number of wavelet coefficients is decreasing on each level and the GCV function is an asymptotical method, not enough wavelet coefficients are available to give a reliable parameter estimation. Therefore the number of decomposition levels must be chosen such that sufficiently many wavelet coefficients are available. Additionally, a global thresholding parameter could be used on the remaining levels. However, in our examples this did not lead to an improvement of the results.

§3. Wavelet Denoising using a Regularization Method

The application of the classical regularization method to wavelets has been studied by Amato and Vazu in [1]. There, the following minimization problem is considered:

$$\min_f S(\delta), \ \ S(\delta) := \|y - f\|_{L^2}^2 + \delta \|f - Ef\|_{H^p}^2$$

with $Ef := \langle f, 1 \rangle_{H^p}$, and H^p being the Hardy space. Using the wavelet expansion for y and f and switching to equivalent discrete norms (see [5]) yields

$$\min_{\underline{f}} \hat{S}(\delta), \quad \hat{S}(\delta) := \sum_{i=0}^{m_\ell-1} (\hat{y}_i - \hat{f}_i)^2 + \sum_{i=m_\ell}^{N-1} (\hat{y}_i - \hat{f}_i)^2 + \delta \sum_{j=m-\ell}^{m-1} 2^{2jp} \sum_{i=0}^{2^j-1} \hat{f}_{2^j+i}^2$$

which has the solution

$$\hat{f}_i = \hat{y}_i \quad (i = 0, \ldots, m_\ell - 1),$$

$$\hat{f}_{2^j+i} = \frac{\hat{y}_{2^j+i}}{1 + \delta 2^{2jp}} \quad (j = m - \ell, \ldots, m - 1, \; i = 0, \ldots, 2^j - 1).$$

Thus the modification of the wavelet coefficients can be expressed by

$$\underline{\hat{y}_\delta} = R(\delta)\underline{\hat{y}}$$

with (for $\ell = m$)

$$R = \text{diag}\left(1, \frac{1}{1+\delta}, \frac{1}{1+\delta}, \frac{1}{1+2^{2p}\delta}, \cdots, \frac{1}{1+2^{2p}\delta}, \cdots, \right.$$

$$\left. \frac{1}{1+2^{2(m-1)p}\delta}, \cdots, \frac{1}{1+2^{2(m-1)p}\delta} \right).$$

The subtraction of Ef in the smoothing term of the regularization functional ensures the preservation of the clean signals mean.

The smoothing parameter δ can again be obtained by means of the GCV function. Here we have $A = R$ and $\tilde{A} = A = R$ because R is independent of the data $\hat{y}$. For the GCV function we obtain

$$GCV(\delta) = \frac{\frac{1}{N} \sum_{j=m-\ell}^{m-1} \sum_{i=0}^{2^j-1} \left(1 - \frac{1}{1+2^{2jp}\delta}\right)^2 \hat{y}_{2^j+i}^2}{\left[\frac{1}{N} \sum_{j=m-\ell}^{m-1} 2^j \left(1 - \frac{1}{1+2^{2jp}\delta}\right)\right]^2}.$$

The proof of the optimality result (1) can be found in [1].

It may be interesting to see that also the wavelet (soft) thresholding method is the solution of a regularization problem (see [1]). Therefore consider the problem

$$\min_{\underline{f}} \hat{S}(\delta), \quad \hat{S}(\delta) := \sum_{i=0}^{m_\ell-1} (\hat{y}_i - \hat{f}_i)^2 + \sum_{i=m_\ell}^{N-1} (\hat{y}_i - \hat{f}_i)^2 + 2\delta \sum_{j=m-\ell}^{m-1} 2^{2jp} \sum_{i=0}^{2^j-1} |\hat{f}_{2^j+i}|$$

which has the solution

$$\hat{f}_i = \hat{y}_i \quad (i = 0, \ldots, m_\ell - 1),$$

$$\hat{f}_i = tr_{1,\,\delta}(\hat{y}_i) = \begin{cases} 0 & \text{if } |\hat{y}_i| \leq \delta, \\ \left(1 - \frac{\delta}{|\hat{y}_i|}\right)\hat{y}_i & \text{else,} \end{cases} \quad (i = m_\ell, \ldots, N - 1).$$

Using the wavelet expansion for y and f and switching to equivalent continuous norms (see [5]) yields the respective continuous regularization problem

$$\min_f S(\delta), \quad S(\delta) := \|y - f\|_{L^2}^2 + 2\delta \|f - Ef\|_{B_{1,1}^1}$$

with $Ef := \langle f, 1 \rangle_{B_{1,1}^1}$, and $B_{1,1}^1$ being the Besov space $B_{p,q}^s$ for $p = q = s = 1$ (see [5]).

§4. Numerical Results

We have tested the thresholding and regularization method with the test functions from [3] using 2048 data and (periodized) Daubechies wavelets of order 4. For the experiments we generated pseudo-random (white Gaussian) noise such that the signal-to-noise ratio (SNR) was 17dB. The SNR (in dB) was computed according to $SNR := 10 \log_{10} \left(\sum_{i=0}^{N-1} f_i^2 / \sum_{i=0}^{N-1} (y_{\delta,i} - f_i)^2 \right)$. For the thresholding method we used level independent ($\ell = 11$) and level dependent parameter selection ($\ell = 5$). For the regularization method only level independent parameter estimation worked. The resulting normalized ℓ^2 errors $NR(\delta) := \left(\sum_{i=0}^{N-1} (y_{\delta,i} - f_i)^2 / \sum_{i=0}^{N-1} f_i^2 \right)^{1/2}$ are displayed in Table 1. Note that the regularization method works well for the smooth test functions and performs poor in the presence of discontinuities.

In Table 2 we present the numerical results for both methods applied to two-dimensional data (e.g., images). As test data, we used the 512×512 Lena picture. The wavelet transformation was performed with biorthogonal wavelets with 6 lowpass and 10 highpass coefficients. The number of decomposition levels was $\ell = 9$ for the level independent parameter estimation and $\ell = 4$ for the level dependent case. In the level dependent case, we computed two separate thresholding parameters on each level, one for the wavelet coefficients representing the horizontal and vertical details and one for the wavelet coefficients representing the diagonal details.

Table 2 suggests that the level dependent parameter estimation is preferable for low SNR's. It may also be interesting to note that the magnitude of the error for the SNR of 17dB is about the same as for the step function in Table 1.

Although the denoised images are not displayed we want to remark that the regularization method had problems with the sharp image edges resulting in a more blurred picture than the thresholding method.

Tab. 1. Normalized ℓ^2 Error for the Test Functions.

Function	Regularization		Thresholding			
	IReg	Reg	IThresh	Thresh	LIThresh	LThresh
Step	0.0893	0.0917	0.0742	0.0782	0.0648	0.0665
Bumps	0.0642	0.0642	0.0731	0.0775	0.0551	0.0588
Doppler	0.0688	0.0689	0.0621	0.0631	0.0501	0.0569
HSine	0.0439	0.0440	0.0472	0.0472	0.0359	0.0374

Tab. 2. Signal-to-Noise-Ratio in dB for the Lena Picture.

SNR	Regularization		Thresholding			
	IReg	Reg	IThresh	Thresh	LIThresh	LThresh
30	30.58	25.90	31.45	30.39	31.48	28.93
20	23.38	23.36	24.56	24.56	24.82	24.20
17	21.86	21.86	22.82	22.80	23.22	22.85
10	18.49	17.28	19.05	18.67	19.81	19.42
0	14.83	12.44	14.25	12.11	15.48	15.33
-3	13.47	10.02	12.95	9.94	14.29	14.02

Legend

IReg: Regularization method, using ideal parameter

Reg: Regularization method, using GCV parameter

IThresh: Thresholding method, using ideal parameter (level independent)

Thresh: Thresholding method, using GCV parameter (level independent)

LIThresh: Thresholding method, using ideal parameter (level dependent)

LThresh: Thresholding method, using GCV parameter (level dependent)

References

1. Amato, U. and D. T. Vuza, Wavelet Regularization for Smoothing Data, Technical Report, 1994.

2. Donoho, D. L. and I. M. Johnstone, Adapting to unknown smoothness via wavelet shrinkage, J. Amer. Statistical Assoc. **90** (1995).

3. Donoho, D. L. and I. M. Johnstone, Ideal spatial adaptation by wavelet shrinkage, Biometrika **81** (1995), 425–455.

4. Jansen, M., *Minimization of Noise in Digital Images with Automatic Selection of Thresholds for Wavelet Coefficients*, Master's Thesis, Department of Computer Science, K. U. Leuven, Belgium, 1995.

5. Meyer, Y., *Wavelets and Operators*, Cambridge University Press, 1992.

6. Nason, G. P., Wavelet function estimation using cross-validation, in *Wavelets and Statistics* A. Antoniadis and G. Oppenheim (eds.), Springer, New York, 1995, pp. 261–280.

7. Wahba, G., *Spline Models for Observational Data*, SIAM, Philadelphia, 1990.

8. Weyrich, N. and G. T. Warhola, De-noising using wavelets and cross validation, in *Recent Developments in Approximation Theory, Wavelets and Applications* S. P. Singh (ed.), Kluwer Academic Publishers, Dordrecht, 1995, pp. 523–532.

9. Weyrich, N. and G. T. Warhola, De-noising by wavelet shrinkage and generalized cross validation with applications to speech, in *Approximation Theory VIII, Vol. 2: Wavelets*, C. K. Chui and L. L. Schumaker (eds.), World Scientific Publishing Co., Inc., Singapore, 1995, pp. 407–414.

10. Weyrich, N. and G. T. Warhola, Wavelet shrinkage and generalized cross validation for image de-noising, IEEE Trans. Image Processing **7** (1998), 82–90.

Norman Weyrich
Synopsys, Inc.
Kaiserstr. 100
52134 Herzogenrath, Germany
weyrich@synopsys.com

Mortar Mixed Finite Element Approximations for Elliptic and Parabolic Equations

Mary F. Wheeler and Ivan Yotov

Abstract. We consider mortar mixed finite element approximations for modeling elliptic and parabolic partial differential equations on multiblock non-matching grids. Mortar finite element spaces are introduced on interfaces to impose physically meaningful matching conditions in a stable and accurate fashion. We consider both linear elliptic and degenerate nonlinear parabolic equations and provide theoretical convergence bounds for the discretization error in terms of optimal approximation error. Superconvergence is also observed in the linear case at certain discrete points. Computational results confirming the theory are presented.

§1. Introduction

In this paper we describe mortar mixed finite element approximations for modeling elliptic and parabolic partial differential equations on multiblock domains. The motivation for these formulations arises from the need of local mass conservation as well as being able to employ non-matching grids to treat problems with irregular geometry. Additional benefits include the use of efficient, parallel scalable domain decomposition solvers and local hierarchical adaptive grids. Multiblock (also known as macro-hybrid) formulations provide numerical models consistent with the physical/engineering description of the underlying equations: that is, the equations hold with their usual meaning on the subdomains, which have physically meaningful interface matching conditions between them. Difficulties with this approach involve imposing these conditions in a stable and accurate fashion. This is achieved through introducing a specially chosen mortar finite element space, in which the scalar variable is approximated and the flux-matching conditions are imposed. The following features make this methodology computationally attractive.

In many cases geometrically highly irregular domains can be described as unions of relatively simple blocks. Each block is independently covered by a relatively simple (e.g. logically rectangular) grid. The local grid structure

Approximation Theory IX, Volume 2: Computational Aspects   377
Charles K. Chui and Larry L. Schumaker (eds.), pp. 377–392.
Copyright ⓒ 1998 by Vanderbilt University Press, Nashville, TN.
ISBN 0-8265-1326-3.

allows for more efficient and accurate discretization techniques to be employed. Moreover, unstructured grid could be used on a given block, if its geometry is very irregular.

Since the numerical grids may be non-matching across interfaces, they can be constructed to follow internal boundaries such as faults and heterogeneous layers in subsurface modeling.

In addition to geometrical considerations, multiblock decomposition can be induced by differences in the physical or discretization models in different parts of the simulation domain.

Dynamic grid adaptivity can be performed locally on each block. This is very convenient for the fast reconstruction of grids and calculation of the corresponding stiffness matrices in time-dependent problems. Mortar degrees of freedom may also vary, providing an additional degree of adaptivity.

The multiblock structure of the algebraic systems of equations allows for the design and use of efficient domain decomposition solvers and preconditioners, which maximize data and computation locality and minimize communication overhead.

The multiblock paradigm is very general and has a broad field of application. Mortar finite elements have been successfully applied for standard and spectral finite element discretizations on non-matching grids (see, e.g. [11, 10]). Motivated by subsurface flow and transport applications, in this paper we consider mixed finite element (finite volume) methods for subdomain discretizations. Mixed methods provide accurate approximation for two variables of physical interest - the scalar variable (pressure) and its flux (velocity). The computed velocity field is locally mass conservative and is continuous in the normal direction across element faces (edges), which is critical for transport. Theoretical and numerical results for single phase flow indicate mortar mixed finite element methods are highly accurate (superconvergent) for both pressure and velocity [35, 4, 5, 8, 37]. A parallel non-overlapping domain decomposition implementation, based on an algorithm originally developed by Glowinski and Wheeler [23, 19, 18], reduces the saddle point problem to a positive definite interface problem, providing an efficient scalable solution technique [35]. Some efficient preconditioners have also been developed [24].

An extension of the method to a degenerate parabolic equation arising in two phase flow is presented in [36], where optimal convergence is shown. Recent work on two phase flow simulation indicates that the multiblock mortar approach works very well for coupled systems of transient highly non-linear differential equations (see [32, 28, 34] for reservoir simulation and [38] for groundwater remediation).

The rest of the paper is organized as follows. In Section 2 we formulate mortar mixed finite element methods for second order elliptic equations and give theoretical and numerical convergence results. In Section 3 we consider degenerate nonlinear parabolic equations which arise in some problems of flow and transport in porous media. We provide error bounds and illustrate the method with a simulation of two-phase flow in porous media.

§2. Mortar Mixed Finite Element Approximations for Elliptic Equations

In this section we consider second order linear elliptic equations which in porous medium applications model single phase Darcy flow. We solve for the pressure p and the velocity $\mathbf{u}$ satisfying

$$\mathbf{u} = -K\nabla p \text{ in } \Omega, \tag{1.1}$$

$$\nabla \cdot \mathbf{u} = f \quad \text{in } \Omega, \tag{1.2}$$

$$p = g \qquad \text{on } \partial\Omega, \tag{1.3}$$

where $\Omega \subset \mathbf{R}^d$, $d = 2$ or 3, is a multiblock domain and K is a symmetric, uniformly positive definite tensor with $L^\infty(\Omega)$ components representing the permeability divided by the viscosity. Dirichlet boundary conditions are considered here only for simplicity. Equation (1.1) is known as Darcy's law and (1.2) is the mass conservation equation. This formulation is commonly referred to as a mixed formulation.

2.1. Formulation and Discretization

In the weak formulation of (1.1)–(1.3) we seek a pair $(\mathbf{u}, p) \in H(\text{div}; \Omega) \times L^2(\Omega)$ such that

$$(K^{-1}\mathbf{u}, \mathbf{v}) = (p, \nabla \cdot \mathbf{v}) - \langle g, \mathbf{v} \cdot \nu \rangle_{\partial\Omega}, \qquad \mathbf{v} \in H(\text{div}; \Omega), \tag{2.1}$$

$$(\nabla \cdot \mathbf{u}, w) = (f, w), \qquad\qquad\qquad w \in L^2(\Omega). \tag{2.2}$$

Here $(\cdot, \cdot)_S$ denotes the $L^2(S)$ inner product and $\langle \cdot, \cdot \rangle_{\partial S}$ denotes the $L^2(\partial S)$ inner product or a duality pairing. We omit S if $S = \Omega$. It is well known (see, e.g., [15]) that (2.1)–(2.2) has a unique solution.

Let $\Omega = \cup_{i=1}^{n_b}\Omega_i$ be decomposed into n_b non-overlapping subdomains Ω_i, and let $\Gamma_{i,j} = \partial\Omega_i \cap \partial\Omega_j$, $\Gamma = \cup_{1 \le i < j \le n_b}\Gamma_{i,j}$, and $\Gamma_i = \partial\Omega_i \cap \Gamma = \partial\Omega_i \backslash \partial\Omega$. Let

$$\mathbf{V}_i = H(\text{div}; \Omega_i), \qquad \mathbf{V} = \bigoplus_{i=1}^{n_b}\mathbf{V}_i,$$

and

$$W_i = L^2(\Omega_i), \qquad W = \bigoplus_{i=1}^{n_b}W_i = L^2(\Omega).$$

If the solution $(\mathbf{u}, p)$ of (2.1)–(2.2) belongs to $H(\text{div}; \Omega) \times H^1(\Omega)$, it is easy to see that it satisfies, for $1 \le i \le n_b$,

$$(K^{-1}\mathbf{u}, \mathbf{v})_{\Omega_i} = (p, \nabla \cdot \mathbf{v})_{\Omega_i} - \langle p, \mathbf{v} \cdot \nu_i \rangle_{\Gamma_i} - \langle g, \mathbf{v} \cdot \nu_i \rangle_{\partial\Omega_i \cap \partial\Omega}, \quad \mathbf{v} \in \mathbf{V}_i, \tag{2.3}$$

$$(\nabla \cdot \mathbf{u}, w)_{\Omega_i} = (f, w)_{\Omega_i}, \qquad w \in W_i, \tag{2.4}$$

where ν_i is the outer unit normal to $\partial\Omega_i$.

Let $\mathcal{T}_{h,i}$ be a conforming, quasi-uniform finite element partition of Ω_i, $1 \leq i \leq n_b$, with $\mathcal{T}_{h,i}$ and $\mathcal{T}_{h,j}$ possibly non-matching on $\Gamma_{i,j}$. Let $\mathcal{T}_h = \cup_{i=1}^{n_b} \mathcal{T}_{h,i}$. Let

$$\mathbf{V}_{h,i} \times W_{h,i} \subset \mathbf{V}_i \times W_i$$

be any of the usual mixed finite element spaces, (i.e., the RT spaces [31, 29, 25]; BDM spaces [14]; BDFM spaces [13]; BDDF spaces [12], or CD spaces [17]). For simplicity we assume that the order of the spaces is the same on every subdomain. Let

$$\mathbf{V}_h = \bigoplus_{i=1}^{n_b} \mathbf{V}_{h,i}, \qquad W_h = \bigoplus_{i=1}^{n_b} W_{h,i}.$$

All of the spaces above satisfy

$$\nabla \cdot \mathbf{V}_{h,i} = W_{h,i},$$

and that there exists a projection Π_i onto $\mathbf{V}_{h,i}$, such that for any $\mathbf{q} \in (H^{1/2+\varepsilon}(\Omega_i))^d \cap \mathbf{V}_i$,

$$(\nabla \cdot (\Pi_i \mathbf{q} - \mathbf{q}), w)_{\Omega_i} = 0, \qquad w \in W_{h,i} \tag{2.5}$$

$$\langle (\mathbf{q} - \Pi_i \mathbf{q}) \cdot \nu_i, \mathbf{v} \cdot \nu_i \rangle_{\partial\Omega_i} = 0, \quad \mathbf{v} \in \mathbf{V}_{h,i}. \tag{2.6}$$

Note that, since $\mathbf{q} \in (H^{1/2+\varepsilon}(\Omega_i))$, $\mathbf{q} \cdot \nu|_e \in H^\varepsilon(e)$ for any element face (edge) e; therefore $\Pi_i \mathbf{q}$ is well defined.

Let $\mathcal{T}_{h,i,j}$ be a quasi-uniform finite element partition of $\Gamma_{i,j}$. Denote by $M_{h,i,j} \subset L^2(\Gamma_{i,j})$ the space of either continuous or discontinuous piecewise polynomials of degree $k+1$ on $\mathcal{T}_{h,i,j}$, where k is associated with the degree of the polynomials in $\mathbf{V}_h \cdot \nu$. More precisely, if $d = 3$, on any boundary element K, $M_{h,i,j}|_K = P_{k+1}(K)$, if K is a triangle, and $M_{h,i,j}|_K = Q_{k+1}(K)$, if K is a rectangle. Let

$$M_h = \bigoplus_{1 \leq i < j \leq n_b} M_{h,i,j}.$$

An additional assumption on M_h and $\mathcal{T}_{h,i,j}$ will be made below in (2.10) and (2.18).

In the mortar mixed finite element approximation of (2.1)–(2.2), we seek $\mathbf{u}_h \in \mathbf{V}_h$, $p_h \in W_h$, and $\lambda_h \in M_h$ such that, for $1 \leq i \leq n_b$,

$$(K^{-1}\mathbf{u}_h, \mathbf{v})_{\Omega_i} = (p_h, \nabla \cdot \mathbf{v})_{\Omega_i} - \langle \lambda_h, \mathbf{v} \cdot \nu_i \rangle_{\Gamma_i}$$
$$\qquad\qquad - \langle g, \mathbf{v} \cdot \nu_i \rangle_{\partial\Omega_i \cap \partial\Omega}, \qquad \mathbf{v} \in \mathbf{V}_{h,i}, \tag{2.7}$$

$$(\nabla \cdot \mathbf{u}_h, w)_{\Omega_i} = (f, w)_{\Omega_i}, \qquad\qquad w \in W_{h,i}, \tag{2.8}$$

$$\sum_{i=1}^{n_b} \langle \mathbf{u}_h \cdot \nu_i, \mu \rangle_{\Gamma_i} = 0, \qquad\qquad \mu \in M_h. \tag{2.9}$$

Existence and uniqueness of a solution to (2.7)–(2.9) is shown in [35, 4] under the assumption that, for any $\phi \in M_{h,i,j}$,

$$\mathcal{Q}_{h,i}\phi|_{\Gamma_{i,j}} = \mathcal{Q}_{h,j}\phi|_{\Gamma_{i,j}} = 0 \text{ implies that } \phi|_{\Gamma_{i,j}} = 0, \tag{2.10}$$

where $\mathcal{Q}_{h,i} : L^2(\Gamma_i) \rightarrow \mathbf{V}_{h,i} \cdot \nu_i|_{\Gamma_i}$ is the L^2-projection satisfying for any $\phi \in L^2(\Gamma_i)$

$$\langle \phi - \mathcal{Q}_{h,i}\phi, \mathbf{v} \cdot \nu_i \rangle_{\Gamma_i} = 0, \quad \mathbf{v} \in \mathbf{V}_{h,i}. \tag{2.11}$$

2.2. Convergence Results

The approximation properties of the finite element spaces are critical in the convergence analysis. It is convenient to express these properties in terms of approximability of projection operators which are later used in the analysis.

Let $\mathcal{P}_h$ be the $L^2(\Gamma)$ projection onto M_h satisfying for any $\psi \in L^2(\Gamma)$

$$\langle \psi - \mathcal{P}_h\psi, \mu \rangle_\Gamma = 0, \quad \mu \in M_h.$$

For any $\varphi \in L^2(\Omega)$, let $\hat{\varphi} \in W_h$ be its $L^2(\Omega)$ projection satisfying

$$(\varphi - \hat{\varphi}, w) = 0, \quad w \in W_h.$$

The above defined projections have the following approximation properties, wherein l is associated with the degree of the polynomials in W_h:

$$\|\psi - \mathcal{P}_h\psi\|_{-s,\Gamma} \leq C\|\psi\|_{r,\Gamma} h^{r+s}, \quad 0 \leq r \leq k+2, \ 0 \leq s \leq k+2, \tag{2.12}$$

$$\|\varphi - \hat{\varphi}\|_0 \leq C\|\varphi\|_r h^r, \quad 0 \leq r \leq l+1, \tag{2.13}$$

$$\|\mathbf{q} - \Pi_i \mathbf{q}\|_{0,\Omega_i} \leq C\|\mathbf{q}\|_{r,\Omega_i} h^r, \quad 1/2 < r \leq k+1, \tag{2.14}$$

$$\|\nabla \cdot (\mathbf{q} - \Pi_i \mathbf{q})\|_{0,\Omega_i} \leq C\|\nabla \cdot \mathbf{q}\|_{r,\Omega_i} h^r, \quad 1 \leq r \leq l+1, \tag{2.15}$$

$$\|\psi - \mathcal{Q}_{h,i}\psi\|_{-s,\Gamma_i} \leq C\|\psi\|_{r,\Gamma_i} h^{r+s}, \quad 0 \leq r \leq k+1, \ 0 \leq s \leq k+1, \tag{2.16}$$

$$\|(\mathbf{q} - \Pi_i \mathbf{q}) \cdot \nu_i\|_{-s,\Gamma_i} \leq C\|\mathbf{q}\|_{r,\Gamma_i} h^{r+s}, \quad 0 \leq r \leq k+1, \ 0 \leq s \leq k+1, \tag{2.17}$$

where $\| \cdot \|_r$ is the H^r-norm and $\| \cdot \|_{-s}$ is the norm of H^{-s}, the dual of H^s. The proof of the following theorem can be found in [4].

Theorem 2.1. *For the solution of the mixed method (2.7)–(2.9), if*

$$\|\mu\|_{0,\Gamma_{i,j}} \leq C(\|\mathcal{Q}_{h,i}\mu\|_{0,\Gamma_{i,j}} + \|\mathcal{Q}_{h,j}\mu\|_{0,\Gamma_{i,j}}), \quad \forall \mu \in M_h, \ 1 \leq i < j \leq n_b,$$
(2.18)

then there exists a positive constant C independent of h such that

$$\|\mathbf{u} - \mathbf{u}_h\|_0 \leq C \sum_{i=1}^{n_b} (\|p\|_{s+1,\Omega_i} h^s + \|\mathbf{u}\|_{r,\Omega_i} h^r),$$
(2.19)

$$\|\hat{p} - p_h\|_0 \leq C \sum_{i=1}^{n_b} (\|p\|_{s+1,\Omega_i} h^{s+1} + \|\mathbf{u}\|_{r,\Omega_i} h^{r+1} + \|\nabla \cdot \mathbf{u}\|_{t,\Omega_i} h^{t+1}),$$
(2.20)

$$\|p - p_h\|_0 \leq C \sum_{i=1}^{n_b} (\|p\|_{s+1,\Omega_i} h^s + \|\mathbf{u}\|_{r,\Omega_i} h^{r+1} + \|\nabla \cdot \mathbf{u}\|_{t,\Omega_i} h^{t+1}),$$
(2.21)

$$\|\nabla \cdot (\mathbf{u} - \mathbf{u}_h)\|_0 \leq C \sum_{i=1}^{n_b} \|\nabla \cdot \mathbf{u}\|_{t,\Omega_i} h^t,$$
(2.22)

where $0 \leq s \leq l+1$, $1/2 < r \leq k+1$, and $0 \leq t \leq l+1$.

Moreover, if the tensor K is diagonal and the mixed finite element spaces are the Raviart-Thomas spaces on rectangular type grids,

$$\||\mathbf{u} - \mathbf{u}_h\|| \leq C \sum_{i=1}^{n_b} (\|p\|_{s+3/2,\Omega_i} h^{s+1/2} + \|\mathbf{u}\|_{r+1/2,\Omega_i} h^{r+1/2}),$$
(2.23)

where $0 \leq s \leq l+1$, $1/2 < r \leq k+1$, and $\|| \cdot \||$ is a discrete approximation to the L^2-norm involving integration along Gaussian lines (see [20, 21, 4] for the exact definition).

Remark 2.1. The condition (2.18) (and subsequently (2.10)) on the mortar grids and spaces is easily satisfied in practice. It does not allow for the mortar space to be too rich compared to the subdomain grids. It can be shown theoretically [35] that it always holds for either continuous or discontinuous mortar spaces, if the the mortar grid on each interface is a coarsening by two in each direction of the trace of either one of the subdomain grids. This choice is reminiscent of the one in the case of standard or spectral finite element subdomain discretizations [11, 10].

Remark 2.2. The proof of Theorem 2.1 is based on reduction of the discretization error to approximation error and use of the approximation properties (2.12)–(2.17). Bounds (2.19), (2.21), and (2.22) imply optimal convergence, bound (2.20) implies superconvergence for the pressure at the Gaussian points, and bound (2.23) implies superconvergence for the velocity along the Gaussian lines, with a slightly higher solution regularity requirement.

Remark 2.3. The choice of the mortar finite element space M_h consisting of piecewise polynomials of degree $k + 1$, i.e., one degree higher than the polynomials in $\mathbf{V}_h \cdot \nu$, is essential for the optimal convergence and superconvergence of the method. If only polynomials of degree k are used, which is the standard Lagrange multiplier choice for mixed methods on conforming grids, no superconvergence for the pressure, and only sub-optimal convergence for the velocity (with a loss of $O(h^{1/2})$), can be shown. This is due to the approximation error on the interfaces. These theoretical estimates are in accordance with the numerically observed convergence rates.

2.3. A Computational Example

To verify the theoretical bounds from the previous subsection, we present a numerical experiment on a problem with a given analytical solution. We take Ω to be the unit square in $\mathbf{R}^2$ and solve (1.1)–(1.2) with

$$p = \cos(9xy)\sin(4x)$$

and

$$K = \begin{pmatrix} x^2 + 1 & 0 \\ 0 & 10y^2 + 1 \end{pmatrix}.$$

Dirichlet boundary conditions are imposed on the left and right edges of the boundary and Neumann boundary conditions are imposed on the rest of the boundary. The domain is divided into four subdomains independently covered by non-matching grids. The subdomains are discretized by the lowest order Raviart-Thomas mixed method. We estimate the convergence rate by solving the problem on a sequence of grids, each time refining by two both the subdomain and the mortar grids. The coarsest subdomain grids are 5×5, 4×4, 7×4, and 6×5, with a 3 element mortar grid on each interface. It is easy to see that (2.18) holds in this case. In Table 1 we report the discrete L^2-norm errors and the estimated by a least squares fit convergence rates. Here $\|\cdot\|_M$ is the discrete L^2-norm induced by the midpoint quadrature rule on $\mathcal{T}_h$. We observe superconvergence in the two linear mortar cases for both pressure and velocity, as predicted by the theory. In the case of piecewise constant mortars, only $O(h)$ rate for the pressure and $O(h^{1/2})$ rate for velocity are obtained, also in agreement with the theory. The computed pressure and velocity on the second level of refinement are shown in Figure 1, and the magnified discretization error is shown in Figure 2.

Table 1. Pressure and velocity errors. Mortar 1 - continuous piecewise linears; Mortar 2 - discontinuous linears; Mortar 3 - piecewise constants.

	Mortar 1		Mortar 2		Mortar 3	
$1/h$	$\|p - p_h\|_M$	$\|\|\mathbf{u} - \mathbf{u}_h\|\|$	$\|p - p_h\|_M$	$\|\|\mathbf{u} - \mathbf{u}_h\|\|$	$\|p - p_h\|_M$	$\|\|\mathbf{u} - \mathbf{u}_h\|\|$
8	0.24E-1	0.83E0	0.25E-1	0.82E0	0.28E-1	1.41E0
16	0.58E-2	0.21E0	0.59E-2	0.22E0	0.98E-2	0.73E0
32	0.15E-2	0.53E-1	0.15E-2	0.57E-1	0.47E-2	0.48E0
64	0.36E-3	0.14E-1	0.36E-3	0.16E-1	0.24E-2	0.33E0
128	0.91E-4	0.37E-2	0.91E-4	0.46E-2	0.12E-2	0.23E0
Rate	$1.6\,h^{2.01}$	$47\,h^{1.95}$	$1.6\,h^{2.01}$	$39\,h^{1.86}$	$0.2\,h^{1.11}$	$4.6\,h^{0.63}$

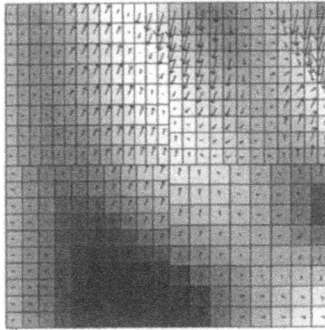

(a) Continuous linears   (b) Discontinuous constants

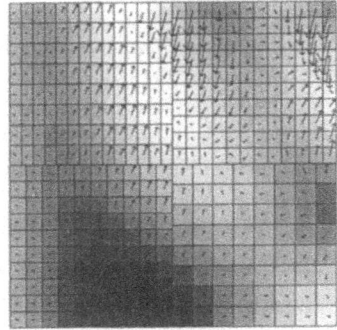

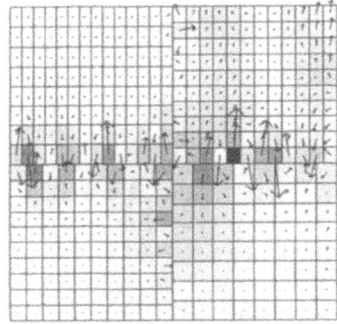

Fig. 1. Computed pressure (shade) and velocity (arrows).

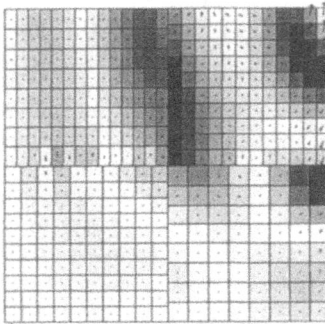

(a) Continuous linears   (b) Discontinuous constants

Fig. 2. Pressure and velocity error.

We notice that, although the computed solutions for the two cases look very similar, there is much larger velocity error on the interfaces in the piecewise constant mortar case. This interface error does not decrease with the refinement of the grids and causes a deterioration in the global convergence.

§3. A Degenerate Nonlinear Parabolic Equation

The results from the previous section can be easily extended to linear and quasi-linear parabolic equations. However, they do not carry directly to degenerate nonlinear advection-diffusion equations. Such equations arise for example in modeling two-phase flow in porous media [16, 9, 2, 7]. Consider the advection-diffusion equation

$$\phi\frac{\partial s}{\partial t} + \nabla \cdot (\beta(s)\mathbf{u} - \sigma(s)K\nabla s) = q, \tag{3.1}$$

where $0 \le s(x,t) \le 1$ is the wetting fluid phase saturation, ϕ is the porosity, $\sigma(s) = -\alpha(s)\frac{\partial p_c}{\partial s}$, $p_c(s)$ is the capillary pressure function, $\alpha(s)$ and $\beta(s)$ are related to the phase mobilities, K is the absolute permeability tensor, and q is the source term. This equation is degenerate, since the diffusion term $\sigma(s)$ vanishes at $s = 0$ and $s = 1$ (see, e.g. [9]). Consequently, the solution has a very low regularity [1, 2, 3]:

$$s \in L^\infty(0,T; L^1(\Omega)), \quad \frac{\partial s}{\partial t} \in L^2(0,T; H^{-1}(\Omega)). \tag{3.2}$$

The degeneracy in the diffusion can be treated analytically via the Kirchhoff transformation [30, 22, 7]. Let

$$D(s) = \int_0^s \sigma(\zeta)\,d\zeta.$$

Then

$$\nabla D(s) = \sigma(s)\nabla s,$$

and (3.1) becomes

$$\phi\frac{\partial s}{\partial t} + \nabla \cdot (\beta(s)\mathbf{u} - K\nabla D(s)) = q. \tag{3.3}$$

Equation (3.3) is considered in a space-time domain $\Omega \times (0,T]$. The model is completed with a no flow boundary condition

$$(\beta(s)\mathbf{u} - K\nabla D(s)) \cdot \nu = 0 \quad \text{on } \partial\Omega \times [0,T]$$

and initial condition $s(x,0) = s_0(x)$.

3.1. Formulation and Discretization

In the standard mixed variational formulation, equation (3.3) is multiplied by a test function $w \in L^2(\Omega)$ and integrated in space. In this case however, because of (3.2), the integral $(\frac{\partial s}{\partial t}, w)$ is not well defined. We integrate (3.3) in time from 0 to t [26, 7, 36] to obtain the equivalent equation (ϕ is considered a constant and is omitted for simplicity in the rest of the section)

$$s(x,t) + \nabla \cdot \int_0^t \psi\,d\tau = \int_0^t q(s)\,d\tau + s_0(x), \tag{3.4}$$

where $\psi = \beta(s)\mathbf{u} - K\nabla D(s)$. Let, for some $0 < \varepsilon < 1/2$, $M = H^{1/2-\varepsilon}(\Gamma)$,

$$\mathbf{V}_i = \{\mathbf{v} \in (H^\varepsilon(\Omega_i))^d \cap H(\text{div};\Omega_i) : \mathbf{v} \cdot \nu = 0 \text{ on } \partial\Omega\}, \qquad \mathbf{V} = \bigoplus_{i=1}^{n_b} \mathbf{V}_i,$$

and let W be as defined in Section 2. Let $a = K^{-1}$ and let γ be the trace of $D(s)$ on Γ. In the multiblock variational formulation of (3.4), we have, for every time $t \in [0, T]$ and $1 \le i \le n_b$,

$$(a\psi, \mathbf{v})_{\Omega_i} = (D(s), \nabla \cdot \mathbf{v})_{\Omega_i} - \langle \gamma, \mathbf{v} \cdot \nu_i \rangle_{\Gamma_i} + (a\beta(s)\mathbf{u}, \mathbf{v})_{\Omega_i}, \mathbf{v} \in \mathbf{V}_i, \quad (3.5)$$

$$(s, w)_{\Omega_i} + \left(\nabla \cdot \int_0^t \psi \, d\tau, w \right)_{\Omega_i}$$

$$= \left(\int_0^t q(s) \, d\tau, w \right)_{\Omega_i} + (s_0, w)_{\Omega_i}, \qquad w \in W_i, \quad (3.6)$$

$$\sum_{i=1}^{n_b} \left\langle \int_0^t \psi \cdot \nu_i \, d\tau, \mu \right\rangle_{\Gamma_i} = 0, \qquad\qquad \mu \in M. \quad (3.7)$$

The boundary integrals in (3.5) and (3.7) are well defined (see [36]).

We consider space discretizations as defined in Section 2, and a backward Euler time discretization. Let $\{t_n\}_{n=0}^N$ be a monotone partition of $[0, T]$ with $t_0 = 0$ and $t_N = T$, let $\Delta t^n = t_n - t_{n-1}$, and let $f^n = f(t_n)$. We seek, for any $0 \le n \le N$, $\psi_h^n \in \mathbf{V}_h$, $s_h^n \in W_h$, and $\gamma_h^n \in M_h$ such that, for $1 \le i \le n_b$,

$$(a\psi_h^n, \mathbf{v})_{\Omega_i} = (D(s_h^n), \nabla \cdot \mathbf{v})_{\Omega_i}$$
$$- \langle \gamma_h^n, \mathbf{v} \cdot \nu_i \rangle_{\Gamma_i} + (a\beta(s_h^n)\mathbf{u}^n, \mathbf{v})_{\Omega_i}, \qquad \mathbf{v} \in \mathbf{V}_{h,i}, \quad (3.8)$$

$$(s_h^n, w)_{\Omega_i} + \left(\nabla \cdot \sum_{j=1}^n \psi_h^j \Delta t^j, w \right)_{\Omega_i}$$

$$= \left(\sum_{j=1}^n q(s_h^j) \Delta t^j, w \right)_{\Omega_i} + (s_{0,h}, w)_{\Omega_i}, w \in W_{h,i}, \quad (3.9)$$

$$\sum_{i=1}^{n_b} \left\langle \sum_{j=1}^n \psi_h^j \cdot \nu_i \Delta t^j, \mu \right\rangle_{\Gamma_i} = 0, \qquad\qquad \mu \in M_h \quad (3.10)$$

3.2. Convergence Results

In the convergence analysis we make the following assumptions on the coefficients of (3.3):

$$\sigma(s) \ge \begin{cases} \beta_1 |s|^{\nu_1}, & 0 \le s \le \alpha_1, \\ \beta_2, & \alpha_1 \le s \le \alpha_2, \\ \beta_3 |1 - s|^{\nu_2}, & \alpha_2 \le s \le 1, \end{cases} \quad (3.11)$$

where β_i, $1 \le i \le 3$, are positive constants, and α_i and ν_i, $i = 1, 2$, satisfy

$$0 < \alpha_1 < 1/2 < \alpha_2 < 1, \quad 0 < \nu_i \le 2.$$

We also assume that there exists a positive constant C such that

$$0 \le \frac{\partial D}{\partial s}(x, t; s) \le C \quad \text{for } (x, t) \in \Omega \times [0, T], \; 0 \le s \le 1. \quad (3.12)$$

Bounds (3.11) and (3.12) are justified by the physical behavior of the capillary pressure and the relative permeabilities [9, 3, 22, 36]. We also assume that, for $0 \le s_1, s_2 \le 1$,

$$|\beta(s_1) - \beta(s_2)|^2 \le C(D(s_1) - D(s_2))(s_1 - s_2), \qquad (3.13)$$

$$|q(s_1) - q(s_2)|^2 \le C(D(s_1) - D(s_2))(s_1 - s_2), \qquad (3.14)$$

which are also physically reasonable (see [36, 22] for details).

The proof of following theorem is given in [36].

Theorem 3.1. *Assume that (3.11)–(3.14) and (2.18) hold, and that there exists a constant $C_1 > 0$ such that*

$$\Delta t^{j+1} \le C_1 \Delta t^j, \quad 1 \le j \le N - 1.$$

For the fully discrete mixed finite element approximation (3.8)–(3.10) of (3.3), there exists a positive constant C such that, for any $1 \le n \le N$,

$$\sum_{j=1}^{n} (s^j - s_h^j, D(s^j) - D(s_h^j)) \Delta t^j + \left\| \sum_{j=1}^{n} (\psi^j - \psi_h^j) \Delta t^j \right\|_0^2$$

$$+ \sum_{j=1}^{n} \|(\psi^j - \psi_h^j)\Delta t^j\|_0^2$$

$$\le C \sum_{j=1}^{n} \left\{ \|\widehat{s^j} - s^j\|_0^2 + \left\| \sum_{l=1}^{j} \left(\int_{t^{l-1}}^{t^l} q(s)\, d\tau - q(s^l)\Delta t^l \right) \right\|_0^2 \right.$$

$$+ \left\| \frac{1}{\Delta t^j} \left(\Pi \int_{t^{j-1}}^{t^j} \psi\, d\tau - \int_{t^{j-1}}^{t^j} \psi\, d\tau \right) \right\|_0^2 + \left\| \frac{1}{\Delta t^j} \left(\int_{t^{j-1}}^{t^j} \psi\, d\tau - \psi_j \Delta t^j \right) \right\|_0^2$$

$$+ \left\| \frac{1}{\Delta t^j} \left(\Pi \int_{t^{j-1}}^{t^j} \psi\, d\tau - \int_{t^{j-1}}^{t^j} \psi\, d\tau \right) \cdot \nu \right\|_{0,\Gamma}^2 + h^{-1}\|\mathcal{P}_h \gamma^j - \gamma^j\|_{0,\Gamma}^2 \right\} \Delta t^j.$$

Moreover,

$$\|s^n - s_h^n\|_{-1}^2 \le C \sum_{j=1}^{n} \left\{ \|\widehat{s^j} - s^j\|_0^2 \right.$$

$$+ \left\| \sum_{l=1}^{j} \left(\int_{t^{l-1}}^{t^l} q(s)\, d\tau - q(s^l)\Delta t^l \right) \right\|_0^2 + h^{-1}\|\mathcal{P}_h \gamma^j - \gamma^j\|_{0,\Gamma}^2$$

$$+ \left\| \frac{1}{\Delta t^j} \left(\Pi \int_{t^{j-1}}^{t^j} \psi\, d\tau - \int_{t^{j-1}}^{t^j} \psi\, d\tau \right) \right\|_0^2 + \left\| \frac{1}{\Delta t^j} \left(\int_{t^{j-1}}^{t^j} \psi\, d\tau - \psi_j \Delta t^j \right) \right\|_0^2$$

$$+ \left\| \frac{1}{\Delta t^j} \left(\Pi \int_{t^{j-1}}^{t^j} \psi\, d\tau - \int_{t^{j-1}}^{t^j} \psi\, d\tau \right) \cdot \nu \right\|_{0,\Gamma}^2 \right\} \Delta t^j + Ch^2\|\widehat{s^n} - s^n\|_0^2.$$

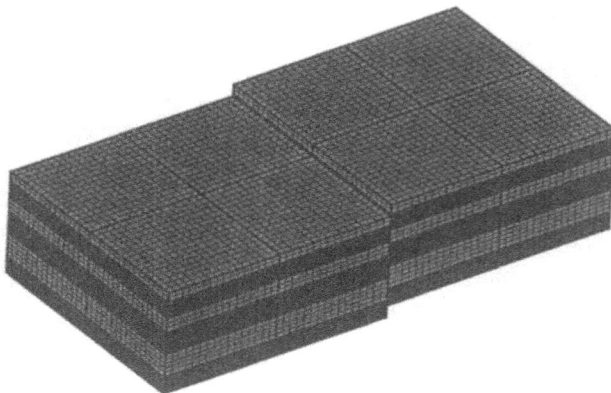

Fig. 3. Geological layers and numerical grids The dark layers (400 md) are eight times more permeable than the light layers.

Remark 3.1. The first estimates bounds the size of $\|D(s) - D(s_h)\|_0$ by (3.12). It is used in the proof of the $\|s - s_h\|_{-1}$ estimate. Both estimates bound the discretization error by optimal order approximation terms in time or space. The term $h^{-1/2}\|\gamma - \mathcal{P}_h\gamma\|_{0,\Gamma}$ provides approximation of order $h^{1/2}$ higher then the other terms, assuming enough regularity of γ, since the functions in M_h are piecewise polynomials of one degree higher than these in $\mathbf{V}_h \cdot \nu$.

3.3. A Computational Example

We illustrate the numerical method described above with a simulation of oil-water flow in a faulted heterogeneous reservoir. A fault cuts through the middle of the domain and divides it into two blocks. The numerical grids follow the geological layers and are non-matching across the fault (see Figure 3). Each block is covered by a $32 \times 32 \times 20$ grid and is discretized by the lowest order RT mixed method, reduced to cell-centered finite differences. Continuous piecewise linear mortar finite elements are used along the fault on a 14×16 grid. The simulation was performed on eight processors on IBM SP2, each block distributed among four processors. Oil concentration contours after 0.25 pore volume water injected (360 days) are given on Figure 4 (water is injected at the right front corner and oil is produced at the left back corner).

Acknowledgments. This work was partially supported by the United States Department of Energy and the National Science Foundation.

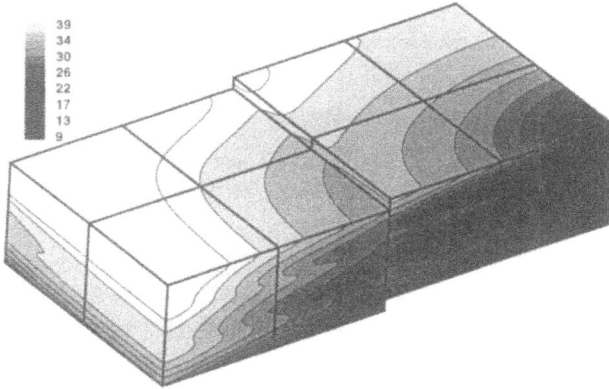

Fig. 4. Oil concentration contours at 360 days.

References

1. Alt, H. W. and S. Luckhaus, Quasilinear elliptic-parabolic differential equations, Math. Z. **183** (1983), 311–341.

2. Alt, H. W. and E. DiBenedetto, Nonsteady flow of water and oil through inhomogeneous porous media, Ann. Scuola Norm. Sup. Pisa Cl. Sci. **12** (1985), 335–392.

3. Arbogast, T., The existence of weak solutions to single porosity and simple dual-porosity models of two-phase incompressible flow, Nonlinear Analysis, Theory, Methods and Applications **19** (1992), 1009–1031.

4. Arbogast, T., L. C. Cowsar, M. F. Wheeler, and I. Yotov, *Mixed Finite Element Methods on Non-Matching Multiblock Grids*, Tech. Report TICAM 96-50, Texas Inst. Comp. Appl. Math., University of Texas at Austin, 1996, submitted to SIAM J. Num. Anal.

5. Arbogast, T., M. F. Wheeler, and I. Yotov, Logically rectangular mixed methods for flow in irregular, heterogeneous domains, in *Computational Methods in Water Resources XI*, A. A. Aldama et al. (eds.), Computational Mechanics Publications, Southampton, 1996, pp. 621–628.

6. Arbogast, T., M. F. Wheeler, and I. Yotov, Mixed finite elements for elliptic problems with tensor coefficients as cell-centered finite differences, SIAM J. Numer. Anal. **34** (1997), 828–852.

7. Arbogast, T., M. F. Wheeler, and N. Zhang, A nonlinear mixed finite element method for a degenerate parabolic equation arising in flow in porous media, SIAM J. Numer. Anal. **33** (1996), 1669–1687.

M. F. Wheeler and I. Yotov

8. Arbogast, T. and I. Yotov, A non-mortar mixed finite element method for elliptic problems on non-matching multiblock grids, Comput. Meth. Appl. Mech. Eng. **149** (1997), 255–265.

9. Bear, J., *Dynamics of Fluids in Porous Media*, Dover, New York, 1972.

10. Belgacem, F. Ben and Y. Maday, The mortar element method for three-dimensional finite elements, RAIRO Mod. Math. Anal. Num. **31** (1997), 289–302.

11. Bernardi, C., Y. Maday, and A. T. Patera, A new nonconforming approach to domain decomposition: the mortar element method, in *Nonlinear Partial Differential Equations and Their Applications*, H. Brezis and J. L. Lions (eds.), Longman Scientific & Technical, UK, 1994.

12. Brezzi, F., J. Douglas, Jr., R. Duràn, and M. Fortin, Mixed finite elements for second order elliptic problems in three variables, Numer. Math. **51** (1987), 237–250.

13. Brezzi, F., J. Douglas, Jr., M. Fortin, and L. D. Marini, Efficient rectangular mixed finite elements in two and three space variables, RAIRO Modèl. Math. Anal. Numèr. **21** (1987), 581–604.

14. Brezzi, F., J. Douglas, Jr., and L. D. Marini, Two families of mixed elements for second order elliptic problems, Numer. Math. **88** (1985), 217–235.

15. Brezzi, F. and M. Fortin, *Mixed and Hybrid Finite Element Methods*, Springer-Verlag, New York, 1991.

16. Chavent, G. and J. Jaffre, *Mathematical Models and Finite Elements for Reservoir Simulation*, North-Holland, Amsterdam, 1986.

17. Chen, Z. and J. Douglas, Jr., Prismatic mixed finite elements for second order elliptic problems, Calcolo **26** (1989), 135–148.

18. Cowsar, L. C., J. Mandel, and M. F. Wheeler, Balancing domain decomposition for mixed finite elements, Math. Comp. **64** (1995), 989–1015.

19. Cowsar, L. C. and M. F. Wheeler, Parallel domain decomposition method for mixed finite elements for elliptic partial differential equations, in *Fourth International Symposium on Domain Decomposition Methods for Partial Differential Equations*, R. Glowinski, Y. Kuznetsov, G. Meurant, J. Periaux, and O. Widlund (eds.), SIAM, Philadelphia, 1991.

21. Ewing, R. E., R. D. Lazarov, and Junping Wang, Superconvergence of the velocity along the Gauss lines in mixed finite element methods, SIAM J. Numer. Anal. **28** (1991), 1015–1029.

22. Fadimba, K. and R. Sharpley, A priori estimates and regularization for a class of porous medium equations, Nonlinear World **2** (1995), 13–41.

23. Glowinski, R. and M. F. Wheeler, Domain decomposition and mixed finite element methods for elliptic problems, in *First International Symposium on*

Domain Decomposition Methods for Partial Differential Equations, R. Glowinski, G. H. Golub, G. A. Meurant, and J. Periaux (eds.), SIAM, Philadelphia, 1988, pp. 144–172.

24. Kuznetsov, Y. A. and M. F. Wheeler, Optimal order substructuring preconditioners for mixed finite element methods on non-matching grids, East-West J. Numer. Math. **3** (1995), 127–143.

25. Nedelec, J. C., Mixed finite elements in $\mathbf{R}^3$, Numer. Math. **35** (1980), 315–341.

26. Nochetto, R. N., Error estimates for the multidimensional singular parabolic problems, Japan J. Appl. Math. **4** (1987), 111-138.

27. Parashar, M. and J. C. Browne, *An Infrastructure for Parallel Adaptive Mesh-Refinement Techniques*, Tech. Report, Department of Computer Science, University of Texas at Austin, 1995.

28. Parashar, M., J. A. Wheeler, G. Pope, K. Wang, and P. Wang, A new generation eos compositional reservoir simulator, part II: Framework and multiprocessing, in *Fourteenth SPE Symposium on Reservoir Simulation*, Dallas, Texas, Society of Petroleum Engineers, June 1997, pp. 31–38.

29. Raviart, R. A. and J. M. Thomas, A mixed finite element method for 2nd order elliptic problems, in *Mathematical Aspects of the Finite Element Method*, Lecture Notes in Mathematics (New York), vol. **606**, Springer-Verlag, New York, 1977, pp. 292–315.

30. Rose, M., Numerical methods for flows through porous media. I, Math. Comp. **40** (1983), 435–467.

31. Thomas, J. M., *These de doctorat d'etat*, 'a l'Universite Pierre et Marie Curie, 1977.

32. Wang, P., I. Yotov, M. F. Wheeler, T. Arbogast, C. N. Dawson, M. Parashar, and K. Sepehrnoori, A new generation eos compositional reservoir simulator, part I: Formulation and discretization, in *Fourteenth SPE Symposium on Reservoir Simulation*, Dallas, Texas, Society of Petroleum Engineers, June 1997, pp. 55–64.

33. Wheeler, M. F., C. N. Dawson, S. Chippada, H. C. Edwards, and M. L. Martinez, Surface flow modeling of bays, estuaries and coastal oceans, Parallel Computing Research **4** (1996), 8–9.

34. Wheeler, M. F. and I. Yotov, Physical and computational domain decompositions for modeling subsurface flows, in *Tenth International Conference on Domain Decomposition Methods, Contemporary Mathematics*, Jan Mandel et al. (eds.), AMS, 1998.

35. Yotov, I., *Mixed Finite Element Methods for Flow in Porous Media*, Ph.D. Thesis, Rice University, Houston, Texas, 1996, TR96-09, Dept. Comp. Appl. Math., Rice University and TICAM Report 96-23, University of Texas at Austin.

36. Yotov, I., A mixed finite element discretization on non-matching multi-block grids for a degenerate parabolic equation arising in porous media flow, East-West J. Numer. Math. **5** (1997), 211–230.

37. Yotov, I., Mortar mixed finite element methods on irregular multiblock domains, in *Third IMACS International Symposium on Iterative Methods in Scientific Computation*, B. Chen, T. Mathew, and J. Wang (eds.), Academic Press, 1997, to appear.

38. Yotov, I. and M. F. Wheeler, Domain decomposition mixed methods for multiphase groundwater flow in multiblock aquifers, in *Computational Methods in Water Resources XII*, V. Burganos et al. (eds.), Computational Mechanics Publications, 1998, to appear.

Mary F. Wheeler and Ivan Yotov
Texas Institute for Computational and Applied Mathematics
The University of Texas at Austin
Austin, Texas 78712
mfw@ticam.utexas.edu and yotov@ticam.utexas.edu